W0051389

Deformation Processes in Minerals, Ceramics and Rocks

TITLES OF RELATED INTEREST

Deformation Processes in Minerals, Ceramics and Rocks

Edited by

D. J. Barber

University of Essex

P. G. Meredith

University College London

Published in association with
The Mineralogical Society of Great Britain and Ireland

London
UNWIN HYMAN
Boston   Sydney   Wellington

© D. J. Barber, P. G. Meredith & contributors, 1990
Softcover reprint of the hardcover 1st edition 1990

This book is copyright under the Berne Convention. No
reproduction without permission. All rights reserved.

Published by the Academic Division of
Unwin Hyman Ltd
15/17 Broadwick Street, London W1V 1FP, UK

Unwin Hyman Inc.,
8 Winchester Place, Winchester, Mass. 01890, USA

Allen & Unwin (Australia) Ltd,
8 Napier Street, North Sydney, NSW 2060, Australia

Allen & Unwin (New Zealand) Ltd in association with the
Port Nicholson Press Ltd,
Compusales Building, 75 Ghuznee Street, Wellington 1, New Zealand

First published in 1990

British Library Cataloguing in Publication Data
Deformation processes in minerals, ceramics and
rocks.
1. Materials. Deformation. Mechanics
I. Barber, D. J. (David J.)   II. Meredith, Philip G.   III.
Series
620.1'123

ISBN-13: 978-94-011-6829-8

Library of Congress Cataloging in Publication Data
Deformation processes in minerals, ceramics, and rocks / edited by
David J. Barber and Philip G. Meredith.
    p. cm. — (The Mineralogical Society series ; 1)
    Includes bibliographical references.
    ISBN-13: 978-94-011-6829-8   e-ISBN-13: 978-94-011-6827-4
    DOI: 10.1007/978-94-011-6827-4
    1. Rock deformation—processes. 2. Deformation (Mechanics)—
processes. 3. Ceramic materials—processes. 4. Mineralogy—
processes. I. Barber, D. J. (David J.) II. Meredith, P. G.
III. Series.
QE604.D45 1990
551.8—dc20
                                                        89-70611
                                                        CIP

Preface

This monograph has its origins in a two-day meeting with the same title held in London, England in the spring of 1987. The idea for the meeting came from members of the UK Mineral and Rock Physics Group. It was held under the auspices of, and made possible by the generous support of, the Mineralogical Society of Great Britain and Ireland. Additional financial assistance was provided by ECC International plc and the Cookson Group plc.

The aims of the London meeting were to survey the current state of knowledge about deformation processes in non-metallic materials and to bring together both experts and less experienced Earth scientists and ceramicists who normally had little contact but shared common interests in deformation mechanisms. This monograph has similar aims and, indeed, most of its authors were keynote speakers at the meeting. Consequently, most of the contributions contain a review element in addition to the presentation and discussion of new results. In adopting this format, the editors hope that the monograph will provide a valuable state-of-the-art sourcebook, both to active researchers and also to graduate students just starting in the relevant fields.

This book is the first of a new series of monographs published by Unwin Hyman, as part of an arrangement made with the Mineralogical Society through its Publications Manager, Dr David Price. The intention of the series is to address active areas of geological and materials research that are not well covered by existing textbooks; a situation that certainly exists for this first topic area. We express our sincere thanks to those who made the book possible. In particular, we thank the authors and other participants in the original very successful meeting, numerous colleagues who kindly assisted us by providing constructive reviews of the papers, the Mineralogical Society and industrial sponsors mentioned above for their financial support, and Roger Jones of Unwin Hyman for his patience and forbearance.

David Barber & Philip Meredith

Contents

List of tables

List of Tables

List of contributors

Michael F. Ashby, Engineering Department, University of Cambridge, Trumpington Street, Cambridge CB2 1PZ, UK

David J. Barber, Department of Physics, University of Essex, Wivenhoe Park, Colchester, Essex CO4 3SQ, UK

Solange Beauchesne, Institut de Physique du Globe de Paris, 4 place Jussieu, 75252 Paris Cedex 05, France

Gilles R. Canova, Laboratoire de Physique et Mécanique des Matériaux, Université, Ile de Sauley, 5700 Metz Cedex, France

Jean-Jacques Couderc, Laboratoire de Physique des Solides, associé au CNRS, UPS et INSA, Avenue de Rangeuil, F-31077 Toulouse Cedex, France

Michel Darot, Laboratoire de Physique des Matériaux, Ecole et Observatoire de Physique du Globe de Strasbourg, 5 rue Réné Déscartes, 67084 Strasbourg Cedex, France

Brian Derby, Department of Metallurgy and Science of Materials, University of Oxford, Parks Road, Oxford OX1 3PH, UK

Stephen W. Freiman, Ceramics Division, National Institute of Standards and Technology, Gaithersburg, Maryland 20899, USA

Yves Gueguen, Laboratoire de Physique des Matériaux, Ecole et Observatoire de Physique du Globe de Strasbourg, 5 rue Réné Déscartes, 67084 Strasbourg Cedex, France

Sheila D. Hallam (née Cooksley), BP Research International, Sunbury Research Centre, Chertsey Road, Sunbury-on-Thames, Middlesex TW16 7LN, UK

Christa Hennig-Michaeli, Institut für Mineralogie und Lagerstättenlehre der RWTH Aachen, Wüllnerstrasse 2, D-5100 Aachen, FRG

Arthur H. Heuer, Department of Metallurgy and Materials Science, Case Western Reserve University, Cleveland, Ohio 44106, USA

Robert J. Keller, Precision Castparts Corporation, 731 Beta Drive, Cleveland, Ohio 44143, USA

Robert J. Knipe, Department of Earth Sciences, University of Leeds, Leeds LS2 9JT, UK

Florian K. Lehner, Koninlijke/Shell Exploratie en Produktie Laboratorium, Rijswijk Z.H., The Netherlands

Philip G. Meredith, Rock Physics Laboratory, Department of Geological Sciences, University College London, Gower Street, London WC1E 6BT, UK

Terry E. Mitchell, Los Alamos National Laboratory, Los Alamos, New Mexico 87544, USA

Alain Molinari, Laboratoire de Physique et Mécanique des Matériaux, Université, Ile de Sauley, 5700 Metz Cedex, France

Stanley A. F. Murrell, Department of Geological Sciences, University College London, Gower Street, London WC1E 6BT, UK

Jean-Paul Poirier, Institut de Physique du Globe de Paris, Departement des Géomatériaux, 4 place Jussieu, 75252 Paris Cedex 05, France

Thierry Reuschlé, Laboratoire de Physique des Matériaux, Ecole et Observatoire de Physique du Globe de Strasbourg, 5 rue Réné Déscartes, 67084 Strasbourg Cedex, France

David C. Rubie, Bayerisches Forschungsinstitut für Experimentelle Geochemie und Geophysik, Universität Bayreuth, Postfach 10 12 51, 8580 Bayreuth, FRG

Peter M. T. M. Schutjens, Department of Geology, Institute of Earth Sciences, Rijksuniversiteit Utrecht, PO Box 80.021, 3508 TA Utrecht, The Netherlands

Christophe Sotin, Laboratoire de Géophysique et Géodynamique Interne, Université de Paris-Sud, 91405 Orsay Cedex, France.

Christopher J. Spiers, Department of Geology, Institute of Earth Sciences, Rijksuniversiteit Utrecht, PO Box 80.021, 3508 TA Utrecht, The Netherlands

Peter L. Swanson, US Bureau of Mines, Denver Research Center, PO Box 25086, Denver Federal Center, Denver, Colorado 80225, USA

Toru Takeshita, Department of Geology and Geophysics, University of California, Berkeley, California 94720, USA

Jan Tullis, Department of Geological Sciences, Brown University, Providence, Rhode Island 02912, USA

Hans-Rudolf Wenk, Department of Geology and Geophysics, University of California, Berkeley, California 94720, USA

An introduction

David J. Barber & Philip G. Meredith

Most of the chapters comprising this monograph are based on oral presentations delivered at the meeting held in London in April 1987, to which we refer in the Preface. The additional contributions were judged to be needed in order to make the monograph more comprehensive and self-sufficient. In particular, there are two overviews written by the individual editors, which are intended to give the necessary frameworks and references to relevant previous work in the two broad and rather poorly defined fields of fracture and ductile deformation, respectively. These overviews by Meredith and Barber also give limited introductions to current problems in these fields and provide links to and between the contributed chapters, which are also grouped under these headings.

The contributors represent a fairly wide cross-section of Earth and materials scientists, who are concerned with various aspects of the deformation of non-metals, whether to do with theory, experimental testing, or naturally occurring deformation, and who address problems concerned with solids ranging from single mineral phases to rocks, synthetic aggregates, ceramics, etc. The term 'deformation processes' embraces a very large spectrum of régimes, size scales, and timescales (earthquakes and plate tectonics are extreme examples), and no monograph based on a two-day meeting can hope to be fully comprehensive or even representative of present-day research activity in this diverse field. However, we concentrate on the microprocesses and micromechanisms which can be important in controlling events on much larger scales, and endeavour to illustrate both the current state of the art and some of the problems that remain to be solved.

It will be tempting for readers to see the book as being in two parts, and the papers in these parts as being completely separate and unconnected. Historically speaking, the divide between the study of fracture behaviour and plastic behaviour has been very large, with developments in both fields tending to occur in ignorance of what has been happening in the other. However, as investigations of deformation processes have broadened in scope and depth, it has become apparent that the distinctions between forms of behaviour that were once respectively termed 'brittle' and 'plastic' (or 'ductile') can be very artificial and misleading. It is recognized that the response of even a model solid to an applied stress can be very sensitive to various environmental parameters, in addition to being influenced by the rate of deformation and such obvious state variables as temperature and pressure. (Probably the first example

1

noted was the marked change in plasticity of single crystals of halite, NaCl, under the influence of water, an effect now interpreted in terms of the blunting or removal of stress-concentrating surface defects.) Mostly as a result of our growing appreciation of environmental and strain-rate sensitive effects, the phenomenological terms 'brittle' and 'ductile' have become words of ill repute.

Awareness of progress in 'parallel' topics in science often assists understanding and interpretation of, or at least provides a stimulus for, the use of relevant new approaches and methods. It is clear that since research on the micromechanisms of both brittle and ductile deformation in non-metals gathered pace, it has benefitted from the methods and been inspired by the findings of similar research on metals, and the large volume of literature on that subject. Without wishing to diminish the value of received knowledge in this case, differences in the behaviour of metals and non-metals (in creep, for example) are now recognized, and it is increasingly necessary that those researching on non-metallic systems be aware of activities in the related non-metals fields, whether these be geologically or commercially important. One hopes that this monograph, which has an interdisciplinary flavour, will make this task easier.

For those coming new to the deformation behaviour of non-metals, it must be emphasized that this monograph is no substitute for several good basic textbooks, and was never intended to fulfil such a function. We assume that our readers will be postgraduate students and researchers of greater years and experience. The latter will perhaps include both scientists to whom deformation processes are relevant but somewhat peripheral to their main interests, and those who are participants in such research, and will, we hope, find this monograph a useful update, stimulus, and source of references.

The level at which the chapters are written therefore assumes the reader to be familiar with crystals and polycrystals (either rocks or ceramics), crystallography, and basic mineralogy, and the principles underlying effects such as diffusion, exsolution, etc. Most undergraduate courses at least give an introduction to crystal defects, including the role of dislocations in plastic deformation, and stress-concentrating flaws in fracture. For any reader who has missed out on one or more of these topics, textbooks which provide suitable coverage include Battey (1975), Hull (1965), Kelly & Groves (1970), Putnis & McConnell (1980), Wyatt & Dew-Hughes (1974), Jaeger & Cook (1976), and Lawn & Wilshaw (1975). Our authors also tacitly assume that readers have some familiarity with modern laboratory methods of characterizing materials, especially optical petrography, X-ray diffraction, electron probe micro-analysis, and electron microscopy. Here again there is no shortage of suitable books to assist those who feel that their background knowledge is in need of some reinforcement. Appropriate and useful texts are Cullity (1978), Goldstein *et al.* (1981), Thomas & Goringe (1979), Goodhew & Humphreys (1988), and Phillips (1971). We imagine that most serious students and researchers will have their own favourite sources for such information.

The transmission electron microscope (TEM) features quite strongly in several of the papers. However, TEM has only recently become an important tool in the Earth sciences, so that many researchers and students are still unfamiliar with its application and potential benefits. Those who lack background in the application of TEM methods to microstructural analysis, and to the deformation of non-metals in particular, should perhaps start by consulting the books by Wenk (1976), and Nicholas & Poirier (1976). We assume that materials scientists and ceramicists will universally have some acquaintance with TEM. Most materials scientists are somewhat unfamiliar with both the methods and the terminology used by geologists studying deformation in the field, and here there is a difficult gap to bridge. However, the overcoming of such problems can bring considerable benefits to workers in the separate disciplines, and in editing this monograph we have endeavoured to see that problems resulting from somewhat different approaches and terminologies in the separate disciplines do not make crossing this gap unduly difficult. Nonetheless, materials scientists interested in learning more about the basis of structural geology and aspects of rock deformation could start by consulting texts such as Hobbs, Means & Williams (1976), Paterson (1978), Ramsay (1967), and Atkinson (1987).

References

Atkinson, B. K. (ed.) 1987. *Fracture mechanics of rock*. London: Academic Press.

Battey, M. H. 1975. *Mineralogy for students*. London: Longman.

Cullity, B. D. 1978. *Elements of X-ray diffraction*. Reading, Mass.: Addison-Wesley.

Goldstein, J. I., D. E. Newbury, P. Echlin, D. C. Joy, C. Fiori & E. Lifshin (eds) 1981. *Scanning electron microscopy and X-ray analysis*. New York: Plenum Press.

Goodhew, P. J. & F. J. Humphreys 1988. *Electron microscopy and analysis*, 2nd edn. London: Taylor and Francis.

Hobbs, B. E., W. D. Means & P. E. Williams 1976. *An outline of structural geology*. New York: Wiley.

Hull, D. 1965. *Introduction to dislocations*. Oxford: Pergamon Press.

Jaeger, J. C. & N. G. W. Cook 1976. *Fundamentals of rock mechanics*, 2nd edn. London: Chapman and Hall.

Kelly, A. & G. W. Groves 1970. *Crystallography and crystal defects*. London: Longman.

Lawn, B. R. & T. R. Wilshaw 1975. *Fracture of brittle solids*. Cambridge: Cambridge University Press.

Nicolas, A. & J.-P. Poirier 1976. *Crystalline plasticity and solid state flow in metamorphic rocks*. New York: Wiley.

Paterson, M. S. 1978. *Experimental rock deformation − the brittle field*. Berlin: Springer.

Phillips, W. R. 1971. *Mineral optics: principles and techniques*. New York: W. H. Freeman.

Putnis, A. & J. D. C. McConnell 1980. *Principles of mineral behaviour*. Oxford: Blackwell Scientific.

Ramsay, J. G. 1967. *Folding and fracturing of rocks*. New York: McGraw-Hill.

Thomas, G. & M. Goringe 1979. *Transmission electron microscopy of materials*. New York: Wiley.

Wenk, H.-R. 1976. *Electron microscopy in mineralogy*. Berlin: Springer.

Wyatt, O. H. & D. Dew-Hughes 1974. *Metals, ceramics and polymers*. Cambridge: Cambridge University Press.

CHAPTER ONE

Fracture and failure of brittle polycrystals: an overview

Philip G. Meredith

1.1 Introduction

The brittle failure of many solids at low homologous temperatures is often attributed to the concept of a critical stress being applied to the material. Traditionally, this failure process has been studied by means of experimentally determined stress–strain relations, with the critical applied stress, or 'fracture strength', normally defined as the peak stress sustained by the material under the particular conditions of the experiment. This continuum, strength-of-materials approach has been successful in identifying a number of important brittle phenomena, such as dilatancy, stick-slip, and the mechanical rôle of pressurized pore fluids (for a comprehensive review of brittle phenomena associated with rock deformation, see Paterson 1978), but has met with only limited success when attempts have been made to extrapolate results outside of the confines of laboratory testing conditions (e.g. to engineering structures or to mechanics of the Earth's crust). This arises in part from an emphasis on phenomenological aspects of macroscopic deformation and failure at the expense of micromechanical aspects of fracture, and in part from the rate, size, and environment dependence of the brittle failure process.

The aim of this chapter is to provide an overview of some recent developments in the understanding of fracture and brittle failure in polycrystalline materials, and hence also to set the scene for following chapters which deal in more detail and in greater depth with specific aspects of brittle failure and the transition from brittle to ductile behaviour.

One of the most important advances in our understanding of the micromechanics of fracture has been through the development of the concept of 'fracture mechanics', and its application in describing crack propagation in brittle materials. Fracture mechanics concerns the study of localized stress concentrations caused by pre-existing 'flaws' in imperfect solids, and the conditions for the extension and propagation of such flaws to become macrocracks. It thus recognizes the crucial rôle in the failure process played by the pre-existing microcracks and pores that are ubiquitous in polycrystalline

5

aggregates such as ceramics and rocks (a review of microcracks in rocks has been provided by Kranz 1983). Hence, the first order 'critical stress' criterion for brittle fracture is replaced by the second order concept of a 'critical stress concentration'. In his pioneering study of cracks in stressed solids, Griffith (1920) invoked an energy criterion to determine whether a crack will tend to close, remain the same length, or propagate. Irwin (1958) developed this idea further, and showed that by measuring the force required to cause unstable propagation of a crack of known length and well defined geometry, it was possible to determine the resistance to fracture or 'fracture toughness' of the material. Fracture toughness is considered to be an intrinsic material property, and can therefore be used to predict failure for widely different geometries and on a much larger scale than that used for its determination.

Failure under many applied stress conditions, however, is not caused by the propagation of a single macrocrack, but by the nucleation, growth, and interaction of arrays of microcracks. In particular, under compressive loading brittle materials can fail progressively, accumulating microcrack damage until a critical level is reached, resulting in macroscopic failure. In recent years a new body of theory, known as 'damage mechanics', has been developed in an attempt to provide complete constitutive relations to describe the mechanical behaviour of brittle materials in complex stress states that induce microcrack damage. The aim of this approach is to link the macroscopic with the microscopic by providing a continuum description of brittle failure, based on the mechanics of the growth of individual cracks.

1.2 Linear elastic fracture mechanics (LEFM)

Introductory texts describing the theory of fracture mechanics have been provided by Knott (1973) and Lawn & Wilshaw (1975); and the application of fracture mechanics analyses to specific crack configurations can be studied in the compilations of Sih (1973), Tada *et al.* (1973), and Rooke & Cartwright (1976). Hence, only a brief resumé is provided here, and the reader is referred to the above texts for more detailed information.

1.2.1 Fracture energy

Fracture mechanics theory was developed from the pioneering work of Griffith (1920, 1924), who formulated a criterion for the extension or closure of an isolated elliptical crack in a stressed solid (see also Jaeger & Cook 1976, Murrell 1964 and this volume). For a static crack in an elastic-brittle solid, the total energy (U) is given by

$$U = (-U_p + U_e) + U_s = U_m + U_s \tag{1.1}$$

where U_m is the total mechanical energy of the system, comprising the potential energy of the applied load $(-U_p)$ and the elastic strain energy stored in the solid (U_ε), and U_s is the surface energy. The mechanical energy term favours crack extension, and the surface energy term opposes crack extension since cohesive molecular forces must be overcome in creating new crack surface area. Equilibrium is achieved by balancing the two terms in (1.1) with respect to any incremental change in crack surface area (A), such that

$$dU/dA = dU_m/dA + dU_s/dA = -G_c + G_R = 0 \qquad (1.2)$$

G_c is defined as the 'mechanical energy release rate' (Irwin 1958) and characterizes the crack driving force for unstable crack extension; and G_R ('energy rate resistance') characterizes the crack extension resistance forces associated with the creation of new crack surfaces. In an ideally brittle solid G_R is equal to twice the surface free energy (γ), since crack growth involves the creation of two new surfaces.

Mai & Lawn (1986) point out that equation (1.2) merely defines when a crack is on the verge of extension, and is a sufficient condition for failure only when the equilibrium state is unstable. Many equilibrium geometries are not unstable, however, and in these cases crack extension can occur in a stable manner with increase in applied load (examples given in Sih 1973, Tada *et al.* 1973, Rooke & Cartwright 1976). Instability depends on the second derivative of energy (d^2U/dA^2). In terms of (1.2), a crack system is unstable, neutral, or stable depending on whether dG_c/dA is greater than, equal to, or less than dG_R/dA respectively (Mai & Lawn 1986). This is an important point since, as we shall see later, the fracture energy of real materials is rarely, if ever, independent of crack length (cf. crack surface area).

1.2.2 Stress intensity analysis

Lawn & Wilshaw (1975) present a means whereby the crack driving force can be determined from analysis of the manner in which the presence of a crack modifies the stress and displacement fields in a linear elastic solid. The stress field in the immediate vicinity of a crack tip is analysed using classical linear elasticity theory, with the simplifying approximations that: (1) any region of inelastic behaviour is negligibly small compared with the length of the crack and the dimensions of the cracked body; (2) the crack tip is atomically sharp; and (3) the crack walls remain traction free at all stages of loading. With these assumptions, the general expression for the near-field stress distribution in polar co-ordinates is given by

$$\sigma_{ij} = K_m (2\pi r)^{-1/2} f_{ij}(\theta) \qquad (1.3)$$

where i and j define the components of the stress tensor, r and θ are the radial

7

distance from the crack tip and the angle measured from the plane of the crack respectively, and $f_{ij}(\theta)$ is a well defined function of θ that depends on loading geometry (Lawn & Wilshaw 1975).

The coefficient of the $r^{-1/2}$ term (K_m) is known as the 'stress intensity factor'. This parameter depends only on the applied loading and the crack configuration, and consequently determines the *magnitude* or *intensity* of the local stress field. The remaining terms in (1.3) depend only on position with respect to the crack tip, and hence determine the *distribution* of the stress field. The subscript m denotes the crack-tip displacement mode. There exist three basic modes of crack-tip displacement: mode I (tensile); mode II (in-plane shear); and mode III (anti-plane shear). Displacements in the shear modes II and III bear a certain analogy to the motion of edge and screw dislocations respectively. These modes are further described and illustrated by Gueguen *et al.* (this volume). Superposition of the three basic modes is sufficient to describe any general case of crack-tip displacement and stress field.

In laboratory configurations for the experimental determination of fracture mechanics parameters, and in modelling of crack systems, it is convenient to simplify the analysis by arranging for a two-dimensional crack-tip stress–strain field and uniform loading in a finite body. Under these conditions, the stress intensity factor for a two-dimensional crack of any mode is given by

$$K = Y\sigma_r(\pi a)^{1/2} \tag{1.4}$$

where σ_r is the remotely applied stress, and a is the crack half-length. Y is a well defined dimensionless modification factor to account for geometry and edge effects, and is tabulated for a wide range of crack configurations (e.g. Sih 1973, Tada *et al.* 1973, Rooke & Cartwright 1976).

1.2.3   Equivalence of parameters

Lawn & Wilshaw (1975) show that since the mechanical energy term in equation (1.2) does not depend on the mode of loading, the crack extension force (G) for the two-dimensional case reduces to the rate of change of elastic strain energy (U_e) with respect to crack length; that is,

$$G = -(dU_e/da) \tag{1.5}$$

which defines the strain energy release rate. Strain energy release rates for different modes of crack-tip displacement are additive, such that

$$G = G_I + G_{II} + G_{III} \tag{1.6}$$

and for fracture in each of the fundamental fracture modes under plane strain conditions, the relationships between strain energy release rate and stress intensity are given by

$$G_I = K_I^2(1 - \nu^2)/E \qquad (1.7)$$

$$G_{II} = K_{II}^2(1 - \nu^2)/E \qquad (1.8)$$

$$G_{III} = K_{III}^2(1 + \nu)/E \qquad (1.9)$$

where ν is the Poisson ratio and E is Young's modulus. For plane stress, the factor $(1 - \nu^2)$ in (1.7) and (1.8) is replaced by unity (Atkinson 1987). This property of linear superposition of the different mode components makes G the more useful parameter when considering mixed-mode fracture.

However, we shall see later that many of the simplifying assumptions made in developing the linear elastic fracture mechanics (LEFM) analysis break down when we come to consider real polycrystalline materials. For these materials the stress intensity notation is often more advantageous, since all contributions to the crack-tip value of K for a given mode of fracture are additive (Mai & Lawn 1986), whereas this is not true for G, which is a global rather than a local parameter (Irwin 1958, Lawn & Wilshaw 1975). Thus, if there are inelastic internal contributions (K_i) to the crack driving force, we can linearly superpose these onto the externally applied contribution (K_a) to obtain (Thomson 1983)

$$K_{eff} = K_a + \Sigma K_i \qquad (1.10)$$

where K_{eff} is the effective driving force felt at the crack tip. The K_i terms in (1.10) can be negative or positive depending on whether they have a shielding or anti-shielding influence on the transmission of externally applied stresses to the crack tip (Mai & Lawn 1986; Swanson 1987; Freiman & Swanson, this volume).

Following (1.2), we can write a similar equilibrium equation in terms of the applied stress intensity factor (K_a) and the stress intensity resistance of the material (K_R), such that

$$K_a = K_R = K_c \qquad (1.11)$$

where K_c defines the critical condition, is thought of as a material constant, and is known as the 'fracture toughness'. Again, stability depends on whether dK_a/da is greater than, equal to, or less than dK_R/da. Where there are internal contributions to K, we can write the stability condition in terms of (1.10) as

$$dK_{eff}/da = dK_a/da + d\Sigma K_i/da \qquad (1.12)$$

and by associating the internal contribution with the resistance term rather than with the driving force term, we can define an effective or apparent fracture toughness (K_Q) as

$$K_Q = K_c - \Sigma K_i \tag{1.13}$$

1.2.4 Dynamic fracture

All of the fracture criteria considered in the LEFM approach to failure discussed above predict that an isolated, planar crack with walls that are free of traction will propagate dynamically at some terminal velocity once the critical condition (critical strain energy release rate, fracture toughness, or apparent fracture toughness) has been reached, or exceeded. At values below the critical value, the crack remains stable and stationary. Both theoretical calculations and numerous experimental measurements support the view that fracture propagation velocities can approach the velocity of sound in the medium (see Lawn & Wilshaw 1975). The more ideally brittle the material, the closer the dynamic fracture velocity approaches the sonic velocity. The terminal velocity is generally determined to be less than or equal to the Rayleigh wave velocity because of inertia effects (Aki & Richards 1980), although diffracted body waves propagating ahead of the crack tip can in special circumstances lead to rupture velocities close to the compressional wave velocity (Andrews 1985).

However, it must be noted here that under impulsive or stress wave loading, even a fracture propagating at its terminal velocity may not be sufficient to relieve the applied stress. Under such conditions, crack branching and bifurcation can occur, and the material fails by fragmentation. This is a very important process, especially for rocks, in considering such phenomena as impact cratering and rock breakage by explosive loading. The field of dynamic rock fragmentation is currently being actively researched, and the interested reader is referred to the comprehensive review by Grady & Kipp (1987).

1.3 Quasi-static fracture and failure

These relatively simple dynamic fracture criteria are, however, generally found to be inadequate to describe fully the mechanical behaviour of most oxides and silicates. An important, commonly observed characteristic of real brittle polycrystals is that their mechanical properties and resistance to fracture depend strongly upon the environmental conditions (physical and chemical) under which the deformation takes place, and also to a significant degree upon the rate of deformation, especially when this is low. For example, Costin (1987) has pointed out that the compressive fracture stress of unconfined rock ('uniaxial compressive strength') can decrease by a factor of two or three as the strain rate is reduced from those ordinarily used in laboratory tests

$(10^{-4}-10^{-6} \text{ s}^{-1})$ to that commonly associated with natural tectonic deformation $(10^{-14} \text{ s}^{-1})$. Most rocks also exhibit brittle creep behaviour (time-dependent cracking at constant stress), especially if the applied stress is a significant fraction of the short-term fracture stress (Kranz & Scholz 1977; Kranz 1979, Costin 1987). Importantly, the time-to-failure in creep tests is reduced dramatically by the mere presence of water (Scholz 1972, Kranz 1980). Note that this latter phenomenon is a chemical effect rather than a mechanical, effective stress effect, and is known as 'static fatigue'.

Hence, in systems where brittle materials are subjected to long-term loading, the classical LEFM approach does not provide an adequate description. This is especially so at elevated temperatures and in the presence of reactive environmental species. A considerable body of experimental evidence now exists to support the idea that cracks can propagate in a stable, quasi-static manner at values of K and G that are substantially below the critical values (K_c, G_c); albeit at velocities that are orders of magnitude lower than the terminal velocities associated with catastrophic, dynamic fracture. This phenomenon is called 'subcritical crack growth' and has been observed experimentally in many brittle materials, including glass (Wiederhorn 1978), ceramics (Wiederhorn 1974), and minerals and rocks (Atkinson 1984; Atkinson & Meredith 1987a, b). A whole range of micromechanisms has been suggested to account for this phenomenon, including atomic diffusion, dissolution, ion exchange, microplasticity, stress corrosion, and cyclic fatigue. The reader is referred to recent reviews by Atkinson (1984) and Atkinson & Meredith (1987a) for a detailed discussion of these mechanisms, and the stress levels and environmental conditions under which any of them might be expected to dominate subcritical crack extension.

However, it should also be noted that subcritical cracking is possible even in the absence of reactive environmental species. Close to the critical stress intensity for dynamic fracture there exists a range of stress intensities and crack-tip positions for which the crack is stable due to the discrete nature of the crystalline lattice. The crack is said to be 'lattice-trapped', and a sufficient input of thermal energy can allow the crack to extend from one position of stability to another (Lawn & Wilshaw 1975). The rôle of any extrinsic, chemically enhanced mechanism may therefore be viewed as an enhanced activation of the intrinsic bond rupture process.

The overwhelming body of experimental and observational evidence suggests that extension of pre-existing cracks and flaws by the mechanism of 'stress corrosion' is likely to dominate subcritical crack growth in brittle materials at low homologous temperatures. Stress corrosion proceeds by the preferential weakening of strained bonds at crack tips through reactions with chemical species in the environment. Although a whole range of surfactants can contribute to stress corrosion (e.g. Dunning *et al.* 1984), the most active reagent for the process appears to be water. Details of the reactions are poorly understood, however, and it is still not clear whether molecular or ionized

11

water is the important reagent (Michalske & Freiman 1982, Atkinson 1984, Freiman 1984).

Regardless of the details of the mechanism, the subcritical crack growth behaviour of many materials approximates to the classical trimodal pattern exhibited by glass in aqueous environments. This pattern of behaviour is illustrated schematically in Figure 1.1, in which an appropriate fracture mechanics parameter is plotted against the logarithm of crack velocity. The details of the subcritical crack growth mechanism and the prevailing environmental conditions will control the details of the diagram, and the particular fracture mechanics parameter of interest will depend upon the constitutive relation invoked to describe the data. In region 1 of Figure 1.1 the crack extension velocity is controlled by the rate of stress corrosion reactions at crack tips. In region 2, crack growth is controlled by the rate of transport of reactive environmental species to crack tips (Lawn & Wilshaw 1975). In region 3 the curve becomes asymptotic to the critical value (denoted by 'c' on Fig. 1.1); crack growth is primarily due to thermally activated bond rupture and is relatively insensitive to the chemical environment (Freiman 1984). A lower limiting threshold is thought to exist, below which no crack growth will occur (denoted by '0' on Fig. 1.1). The existence of such a subcritical crack growth limit has been demonstrated for various glass/water systems (e.g. Wiederhorn & Johnson 1973, Simmons & Freiman 1980), but has not yet been confirmed for polycrystalline ceramics or rocks. Note that because stress corrosion is a chemically enhanced and thermally activated process, the subcritical crack growth curve is shifted to a higher velocity for the same value of K or G if

Figure 1.1 Schematic diagram showing the dependence of crack velocity on stress-intensity factor (K) or strain-energy release rate (G) between the subcritical crack-growth limit (0) and catastrophic rupture (c). The influence of temperature and partial pressure of water is also indicated. The different mechanisms responsible for behaviour in regions 1, 2, and 3 are discussed in the text.

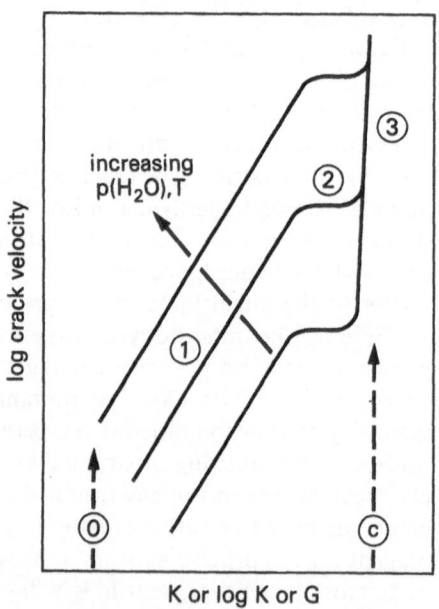

either the partial pressure of the active environmental reagent or the temperature is increased.

Over the years a plethora of constitutive equations have been proposed to describe the relation between various fracture mechanics parameters (particularly K and G) and crack extension velocity (v) during subcritical crack growth, as illustrated in Figure 1.1. A number of the more important theories have been reviewed by Anderson & Grew (1977), Atkinson (1982), and Atkinson & Meredith (1987a). Three formulations in particular have been used extensively to describe the results of subcritical crack growth experiments. These are (a) the Charles (1958) power law, and the (b) Wiederhorn & Bolz (1970) and (c) Lawn (1975) exponential laws:

(a) $$v \propto K^n \qquad (1.14)$$

(b) $$v \propto \exp K \qquad (1.15)$$

(c) $$v \propto \exp K^2 \propto \exp G \qquad (1.16)$$

Note that the final expression in (1.16) arises directly from the relationship between K and G given in (1.7)–(1.9). The Charles (1958) power law is by far the most commonly used expression, even though it has been argued that the exponential laws have stronger theoretical justifications (e.g. Atkinson, 1982; Gueguen et al., this volume). The power law is commonly preferred because it is much more readily integrated to predict analytically the time-dependent failure properties of brittle materials (e.g. Atkinson 1980, Lankford 1981, Sano et al, 1981, Main et al. 1989). In practice, however, it is often not possible to distinguish between any of these expressions in describing experimental data due to the very strong dependence of v on K. For example, Atkinson & Meredith (1987b) have recently provided a compilation of experimental data describing subcritical crack growth in minerals and rocks, in which the value of the exponent (n) in (1.14) is generally found to be in the range 30–60 for polycrystalline rocks.

Gueguen et al. (this volume) have made a direct comparison between expressions (1.14) and (1.16), and concluded that if an exponential relation between v and G is valid then this implies that the value of n in (1.14) depends on the value of K. If this were true, then comparisons between different data sets would lead to erroneous results unless the range of K values considered were identical. They therefore prefer to use the parameter G and expression (1.16) to describe their experimental results. However, they also point out that, due to the narrow range of values of K necessarily investigated in most experiments, it is rarely possible to state definitively which expression fits the data better. The most important conclusion seems to be that any extrapolation of curves outside of the range of the actual data must be treated with extreme caution.

13

Gueguen *et al.* go on to compare their experimental data for glass, quartz, and sandstone with their preferred theoretical model (1.16) (Lawn & Wilshaw 1975, Lawn 1975). Glass behaves as an ideal, homogeneous, isotropic, elastic solid and exhibits the classical, trimodal subcritical crack growth behaviour described above, with a transition from reaction-rate-controlled to transport-controlled crack growth at a velocity of *c.* 10^{-4} m s^{-1}, in broad agreement with Lawn & Wilshaw's (1975) prediction. The other materials, however, are found to deviate to a greater or lesser extent from the ideal case. Although quartz is homogeneous and quasi-elastic, it is not isotropic. Its fracture surface energy depends on crack plane orientation, and this leads to much greater scatter in the data. The situation is even worse for the sandstone, which cannot be considered homogeneous, elastic, or isotropic at the scale tested. For this material the basic assumptions of LEFM become invalid, and not only the slope and position of the subcritical crack growth curve, but also the significance of the parameter G, are not well defined.

Consideration of these deviations from the ideal LEFM case is the subject of the next section.

1.4 Inelastic fracture processes

To a greater or lesser extent, the mechanical behaviour of all real materials deviates from the ideal behaviour assumed in development of the LEFM approach to fracture. Even for mineral single crystals that are considered to be highly brittle, experimentally determined values of the fracture surface energy ($\Gamma = G_c/2$) are generally found to exceed theoretically calculated values of the surface free energy (γ), often by a factor of two or three (see review by Atkinson 1984). For semi-brittle or non-brittle materials, the difference may be several orders of magnitude (e.g., see Murrell, this volume, Table 5.6). This discrepancy occurs because energy dissipative processes other than those associated entirely with the generation of a single, new, planar crack surface are operative in most real materials. Depending on the particular material concerned, these processes can include crack-tip microplasticity, heat generation, and the generation of acoustic waves.

The situation becomes even more complex in polycrystalline materials. Freiman & Swanson (this volume) show how the fracture resistance of single-phase polycrystalline ceramics can be as much as ten times higher than that of single crystals of the same material; and go on to explain this in terms of inelastic processes such as distributed damage by subsidiary microcracking, crack deflections, and phase transformations. Perusal of the data compilations of Atkinson & Meredith (1987b) shows a similar trend for monomineralic rocks. Moreover, fracture toughness values for polymineralic crystalline rocks are found to be significantly higher than those of even the toughest of the rock's constituent minerals. Since fracture energy is related to the square of the fracture toughness, this means that the fracture energy of some rocks can be as much as two orders of magnitude higher than that of any of its constituent minerals. The overall trend is of a general increase in fracture resistance with increase in microstructural complexity. Atkinson's (1984) review of subcritical cracking provides convincing evidence to show that this trend is as true for quasi-static crack growth as it is for dynamic fracture.

1.4.1 Fracture process zone

The power of the LEFM approach to fracture propagation lies in the simplicity with which the magnitude of the crack-tip stress field can be expressed via a single parameter (K) through equation (1.3). However, (1.3) only holds when any zone of inelastic breakdown processes in the region of a macrocrack tip ('process zone') is small compared with the crack length and the dimensions of the cracked body. It has traditionally been considered that the LEFM analysis holds to a good approximation when $r/a < 0.02$; where r is the process zone size and a is the length of the macrocrack or any dimension of the cracked body (Knott 1973). Recent experience has shown that for many polycrystalline ceramics and rocks this basic assumption of small-scale inelasticity is commonly violated (e.g. Hoagland *et al.* 1973; Swanson 1984; Swanson & Spetzler 1984; Freiman & Swanson, this volume). Under these conditions, the magnitude of the crack-tip stress field is not completely characterized by the single parameter K, and the detailed characteristics of the material breakdown processes within the fracture process zone must be considered.

Several candidate mechanisms for process zone inelasticity in brittle polycrystals have been proposed: (i) the formation of a densely microcracked dilatant region ahead of the macrocrack tip (Hoagland *et al.* 1973, Kobayashi & Fourney 1978, Swanson & Spetzler 1984); and (ii) crack interface tractions behind the crack tip (Swanson 1987; Freiman & Swanson, this volume).

1.4.2 Process zone models

The most frequently suggested process zone models postulate a cloud of microcracks surrounding the tip of a macroscopic tensile fracture (e.g. Hoagland *et al.* 1973, Evans 1976, Buresch 1979, Schmidt & Lutz, 1979, Atkinson 1987). Atkinson (1987) describes a model in which the process zone is idealized as a circle centred on the macrocrack tip for the two-dimensional case, and is cylindrical for the three-dimensional case. Other models, which follow directly from the plastic yield zone fracture models originally developed for metals (Irwin 1960, Dugdale 1960), have been applied to rocks (e.g. Schmidt & Lutz 1979). They predict a large double-lobe shaped fracture process zone (e.g. see Swanson 1987) where the size of the zone (r) is given by

$$r \propto (K/\sigma_{\text{crit}})^2 \tag{1.17}$$

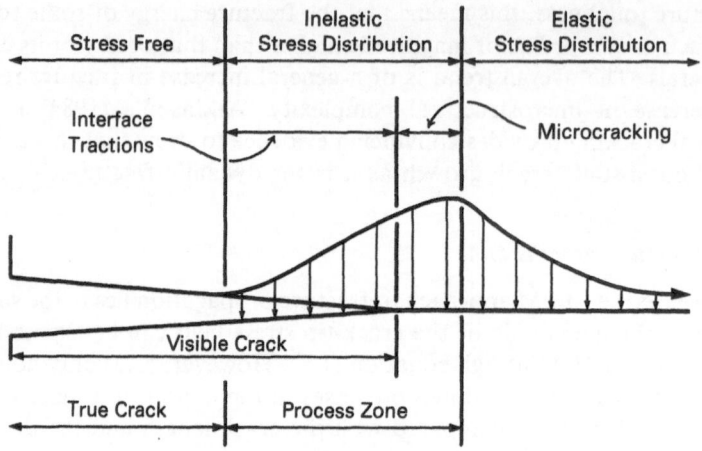

Figure 1.2 Schematic illustration of the hypothesized 'effective' crack that comprises a true crack and a process zone (after Labuz *et al.* 1985, and Ingraffea 1987).

where σ_{crit} is a critical tensile stress for microcracking, and replaces the yield stress used in the plastic yield models for metals. To date, no direct or indirect physical evidence has been presented to support such a process zone shape or size in brittle materials.

More recently, Labuz *et al.* (1985) and Ingraffea (1987) have developed a modified Barenblatt (1962) thin-zone model of tensile fracture. Their model assumes: (i) that a strain-softening constitutive relation exists between applied tensile stress and crack-opening displacement; (ii) that the process zone localizes to a thin zone due to this strain-softening behaviour; (iii) that normal stresses continue to be transferred across a portion of the 'visible' fracture due to grain bridging and the rough, three-dimensional nature of the crack surfaces; and (iv) that the principal stress parallel to the crack front has no influence on the process zone (Mindess & Nadeau 1976, Schmidt & Lutz 1979). Together, these assumptions combine to produce the type of crack model illustrated schematically in Figure 1.2, which considers an 'effective' crack length composed of a true traction-free portion and a process zone that comprises microcracking ahead of, and interface tractions behind, the macro-crack tip.

1.4.3 Experimental observations

The first assumption of the thin-zone model is supported by a number of observations of strain-softening behaviour reported from direct tension tests on a number of different rock types (e.g. Labuz *et al.* 1985, Hashida 1989) and on cementitious composites (e.g. Li & Ward 1989).

Swanson & Spetzler (1984) and Swanson (1987) report results from a series of experiments in which ultrasonic waves were used to probe the fracture process zone during tensile cracking in specimens of Westerly granite (200 mm long × 76 mm wide). In these experiments surface wave and compressional wave ultrasonic transducers were traversed in pairs, both parallel and perpendicular to the direction of macrocrack propagation. Changes in the travel times and amplitudes of waves pulsed between the transducer pairs were monitored as a function of distance from an artificially sawn notch, as the applied load was increased to the level required for macrocrack extension. In summary, the results obtained from traverses parallel to the macrocrack revealed the development of a zone of partial transmission of the order of 30 mm long (equivalent to approximately 40 grain diameters) prior to extension of the macrocrack. The width of this zone of partial transmission was measured during the perpendicular traverses and was found to be in the range 1–4 mm. These observations are entirely consistent with the thin-zone model of the fracture process zone.

Furthermore, Main *et al.* (1990) present data for shear crack growth over a range of scales from 10^{-3} to 10^{+6} m that clearly show evidence of single long macrocracks surrounded by a relatively narrow zone of much shorter subparallel cracks at every scale. Indeed, the basic concept of plate tectonic theory used to describe the dynamics of the Earth's lithosphere inherently relies on a process of strain softening caused by the localization of deformation into long, narrow bands at plate boundaries.

The inclusion of crack interface tractions in the thin-zone model is particularly useful, and may well explain Swanson's (1981) measurement of a significant level of acoustic emission activity generated from points on the crack plane well behind the propagating macrocrack tip in tensile fracture experiments on granite. Moreover, Freiman & Swanson (this volume) present convincing photomicrographic evidence from *in-situ* observations of the cracking process to support their case that crack interface tractions can make a significant contribution to the fracture resistance of brittle polycrystalline ceramics. They have identified two specific traction mechanisms: (i) frictional interlocking between asperities on the rough crack surfaces; and (ii) ligamentary bridging between islands of unbroken material behind the advancing macrocrack front. Both traction mechanisms serve to shield the macrocrack tip from the remotely applied stress, and hence reduce the effective stress intensity factor felt at the crack tip.

1.4.4 Crack extension resistance curves

A substantial body of recent experimental evidence indicates clearly that fracture propagation in many polycrystalline ceramics and rocks is not uniquely related to a single value of the fracture toughness or fracture energy. Rather, the resistance to further crack growth is dependent on the crack extension or crack length already achieved. This type of behaviour, commonly termed R-curve behaviour, has been identified in many different polycrystalline

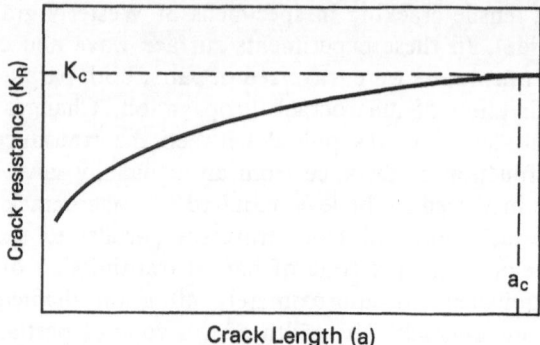

Figure 1.3 Schematic diagram illustrating rising R-curve behaviour up to critical crack length, a_c, observed in many polycrystalline materials.

ceramics (e.g. Hübner & Jillek 1977, Cook *et al.* 1985) and rocks (e.g. Schmidt & Lutz 1979, Ingraffea 1987, Meredith 1989), and is illustrated schematically in Figure 1.3. The existence of R-curve behaviour is closely associated with the development of the fracture process zone during inelastic crack growth. The idea is simply that crack extension resistance increases as crack length increases until the process zone is fully developed, and there is at least a portion of the crack that is traction free. Only at or beyond this critical crack length is a steady state reached, and only then can a fracture toughness value that is truly representative of the bulk material be determined.

1.4.5 Inelastic fracture mechanics

It is now pertinent to consider how a fracture mechanics analysis might be able to accommodate the violation of some of its basic assumptions as discussed above. Swanson (1987) has pointed out that for the assumption of small-scale inelasticity to hold for materials such as Westerly granite, then metre-sized homogeneous test specimens would be required, and this is impracticable in general.

However, the inelastic crack-tip processes described above can be accommodated without sacrificing the simplicity of the LEFM approach, either by considering an 'effective crack length' (e.g. Labuz *et al.* 1985) or an 'effective stress intensity factor' via equation (1.10). In the latter approach, the effective stress intensity factor felt at the crack tip comprises the stress intensity factor due to the applied load and calculated according to LEFM principles minus that value required to overcome the extra, inelastic crack-tip restraining forces. A major difficulty remains in providing quantitative descriptions of the inelastic crack-tip processes as functions of macrocrack length and crack face separation, and the reader is referred to more detailed discussions of specific thin-zone models of the mechanics of fracture in ceramics (Mai & Lawn 1986, 1987), rocks (Labuz *et al.* 1985, Swanson 1987), and concrete (Wittman 1983).

1.5 Fracture and failure prediction

The nucleation and growth of microcracks in stressed brittle materials causes progressive changes in the elastic moduli and stiffness, and hence changes in the mechanical response to the applied stresses. Since crack growth has been shown to be time- and environment-dependent, this leads to a time-, rate-, and environment-dependence of the mechanical properties of brittle polycrystals. Furthermore, since cracks are quasi-planar features, the growth of micro-cracks that are aligned with respect to principal stress directions will result in an initially isotropic material being rapidly transformed into a mechanically anisotropic material (Costin 1987). Eventually, microcrack growth and inter-action leads to macroscopic failure.

1.5.1 Tensile failure

Fracture mechanics predicts that failure in tension will occur by eventual cata-strophic growth of the largest or most critical flaw or microcrack. Hence, tensile failure generally involves the interaction of only a limited number of discrete microcracks in a relatively narrow zone. Experimental observations reviewed earlier support this view. In general, once the applied stress intensity at the tip of a favourably oriented flaw or microcrack becomes high enough, then extension will occur. If the rate of unloading due to crack extension is less than the rate of stress increase, then the crack will accelerate to cata-strophic failure since stress intensity increases with increasing crack length (eqn (1.4)).

In principle, integration of the area under a stress intensity factor − crack velocity curve (see Fig. 1.1) can provide all of the information necessary to pre-dict the time-, rate-, and environment-dependent tensile fracture strength of brittle materials (Evans 1972). For the case of constant remotely applied stress (creep or static fatigue), the time-to-failure (t_f) is given by

$$t_f = \int_{a_i}^{a_c} 1/v \, da \qquad (1.18)$$

where the subscripts i and c denote initial and critical conditions respectively. In order to evaluate (1.18), we need to invoke the relation between applied stress, crack geometry and stress intensity given in (1.4), to yield

$$t_f = 2/\sigma^2 Y^2 \int_{K_{Ii}}^{K_{Ic}} (K_I/v) \, dK_I \qquad (1.19)$$

Since the acceleration to failure of subcritically propagating cracks is a very non-linear process (e.g. Das & Scholz 1981, Main 1988, Meredith *et al.* 1990), the length of time that such cracks will spend in rapid propagation to failure in regions 2 and 3 of the stress intensity factor − crack velocity curve will be negligibly small compared with the length of time spent growing to the critical condition in region 1. Hence, (1.19) may be evaluated numerically from a stress intensity factor − crack velocity curve, or analytically using one of the theoretical relationships for subcritical crack growth given in (1.14), (1.15), and (1.16).

Similarly, Evans & Johnson (1975) have derived an expression to predict the fracture stress (σ_f) for the case of constant applied stress rate (dynamic fatigue), and assuming the Charles power law relationship between K and v, such that

$$\sigma_f = \left[\frac{2\dot{\sigma}(n+1)}{A Y^n(n-2)a_i^{(n-2)/2}}\right]^{1/(n+1)}, \qquad n \neq 2 \qquad (1.20)$$

Further, both Evans & Johnson (1975) and Atkinson (1980) have noted that the constant stress rate expression can also be used to predict failure at constant strain rate ($\dot{\varepsilon}$) simply by substituting: $\dot{\sigma} = E\dot{\varepsilon}$ in (1.20), where E is Young's modulus. Henry & Paquet (1976) and Atkinson (1980) have used (1.19) and (1.20) to predict the time-to-failure at constant stress, and the influence of stressing rate on tensile fracture stress, for carbonate rocks and quartzites respectively.

However, as Costin (1987) has pointed out, the sizes of natural flaws in both ceramics and rocks vary considerably throughout the material. This is especially true for polyphase materials with variations in grain size. It is commonly not possible, therefore, to establish a single value of the initial flaw size that is representative of the bulk material, as required by relationships based on (1.18). One way to overcome this problem is through the use of a suitable statistical distribution of pre-existing flaw sizes. Obviously, since naturally occurring flaws are distributed over a range of sizes and shapes, it follows that the tensile failure stress is not a unique quantity, but will vary statistically in a manner similar to the flaw distribution (Costin 1987). This combination of a highly non-linear constitutive relation with such sensitivity to initial conditions makes the accurate prediction of failure times very difficult to achieve in practice for polycrystalline materials. An unfortunate, but inescapable, corollary of this is that deterministic earthquake prediction may be inherently unreliable, except in special cases.

A statistical distribution commonly used to describe tensile failure stresses is the Weibull distribution (Weibull 1939, 1951), which is based on a 'weakest-link' criterion: that is, it is assumed that propagation of the most deleterious flaw will lead to ultimate failure. The most deleterious flaw will be the one with the most appropriate size and orientation with respect to the applied stress to produce the highest stress intensity at its tip. Under these conditions, the

20

probability of failure at a given stress (σ) is

$$P(\sigma) = 1 - \exp\left[- \int \frac{(\sigma - \sigma_u)^m}{\sigma_0^m} \, dV \right], \qquad \sigma > \sigma_u \qquad (1.21)$$

where σ_0, σ_u, and m are the distribution parameters, and V is the volume of the stressed region. If, as is common for tensile tests, the stress distribution is assumed to be uniform and σ_u is taken to be zero, then (1.21) reduces to

$$P(\sigma) = 1 - \exp\left[- (\sigma/\sigma_0)^m V \right] \qquad (1.22)$$

Experimental methods for evaluating σ_0 and m, and thus determining $P(\sigma)$, are given by Varder & Finnie (1975) and Costin (1987).

Finally, Swan (1980) has shown that Wiebull theory could not explain variations in the failure stress observed in his tensile tests on different sized samples of Stripa granite. The explanation for this discrepancy is simply that traditional Weibull theory takes no account of subcritical crack growth, but assumes that the crack size distribution remains unchanged up to the point of catastrophic failure. However, Wilkins (1980) has shown that it is possible to adapt Weibull theory to predict time-dependent failure, by first performing a series of rapid loading tests to determine the Weibull distribution parameters, and then a series of time-to-failure (static fatigue) tests at constant tensile stress. By ranking the times-to-failure in static fatigue tests with the failure stresses in rapid loading tests, and plotting the resulting data on double logarithmic axes, it is possible to determine the subcritical crack growth index (n) in the Charles power-law relationship (eqn (1.14)).

1.5.2 Compressive failure

Fracture and failure of brittle materials subjected to compressive stresses is generally much more complex than for the tensile case, since compressive failure generally involves the nucleation, propagation, and interaction of a far greater number of microcracks. Direct observation of such stress-induced microcracks in rocks (e.g. Peng & Johnson 1972; Tapponier & Brace 1976; Kranz 1979, 1980, 1983; Wong & Biegel 1985) suggests strongly that they nucleate from pre-existing flaws (pores, inclusions, microcracks, etc.), and propagate primarily as tensile (mode I) cracks in a direction parallel to the maximum principal compressive stress, σ_1 (readers unfamiliar with the concept of principal stresses should consult Paterson (1978) or Jaeger & Cook (1976)). However, the actual initiation mechanism is the subject of much recent debate (e.g. Brace & Bombolakis 1963; Murrell 1964; Murrell & Digby 1970; Tapponier & Brace 1976; Stevens & Holcomb 1980; Horrii & Nemat-Nasser 1986; Ashby & Hallam 1986; Sammis & Ashby 1986; Costin 1983, 1987; Hallam & Ashby, this volume). Gueguen et al. (this volume) point out that under these circumstances an understanding of the behaviour of a single crack may at first sight appear irrelevant, but should be considered as a necessary, although not sufficient, preliminary step.

We have already seen that in a tensile stress field a single crack can grow in an unstable manner and lead to macroscopic failure. The same can also be true for uniaxial compression (e.g. see Hallam & Ashby, this volume). By contrast, the application of even a modest compressive confining stress (σ_3) causes individual cracks to extend in a stable manner. Under these latter circumstances, a microcrack will extend to relieve the local stress concentration caused by an increase in the differential compressive stress ($\sigma_1 - \sigma_3$) and will then arrest. In terms of fracture mechanics, neglecting for a moment time-dependent subcritical crack growth, microcracks extend once the critical tensile stress intensity (K_{Ic}) is exceeded locally. However, since K is a decreasing function of crack length under these conditions (Tada *et al.* 1973, Costin 1987), individual cracks only extend until equilibrium is reached at $K = K_{Ic}$. As the differential compressive stress is increased, an increasing population of microcracks extends until their density and average size is such that they interact with each other to produce macroscopic failure. Hence, the strength of brittle solids in compression is normally greater by at least an order of magnitude than that in tension (Paterson 1978), an empirical observation that was well known to architects and bridge-builders long before a convincing explanation became possible.

Hallam & Ashby (this volume) show how failure may be dominated at very low confining pressures by the propagation of a small number of cracks; at moderate confining pressures by the interaction of numerous microcracks in a relatively narrow fracture zone to form a macroscopic shear failure (i.e. a fault); and at very high confining pressures by distributed microcracking to cause pseudo-ductile deformation by cataclastic flow. Hence the compressive strength of brittle materials is pressure dependent. These different styles of deformation give rise to markedly different stress–strain relations, and these are illustrated in Figure 1.4 (and in more detail in Figure 4.1 of Hallam & Ashby, this volume).

Furthermore, Main *et al.* (1990) have noted that both seismic data from crustal earthquakes and acoustic emission data from laboratory-scale rock fracture experiments (see Section 1.6 for details) are consistent with a feedback model of compressive failure. Where a population of microcracks exists, the overall situation appears to be initially one of negative feedback. Once a crack has grown to relieve the stress locally in a high-stress zone, it becomes a relatively low-stress zone. It is then more likely that further stress relief will be accommodated by growth of a different crack than by further extension of the same crack. Eventually, under conditions of increasing remote stress, a proportion of the original population of cracks will have grown in a stable manner until their lengths are comparable to their spacing, whence the locally perturbed stress fields due to the presence of the cracks interact in a co-

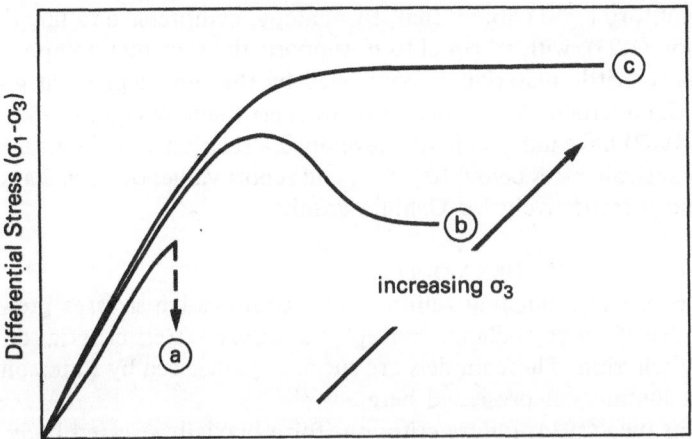

Figure 1.4 Schematic stress–strain curves for a material undergoing deformation over a range of confining pressures (σ_3). (a) At zero or very low confining pressure, the material fails catastrophically in a brittle manner by axial splitting. (b) At intermediate confining pressure, the material has higher strength and fails in a semi-brittle manner by localization of deformation on a shear fault. (c) At very high confining pressure, the material is even stronger and fails in a ductile manner by cataclastic flow. The material may flow at constant stress or exhibit strain hardening.

operative manner. The situation can then flip from one of negative feedback to one of positive feedback, and an instability then develops and the material fails. In terms of the stress–strain relation, this corresponds to a change from strain-hardening to strain-softening behaviour during dilatancy.

So far we have neglected any influence of time-dependent processes such as stress corrosion, but like tensile fracture, failure under compressive loading has long been recognized as a process that is dependent upon both environment and loading rate (see Paterson 1978). Studies of the uniaxial compressive failure of a range of ceramic materials (silicon carbide, silicon nitride, alumina) have shown that the dependence of compressive strength (σ_c) on strain rate ($\dot{\varepsilon}$) follows a relationship of the form (Lankford 1979, 1981)

$$\sigma_c \propto \dot{\varepsilon}^{1/(n^* + 1)} \tag{1.23}$$

where n^* is an environment- and material-dependent constant for strain rates below some critical value ($10^{-2}\,\mathrm{s}^{-1}$ in this case). Lankford noted that the value of n^* was very close (i.e. within experimental accuracy) to the value of n (the subcritical crack growth index for tensile crack growth) for the same material under the same environmental conditions. It is well known that the tensile strength – strain rate relation for these ceramics is identical to equation (1.23) (note, for example, the identical form of (1.23) and (1.20)); and the observed equivalence between the exponent in this relationship and the subcritical crack growth index is accepted as proof that the strain rate dependence of tensile strength is based on thermally activated, tensile microcrack growth (Evans

1974). Lankford (1981) argues that, by analogy, compressive failure described by equation (1.23), with n^* equal to n, supports the view that compressive failure in many brittle materials is controlled by the time-dependent growth of tensile, axial microcracks. Sano and co-workers (Sano & Ogino 1980; Sano et al. 1981, 1982) have independently developed a relationship identical to (1.23) for rocks at strain rates below $10^{-3}\,\mathrm{s}^{-1}$, and report values of 32 ± 2 and 30 ± 5 for n^* and n respectively for Oshima granite.

1.5.2.1 GRIFFITH FAILURE MODELS

Early attempts at modelling failure under compressive stresses grew out of Griffith's (1920) energy balance concept, but utilized stress criteria rather than an energy criterion. These models are succinctly reviewed by Paterson (1978), so only a summary is presented here.

Griffith's own (1924) failure criterion, for a biaxially stressed body is given by

$$(\sigma_1 - \sigma_3)^2 = 8\sigma_t(\sigma_1 + \sigma_3) \qquad \text{for } \sigma_1 + 3\sigma_3 > 0 \qquad (1.24a)$$

$$\sigma_3 = -\sigma_t \qquad \text{for } \sigma_1 + 3\sigma_3 < 0 \qquad (1.24b)$$

where σ_t is the uniaxial tensile strength. This criterion predicts that the uniaxial compressive strength will be eight times the uniaxial tensile strength, which is rather smaller than the ratio commonly measured for rocks (e.g. McLintock & Walsh 1962). Murrell (1964) has suggested that the two-dimensional Griffith criterion can be generalized to the triaxial stress state according to

$$(\sigma_2 - \sigma_3)^2 + (\sigma_3 - \sigma_1)^2 + (\sigma_1 - \sigma_2)^2 = 24\sigma_t(\sigma_1 + \sigma_2 + \sigma_3) \qquad (1.25)$$

This criterion has the advantage that it takes account of intermediate principal stresses and that it provides a more realistic ratio of the uniaxial compressive strength to the uniaxial tensile strength of 12. Whereas the above models provide criteria for failure, Murrell & Digby (1970) have used a similar approach to provide a general three-dimensional solution for fracture initiation from randomly oriented ellipsoidal cavities, that results in

$$(\sigma_1 - \sigma_3)^2 - \alpha\sigma_t(\sigma_1 + \sigma_3) = \beta\sigma_t^2 \qquad (1.26)$$

where α and β are constants that depend on the Poisson ratio and the aspect ratios of the elliptical cavities. For example, for the special case of penny-shaped cracks, $\alpha = 2(2 - \nu)^2$ and $\beta = \nu(4 - \nu)(2 - \nu)^2$. Note that (1.26) implies that crack initiation is independent of the intermediate principal stress.

Two major restrictions are implicit in this classical Griffith approach. First, it is assumed that all cracks are open and, second, it is assumed that no internal pressure (such as a pore fluid pressure) acts within the cracks. In reality, both

24

of these assumptions are commonly violated. The application of even a modest confining pressure will tend to close cracks, and then normal stresses will act across the crack faces, and frictional forces will act along the faces to resist sliding. In addition, any fluid pressure within the cracks acts to reduce the normal stress across the crack according to the law of effective stress (e.g. see Jaeger & Cook 1976, Paterson 1978). Taking account of these effects, and assuming a simple uniform stress distribution on the crack faces, McLintock & Walsh (1962) and Murrell (1964) calculated the condition for fracture initiation in a biaxial stress field, which was later extended to three dimensions by Murrell & Digby (1970). For a body containing only closed penny-shaped cracks under triaxial conditions, the fracture initiation condition is given by

$$[(1 + \mu^2)^{1/2} - \mu](\sigma_1 - \sigma_3) = 2(2 - \nu)\sigma_t[(1 + \sigma^*/\sigma_t) + 2\mu](\sigma_3 - \sigma^*) \quad (1.27)$$

where μ is the coefficient of friction between the crack faces, and σ^* is the critical compressive normal stress required for crack closure.

1.5.2.2 TIME-DEPENDENT FAILURE MODELS

None of the above criteria based on Griffith crack theory make any attempt to take account of time-dependency, nor do they relate directly to the fracture mechanics concept of stress intensity. Amongst the first attempts to include subcritical crack growth involving stress corrosion into theories of compressive deformation were the brittle uniaxial creep models developed for rocks by Scholz (1968a) and Cruden (1970, 1974). Scholz (1968) postulated that microcracks grew in response to local stress concentrations that varied from point to point even under a constant remote stress, due to heterogeneity of the microstructure, and simulated this variation by means of a probability function. Time-dependence was accounted for by assuming that the crack growth rate was controlled by stress corrosion reactions. Additionally, the rock material was assumed to be composed of a number of small zones, each with a fracture strength that also varied statistically. When the local stress in any zone exceeds its fracture strength then that zone is deemed to have failed and can no longer support any of the remote stress. The total stress must then be supported by the remaining intact zones, and overall failure of the material occurs when a sufficient number of zones have failed.

Cruden's (1970, 1974) model was based on the growth of a distribution of microcracks. He determined the crack-tip stress for a crack at any angle to the remote applied stress, computed the crack growth rate using stress corrosion theory (Eqn (1.15)), and then summed the results over all possible angles. In this way, Cruden was able qualitatively to predict the dependence of creep rate on stress, temperature, and time.

Using a rather different approach, a number of workers (Mizutani et al. 1977, Soga et al. 1979, Spetzler et al. 1981) have developed a model of compressive rock failure that attempts to predict fracture strength as a function of time, strain rate, humidity, and temperature, assuming that brittle failure occurs as a result of the interaction of numerous microcracks formed under the combined influence of applied stress and moisture at crack tips. If the applied stress rate is low and the humidity high, then the initial crack population will increase in length and coalesce into a failure plane at low stress. By contrast, if the stress rate is high and the humidity low then the existing crack population will not be able to relieve the applied stress, and many new microcracks will be formed. In this case, failure occurs by the coalescence of a larger number of smaller cracks. Assuming that for a given crack configuration (size, shape, and density) the rock will fail at a stress (σ_f) when the cracks have reached an average critical length, and that crack growth is governed by the mechanism of stress corrosion, then

$$\sigma_f = B[(\ln \dot{\sigma} - \ln T - n \ln P_0 - C)RT + Q] \qquad (1.28)$$

where T is the absolute temperature, n is the stress corrosion index, P_0 is the partial pressure of water, R is the universal gas constant, Q is the activation energy for stress corrosion, and B and C are constants that depend on activation volume, initial crack configuration, and rock type.

In more recent refinements of this approach (Spetzler *et al.* 1982, Brodsky *et al.* 1985), the main assumption has been that the applied stress is supported by the intact material between a population of microcracks. As the microcracks grow, the area of intact supporting material decreases until a critical failure stress is reached. The average crack length ($\bar{a}$) is used as a state variable to describe the state of 'crack damage' at any time, and the variation of $\bar{a}$ with stress and time is again derived from stress corrosion theory.

Costin (1987) has criticized those models which rely on statistical distributions of strengths and crack sizes to describe compressive failure, since the values of the parameters calculated depend on the particular distribution chosen and are therefore not necessarily directly related to the physical process of fracturing. On the other hand, in the models based on microstructural variables such as average crack length, the parameters do have some physical meaning and hence may be related to the fracturing process and its continuum response. However, although some of the models have had some success in predicting rate- and time-dependent failure in uniaxial and axisymmetric compression, there are severe doubts about their applicability to more general compressive loading conditions, and a more versatile approach based on microscopic processes is required (Costin 1987).

1.5.2.3 CONTINUUM DAMAGE MODELS

Since macroscopic brittle failure in compression can occur at stresses up to twice that required for microcrack initiation (e.g. Brace *et al.* 1966, Jones 1988, Meredith *et al.* 1990), a full treatment of failure requires consideration

26

of all phases of crack development including crack nucleation, crack extension, and crack interaction. In recent years a number of authors have contributed to the development of a new body of theory known as 'damage mechanics' that takes account of all these phases of cracking in an attempt to explain the various aspects of non-linear, time-dependent mechanical behaviour referred to earlier by utilizing the concept of a single variable or set of variables to describe the changing microstructural state of a material as it is deformed (e.g. Costin 1983, 1985; Horii & Nemat-Nasser 1986; Ashby & Hallam 1986; Sammis & Ashby 1986; Kemeny & Cook 1987; Hallam & Ashby, this volume). Costin (1987) has pointed out that there are three basic elements required for any damage theory: (1) a definition of the state of damage, (2) an equation to describe damage evolution, and (3) a constitutive law that predicts the relation of damage to stress and strain. Gueguen et al. (this volume) review most of the damage mechanics models proposed to date, and make a detailed comparison of the various model predictions for the uniaxial compressive strength of a micrograined limestone based on their own single tensile crack growth results. Hence only a very brief summary of the models is presented here for completeness.

The model developed by Hallam & Ashby (this volume; see also Ashby & Hallam 1986) is based on a fracture mechanics analysis of the growth of 'wing cracks' from pre-existing microcracks inclined to the direction of the maximum applied compressive stress. This growth of tensile wing cracks from inclined shear cracks was the model first proposed by Brace & Bombolakis (1963) to explain rock dilatancy. Such wing cracks are postulated to grow parallel to the maximum principal compressive stress direction.

The model develops criteria for wing-crack initiation and wing-crack growth based on the fracture toughness of the material, the ratio of the principal stresses, the coefficient of friction on the inclined crack, and the ratio (L) of the length of the wing crack to the half-length of the initial inclined flaw (see Hallam & Ashby, this volume; and Table 2.1 of Gueguen et al., this volume, for details). Physically, the result for crack growth encompasses three régimes: (1) when L is zero, the stress intensity factor for crack initiation is determined by shear of the initial inclined flaw; (2) when L is close to unity, the stress intensity for further growth is dominated by the wedging open of the wing crack due to shear displacement on the inclined flaw; and (3) when L is large, the stress intensity is very dependent on the confining stress (σ_3) since this acts across the entire crack length.

The model has been further developed to consider an array of inclined cracks, and growth of their associated wing cracks (Ashby & Hallam 1986). Eventually, the latter grow and interlink to divide the material into ligaments or beams, through which the remotely applied stresses are transmitted as axial forces and bending moments. The bending caused by wing-crack linkage leads to an additional term in the stress intensity expression (Gueguen et al., this volume, Table 2.1). Finally, the model criteria can be used to define surfaces in principal stress space that describe fracture initiation and fracture propagation for different loading configurations (e.g. plane stress, plane strain, axisymmetry). Yield surfaces for yield or creep behaviour may also be constructed on the same diagram, thus allowing assessment of the different possible failure mechanisms under various states of stress.

Sammis & Ashby (1986) have analysed the compressive failure of poro-elastic materials in a similar manner, but by considering spherical holes rather than sharp inclined flaws as crack nucleation sites.

The models of Horii & Nemat-Nasser (1986) and Kemeny & Cook (1987) are also based on the concept of the growth of wing cracks from the ends of inclined flaws. Kemeny & Cook (1987), for example, consider the growth of wing cracks at an angle to the initial flaw that varies with wing-crack length (see Fig. 2.9a of Gueguen et al., this volume). This variation in wing-crack orientation is introduced to take account of the curved shape of wing cracks observed in fracture experiments on a number of model materials (e.g. Horii & Nemat-Nasser 1986; Hallam & Ashby, this volume). Once again, wing-crack linkage results in separation of the material into axially aligned columns, and failure occurs either within a single column (axial failure) or across adjacent columns (shear failure).

Dilatancy is a fundamental and ubiquitous feature of rock deformation under triaxial compression, and any viable failure model must therefore be able to reproduce all of its observable characteristics. Stevens & Holcomb (1980) have criticized the wing-crack model of deformation since it fails to account for a number of important properties of dilatancy in granite subjected to a cyclic differential stress. For example, elastic wave velocities are observed to increase immediately on unloading, implying that cracks begin to close as soon as unloading commences. If wing cracks growing from the ends of inclined shear cracks were responsible for dilatancy, then hysteresis in elastic wave velocities would be expected since wing cracks should not begin to close until the frictional constraint on the shear cracks was overcome. Further more, they point out that Tapponier & Brace (1976) observed only very few well developed sliding surfaces compared with cracks with preferred orientations close to the maximum principal stress axis in their microscopy study of stress-induced microcracking in Westerly granite. By contrast, Holcomb & Stevens (1980) show that an array of reversible, axial, tensile microcracks ('Griffith cracks') is in agreement with all of the observed properties of dilatancy, and therefore conclude that this is the more viable model.

Costin (1983, 1985, 1987) has developed a damage model for rock based on a fracture mechanics description of the growth of an array of collinear tensile microcracks, oriented parallel to the direction of the maximum principal compressive stress. It is assumed that, even under compressive loading, local regions of tension exist due to the heterogeneous nature of the material. Individual crack extension occurs when the local crack-tip stress intensity due to tensile loading exceeds the fracture toughness, K_{Ic}. It is then assumed that the form of the equation describing the response of an ensemble of cracks is the same as that describing the response of an individual crack, and crack interaction is taken account of by employing the 'pseudo-traction' method of Horii & Nemat-Nasser (1983). This technique, and the resulting stress intensity formulation, are summarized by Gueguen *et al.* (this volume). In Figure 1.5 is shown the relation between normalized stress intensity ($K_I^* = 3\pi K_I/2\sigma\sqrt{\pi\alpha}$) and the ratio of crack length to inter-crack spacing (a/d_i) for interacting penny-shaped cracks in a material subjected to a uniaxial compressive stress (σ). For small a/d_i crack growth is stable, since crack growth results in a decrease in the crack driving force, K^*. However, further crack growth eventually leads to stronger interaction, and eventually to unstable propagation when K^* becomes an increasing function of a/d_i.

Costin's later (1985) development of the model also takes account of time-dependent crack growth by stress corrosion, and concludes with a constitutive equation relating strains to the imposed stresses and the current state of damage. The constitutive equation can be used to construct 'damage surfaces' in differential stress ($\sigma_1 - \sigma_3$) against confining stress (σ_3) space, ranging from the surface describing the initial onset of damage to that describing failure. A series of experiments reported by Holcomb (1983) supports Costin's (1983) assertion that these surfaces should be straight lines. Holcomb assumed that the onset of acoustic emission activity in triaxial tests on Westerly granite samples corresponded to the onset of microcrack damage. He increased the axial load on his samples just up to the onset of acoustic emission, and then increased the confining stress before increasing the axial load further until

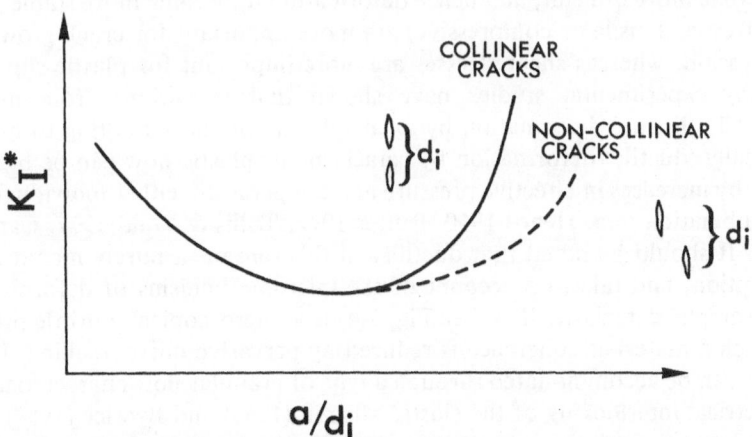

Figure 1.5 The relation between normalized stress-intensity factor (K_I^*) and the ratio of crack length to inter-crack spacing (a/d_i) for an array of equisized penny-shaped cracks (after Costin 1985). The formulation for K_I^* in terms of stress, crack length, and stress intensity is given in the text. The diagram illustrates how crack growth is initially stable, but becomes unstable as cracks become longer and the ratio a/d_i increases.

acoustic emission started once again. In this way he was able to map out a surface describing damage initiation that did indeed turn out to be a straight line.

All of the damage models discussed here make use of very simple crack distributions in order to simplify the complex problem of compressive fracture in polycrystalline materials. In fact, they all assume an array of parallel cracks of equal length. However, a number of workers (see Kranz 1983) have shown that the crack-length distribution function in rocks must be a power law to be compatible with the Gutenberg–Richter frequency–magnitude relation observed both for the earthquake seismicity associated with crustal-scale rupture (Vere-Jones 1976, 1977) and for the acoustic emissions associated with laboratory-scale fracture (Scholz 1968b, Meredith *et al.* 1990). Obviously, the development of continuum damage models is in its infancy, but this approach holds out much promise for enabling physically realistic descriptions to be made of the deformation behaviour of complex heterogeneous materials under different stress régimes.

1.5.3 Brittle-to-ductile transitions

Murrell (this volume) points out that the chemical bonding and crystal structure of most minerals, ceramics, and rocks causes them to be typically brittle at atmospheric pressure and temperature, and commonly to remain brittle even at relatively high homologous temperatures (e.g. Murrell & Chakravarty 1973). The fracture and flow properties of such materials are strongly influenced by the nature of the applied stress system. We have noted earlier how the application of a confining stress causes crack nucleation and propagation to become more difficult, and hence deformation to become more stable. Normal stresses (tensile or compressive) are more important for crack growth or suppression, whereas shear stresses are more important for plastic slip.

Many experimental studies have shown that transitions from macroscopically 'brittle' deformation by axial splitting or shear faulting to macroscopically 'ductile' deformation by cataclastic or plastic flow can be brought about by increases in effective pressure and temperature, either individually or in combination (e.g. Heard 1960, Rutter 1972, Tullis & Yund 1977, Caristan 1982). It should be noted that ductility in this sense is a purely macroscopic description, and takes no account of the micromechanisms of deformation. For example, cataclastic flow (see Fig. 1.4) is a microscopically brittle process in which a materials coherence is reduced by pervasive microcracking. Large strains can be accommodated through a type of granular flow characterized by geometrical interlocking of the clasts. Murrell (1965) and Byerlee (1968) both showed that the transition from shear faulting to cataclastic flow occurs when the confining pressure is sufficiently high for the differential stress for failure (i.e. compressive strength) to equal the differential stress required for frictional sliding. This is clearly illustrated by Murrell (this volume) in his Figure 5.4. Of course, cataclastic flow is pressure-dependent, since it induces dilatancy.

Hence, at very high effective confining pressures cataclastic flow may be supplanted by plastic flow as the dominant deformation mechanism, even in the absence of an elevated temperature, because plastic flow is independent of pressure since it is non-dilatant.

Murrell (this volume) discusses the important deformation mechanisms, and uses deformation mechanism maps (Gandhi & Ashby 1979) and a Mohr diagram (see Paterson 1978) to illustrate the various transitions from fracture to cataclastic and plastic flow. The maps show the differences in behaviour between different 'isomechanical groups' (Frost & Ashby 1982). Isomechanical groups define classes of materials with closely similar mechanical behaviour. The brittle-to-ductile transition temperature is shown to depend on the strain rate, the grain size, the isomechanical class, and also on the applied stress system. For example, an increase in effective confining pressure leads to an increase in strength but a decrease in transition temperature. Hence, a combination of high temperature, high confining pressure, absence of a pore fluid (to ensure high effective pressures), and a low strain rate is required to ensure ductile flow of most minerals, ceramics, and rocks by crystal plasticity.

1.6 Indirect monitoring of fracture

Real-time, *in-situ* observation of crack development during deformation, although desirable, is notoriously difficult, especially in opaque polycrystalline materials such as ceramics and rocks. A few *in-situ* scanning electron microscopy studies have been made (e.g. Lindqvist *et al.* 1984; Swanson 1987; Freiman & Swanson, this volume), and these have been very valuable. However, in this type of study, the sample size is constrained to be very small; and it must necessarily be deformed in a high vacuum, thus precluding any time-dependent influence of active environmental species. More commonly, crack development and statistics are studied by stopping a deformation experiment at some prearranged point, unloading the sample, and then removing it from the test apparatus for examination. Examination also normally involves some sample preparation technique, such as cutting, grinding, or polishing. Both the unloading of the sample and its subsequent preparation for examination are likely to change the crack structure in some unknown way.

Hence, several indirect methods of monitoring crack growth and interaction have been developed. The most common of these are the study of changes in the properties of extrinsic elastic waves pulsed through the sample during deformation, and the monitoring of the intrinsic elastic waves generated by propagating cracks, known as 'acoustic emissions'. Changes in the properties of extrinsic elastic waves reflect the state of accumulated crack damage at any time, whereas acoustic emission statistics reflect the contemporary rate of damage accumulation. However, both techniques sample the three-dimensional crack structure.

31

1.6.1 Elastic wave velocities

The most common property of extrinsic elastic waves that is monitored during deformation is the wave velocity. An example of changes in the velocity of axially pulsed compressional (P) and shear (S) waves during triaxial deformation of a rock sample that failed by shear faulting is shown in Figure 1.6 (after Sammonds *et al.* 1989). The velocity of both P- and S-waves increases during the initial quasi-linear phase of loading, with maximum velocity occurring at about half of the peak differential stress. This behaviour is interpreted as being due to the closure of cracks in response to the increasing applied stress. The most favourably oriented (i.e. those oriented normal or subnormal to the axial stress) and most open cracks close first, and crack closure then becomes progressively more difficult. Above about half the peak stress, new, dilatant cracks begin to propagate, and hence the wave velocities start to decrease. The wave velocities only stabilize during the final phase of deformation – frictional sliding on the fault. This is entirely as expected, since frictional sliding occurs at essentially constant stress, and hence constant state of damage. However, note that during the initial phase of deformation, the relative increase in V_p is substantially higher than the increase in V_s, but that V_s decreases at a higher rate than V_p in response to dilatant crack growth. Since axial P-waves are more sensitive to cracks normal to the axial stress, and

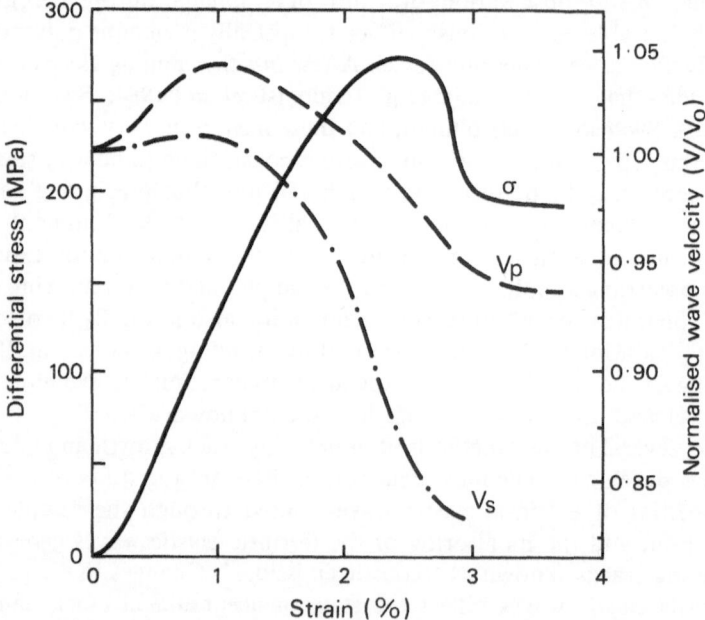

Figure 1.6 Variations in the velocities of compressional (V_p) and shear (V_s) wave velocities in a dry sample of Darley Dale sandstone deformed under a confining pressure of 50 MPa at a strain rate of $10^{-5}\,\mathrm{s}^{-1}$ (after Sammonds *et al.* 1989).

S-waves more sensitive to cracks parallel to the axial stress, this observation supports the concept that dilatancy occurs by the growth of tensile cracks oriented parallel to the maximum compressive stress. It also shows that measurement of changes in both velocities provides a guide to the development of crack anisotropy during progressive deformation.

Elastic wave velocity changes can also provide information about changes in deformation mechanism that might not be apparent from stress–strain relations alone. Data from Jones (1988) for deformation of Solnhofen limestone at temperatures from ambient to 300°C is shown in Figure 1.7. The stress–strain curves are all qualitatively similar, with little indication of any change in deformation mechanism. However, the samples failed by two entirely different flow mechanisms. At room temperature the flow was cataclastic and dilatant, and therefore accompanied by a continuous decrease in wave velocity after the initial slight increase due to crack closure. By contrast, at 300°C deformation is accompanied by a continuous increase in wave velocity that the author attributes to non-dilatant crystal plasticity and plastic pore compaction. At intermediate temperatures it is likely that the velocity curves reflect transitional behaviour between these competing mechanisms.

A number of theories have been proposed to relate changes in wave velocities to change in elastic moduli and thereby to changes in microcrack concentration or density. One of the first of these was by Walsh (1965), who calculated the excess strain energy due to the presence of an isolated penny-

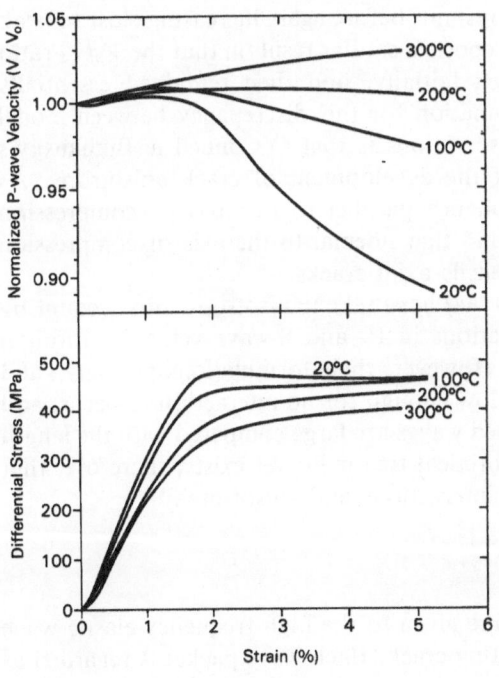

Figure 1.7 Stress and compressional wave velocity as a function of strain and temperature for dry samples of Solnhofen limestone under triaxial loading at a confining pressure of 100 MPa and a strain rate of $10^{-5}\,s^{-1}$ (after Jones 1988). The data illustrate the change from cataclastic flow at ambient temperature to plastic flow at 300°C.

shaped crack in an infinite medium under stress. However, Walsh's formulation did not take account of stress-field interactions between adjacent cracks, and is therefore only applicable to low densities of relatively short cracks. On the other hand, O'Connell & Budiansky (1974) specifically included crack interactions in their model, which assumes that a crack is imbedded in an homogeneous medium the effective elastic moduli of which are those of the cracked material. Not surprisingly, the theory that takes account of crack interactions predicts a much more rapid decrease in moduli with increase in crack density and average crack length than does the formulation that ignores interactions.

Salganik (1973) has developed a slightly different model which considers cracks emplaced in a medium sequentially. The strain energy of the $(N+1)$th crack is calculated as if it were emplaced in a homogeneous medium, the moduli of which were those of the material containing N cracks. Zimmerman (1984) and Zimmerman & King (1985) provide results supporting Salganik's model, in which they use changes in V_p data to predict changes in V_s as a function of pressure in a cracked solid.

O'Connell & Budiansky's model predicts that not only both V_p and V_s, but also the ratio V_p/V_s, would decrease with increasing crack density. However, experimental results reported by Gupta (1973) and Sammonds et al. (1989) do not support this latter prediction. For example, Sammonds et al. note that, for their experimental results (e.g. Fig. 1.6), the V_p/V_s ratio initially increases with increase in differential stress. This initial increase is followed by a period during which the ratio is essentially constant, before again increasing close to peak stress and beyond. Gupta (1973) reports a similar result in that the V_p/V_s ratio in his experiments also increased initially, and then remained essentially constant up to failure. The explanation for this discrepancy between model predictions and experimental observations is that O'Connell & Budiansky's model does not take account of the development of crack anisotropy. As rocks are deformed, their compliance parallel to the axis of compression decreases due to crack closure, and that normal to the axis of compression increases due to the growth of tensile axial cracks.

Hudson (1981) and Crampin (1984) have taken anisotropy into account by calculating theoretically the variations in P- and S-wave velocities through solids containing aligned cracks. However, their formulations are only valid for crack concentrations that are low enough for no interaction to occur, and where the wavelengths of the pulsed waves are large compared with the length of the cracks. No systematic theoretical treatment yet exists, therefore, that takes full account of both crack interactions and anisotropy.

1.6.2 Acoustic emissions

Acoustic emission (AE) is the name given to the high-frequency elastic wave 'packets' generated by a propagating crack. Each wave packet is regarded as an individual AE 'event'. The mechanism of acoustic emission generation is thought to be the release of elastic energy caused by the rapid advance of a crack tip. During deformation experiments, AE activity is detected by transducers bonded to the sample surface. For laboratory-sized rock or ceramic samples, it has been usual for a rather narrow frequency band within the range 50 kHz to 1 MHz to be monitored, although more recently, broader bandwidth (e.g. 100 kHz to 1 MHz) miniature transducers have become available. In most applications it is necessary to compromise between bandwidth and transducer size, since bandwidth generally decreases with decrease in transducer size. The AE technique is in an active state of development, and modern instrumentation can be used to record a wide range of parameters, including: the AE event rate as a function of time, load, strain, etc.; the number of individual vibrations (known as 'counts') per event; the peak amplitude of each event; the time to peak amplitude ('rise time'); the event duration; the energy of the event; and the frequency content, amongst others. However, to date, by far the most commonly recorded parameters have been the event rate, the cumulative number of events, and the distribution of event amplitudes, as functions of load, strain, or time.

The first reports of the use of acoustic emission to detect microcracking in rocks were by Obert (1941) and Obert & Duvall (1942). Since that time, a number of researchers have reported the results of laboratory AE studies of rock deformation under a variety of loading configurations, including tensile loading, uniaxial and triaxial compressive loading, constant stress (creep) loading, and cyclic loading (e.g. Mogi 1968, Scholz 1968b, Gowd 1980, Atkinson & Rawlings 1981, Boyce et al. 1981, Ohnaka & Mogi 1982, Sano et al. 1982, Meredith & Atkinson 1983, Sondergeld et al. 1984, Fonseka et al. 1985, Meredith et al. 1990).

In Figure 1.8 is shown a log–log plot of crack growth rate and AE event rate against stress intensity factor for tensile crack growth in plates of Whin Sill dolerite under different environmental conditions (after Meredith & Atkinson 1983). Note that the AE event rate has an identical functional relation to stress intensity factor as the crack growth rate. Hence, Meredith & Atkinson (1983) conclude that the event rate can be used as a remote monitor of crack velocity, for tensile crack growth at least.

By contrast, in Figure 1.9 are shown both AE event rate and P-wave velocity data from a triaxial deformation experiment on a sample of Darley Dale sandstone (after Jones 1988). Significant AE activity commences at about half the peak stress, corresponds approximately to the peak P-wave velocity, and indicates the onset of dilatant cracking (point 'a' of Fig. 1.9). The AE rate then increases rapidly up to peak stress (point 'b'), whereupon it remains steady for a short period before decreasing dramatically to a period of apparent quiescence at dynamic failure (point 'c'). Immediately following failure the rate recovers before finally decaying during stable sliding on the shear fault (point 'd' onwards). The phenomenon of quiescence close to peak stress has

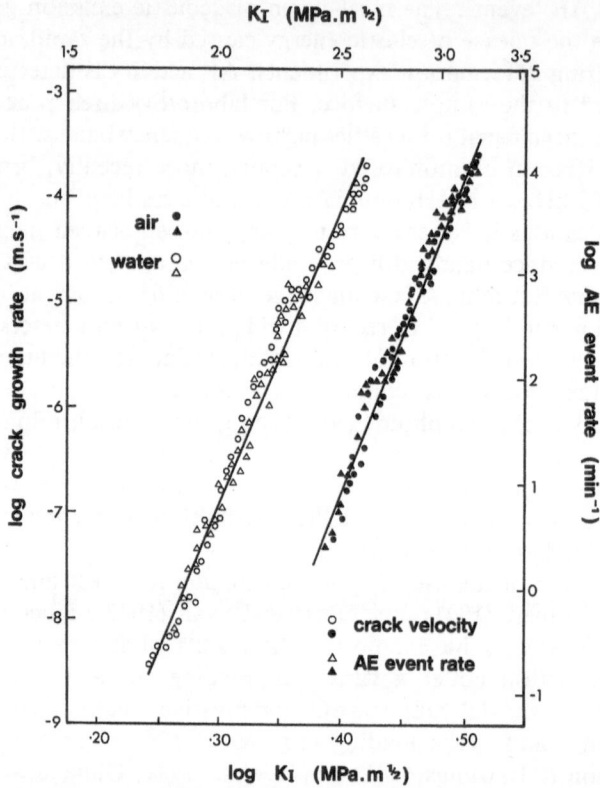

Figure 1.8 Experimental data illustrating the dependence of crack-growth rate and acoustic emission event rate on stress intensity factor (K_I) and crack tip humidity for tensile crack growth in samples of Whin Sill dolerite at $20°C$ (after Meredith & Atkinson 1983). Solid lines are least-squares fits to the data.

been reported, by a number of researchers (e.g. Gowd 1980; Sondergeld *et al.* 1984), to occur close to catastrophic failure, but only when a period of strain softening has preceded failure. Quiescence may be a real or an instrumental phenomenon, and a number of possible explanations is discussed by Meredith *et al.* (1990).

Ohnaka & Mogi (1982) and Sondergeld *et al.* (1984) report changes in the frequency content of emissions during progressive deformation, which they attribute to the changing microstructure. Once dilatancy is well developed, microcracks interact and link to form larger cracks and hence larger source dimensions, and this is associated with an increase in the low-frequency content of AE signals. Under these conditions, the higher-frequency components would also be preferentially attenuated by the open, dilatant cracks. However, as failure is approached, a contrasting enhancement of high-frequency components is observed. Once again, this is interpreted as being due to a change in

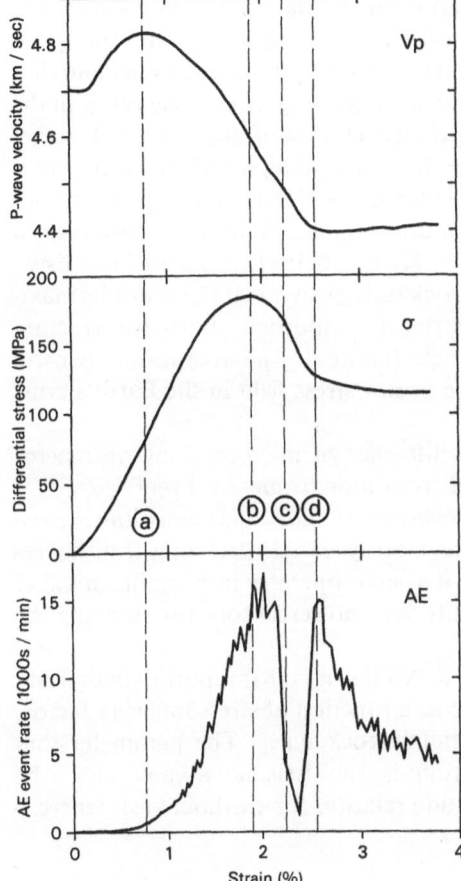

Figure 1.9 Compressional wave velocity (V_p), stress (σ), and acoustic emission event rate (AE) as functions of strain for dry Darley Dale sandstone undergoing triaxial deformation at a confining pressure of 50 MPa and a strain rate of 10^{-5} s^{-1} (after Jones 1988).

source dimension. As crack coalescence relieves the applied stress by producing greater numbers of larger cracks during strain softening, eventually the only sites where further crack growth can occur are the relatively small barriers and ligamentary bridges between the large cracks (see Swanson 1987; Freiman & Swanson, this volume). Crack growth therefore temporarily decreases in dimension, with a concomitant decrease in AE amplitude and increase in high-frequency content. Both of these relative shifts in AE frequency content can therefore be explained as manifestations of the microcracking process leading to macroscopic failure (Sondergeld *et al.* 1984).

We have already noted that Holcomb (1983) has tested some of the predictions of Costin's (1983) damage model by mapping out a damage initiation surface using AE data. He was able to do this because of a particular AE phenomenon known as the 'Kaiser' or stress memory effect. Kaiser (1953) noted that when metal specimens were cyclically loaded, no significant AE was gener-ated on a particular cycle until the load exceeded that on the previous cycle. The concept is simply that an amount of damage is introduced into the speci-men, such as by the growth of microcracks, on the first loading cycle, and that on unloading the cracks will close. Subsequent loading will merely result in the re-opening of existing cracks, without the generation of any further damage, until the previous load is exceeded. Any further loading will then result in new crack growth, and hence AE will re-commence. Several authors have con-firmed that a similar, although rather more complicated, effect is observed in stressed rock (e.g. Kurita & Fujii 1979, Holcomb 1981, Sondergeld & Estey 1981). This observation that cracks in rock only grow when the previous maxi-mum stress has been exceeded is extremely important, both for fracture mechanics analyses of progressive rock failure by microcracking (Costin 1987), and for the interpretation of the *in-situ* stress field in the Earth's crust (e.g. Yoshikawa & Mogi 1981).

In addition to the insights into the influence of microcracking on macro-scopic deformation that can be gained from monitoring AE event rates, it is also now possible to determine the location of individual emissions, given a relatively large sample and a sufficient array of AE detecting transducers (e.g. Sondergeld & Estey 1981). Such a system operates in a similar manner to a seismic network and is a potentially very powerful tool for studying the localization of deformation.

In addition to event rates, Meredith & Atkinson (1983) report data describ-ing the distribution of event amplitudes as a function of stress intensity factor, from tensile tests on a range of crystalline rock types. The parameter that characterizes the amplitude distribution is the 'seismic b-value' (cf. the Gutenberg–Richter frequency–magnitude relation for earthquakes), where b is an empirical constant in the relation

$$\log N = a - bm \qquad (1.29)$$

N is the number of times an event of magnitude (or peak amplitude in dB) m occurs, and a is a constant. The data compiled from a number of tensile tests are shown in Figure 1.10. For these experiments, the b-values are seen to lie in the range 1–3, and are negatively correlated and linearly related to K/K_c, the normalized stress intensity. Physically, a high b-value characterizes crack growth dominated by a large number of relatively small events, whereas a low b-value indicates a larger number of relatively large events. Note also that the b-value is higher in a 'wet' as opposed to a 'dry' environment, especially at low values of K. Changes in b may therefore be due either to changes in stress intensity or to changes in the humidity of the crack-tip environment. Meredith & Atkinson (1983) suggest that this may be related to the observation of a trend from dominantly transgranular cracking at high crack velocity and low humidity, to dominantly intergranular cracking at low velocity and high

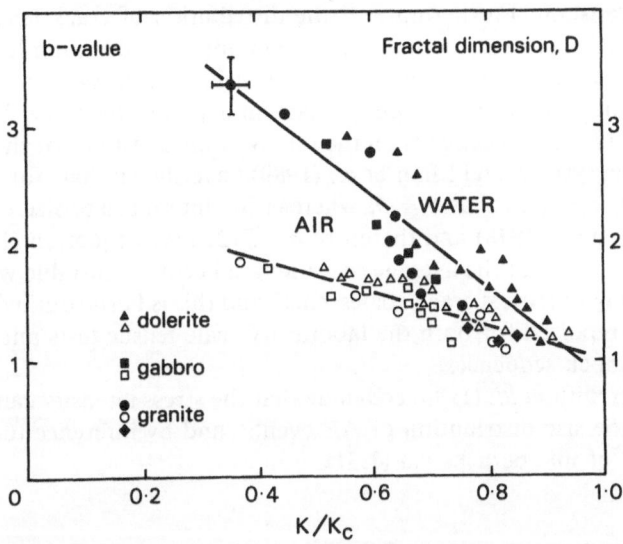

Figure 1.10 Synoptic diagram of acoustic emission b-value and crack fractal dimension (D) as functions of normalized stress-intensity factor (K/K_c) and crack-tip humidity for a range of crystalline rocks (after Meredith & Atkinson 1983, and Main *et al.* 1989b). Data are from tensile crack-growth experiments performed at room temperature and pressure, in air and de-ionized water. Solid lines are least-squares fits to the data.

humidity, since grain boundaries provide the main conduits for the access of water and will therefore stress corrode preferentially. At values of K close to the fracture toughness crack growth will be very rapid, and hence the environment has little effect on either crack growth rate or the b-value. For either environment, the critical b-value at dynamic failure (i.e. $K = K_c$) is unity. Since it has already been shown that a great deal of evidence supports the idea that fracture in compression is dominated by the growth of tensile cracks, a number of authors (e.g. Das & Scholz 1981, Main 1988, Main *et al.* 1989, Meredith *et al.* 1990) have suggested that it is reasonable to use results derived from tensile tests to make predictions about compressive deformation.

In the Earth's crust, b-values are generally found to be in the range 0.5–1.5, and the critical b-value, evaluated from earthquake foreshock sequences, is 0.5 (Von Seggern 1980). The discrepancy between the field and laboratory data can be explained by considering a power-law distribution of crack lengths (a) with exponent D, such that

$$N(a) \propto a^{-D} \qquad (1.30)$$

where several authors (e.g. Caputo 1976, Aki 1981) have shown that

$$D = 3b/c \qquad (1.31)$$

and c is a constant. Furthermore, if the distribution of crack lengths is geometrically scale-invariant between the minimum and maximum lengths, then D is strictly defined as one of the fractal dimensions of the system (Mandelbrot 1982, Turcotte 1989). The more fundamental parameter is D, because the value of c and hence b depends on the relative time constants of the event and the recording system, and Main *et al.* (1989) have shown that, for the data of Figure 1.10, $c = 3$, so that $b = D$; whereas for intermediate-size earthquakes $c = 3/2$ (Kanamori 1978) and therefore $b = D/2$. From equation (1.30), Main (1988) has shown that the average crack length becomes unstable when $D = 1$, corresponding to critical crack coalescence, and this is borne out by the critical values of b reported for both the laboratory-scale tensile tests and the earthquake foreshock sequences.

Thus, Meredith *et al.* (1990) conclude that the stress intensity can be related directly to the size distribution of AE events, and by inference to the length distribution of microcracks via (1.31).

1.7 Concluding remarks

In an overview of this nature it is not possible to do justice to all aspects of a field of research under such rapid and intense development as that of fracture in brittle polycrystalline materials. Therefore this chapter has attempted to concentrate on those aspects of brittle failure that are the most relevant to the following chapters in this volume, possibly at the expense of other equally important areas. Important topics that have been neglected to a greater or lesser extent include: time-dependent crack growth mechanisms other than stress corrosion (reviewed by Atkinson & Meredith 1987a); crack growth by cyclic fatigue (Kim & Mubeen 1981, Costin 1987); mixed-mode failure and, especially, shear failure and slip-weakening (Rudnicki 1984, Rudnicki & Rice 1975, Li 1987); shear localization (Evans & Wong 1985); inelastic fracture mechanics analysis by path-independent integrals (Rice 1968, Li 1987); and crack growth and fracture mechanism maps (Fields & Fuller 1981, Gandhi & Ashby 1979) that are analogous to deformation mechanism maps for plastic deformation. Interested readers are directed to the relevant references listed with these topics for more details.

It will be apparent that our understanding of the processes associated with the nucleation and growth of a single crack in a stress field is relatively well developed, even for complex materials. However, we still have no quantitative way of describing the crack-tip shielding processes that contribute to inelastic fracture propagation. Nevertheless, the new concept broadly classified as damage mechanics leads to some optimism concerning our ability to integrate knowledge of the mechanics of single crack growth to predict the continuum response of a material containing an ensemble of cracks. However, in order for meaningful predictions of mechanical behaviour to be made, more detailed models will be required that accommodate time-dependent cracking and more realistic descriptions of both initial flaw distributions and the developing state of damage. This is likely to be an iterative process that will require input from the techniques that indirectly monitor damage development during progressive deformation.

Acknowledgements

I should like to take this opportunity to thank Barry Atkinson for enthusiastically introducing me to fracture in rocks and minerals, and for sharing a number of years of very enjoyable collaborative research. Stan Murrell, Peter Sammonds and Ian Main are thanked for many helpful and constructive comments on earlier drafts of this overview.

References

Anderson, O. L. & P. C. Grew 1977. Stress corrosion theory of crack propagation with application to geophysics. *Rev. Geophys. Space Phys.* **15**, 77–104.

Aki, K. 1981. A probabilistic synthesis of precursory phenomena, In *Earthquake prediction: an international review*, 566–74. Maurice Ewing Series 4. Washington, DC: Am. Geophys. Union.

Aki, K. & P. G. Richards 1980. *Quantitative seismology*. 2 volumes. 932pp. San Francisco: Freeman.

Andrews, D. J. 1985. Dynamic plane strain shear rupture with a slip-weakening friction law calculated by a boundary integral method. *Bull. Seism. Soc. Am.*, **75**, 1–22.

Ashby, M. F. and S. D. Hallam 1986. The failure of brittle solids containing small cracks under compressive stress states. *Acta Metall.* **34**, 497–510.

Atkinson, B. K. 1980. Stress corrosion and the rate-dependent tensile failure of a fine-grained quartz rock. *Tectonophysics* **65**, 281–90.

Atkinson, B. K. 1982. Subcritical crack propagation in rocks: theory, experimental results and applications. *J. Struct. Geol.* **4**, 41–56.

Atkinson, B. K. 1984. Subcritical crack growth in geological materials. *J. Geophys. Res.* **89**, 4077–114.

Atkinson, B. K. 1987. Introduction to fracture mechanics and its geophysical applications. In *Fracture mechanics of rock*, B. K. Atkinson (ed.), 1–26. London: Academic Press.

Atkinson, B. K. & P. G. Meredith 1987a. The theory of subcritical crack growth with application to minerals and rocks. In *Fracture mechanics of rock*, B. K. Atkinson (ed.), 111–66. London: Academic Press.

Atkinson, B. K. & P. G. Meredith 1987b. Experimental fracture mechanics data for rocks and minerals. In *Fracture mechanics of rock*, B. K. Atkinson (ed.), 477–525. London: Academic Press.

Atkinson, B. K. & R. D. Rawlings 1981. Acoustic emission during stress corrosion cracking in rocks. In *Earthquake prediction: an international review*, 605–16. Maurice Ewing Series 4. Washington, DC: Am. Geophys. Union.

Barenblatt, G. I. 1962. Mathematical theory of equilibrium cracks in brittle fracture. *Adv. Appl. Mech.* **7**, 55–129.

Boyce, G. M., W. M. McCabe & R. M. Koerner 1981. Acoustic emission signatures of various rock types in unconfined compression. In *Acoustic emissions in geotechnical engineering practice*, 142–54. ASTM STP 750. Philadelphia: American Society for Testing and Materials.

Brace, W. F. & E. G. Bombolakis 1963. A note on brittle crack growth in compression. *J. Geophys. Res.* **68**, 3709–13.

Brace, W. F., B. W. Paulding & C. H. Scholz 1966. Dilatancy in the fracture of crystalline rock. *J. Geophys. Res.* **71**, 3939–53.

Brodsky, N. S., I. C. Getting & H. Spetzler 1985. An experimental and theoretical approach to rock deformation at elevated temperature and pressure. In *Measurement of rock properties at elevated pressures and temperatures*, 37–54. ASTM Sp. Tech. Publ., STP 869.

Buresch, F. E. 1979. Fracture toughness testing of alumina, In *Fracture mechanics applied to brittle materials*, 151–65. ASTM Sp. Tech. Publ., STP 678.

Byerlee, J. D. 1968. The brittle–ductile transition in rocks. *J. Geophys. Res.* **73**, 4741–50.

Caputo, M. 1976. Model and observed seismicity represented in a two-dimensional space. *Ann. Geophys. (Rome)* **4**, 277–288.

Caristan, Y. 1982. The transition from high-temperature creep to fracture in Maryland diabase. *J. Geophys. Res.* **87**, 6781–90.

Charles, R. J. 1958. Static fatigue of glass. *J. Appl. Phys.* **29**, 1549–60.

Charles, R. J. & W. B. Hillig 1962. Kinetics of glass failure by stress corrosion. In *Symposium sur la Resistance du Verre et les Moyens de l'Ameliorer*, 511–27. Charleroi: Union Scientifique Continentale du Verre.

Cook. R. F., B. R. Lawn & C. J. Fairbanks 1985. Microstructure–strength properties in ceramics, I: effect of crack size on toughness. *J. Am. Ceram. Soc.* **68**, 604–15.

Costin, L. S. 1983. A microcrack model for the deformation and failure of brittle rock. *J. Geophys. Res.* **88**, 9485–92.

Costin, L. S. 1985. Damage mechanics in the post-failure regime. *Mech. Mat.* **4**, 149–60.

Costin, L. S. 1987. Time-dependent deformation and failure. In *Fracture mechanics of rock*, B. K. Atkinson (ed.), 167–215. London: Academic Press.

Crampin, S. 1984. Effective anisotropic elastic constants for wave propagation through cracked solids. *Geophys. J. R. Astron. Soc.* **76**, 135–45.

Cruden, D. M. 1970. A theory of brittle creep in rock under uniaxial compression. *J. Geophys. Res.* **75**, 3431–42.

Cruden, D. M. 1972. The static fatigue of brittle rock under uniaxial compression. *Int. J. Rock Mech. Min. Sci.* **11**, 67–73.

Das, S. & Scholz, C. H. 1981. Theory of time-dependent rupture in the Earth. *J. Geophys. Res.* **86**, 6039–51.

Dugdale, D. C. 1960. Yielding of steel sheetings containing slits. *J. Mech. Phys. Solids* **8**, 100–4.

Dunning, J. D., W. L. Lewis & D. E. Dunn 1980. Chemomechanical weakening in the presence of surfactants. *J. Geophys. Res.* **85**, 5344–54.

Dunning, J. D., D. Petrovski, J. Schuyler & A. Owens 1984. The effects of aqueous chemical environments on crack propagation in quartz. *J. Geophys. Res.* **89**, 4115–23.

Evans, A. G. 1972. A method for evaluating the time-dependent failure characteristics of brittle materials – and its application to polycrystalline alumina. *J. Mat. Sci.* **7**, 1137–46.

Evans, A. G. 1974. Fracture mechanics determinations. In *Fracture mechanics of Ceramics*, Vol. 1, R. C. Bradt, D. P. H. Hasselman & F. F. Lange (eds), 17–48. New York: Plenum.

Evans, A. G. 1976. On the formation of a crack tip microcrack zone. *Scripta Metall.* **10**, 93–7.

Evans, A. G. & H. Johnson 1975. The fracture stress and its dependence on slow crack growth. *J. Mat. Sci.* **10**, 214–22.

Evans, B. & T.-F. Wong 1985. Shear localization in rocks induced by tectonic deformation. In *Mechanics of geomaterials*, Z. P. Bažant (ed.), 189–210. Chichester: Wiley.

Fields, R. J. & E. R. Fuller Jr 1981. Crack growth mechanism maps. In *Advances in Fracture Research*, D. Francois (ed.), 1313–22. Oxford: Pergamon Press.

Fonseka, G. M., S. A. F. Murrell & P. Barnes 1985. Scanning electron microscope and acoustic emission studies of crack development in rocks. *Int. J. Rock Mech. Min. Sci. & Geomech. Abstr.* **22**, 273–89.

Freiman, S. W. 1984. Effects of chemical environments on slow crack growth in glasses and ceramics. *J. Geophys. Res.* **89**, 4072–6.

Freiman, S. W. & P. L. Swanson 1990. Fracture of polycrystalline ceramics. This volume, 72–83.

Frost, H. J. & M. F. Ashby 1982. *Deformation mechanism maps: the plasticity and creep of metals and ceramics*. Oxford: Pergamon Press.

Gandhi, C. & M. F. Ashby 1979. Fracture mechanism maps for materials which cleave: F.C.C., B.C.C. and H.C.P. metals and ceramics. *Acta Metall.* **27**, 1565–602.

Gowd, T. N. 1980. Factors affecting the acoustic emission response of triaxially compressed rock. *Int. J. Rock Mech. Min. Sci. & Geomech. Abstr.* **17**, 219–23.

Grady, D. E. & M. E. Kipp 1987. Dynamic rock fragmentation. In *Fracture mechanics of rock*, B. K. Atkinson (ed.), 429–75. London: Academic Press.

Griffith, A. A. 1920. The phenomenon of rupture and flow in solids. *Phil. Trans R. Soc. Lond., Ser. A* **221**, 163–98.

Griffith, A. A. 1924. The theory of rupture. In *Proc. 1st Int. Congr. Appl. Mech.* C. B. Biezano & J. M. Burgers (eds), 54–63. Delft: Waltman.

Gueguen, Y., T. Reuschlé & M. Darot 1990. Single-crack behaviour and crack statistics. This volume. 48–71.

Gupta, I. N. 1973. Seismic velocities in rocks subjected to axial loading. *J. Geophys. Res.* **65**, 1083–102.

Hallam, S. D. & M. F. Ashby 1990. Compressive brittle fracture and the construction of multi-axial failure maps. This volume, 84–108.

Hashida, T. 1989. Tension-softening curve measurements for fracture toughness determination in granite. In *Fracture toughness and fracture energy: test methods for concrete and rock*, H. Milashi, H. Takahashi & F. H. Wittmann (eds), 47–56. Rotterdam: Balkema.

Heard, H. C. 1960. Transition from brittle fracture to ductile flow in Solnhofen limestone as a function of temperature, confining pressure and interstitial fluid pressure. In *Rock deformation* D. T. Griggs & J. Handin (eds). *Mem. Geol. Soc. Am.* **79**, 193–226.

Henry, J. P. & J. Paquet 1976. Mecanique de la rupture de roches calcitiques. *Bull. Géol. Soc. Fr.* **7**, Ser. 18, 1573–82.

Hoagland, R. G., G. T. Hahn & R. A. Rosenfield 1973. Influence of microstructure on fracture propagation in rocks. *Rock Mech.* **5**, 77–106.

Holcomb, D. J. 1981. Memory, relaxation, and microfracturing in dilatant rock. *J. Geophys. Res.* **86**, 6235–48.

Holcomb, D. J. 1983. *Proc. Am. Soc. Mech. Engng. Symp. Mech. Rocks, Soils and Ice*, Houston, Texas, 11–21. New York: ASME.

Holcomb, D. J. & J. L. Stevens 1980. The reversible Griffith crack: a viable model for dilatancy. *J. Geophys. Res.* **85**, 7101–7.

Horii, H. & S. Nemat-Nasser 1983. Estimate of stress intensity factors for interacting cracks. In *Advances in Aerospace Structures and Materials*, R. M. Laurenson & U. Yuceoglu (eds), 111–17. AD Publ. Am. Soc. Mech. Engng, Aerospace Div., AD 06.

Horii, H. & S. Nemat-Nasser 1986. Brittle failure in compression: splitting, faulting and brittle–ductile transition. *Phil Trans R. Soc. Lond., Ser. A* **319**, 337–374.

Hübner, H. & W. Jillek 1977. Sub-critical crack extension and crack resistance in polycrystalline alumina. *J. Mat. Sci.* **12**, 117–25.

Hudson, J. A. 1981. Wave speeds and attenuation of elastic waves in material containing cracks. *Geophys. J. R. Astron. Soc.* **64**, 133–50.

Ingraffea, A. R. 1987. Theory of crack initiation and propagation in rock. In *Fracture mechanics of rock*, B. K. Atkinson (ed.), 71–110. London: Academic Press.

Irwin, G. R. 1958. Fracture. In *Handbüch der Physik*, S. Flügge (ed.), Vol. 6, 551–90. Berlin: Springer.

Irwin, G. R. 1960. Plastic zone near a crack and fracture toughness. *Proc. 7th Sagamore Conf.*, iv–63.

Jaeger, J. C. & N. G. W. Cook 1976. *Fundamentals of rock mechanics*, 2nd edn. London: Chapman and Hall.

Jones, C. 1988. An experimental study of the relationships between compressional wave velocity, acoustic emission and deformation in rocks under stress. Ph.D. thesis, University of London.

Kaiser, J. 1953. Erkenntnisse und Folgerungen aus der Messung von gerauschen bei Zugbeanspruchung von Metallischen Werkstoffen. *Arch. Eisenhütten.* **24**, 43–45.

Kanamori, H. 1978. Quantification of earthquakes. *Nature* **271**, 411–14.

Kemeny, J. M. & N. G. W. Cook 1987. Crack models for the failure of rocks in compression. *Proc. 2nd Int. Conf. on Constitutive Laws for Engineering Materials*, Tucson, Arizona.

Kim, K. & A. Mubeen 1981. Relationship between differential stress intensity factor and crack growth rate in cyclic tension in Westerly granite. In *Fracture mechanics for ceramics, rock and concrete*, 157–68. ASTM Sp. Tech. Publ., STP 745.

Kobayashi, T. & W. L. Fourney 1978. Experimental characterization of the development of the microcrack process zone at a crack tip in rock under load. *Proc. 19th US Symp. Rock Mech.*, 243–50.

Knott, J. F. 1973. *Fundamentals of fracture mechanics*. London: Butterworth.

Kranz, R. L. 1979. Crack growth and development during creep of Barre granite. *Int. J. Rock Mech. Min. Sci. & Geomech. Abstr.* **16**, 23–35.

Kranz, R. L. 1980. The effects of confining pressure and stress difference on static fatigue of granite. *J. Geophys. Res.* **85**, 1854–66.

Kranz, R. L. 1983. Microcracks in rocks: a review. *Tectonophysics* **100**, 449–80.

Kranz, R. L. & C. H. Scholz 1977. Critical dilatant volume at the onset of tertiary creep. *J. Geophys. Res.* **82**, 4893–8.

Kurita, K. & N. Fujii 1979. Stress memory of crystalline rocks in acoustic emission. *Geophys. Res. Lett.* **6**, 9–12.

Labuz, J. F., S. P. Shah & C. H. Dowding 1985. Experimental analysis of crack propagation in granite. *Int. J. Rock Mech. Min. Sci. & Geomech. Abstr.* **22**, 85–98.

Lankford, J. 1979. Uniaxial compressive damage in α-SiC at low homologous temperatures. *J. Am. Ceram. Soc.* **62**, 310–12.

Lankford, J. 1981. The role of tensile microfracture in the strain rate dependence of compressive strength of fine-grained limestone – analogy with strong ceramics. *Int. J. Rock Mech. Min. Sci. & Geomech. Abstr.* **18**, 173–5.

Lawn, B. R. 1975. An atomistic model of kinetic crack growth in brittle solids. *J. Mat. Sci.* **10**, 469–80.

Lawn, B. R. & T. R. Wilshaw 1975. *Fracture of brittle solids*. Cambridge: Cambridge University Press.

Li, V. C. 1987. Mechanics of shear rupture applied to earthquake zones. In *Fracture mechanics of rock*, B. K. Atkinson (ed.), 351–428. London: Academic Press.

Li, V. C. & R. J. Ward 1989. A novel testing technique for post-peak tensile behaviour of cementitious materials. In *Fracture toughness and fracture energy: test methods for concrete and rock*, H. Mihashi, H. Takahashi & F. H. Wittman (eds), 183–96. Rotterdam: Balkema.

Mai, Y.-W. & B. R. Lawn 1986. Crack stability and toughness characteristics in brittle materials. *Ann. Rev. Mat. Sci.* **16**, 415–39.

Mai, Y.-W. & B. R. Lawn 1987. Crack-interface grain bridging as a fracture resistance mechanism in ceramics, II: theoretical fracture mechanics model. *J. Am. Ceram. Soc.* **70**, 289–94.

Main, I. G. 1988. Prediction of failure times in the Earth for a time-varying stress. *Geophys. J.* **92**, 455–64.

Main, I. G., P. G. Meredith & C. Jones 1989. A reinterpretation of the precursory seismic *b*-value anomaly using fracture mechanics. *Geophys. J.* **96**, 131–8.

Main, I. G., S. Peacock & P. G. Meredith 1990. Scattering attenuation and the fractal geometry of fracture systems. *Pageoph.* (in press).

Mandelbrot, B. B. 1982. *The fractal geometry of nature*. New York: W. H. Freeman.

McLintock, F. A. & J. B. Walsh 1962. Friction on Griffith cracks in rocks under pressure. *Proc. 4th US Nat. Congr. Appl. Mech.*, Vol. II, 1015–21. New York: ASME.

Meredith, P. G. 1989. Comparative fracture toughness testing of rocks. In *Fracture toughness and fracture energy: test methods for concrete and rock*, H. Mihashi, H. Takahashi & F. H. Wittmann (eds), 265–77. Rotterdam: Balkema.

Meredith, P. G. & B. K. Atkinson 1983. Stress corrosion and acoustic emission during tensile crack propagation in Whin Sill dolerite and other basic rocks. *Geophys. J. R. Astron. Soc.* **75**, 1–21.

Meredith, P. G., I. G. Main & C. Jones 1990. Temporal variations in seismicity during quasi-static and dynamic rock failure. *Tectonophysics* (in press).

Michalske, T. A. & S. W. Freiman 1982. A molecular interpretation of stress corrosion in silica. *Nature* **295**, 511–12.

Mindess, S. & J. S. Nadeau 1976. Effect of notch width on K_{Ic} for mortar and concrete. *Cement Concrete Res.* **6**, 529–34.

Mizutani, H., H. Spetzler, I. C. Getting, R. J. Martin & N. Soga 1977. The effect of outgassing upon the closure of cracks and the strength of lunar analogues. *Proc. 8th Lunar Sci. Conf.*, 1235–48.

Mogi, K. 1968. Source locations of elastic shocks in the fracturing process in rocks. *Bull. Earthq. Res. Inst., Tokyo Univ.* **46**, 1103–25.

Murrell, S. A. F. 1964. The theory of the propagation of elliptical Griffith cracks under various conditions of plane strain or plane stress. *Br. J. Appl. Phys.* **15**, 1195–223.

Murrell, S. A. F. 1965. The effect of triaxial stress systems on the strength of rocks at atmospheric temperatures. *Geophys. J. R. Astron. Soc.* **10**, 231–81.

Murrell, S. A. F. 1989. Brittle to ductile transitions in polycrystalline non-metallic materials. This volume, 109–37.

Murrell, S. A. F. & S. Chakravarty 1973. Some new rheological experiments on igneous rocks at temperatures up to 1120°C. *Geophys. J. R. Astron. Soc.* **34**, 211–50.

Murrell, S. A. F. & P. J. Digby 1970. The theory of brittle fracture initiation under triaxial stress conditions, Parts I and II. *Geophys. J. R. Astron. Soc.* **19**, 309–34, 499–512.

Paterson, M. S. 1978. *Experimental rock deformation – the brittle field*. Berlin: Springer.

Peng, S. S. & A. M. Johnson 1972. Crack growth and faulting in cylindrical specimens of Chelmsford granite. *Int. J. Rock Mech. Min. Sci.* **9**, 37–86.

O'Connell, R. J. & B. Budiansky 1974. Seismic velocities in dry and saturated cracked solids. *J. Geophys. Res.* **79**, 5412–26.

Ohnaka, M. & K. Mogi 1982. Frequency characteristics of acoustic emission in rocks under uniaxial compression and its relation to the fracturing process to failure. *J. Geophys. Res.* **87**, 3873–84.

Rice, J. R. 1968. A path independent integral and the approximate analysis of strain concentration by notches and cracks. *J. Appl. Mech.* **35**, 379–86.

Rooke, D. P. & D. J. Cartwright 1976. *Compendium of stress intensity factors*. Procurement Executive, Ministry of Defence. London: HMSO.

Rudnicki, J. W. 1984. Effects of dilatant hardening on the development of concentrated shear deformation in fissured rock masses. *J. Geophys. Res.* **89**, 9259–70.

Rudnicki, J. W. and J. R. Rice 1975. Conditions for the localization of deformation in pressure-sensitive dilatant materials. *J. Mech. Phys. Solids* **23**, 371–91.

Rutter, E. H. 1972. The effects of strain rate changes on the strength and ductility of Solnhofen limestone at low temperatures and confining pressures. *Int. J. Rock Mech. Min. Sci.* **9**, 183–9.

Salganik, R. L. 1973. Mechanics of bodies with many cracks. *Mech. Solids* **8**, 135–43.

Sammis, C. G. & M. F. Ashby 1986. The failure of brittle porous solids under compressive stress states. *Acta Metall.* **34**, 511–26.

Sammonds, P. R., M. R. Ayling, C. Jones, P. G. Meredith & S. A. F. Murrell 1989. A laboratory investigation of acoustic emission and elastic wave velocity changes during rock failure under triaxial stresses. In Rock at great depth, V. Maury & D. Fourmaintraux (eds), Vol. 1, 233–40. Rotterdam: Balkema.

Sano, O. & S. Ogino 1980. Acoustic emission during slow crack growth. *Tech. Rep., Yamaguchi Univ.* **2**, 381–8.

Sano, O., I. Ito & M. Terada 1981. Influence of strain rate on dilatancy and strength of Oshima granite under uniaxial compression. *J. Geophys. Res.* **86**, 9299–311.

Sano, O., M. Terada & S. Ehara 1982. A study on the time-dependent microfracturing and strength of Oshima granite. *Tectonophysics* **84**, 343–62.

Schmidt, R. A. & T. J. Lutz 1979. K_{Ic} and J_{Ic} of Westerly granite – effects of thickness and in-plane dimensions. In *Fracture mechanics applied to brittle materials*, 166–82. ASTM Sp. Tech. Publ., STP 678.

Scholz, C. H. 1968a. Mechanism of creep in brittle rock. *J. Geophys. Res.* **73**, 3295–302.

Scholz, C. H. 1968b. The frequency–magnitude relation of microfracturing in rock and its relation to earthquakes. *Bull. Seismol. Soc. Am.* **58**, 399–415.

Scholz, C. H. 1972. Static fatigue of quartz. *J. Geophys. Res.* **77**, 2104–14.

Sih, G. C. 1973. *Handbook of Stress Intensity Factors for Researchers and Engineers*. Institute of Fracture and Solid Mechanics, Lehigh University.

Simmons, C. J. & S. W. Freiman 1983. Effect of corrosion processes on subcritical crack growth in glass. *J. Am. Ceram. Soc.* **64**, 683–6.

Soga, N., H. Spetzler & H. Mizutani 1979. Comparison of single crack propagation in lunar analogue glass and the failure strength of rocks. *Proc. 10th Lunar Sci. Conf.*, 33–41.

Sondergeld, C. H. & L. H. Estey 1981. Acoustic emission study of microfracturing during the cyclic loading of westerly granite. *J. Geophys. Res.* **86**, 2915–24.

Sondergeld, C. H., L. A. Granryd & L. H. Estey 1984. Acoustic emissions during compression testing of rock. In *Proc. 3rd Conf. on Acoustic Emission/Microseismic Activity in Geologic Structures and Materials*, 131–45. Clausthal: Trans Tech.

Spetzler, H., C. Sondergeld & I. C. Getting 1981. The influence of strain rate and moisture content on rock failure. In *Anelasticity in the Earth*, F. D. Stacey, M. S. Paterson & A. Nicolas (eds), 105–12. Geodynamics Ser. 4. Washington, DC: Am. Geophys. Union. Boulder: Geol. Soc. Am.

Spetzler, H., H. Mizutani & F. Rummel 1982. A model for time-dependent rock failure. In. *High-pressure researches in geoscience*, W. Schreyer (ed.), 85–94. Stuttgart: E. Schweizerbart'sche.

Stevens, J. L. & D. J. Holcomb 1980. A theoretical investigation of the sliding crack model of dilatancy. *J. Geophys. Res.* **85**, 7091–100.

Swan, G. 1980. Fracture stress scale effects. *Int. J. Rock Mech. Min. Sci. & Geomech. Abstr.* **17**, 239–43.

Swanson, P. L. 1981. Subcritical crack propagation in Westerly granite: an investigation into the double torsion method. *Int. J. Rock Mech. Min. Sci. & Geomech. Abstr.* **18**, 445–9.

Swanson, P. L. 1984. Subcritical crack growth and other time- and environment-dependent behavior in crustal rocks. *J. Geophys. Res.* **89**, 4137–52.

Swanson, P. L. 1987. Tensile fracture resistance mechanisms in brittle polycrystals: an ultrasonics and *in-situ* microscopy investigation. *J. Geophys. Res.* **92**, 8015–36.

Swanson, P. L. & H. Spetzler 1984. Ultrasonic probing of the fracture process zone in rock using surface waves. *Proc. 25th US Symp. Rock Mech.*, 67–76.

Tada, H., P. C. Paris & G. R. Irwin 1973. *The stress analysis of cracks handbook*. Hellertown, PA: Del Research Corp.

Tapponier, P. and W. F. Brace 1976. Development of stress-induced microcracks in Westerly granite. *Int. J. Rock Mech. Min. Sci. & Geomech. Abstr.* **13**, 103–12.

Thomson, R. M. 1983. Fracture. In *Physical metallurgy*. R. W. Cahn & P. Haasen (eds), 1487–551. Amsterdam: Elsevier.

Tullis, J. & R. A. Yund 1977. Experimental deformation of dry Westerly granite. *J. Geophys. Res.* **82**, 5705–18.

Turcotte, D. L. 1989. Fractals in geology and geophysics. *Pure Appl. Geophys.*, **131**, 171–96.

Vardar, Ö, & I. Finnie 1975. An analysis of the Brazilian disk fracture test using the Weibull probabilistic treatment of brittle strength. *Int. J. Fract.* **11**, 495–508.

Vere-Jones, D. 1976. A branching model for crack propagation. *Pageoph.* **114**, 711–25.

Vere-Jones, D. 1977. Statistical theories of crack propagation. *Math. Geol.* **9**, 455–81.

Von Seggern, D. 1980. A random stress model for seismicity statistics and earthquake prediction. *Geophys. Res. Lett.* **7**, 540–637.

Walsh, J. B. 1965. The effect of cracks on the compressibility of rocks. *J. Geophys. Res.* **70**, 381–9.

Weibull, W. A. 1939. A statistical theory of the strength of materials. *Ingvetenskakad Handl., Stockholm* **151**, 5–45.

Weibull, W. A. 1951. A statistical distribution function of wide applicability. *J. Appl. Mech.* **18**, 293–7.

Wiederhorn, S. M. 1974. Subcritical crack growth in ceramics. In *Fracture mechanics of ceramics*, R. C. Bradt, D. P. H. Hasselman & F. F. Lange (eds), Vol. 2, 613–46. New York: Plenum.

Weiderhorn, S. M. 1978. Mechanisms of subcritical crack growth in glass. In *Fracture mechanics of ceramics*, R. C. Bradt, D. P. H. Hasselman & F. F. Lange (eds), Vol. 4, 549–80. New York: Plenum.

Wiederhorn, S. M. & L. H. Bolz 1970. Stress corrosion and static fatigue of glass. *J. Am. Ceram. Soc.* **53**, 543–8.

Wiederhorn, S. M. & H. Johnson 1973. Effect of electrolyte pH on crack propagation in glass. *J. Am. Ceram. Soc.* **56**, 192–7.

Wilkins, B. J. S. 1980. Slow crack growth and delayed failure of granite. *Int. J. Rock Mech. Min. Sci. & Geomech. Abstr.* **17**, 365–9.

Wittman, F. H. (ed.) 1984. *Fracture mechanics of concrete*. New York: Elsevier Applied Science.

Wong, T.-F. & R. Biegel 1985. Effects of pressure on the micromechanics of faulting in San Marcos gabbro. *J. Struct. Geol.* **7**, 737–49.

Yoshikawa, S. & K. Mogi 1981. A new method for estimation of the crustal stress from cored rock samples: laboratory study in the case of uniaxial compression. *Tectonophysics* **74**, 323–39.

Zimmerman, R. W. 1984. Elastic moduli of a solid with spherical pores: a new self-consistent method. *Int. J. Rock Mech. Min. Sci. & Geomech. Abstr.* **21**, 339–43.

Zimmerman, R. W. & M. S. King 1985. Propagation of acoustic waves through cracked rock. *Proc. 26th US Symp. Rock Mech., Rapid City, SD*, 739–45.

CHAPTER TWO

Single-crack behaviour and crack statistics

Yves Gueguen, Thierry Reuschlé & Michel Darot

2.1 Introduction

Physical and mechanical properties of rocks are controlled by their crack content and crack behaviour. In general, rocks contain a random population of cracks, not necessarily isotropic and homogeneous, more or less interacting, so that the physical properties are those of a complex structure. In such a situation, understanding of the behaviour of a single crack may appear to be irrelevant. It should be looked at rather as a necessary, although not sufficient, preliminary step. With some approximations, fracture mechanics appears to be a useful tool with which to understand the behaviour of a single crack (Rudnicki 1980; Atkinson 1984, 1987; Swanson 1984). Fracture mechanics gives a basic rationale for the various regimes of crack propagation; crack healing, subcritical crack growth, and dynamic (overcritical) crack growth. It points to the importance of extensional cracks at the microscopic level (Lawn & Wilshaw 1975, Section 3.3). Macroscopic behaviour is to be interpreted in terms of microscopic processes. As far as cracks are concerned, important microscopic processes are nucleation and propagation of extensional cracks. However, in order to take into account the fact that a real rock contains a population of cracks, statistical methods are required. Statistical crack mechanics was developed by Dienes (1978, 1982) to calculate the permeability of a cracked rock. Percolation theory also appears to be useful to discuss permeability in terms of crack concentration (Dienes 1982, Gueguen *et al*. 1986). Both statistics and percolation contribute to improving our understanding of transport properties. On the other hand, they could also be used to discuss fracture (Chelidze 1982). In this last case, some important difficulties remain, as our understanding of the fracture of real rocks is far from being satisfactory.

2.2 Single-crack behaviour

We consider the propagation of a well developed single crack in a linear, elastic, homogeneous, isotropic material. The above assumptions are restrictive; they are not always valid for real rocks and minerals. However, they allow us to discuss crack propagation within the framework of fracture mechanics, which is an important advantage. Examples of failure of one or several of these assumptions will be considered below. We examine successively the basis of fracture mechanics, the subcritical crack-propagation régime, and the overcritical crack-propagation régime.

2.2.1 Fundamentals of fracture mechanics

It is traditional to distinguish three basic 'modes' of crack-surface displacement (Fig. 2.1). Mode I (opening mode) corresponds to a displacement perpendicular to the crack surface under the action of tensile stress. Mode II (sliding mode) corresponds to a displacement in the crack plane, in a direction normal to the crack front, under the action of a shear stress. Mode III (tearing mode) corresponds to a displacement in the crack plane, in a direction parallel to the crack front, again under the action of a shear stress.

From linear elasticity, the components of the stress tensor near the crack tip are found to be

$$\sigma_{ij} = \frac{K}{\sqrt{2\pi r}}\, f_{ij}(\theta)$$

in polar co-ordinates, where K is the stress intensity factor. In the case of homogeneous loading where the external stress is applied remote from the crack, $K = \sigma\sqrt{\pi a}$ where σ is the remote stress (tensile stress perpendicular to the crack plane for mode I; shear stress parallel to the crack plane for modes II and III) and $2a$ is the crack length.

By definition, the crack extension force (or strain energy release rate) is $G = -\partial U_m/\partial a$, where U_m is the total mechanical energy (sample + loading system). Crack extension is considered per unit of crack front.

Figure 2.1 The three modes of fracture: I, opening mode; II, sliding mode; III tearing mode.

In the case of plane stress, one obtains:

$$G = K^2/E \quad \text{(modes I and II)}, \qquad G = K^2(1 + \nu)/E \quad \text{(mode III)}$$

where E is Young's modulus and ν is the Poisson ratio.

Fracture mechanics generalizes the Griffith criterion by using the energy-balance concept for any situation (mode I, II or III). The limit of stability for a given crack is obtained when G reaches a critical value, G_c such that $G_c = 2\gamma$, where γ is the specific surface energy.

This is a basic criterion for fracture of ideally brittle solids. In the case of rocks, which are not ideally brittle solids, the Irwin–Orowan generalization of the Griffith equilibrium criterion is $G_c = 2\Gamma$, where the fracture resistance term (fracture surface energy) includes dissipative processes which are operative within a small zone near the crack tip. Linear elasticity is no longer valid in this zone. The size of this zone is assumed to be small ('small-scale zone' approximation). If this is not the case, then fracture mechanics is no longer valid.

The reader is referred to standard textbooks on fracture mechanics (e.g. Lawn & Wilshaw 1975, Atkinson 1987) and the overview at the start of this volume for a more complete treatment of this subject.

2.2.2 Subcritical crack-propagation régime

According to the previous analysis, no crack propagation should be observed at $G < G_c = 2\Gamma$. This is not found to be true: subcritical crack growth has been documented in glasses, ceramics, and rocks for a long time (Hillig & Charles 1965). Rice (1978) has shown that quasi-static healing of a Griffith crack is possible for $G < 2\gamma$ as long as the physical processes are assumed to be thermodynamically reversible. Symmetrically, quasi-static growth is possible for $G > 2\gamma$.

Various possible derivations of the kinetics laws relative to crack growth exist (Law & Wilshaw 1975, Ch. 8; Wiederhorn *et al.* 1980). These different laws have been discussed in several important review papers during the past decade (Anderson & Grew 1977; Atkinson 1984; Atkinson & Meredith 1987a; see also Meredith, this volume), so that a further general review would not be useful and is not given here. Let us concentrate here on the two most frequently used relationships:

$$v = A \exp(\alpha G) \quad \text{and} \quad v = BK^n$$

where v is the crack velocity. The exponential law has been used recently by Darot & Guegon (1986) and the power law has been used by Atkinson (1982), Meredith & Atkinson (1982, 1983, 1985), and Swanson (1984), amongst others. Although it is easier theoretically to justify the exponential law, for practical reasons the power law is more frequently used. It can be shown from a microscopic model that when an elementary crack-front length l advances over a distance b, $\alpha = bl/2\mathbf{k}T$, where T is the absolute temperature and $\mathbf{k}$ is the Boltzmann constant (Lawn & Wilshaw 1975, Section 8.3; Darot & Gueguen 1986). Let us compare the two laws, with the additional assumption that $G = K^2/E$.

It follows from the first law that $\log v \propto K^2$, and from the second that $\log v \propto n \log K$. This implies that $n \propto K^2/\log K$, which means that n depends on K: hence the greater the K values are, the larger n will be. It follows that if one accepts the exponential law as being valid, the power law $v = BK^n$ can lead to erroneous comparisons between different sets of data when the investigated K domains are not identical. Variations in n could thus be artefacts in many cases, and n data should be examined with caution. However, due to the narrow range of K associated with large changes in crack velocity, it can rarely be stated that only one of either the exponential or power law forms fit the data better than the other. Extrapolations of the fitted curves outside the range of actual data should thus be considered as highly uncertain.

How do the experimental data compare with theoretical models? We will examine this briefly in the following three situations corresponding to increasing complexity. The first situation is that of glass, a material which does not violate any of our three basic assumptions (elastic, homogeneous, isotropic material). The second is that of quartz, a material which violates the assumption of isotropy. The third is that of sandstone, a material which violates the assumption of perfect elasticity. Many subcritical crack growth data exist in the literature: our purpose here is not to give an extensive review, but rather to give a critical introduction to the subject. In order to do so, we will focus on the three above cases and use selected results which have been obtained in our laboratory. These results are similar to those obtained by many other authors. For an extensive review of experimental data, the reader is referred to Atkinson (1984) and Atkinson & Meredith (1987b).

2.2.2.1 DATA ON GLASS

Experimental data on subcritical propagation in glass, quartz, and sandstones are shown in Figures 2.2, 3 & 4. The results on glass in ambient air (Fig. 2.2) display the well known three domains. Subcritical growth is controlled by chemical reactions (adsorption) in domain I. Using very low velocity data (Double Torsion test), for the limit $v \to 0$ one obtains $\gamma = 1.1 \ \mathrm{J\,m^{-2}}$. In domain I, one obtains $bl = 3.2 \times 10^{-20} \ \mathrm{m^2}$ or $n = 22$. This last result compares well with Swanson's (1984) value ($n = 18.7 \pm 1.8$). Domain II is a transitional régime: crack propagation is so fast that fluid is no longer available at the crack tip. The transition occurs for a critical velocity, v_t, above which crack velocity is transport-limited. Accurate prediction of v_t is difficult because fluid advance is difficult to model at the crack tip. Lawn & Wilshaw (1975, Section 8.3) have shown that $v_t \ll 10^{-2} \ \mathrm{m\,s^{-1}}$ in normal conditions. Experimentally one obtains $v_t \simeq 10^{-4} \ \mathrm{m\,s^{-1}}$. Domain III corresponds to subcritical

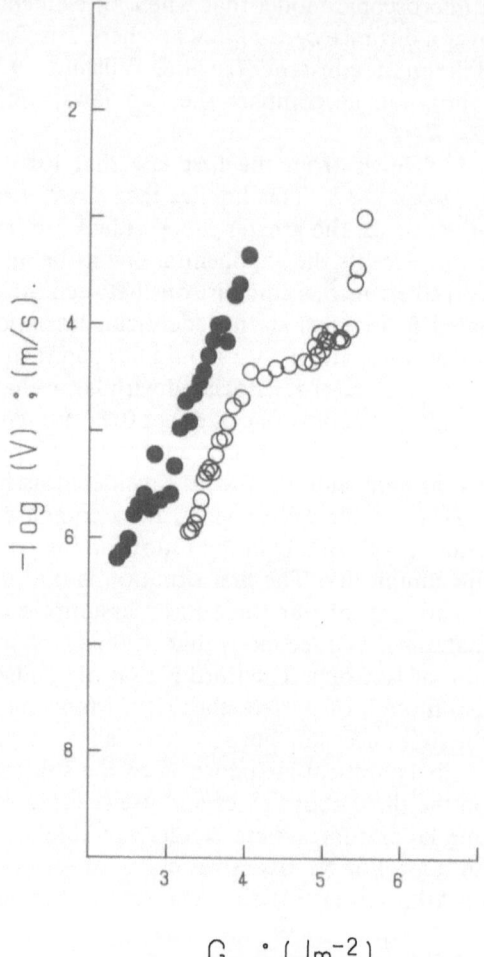

Figure 2.2 Subcritical crack propagation in glass at room temperature and pressure (Double Cantilever Beam test). •, Water; ○, air. Data in air exhibit the three régimes (I, II, and III) corresponding to low, intermediate, and high velocities. Data in water exhibit only régime I.

propagation in vacuum ($\gamma = \gamma_0$). The presence of liquid water increases v_t so that only domain I is observed (Fig. 2.2) in that case. Glass is the most convenient material with which to investigate subcritical crack growth: it is homogeneous, isotropic and elastic (in the usual sense of solid mechanics). For this reason, tests on glass lead to the almost ideal picture discussed above. They also allow a check of the accuracy of experimental devices and of the reliability of the test, which is also of great interest. In the case of Double Torsion (DT) tests for instance, theoretical and experimental compliances agree to better than 5 per cent (Darot *et al.* 1985).

Results on glass can be used to point out an important difficulty with subcritical crack-growth experiments: a variation of v values over six orders of

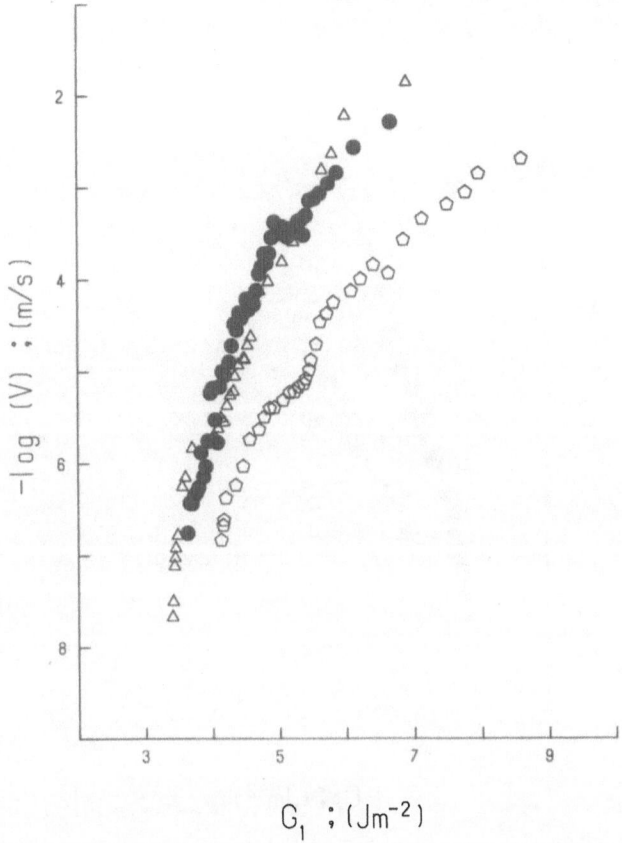

Figure 2.3 Subcritical crack propagation in quartz at room temperature and pressure (Double Torsion test). •, Water; △, water + DTAB 10^{-3}M; ○, water + DTAB 10^{-6}M.

magnitude results from a variation of G which is less than one order of magnitude. An extraordinary accuracy in G is then required.

2.2.2.2 DATA ON SYNTHETIC QUARTZ

Results on synthetic quartz present important differences with those on glass. Quartz is homogeneous and elastic, but it is anisotropic: surface energy depends on crack-plane orientation (Ball & Payne 1976). For this reason, it is difficult to produce crack propagation in a plane which is not a cleavage plane. In such a situation, the crack path will be irregular so that data scatter will be large and the accuracy in G will be poor.

Forty experiments were conducted with crack planes normal to $[a]$, and identical experiments were repeated several times. This orientation resulted from the shape of the single crystals, and is not that of cleavage planes (rhombohedral planes $(01\bar{1}1)$ and $(10\bar{1}1)$). The γ value deduced from

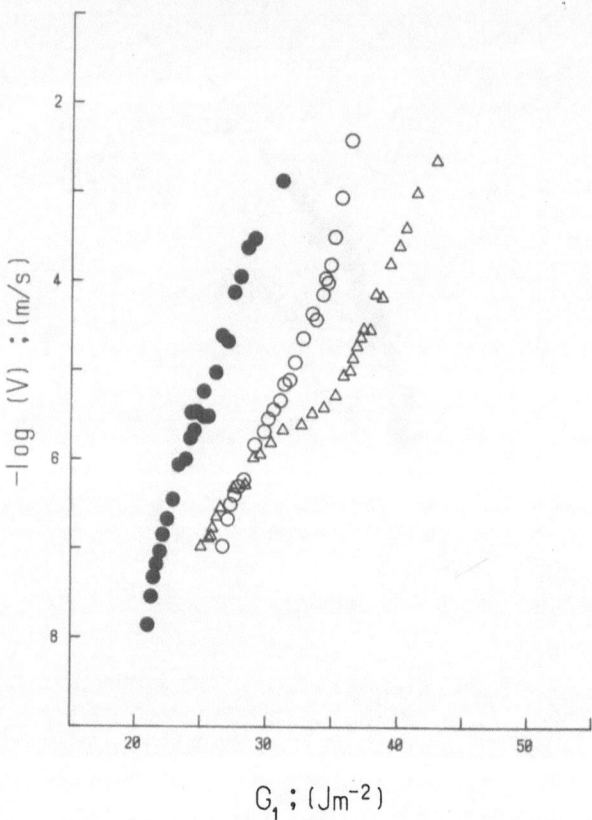

Figure 2.4 Subcritical crack propagation in Fontainebleau sandstones (Double Torsion test). •, Water; △, water + DTAB 10^{-3}M; ○ water + DTAB 10^{-6}M.

Figure 2.3 is $\gamma \simeq 2 \, \mathrm{J\,m^{-2}}$, which is higher than that obtained for powder ($\gamma \simeq 0.4 \, \mathrm{J\,m^{-2}}$; Parks 1984). Anisotropy of quartz is likely to be the main cause of the observed scatter (Fig. 2.3). An additional cause may be the existence of a stress gradient in DT specimens. The result of this data scatter is that the effect of surfactant adsorption on the surface energy of quartz is barely discernible. Surfactant effects on crack propagation probably exist (Dunning *et al.* 1984) but our DT tests were not accurate enough to clearly display it. As pointed out by Parks (1984), the decrease in quartz surface energy is mainly due to water adsorption, and any additional decrease due to surfactant adsorption is much smaller. Nevertheless, the slopes measured in Figure 2.3 lead to an averaged bl value of $4 \times 10^{-20} \, \mathrm{m^2}$. The parameter $d = (bl)^{1/2} \simeq 2 \, \mathrm{\AA}$ is close to the SiO bond length ($a_0 \simeq 1.6 \, \mathrm{\AA}$). This can tentatively be interpreted in terms of a propagation model in which the elementary step for crack growth is the rupture of one atomic bond, as suggested also by the model of Michalske & Freiman (1982). In terms of n values, from

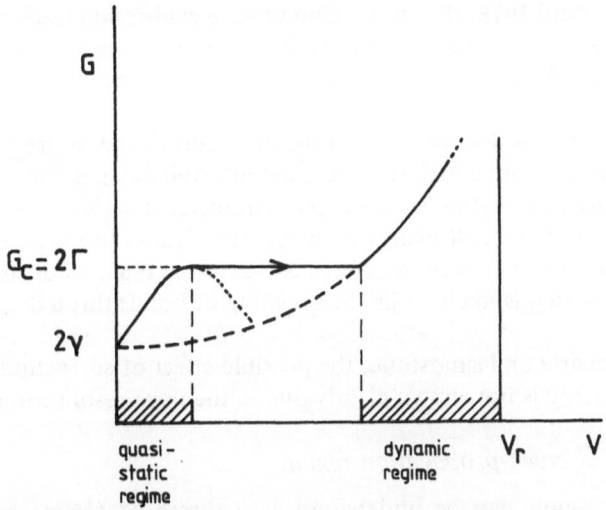

Figure 2.5 G as a function of velocity V: transition from quasi-static régime (subcritical crack growth) to dynamic régime (dynamic rupture). Dotted curve, friction; dashed curve, ideal solid. G is obtained as a function of V by addition.

Figure 2.3 one obtains $n \simeq 40$, which is not in agreement with Atkinson's (1984) data ($n \simeq 20$). The discrepancy can be attributed to differences in experimental procedures and to the fact that different K_I domains were investigated. In any case, the uncertainties in the data are important, as discussed above.

2.2.2.3 DATA ON SANDSTONES

The situation is worse for rocks. Swanson (1984) has previously shown that it is difficult to obtain reproducible data in DT tests with rocks. The difficulties arise from the fact that rocks are frequently not homogeneous and not perfectly elastic. They can also, of course, be anisotropic but on average their anisotropy is weak. The departure from the basic assumptions of fracture mechanics which results is not negligible. The accuracy on absolute values of G is reduced. We have investigated subcritical growth in almost pure quartz sandstones (Fontainebleau sandstones) which are cemented by quartz. Interpretation of the DT results is difficult, mainly because of the imperfect elastic behaviour of the rocks (Darot *et al.* 1985). A large permanent deformation is observed when the sample is unloaded. This deformation can be as high as one third of the total deformation.

In addition, there is an intrinsic relaxation of the rock which is neglected. This relaxation occurs without any crack propagation and leads to erroneous $v(G)$ slopes, mainly at the beginning of the relaxation test. Another problem is that of Young's modulus. It has been shown that in rocks Young's modulus in compression often differs significantly from Young's modulus in tension

(Lama & Vutukuri 1978, II, Ch. 6). Compliance calibration leads to a Young's modulus value of 5.2 GPa, while uniaxial compression tests lead to values of 87 GPa for the Fontainebleau sandstone for which results are presented in Figure 2.4.

Therefore, the absolute position of the $v(G)$ curves and, more generally, the significance of G in such inelastic rocks are not well defined. Nevertheless, the slopes measured on Figure 2.4 lead to a parameter $d = (bl)^{1/2}$ of about 1 Å, which is close to the result observed for quartz. Cracks propagate along grain boundaries at low velocities (Meredith & Atkinson 1983, Swanson 1984) and the elementary step is likely to be the breaking of bonds through grain boundaries.

For both quartz and sandstone, the possible effect of surfactant adsorption on surface energy is not shown clearly due to the poor resolution in G values.

2.2.3 Critical crack-propagation régime

The critical régime can be understood as a threshold régime between subcritical propagation and dynamic rupture.

Maugis (1985) derived a model which appears to be complementary to the previous ones in that it relates the quasi-static régime to the dynamic one, starting from the same first principles. He used the energy balance contained in the Griffith concept to obtain an interpretation of stick-slip and acoustic emission. When the unit length of crack front advances at a velocity v, the energy balance is:

$$(G - 2\gamma)v = f(v)$$

where $f(v)$ is an empirical function which represents the energy dissipation at the crack tip (plasticity, inelasticity, internal frictions).

Maugis (1985) assumed that $f(v)$ increased with v up to a critical value, v_c. Above v_c, friction is reduced because the régime is no longer that of static friction but of dynamic friction: v then jumps from v_c to a much higher velocity (Fig. 2.5).

The critical value, v_c, corresponds to the critical crack-extension force $G_c = 2\Gamma$ introduced earlier. Slow crack growth is restricted to the domain $(2\gamma, G_c)$. Thus G_c should be considered as the upper limit of the $v(G)$ curves computed from Double Torsion or Double Cantilever Beam (DCB) tests (for example, the DCB curve for glass in ambient air (Fig. 2.2) leads to $G_c \simeq 5.5 \text{ J m}^{-2}$). In fact, G_c is usually measured by rapidly loading the DCB or DT specimen to avoid any environmental effects on crack propagation. Values obtained in this latter way are always higher than those obtained in the former way (for example, $G_c \simeq 8 \text{ J m}^{-2}$ in glass by DT and DCB).

Calculated G_c values for quartz in different environments again display some scatter. The average value is $G_c \simeq 10 \pm 1 \text{ J m}^{-2}$, in agreement with Atkinson's (1984) data. Experiments were conducted at high temperature up

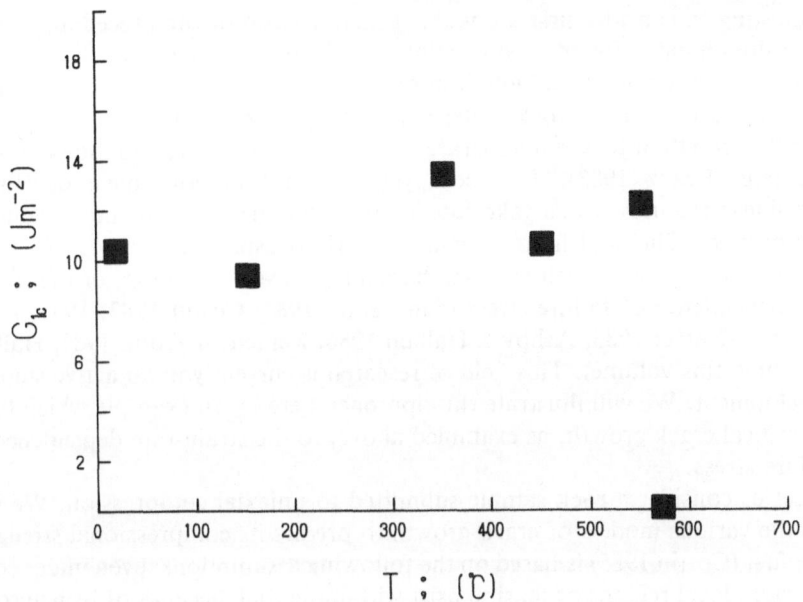

Figure 2.6 Critical crack-extension force G_c in quartz as a function of temperature (Double Torsion test).

to the α–β transition (573°C). The G_c values obtained by the Double Torsion method are displayed in Figure 2.6. The onset of plasticity above 350°C leads to high G_c values up to 553°C. A drastic decrease in G_c is observed in the close vicinity of the transition point. When the temperature crosses the transition point, the specimen breaks down under its own weight without the addition of any external load.

The scatter is much greater in sandstones ($G_c = 56 \pm 11$ J m^{-2}). This is due partly to inelasticity, which leads to an apparent decrease of G_c when the crack length increases (Darot *et al.* 1985), and partly to uncertainties in the value of Young's modulus. The dependence of G_c on crack length can be interpreted as follows: the inelastic zone ahead of the crack is large and is limited by the end of the specimen, so that as the crack advances the size of this zone decreases. Thus the amount of energy dissipated in the specimen by inelastic processes also decreases. Therefore, in rocks such as sandstones, G_c, as measured from most tests, cannot be considered to be an absolute intrinsic parameter.

2.3 Multiple-crack behaviour

Mechanical properties of real rocks are controlled by the development of a population of cracks and not by the existence of a single crack. Thus, it is interesting to consider first a possible generalization of the preceding theory: what does happen for the propagation of a family of cracks? Each crack will be treated individually, although interactions between cracks are possible.

In the case of brittle rock under compressive loading, it is well established that the growth of tensile microcracks results in inelastic deformation and failure (e.g. Kranz 1983). In recent years several authors have developed non-linear theories which take into account the growth and interactions of microcracks. The goal is to explain the various aspects of non-linear, time-dependent behaviour such as strain hardening, strain softening, or the strain-rate-dependence of failure stress (Sano *et al.* 1981; Costin 1983, 1985; Horii & Nemat-Nasser 1985; Ashby & Hallam 1986; Kemeny & Cook 1987; Hallam & Ashby, this volume). This field of research is currently in an active state of development. We will illustrate this approach here by an example which links subcritical crack growth, as examined above, to the strain-rate dependence of failure stress.

Let us consider a rock sample submitted to uniaxial compression. We will use the various models of crack growth to predict its compressional strength. The first (Costin 1985) is based on the following assumptions: even under compression, local regions of tension exist within the rock because of its heterogeneous nature; the magnitude of the stress at a microcrack tip within a local region of tension is proportional to the applied deviatoric stress; the compressive mean stress has an inhibiting effect on crack growth; and microcracks are assumed to grow in their own planes and in directions normal to positive components of the deviatoric stress. Thus microcrack growth occurs parallel to the direction of applied compressive stress and, as the crack grows, the local stress is relieved in a way illustrated in Figure 2.7a. Microcrack interactions are taken into account by using an approximation of the 'pseudo-traction' method given by Horii & Nemat-Nasser (1983). This method can be summarized as follows. Let us consider an array of cracks parallel to the applied compressive stress (Fig. 2.7b). When the cracks grow, their average spacing decreases and some interaction takes place. The solution of this complex problem is obtained by the superposition of the solution of subproblems, each of them involving an infinite homogeneous solid containing one single crack, and of the solution of the problem of an infinitely extended homogeneous solid under applied far-

Figure 2.7 Schematic illustration of Costin's (1985) model. (a) Crack-tip stress (σ_l) relieves when length of crack (a) increases: d is a measure of the local region of tension. (b) Crack interactions model: cracks are parallel to the applied compressive stress σ. d_1 is the initial crack spacing.

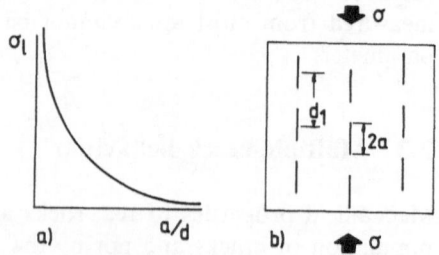

field stresses. For each subproblem, there are unknown quantities ('pseudo-tractions') which are the tractions exerted by the other cracks on the crack considered in the subproblem. The requirement that the sum of the sub-problems must be equivalent to the original problem leads to consistency conditions: K_I is numerically calculated from these conditions. Costin's K_I value (Table 2.1) is an analytical approximation of Horii & Nemat-Nasser's numerical results.

The second model (Ashby & Hallam 1986; Hallam & Ashby, this volume) is based on the growth of wing cracks from initial flaws inclined with respect to the applied compressive stress (Fig. 2.8a). The wing cracks grow parallel to the applied remote stress. The stress intensity factor K_I at the wing-crack tip is the superposition of a term due to the stress field of the initial angled flaw and a term due to the opening force caused by the sliding displacement of the inclined flaw. The angle ψ is calculated so as to maximize K_I. Now consider an array of angled cracks and the associated wing cracks (Fig. 2.8b). When the latter grow, they will interpenetrate, dividing the solid into beams (Fig. 2.8b) which become longer and narrower as the cracks extend. The remote compressive stress will then cause bending deflections in these beams. This

Table 2.1 K_1 results used for axial failure stress prediction.

Crack-growth model	Crack-tip stress-intensity factor
Costin (1985)	$K_1 = \dfrac{2}{\pi}\sqrt{\pi a}\,\dfrac{\sigma}{3}\left\{-1 + f\left[\dfrac{d}{a}\dfrac{1}{[1-(a/d_1)^2]^{1/2}}\right]\right\}$
	f = constant of proportionality between the applied deviatoric stress and the local tensile stress
	σ, a, d, and d_1 are defined in Figure 2.7
Ashby & Hallam (1986), Hallam & Ashby (this volume)	$K_1 = \sigma\sqrt{\pi a}\left\{\dfrac{[(1+\mu^2)^{1/2}-\mu]\{0.23L + [1/\sqrt{3(1+L)}]\}}{(1+L)^{3/2}}\right.$
	$\left.+\dfrac{(L+\alpha)^{1/2}D_0^{1/2}\sqrt{2}}{\pi}\right\}$
	$L = l/a$, $\alpha = 1/\sqrt{2}$
	$D_0 = \pi a^2 N_a$, where $N_a = 1/[2t(l+a)]$ is the number of initial, angled, cracks per unit area
	μ = coefficient of friction
	σ, a, and l are defined in Figure 2.8
Kemeny & Cook (1987)	$K_1 = \dfrac{2l_0\tau\sin\beta}{\sqrt{b}\,\sin(\pi(l+l^*)/b)} - \dfrac{\sqrt{b}\,\tan(\pi l/b)}{2}[1+\cos 2(\beta+\theta)]\sigma$
	where $\tau = \frac{1}{2}\sigma[\sin 2\theta - \mu(1+\cos 2\theta)]$
	$l^* = 0.27l_0$
	l_0, β, b, l, and θ are defined in Figure 2.9

Figure 2.8 Schematic illustration of wing-crack model (after Ashby & Hallam 1986, and Hallam & Ashby, this volume). (a) Wing cracks grow from an angled sliding crack under uniaxial compression σ. The wing cracks are parallel to σ. (b) Crack interactions model. The solid is divided into beams (one is shaded). As beams become longer and narrower, bending takes place simulating crack interactions (t is the smallest distance between wing cracks).

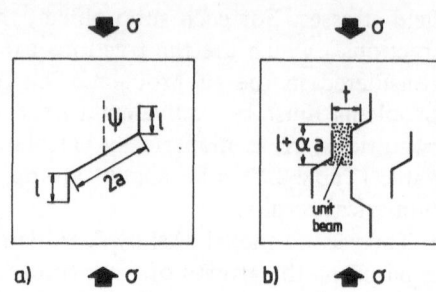

bending corresponds to crack interaction and leads to an additional term in the expression for K_I. The final K_I value is given in Table 2.1.

The third model (Kemeny & Cook 1987) is based on a similar crack-growth geometry. Wing cracks are initiated from an angled flaw, and deviate from it by an angle β which varies with the wing-crack length (Fig. 2.9a). This dependence is introduced to take account of the curved shape of wing cracks in real situations (Horii & Nemat-Nasser 1985). β is again chosen so as to maximize K_I. As the wing cracks grow further, they will orient themselves parallel to the applied compressive stress. Interactions are taken into account by extending the K_I calculation to the case of an axially aligned column of sliding cracks (Fig. 2.9b). Crack interactions take place either within a single column (axial failure) or between columns (shear failure). The K_I value is given in Table 2.1.

To test these different models we use single-crack growth results obtained on a micrograined limestone using the double torsion technique (Fig. 2.10). The $v(G_I)$ curve is divided into two linear régimes, and an exponential law $v(G_I)$ is fitted to both régimes. A third linear régime is added to the experimental data to take account of the final crack acceleration when G_I approaches the critical crack-extension force G_{Ic}.

Assuming a constant deformation rate, $\dot{\varepsilon}$, the stress is increased incrementally with, at each increment, $\Delta\sigma = E\Delta\varepsilon$ (uniaxial compression). After each stress increment, K_I and hence $G_I = K_I^2/E$ are calculated at the tips of the growing cracks, using the formulae of Table 2.1. We can then calculate the

Figure 2.9 Schematic illustration of wing-crack model (after Kemeny & Cook 1987). (a) Wing cracks deviate from the sliding crack by an angle β chosen so as to maximise K_I. They orient themselves in the σ direction when growing further. (b) Crack interactions model: axially aligned columns of sliding cracks are used. Crack interactions take place either within a single column (axial failure) or between columns (shear failure).

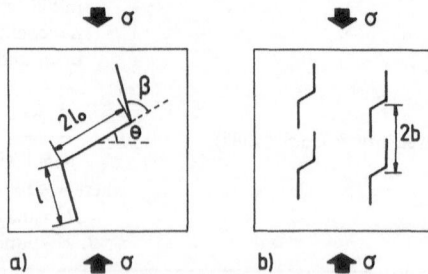

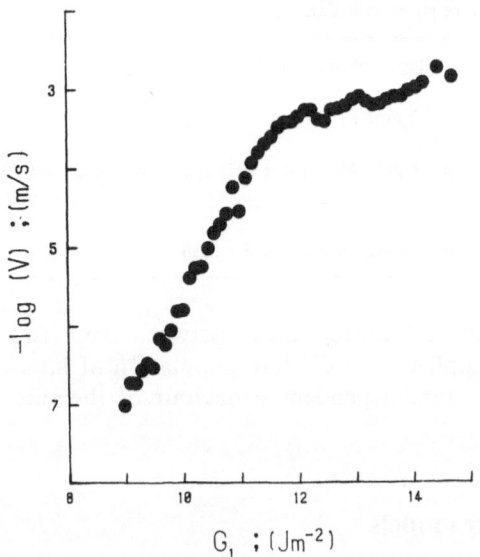

Figure 2.10 Subcritical crack propagation in micrograined limestone (Soultz micrite) at room temperature and pressure (de-ionized water) (Double Torsion test).

Figure 2.11 (below) Failure strength, σ_m, of micrograined limestone (Soultz micrite) as a function of strain rate $\dot\varepsilon$ (uniaxial compression). +, Experimental results; solid line, failure strength prediction after Costin's (1985) model; dashed line, failure strength prediction after Ashby & Hallam's (1986) model; mixed line, failure strength prediction after Kemeny & Cook's (1987) model.

crack velocity at this step, and the new crack length. Failure is assumed to take place when the cracks have grown over a distance equal to half the initial crack spacing. In Figure 2.11 are displayed the three theoretical curves relating the uniaxial compressional strength, according to the above three models, and the deformation rate. Superimposed are the experimental results obtained on uniaxial compression of cylinders of the same rock. The fit is equally good for all three curves. Material parameters are given in Table 2.2. The values of the microstructural parameters ($a_0, d_1, a, l_0, 2b$) have to be compared with the limestone grain size (1–10 μm). It is noteworthy that the compressional strength decreases with the deformation rate down to a limiting value where slow crack growth no longer has any effect. This means that, according to these models, the rock can sustain a load lower than some threshold value indefinitely. At the other end of the deformation rate scale, crack growth is extremely rapid. In this case time-dependence disappears again and the

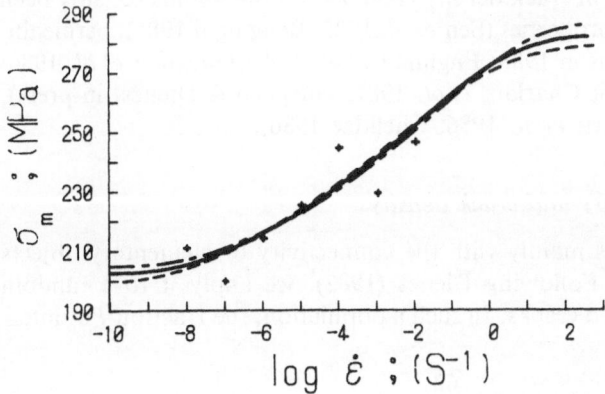

Table 2.2 Material parameters used for failure stress calculations.

Crack-growth model	Material parameters
Costin (1985)	$a = 3$ μm, $d_1 = 9$ μm, $d = 23$ μm
Ashby & Hallam (1986), Hallam & Ashby (this volume)	$a = 6$ μm, $N_a = 1.4 \times 10^{10}$ cracks per m^2, $\mu = 0.2$
Kemeny & Cook (1987)	$l_0 = 12$ μm, $b = 5$ μm, $\mu = 0.35$

strength again reaches an independent limiting value. Between these two limits, single-crack growth results applied to a discrete population of inter-acting cracks appear to explain the time-dependent behaviour of the rock reasonably well.

2.4 Crack statistics: percolation models

In the above models multiple cracks are considered and it becomes possible to calculate some macroscopic properties of real rocks from knowledge of single-crack behaviour. The models correspond to an intermediate step between the single-crack problem and the statistical problem since they take crack interactions into account, albeit in a very crude way. We will now examine another approach, which is a statistical one. In this approach, crack density is a param-eter which plays a crucial rôle, and percolation theory is the starting point. Crack density is known to be an important parameter in rocks (e.g. Brace *et al.* 1966, Paterson 1978). For low crack density, cracks are generally iso-lated and primarily produce a modification of elastic properties (Walsh 1965). When crack density increases, a threshold can be reached where crack inter-connection takes place (Fig. 2.12). As a result, permeability is strongly modified. At even higher crack density, crack interactions result in the devel-opment of a macroscopic fracture. Percolation theory appears to be an appro-priate framework in which to discuss this evolution of mechanical and physical properties with increase in crack density. Percolation theory has recently been used to examine elastic properties (Sen *et al.* 1985, Benguigui 1985), permeabi-lity (Dienes 1982; Robinson 1983; Englman *et al.* 1983; Gueguen *et al.* 1986; Katz & Thompson 1986; Charlaix *et al.* 1987; Gueguen & Dienes, in press), and fracture (Chakrabarti *et al.* 1986, Chelidze 1986).

2.4.1 Percolation theory and crack density

Percolation theory deals mainly with the connectivity of elementary objects (Stauffer 1985, Ch. 2). Following Dienes (1982), we apply it to a random population of disc-shaped cracks. In such a population, the fraction f of inter-

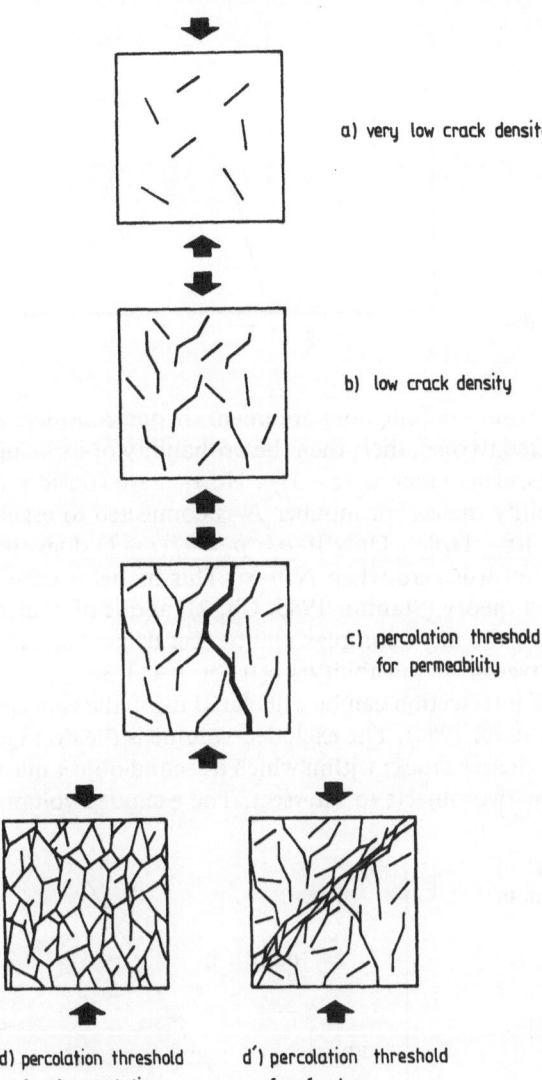

a) very low crack density

b) low crack density

c) percolation threshold
   for permeability

d) percolation threshold
   for fragmentation

d') percolation threshold
    for fracture
    [anisotropy ; interaction]

Figure 2.12 Crack-density evolution and percolation thresholds: (a) very low crack density: isolated cracks (b) low crack density: development of a few clusters of connected cracks. (c) percolation threshold for permeability. (d) percolation threshold for fragmentation. (d') percolation threshold for fracture: anisotropy and correlation.

connected cracks is variable. Some cracks are isolated, while others are connected to form a finite 'cluster'. When the probability of crack intersection, p, is higher than a critical value, p_c, there exists an 'infinite cluster'; that is, there exists a continuous network of connected cracks throughout the medium. Only in this last case is the overall permeability not equal to zero. The permeability is proportional to the fraction f of connected cracks and varies accordingly (Fig. 2.13). Let us assume that crack centres are distributed randomly and that each crack has z neighbours: then the critical probability

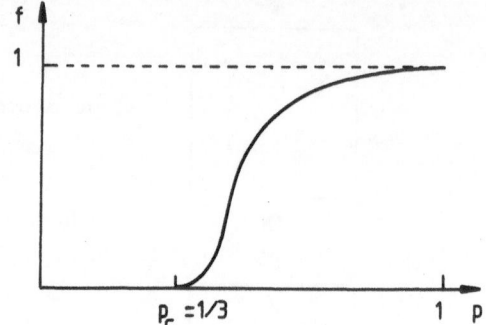

Figure 2.13 Variation of the fraction of connected cracks, f, as a function of probability p of crack intersection. Permeability is proportional to f.

is $p_c = 1/(z-1)$. This results from the following argument: if one considers a crack which is already connected to one other, then the probability of its being connected to a second neighbouring crack is $(z-1)p$. Now, if we consider a chain of N steps, the probability that crack number N is connected to crack number 1 is proportional to $[(z-1)p]^N$. Only if $p = p_c = 1/(z-1)$ does the above quantity not decrease towards zero when $N \to \infty$. This model is called the Bethe lattice in percolation theory (Stauffer 1985, Ch. 2), and is of course only a simple approximation of the more complex reality. Let us now assume $z = 4$, so that $p_c = 1/3$ and 'overall' permeability $k = 0$ for $p \leqslant 1/3$.

The probability, p, of crack intersection can be calculated using the concept of excluded volume (Charlaix *et al.* 1984). The excluded volume is the average volume around a given object (here a crack) within which a second object must have its centre in order for the two objects to intersect. The excluded volume

Figure 2.14 Percolation domain in (l, a) space. Path a, crack nucleation; path b, crack growth.

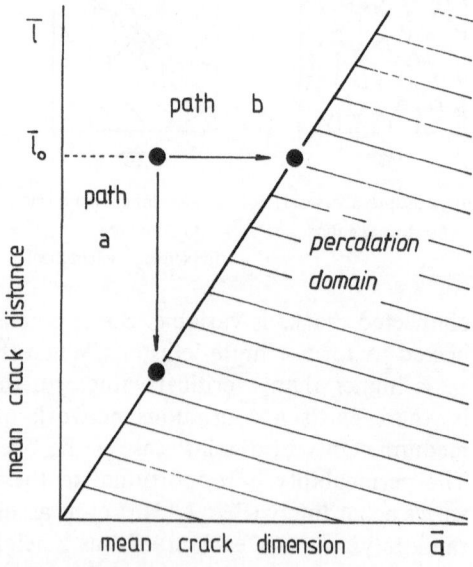

64

is thus a statistical volume. In the case of discs of radius a one obtains $V_{e'} = \pi^2 a^3$. Let us define the crack density n as $n = 1/l^3$, where l is the average crack spacing.

The probability is then given by $p = (\pi^2 a^3/4l^3)$ since $z = 4$. It is thus expressed in terms of two microvariables, a (crack radius) and l (crack spacing). As expected, p is low if cracks are very small or if their density is very small (small a or large l). Non-zero permeability is predicted for $p \geqslant 1/3$; that is, $a/l > 0.5$. There are thus basically two distinct paths in an $l-a$ plot by which to reach the percolation domain (Fig. 2.14): the first path corresponds to crack nucleation and the second to crack growth. Both crack nucleation and crack growth are mainly controlled by deviatoric stress.

Near the percolation threshold $p_c = 1/3$, and f varies rapidly with p. The approximation of the Bethe lattice with $z = 4$ leads to $f \simeq 54(p - p_c)^2$. This result is obtained from equation (27) in Dienes (1982). More generally, percolation theory predicts that f varies as $(p - p_c)^n$ with $n \simeq 1.9$ (Stauffer 1985, Ch. 2). Well above the threshold, f is close to 1.

2.4.2 Application to permeability

The previous results lead directly to a discussion of permeability. Two different approaches can be used at this stage to derive permeability. The first seems to be more appropriate when the statistical distributions of cracks parameters (apertures, radii) are not very broad. The second seems to be more appropriate when those distributions are very broad.

Let us consider the first situation. It has been shown by Dienes (1982), from a statistical calculation, that an isotropic distribution of cracks with radius a, number density n, and aspect ratio $A = w/a$ results in a permeability:

$$k = \frac{4\pi}{15} A^3 n \overline{a^5} \theta f$$

where $n\overline{a^5}$ is the fifth moment of the crack number density. The factor θ accounts for the hydrodynamics of flow through a system of cracks with varying thickness, but seems to differ little from unity in the cases examined. We introduce the three following microvariables; $\bar{a}$ (average crack radius), $\bar{w}$ (average crack aperture), and $\bar{l}$ (average crack spacing) (Fig. 2.15). We restrict our attention to the isotropic case and assume that the statistical distributions are narrow, so that $(\bar{a})^5$ is close to $(\overline{a^5})$. Then:

$$k = \frac{4\pi}{15} f \frac{\bar{w}^3 \bar{a}^2}{\bar{l}^3}$$

The factor f has been discussed previously, and has been shown to be a function of $\bar{a}$ and $\bar{l}$. Above the percolation threshold, k can vary by a large amount

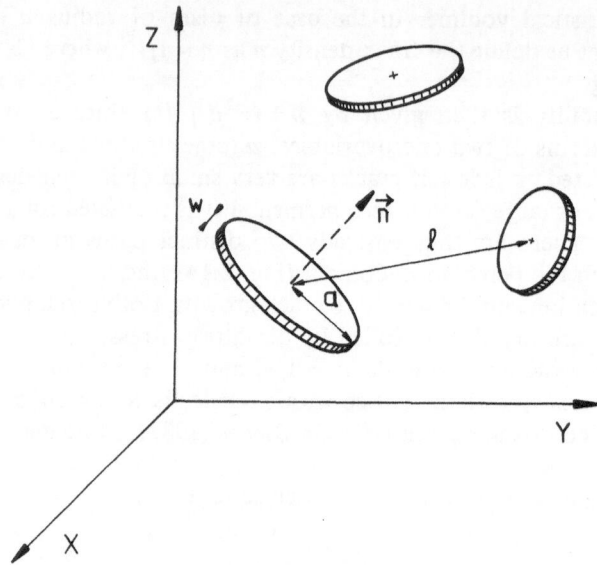

Figure 2.15 Crack model: isotropic distribution of cracks. Each crack is characterized by its radius a and aperture w; $\vec{n}$ is the normal to a crack and l is the crack spacing.

if $\bar{w}$, $\bar{a}$, or $\bar{l}$ vary. Crack propagation (increase of $\bar{a}$) or crack nucleation (decrease of $\bar{l}$) will produce a strong increase of k, a result which is similar to the effect discussed previously (Fig. 2.14). In addition, however, k appears to be strongly dependent on crack aperture, which is itself strongly dependent on effective pressure.

A second permeability model can be derived for the case in which statistical distributions of crack parameters are very broad. Such a model has been published recently by Katz & Thompson (1986) and Charlaix *et al.* (1987). It results from the direct application of the theory of Ambegaokar *et al.* (1981), first derived for the calculation of the electrical conductivity of a population of widely variable resistances. If one considers, for instance, cracks which have very variable apertures, it is possible to classify those cracks into subnetworks made of equal-aperture cracks. Cracks with small apertures are numerous, so that their subnetworks are continuously connected. But their overall contribution to flow is negligible (their aperture w is small, and k varies as w^3). For the same reason, cracks with large apertures have a high individual permeability. They are not numerous, however, so their subnetworks are not continuously connected. It therefore results that their contribution to flow is zero. The system behaves approximately as if there were only one size of crack aperture, which is the aperture of the most permeable connected subnetwork. By definition, this subnetwork is the one which has the largest crack aperture amongst those which are connected continuously. Then:

$$k = k_l(p - p_c)^2$$

where k_l is the elementary permeability of any crack of the considered subnetwork and $(p - p_c)^2$ is the factor which weights the 'size' of this subnetwork.

2.4.3 Application to fracture

Fracture of a solid is generally characterized by the development of a macroscopic discontinuity. The process is thus essentially non-homogeneous and localized. Fracture should not be confused with fragmentation, which could be looked on as an 'homogeneous' fracture, in the sense that there is no local-·ization. We have seen above that, when crack density is high enough, the crack network becomes connected (at $p = p_c$). In this case there exists an 'infinite cluster' of cracks. This infinite cluster is indeed an infinite path made of connected cracks (Fig. 2.12c): but, as far as the rock is concerned, it remains solid and intact with respect to mechanical resistance since most of the cracks do not belong to the 'infinite cluster'.

For higher crack densities, more and more infinite paths exist, so that above a second critical crack density (at $p = p'_{c'} > p_c$) it will not be possible to go through the rock without crossing such a path. This stage corresponds to fragmentation (Fig. 2.12d). Most of the cracks then belong to the infinite cluster. It results from this analysis that percolation leads directly to criteria that are relevant to fragmentation but not to fracture. The fragmentation criteria would again be of the type $(a^3/l^3) = \alpha$, where α is an appropriate constant (Chelidze 1986). In the case of homogeneous crack nucleation, fragmentation should indeed be expected. Homogeneous nucleation should occur when stress distribution is homogeneous. Two examples of such a situation can be given. The first is that of explosion, in which a stress pulse travels through the rock: for a short time interval, δt, each part of the rock is submitted to the same high stress pulse $\Delta\sigma$. The second example is that of a highly porous rock submitted to stress. A narrow size distribution of pores provides an almost equal stress concentration at each pore, so that again nucleation is homogeneous.

In general, however, the stress distribution resistance to crack nucleation within the solid is not homogeneous, so that nucleation is not homogeneous. In that case, there are two other important factors which will play a major rôle. The first is that cracks are not isotropic objects, and the second is that they interact. The problem is consequently a percolation problem with anisotropy and correlation (Kolesnikov & Chelidze 1985). Anisotropy means that the probability of having a second crack close to the first one depends on the orientation. Correlation means that the probability of having a second crack close to a first one is higher than the probability of it occurring at an isolated site. More specifically, in a sufficiently dense and homogeneously cracked sample, the probability of local crack extension will increase as cracks start to interact: the latter phenomenon, extension rather than nucleation, would perhaps serve as an appropriate correlation parameter in the percolation problem. Correlation localizes the cracks, and anisotropy decreases the 'tortuosity' of their connections. Because of anisotropy and correlation, localization and development of a planar discontinuity will take place. Unfortunately, it is very difficult to express this in terms of simple equations. Anisotropic percolation has received virtually no attention up to now: and the same applies to correlation in percolation. This research direction does, however, offer a promising way in which to analyse the fracture process.

2.5 Conclusions

The behaviour of a single tensile crack propagating in a rock sample has received much attention in recent years. For instance, effects such as subcritical crack growth have been documented in detail and are presently well known. If some uncertainties remain in that field, they result mainly from experimental difficulties. The physical processes related to single tensile cracks are reasonably well understood from a theoretical point of view. However, real rocks do contain a population of cracks, and theoretical difficulties are bound up with this basic fact. Multiple cracks can be dealt with using models which take into account the growth and interactions of microcracks. From such models it is possible to calculate some macroscopic properties of real rocks using single crack data as input. Statistical theories would be preferable, however, given that crack density appears to be a very important parameter. Percolation theory and random media theory are useful in order to gain some insights into the physics of these statistical processes. It has been shown that a key parameter is the crack intersection probability, which can be expressed in terms of two microvariables (crack radius and crack spacing). An appropriate theory for fracture appears to be anisotropic correlated percolation. However, this tool has not received much attention to date. Cracks are strongly anisotropic, and they interact when sufficiently close to each other. These two factors determine the localization of the fracture process.

References

Ambegaokar, V. B., R. I. Halperin & J. S. Langer 1971. Hopping conductivity in disordered systems. *Phys. Rev. B* **4**, 2612–20.

Anderson, O. L. & P. C. Grew 1977. Stress corrosion theory of crack propagation with application to geophysics. *Rev. Geophys. Space Phys.* **17**, 77–104.

Ashby, M. F. & S. D. Hallam 1986. The failure of brittle solids containing small cracks under compressive stress states. *Acta Metall.* **34**, 497–510.

Atkinson, B. K. 1982. Subcritical crack propagation in rocks: theory, experimental results and applications. *J. Struct. Geol.* **4**, 41–56.

Atkinson, B. K. 1984 . Subcritical crack growth in geological materials. *J. Geophys. Res.* **89**(B6), 4077–114.

Atkinson, B. K. (ed.) 1987. *Fracture mechanics of rock*. London: Academic Press.

Atkinson, B. K. & P. G. Meredith 1987a. The theory of subcritical crack growth with application to minerals and rocks. In *Fracture mechanics of rock*, B. K. Atkinson (ed.), 111–66. London: Academic Press.

Atkinson, B. K. & P. G. Meredith 1987b. Experimental fracture mechanics data for rocks and minerals. In *Fracture mechanics of rock*, B. K. Atkinson (ed.), 477–525. London: Academic Press.

Ball, A. & B. W. Payne 1976. The tensile fracture of crack crystals. *J. Mat. Sci.* 11, 731–40.

Benguigui, L. 1985. Elasticity of percolation systems. In *Physics of finely divided matter*, N. Boccara & M. Daoud (eds), 188–92. Berlin: Springer.

Brace, W. F. & E. G. Bombolakis 1963. A note on brittle crack growth in compression. *J. Geophys. Res.* 68, 3709–13.

Brace, W. F., B. W. Paulding and C. Scholz 1966. Dilatancy of crystalline rocks. *J. Geophys. Res.* 71, 3939–53.

Bruce, J. G. & B. G. Koepke 1977. Evaluation of K_{1c} by the double torsion technique. *J. Am. Ceram. Soc.* 60(5–6), 284–5.

Chakrabarti, B. K., C. Debashish & D. Stauffer 1986. Fracture in 2D disordered Lennard–Jones solid: two different thresholds. In *Fragmentation, form and flow in fractured media*, 251–54. Annals of Israel Physical Society, vol. 8.

Charlaix, E., E. Guyon & N. Rivier 1984. A criterion for percolation threshold in a random array of plates. *Solid State Communs* 50(11), 999–1002.

Charlaix, E., E. Guyon & S. Roux 1987. Permeability of a random array of fractures of widely varying apertures. *Transport in Porous Media* 2, 31–43.

Chelidze, T. 1982. Percolation and fracture. *Phys. Earth Planet. Inter.* 28, 93–101.

Chelidze, T. 1986. Percolation theory as a tool for imitation of fracture processes in rocks. *Pure Appl. Geophys.* 124, 731–48.

Costin, L. S. 1983. A microcrack model for the deformation and failure of brittle rock. *J. Geophys. Res.* 88, 9485–92.

Costin, L. S. 1985. Damage mechanics in the post-failure regime. *Mech. Mat.* 4, 149–60.

Darot, M. & Y. Gueguen 1986. Slow crack growth in minerals and rocks: theory and experiments. *Pure Appl. Geophys.* 124(4–5), 677–92.

Darot, M., T. Reuschlé & Y. Gueguen 1985. Fracture parameters of Fontainebleau sandstones: experimental study using a high temperature controlled atmosphere double torsion apparatus. In *Research and engineering applications in rock masses*, E. Ashworth (ed.), 463–9. Rotterdam: A. A. Balkema.

Dienes, J. K. 1978. A statistical theory of fragmentation. In *Proceedings of the 19th US Rock Mechanics Symposium*, Y. S. Kim (ed.), 51–5. Conferences and Institutes, Extend Programs and Continuing Education, University of Nevada, Stateline, Nevada.

Dienes, J. K. 1982. Permeability, percolation and statistical crack mechanics. In *Issues in rock mechanics*, R. E. Goodman & F. E. Heuze (eds), 86–94. New York: American Institute of Mining, Metallurgical, and Petroleum Engineers.

Dunning, J. D., D. Petrovsky, J. Schuler & A. Owens 1984. The effect of aqueous chemical environments on propagation on quartz. *J. Geophys. Res.* 89(B6), 4115–23.

Englman, R., Y. Gur & Z. Jaeger 1983. Fluid flow through a crack network in rocks. *J. Appl. Mech.* 50, 707–11.

Evans, A. G. 1972. A method for evaluating the time dependent characteristics of brittle materials and its applications to polycristalline alumina. *J. Mat. Sci.* 7, 1137–46.

Gueguen, Y., C. David & M. Darot 1986. Models and time constants for permeability evolution. *Geophys. Res. Letts* 13(5), 460–3.

Gueguen, Y. & J. Dienes, 1989. Transport properties of rocks from statistics and percolation. *Math. Geol.* **29**, 1–13.

Hallam, S. D. & M. F. Ashby 1989. Compressive brittle fracture and the construction of multi-axial failure maps. This volume, 84–108.

Hillig, W. S. & R. J. Charles 1965. Surfaces, strain dependent reactions and strength. In *High strength materials*, F. Zackey (ed.), 682–705. New York: Wiley.

Horii, H. & S. Nemat-Nasser 1983. Estimate of stress intensity factors for interacting cracks. In *Advances in aerospace structures and materials*, R. M. Laurenson & U. Yuceoglu (eds), 111–17. AD Publ. Am. Soc. Mech. Engrs. Aerospace Div., AD 06.

Horii, H. & S. Nemat-Nasser 1985. Compression-induced microcrack growth in brittle solids: axial splitting and shear failure. *J. Geophys. Res.* **90**, 3105–25.

Katz, A. J. & A. H. Thompson 1986. Quantitative prediction of permeability in porous rocks. *Phys. Rev. B.* **34**(11), 8179–81.

Kemeny, J. M. & N. G. W. Cook 1987. Crack models for the failure of rocks in compression. In *Proc. 2nd Int. Conf. on Constitutive Laws for Engineering Materials*, Tucson, Arizona.

Kolesnikov, Y. & T. L. Chelidze 1985. The anisotropic correlation in percolation theory. *J. Phys. A* **18**, L273.

Kranz, R. L. 1983. Microcracks in rocks: a review. *Tectonophysics* **100**, 449–80.

Lama, R. D. & V. S. Vukuturi 1972. Static elastic constants of rocks. Ch. 6 in *Handbook on mechanical properties of rocks. Testing techniques and results*, vol. II. Clausthal: Trans. Techn. Publications.

Lawn, B. R. & T. R. Wilshaw 1975. *Fractures of brittle solids*. Cambridge: Cambridge University Press.

Maugis, D. 1985. Subcritical crack growth, surface energy and fracture toughness of brittle materials. In *Fracture mechanics of ceramics*, R. C. Bradt, A. G. Evans, D. P. H. Hasselman & F. F. Lange (eds). New York: Plenum.

Meredith, P. G. 1989. Fracture and failure of polycrystals: an overview. This volume, 5–47.

Meredith, P. G. & B. K. Atkinson 1982. High temperature tensile crack propagation in quartz: experimental results and applications to time-dependent earthquake rupture. *Earth Predict. Res.* **1**, 377–91.

Meredith, P. G. & B. K. Atkinson 1983. Stress corrosion and acoustic emission during tensile crack propagation in Whin Sill dolerite and other basic rocks. *Geophys. J. R. Astron. Soc.* **75**, 1–21.

Meredith, P. G. & B. K. Atkinson 1985. Fracture toughness and subcritical crack growth during high-temperature tensile deformation of Westerly granite and Black gabbro. *Phys. Earth Planet. Inter.* **39**, 33–51.

Michalske, T. A. & S. W. Freiman 1982. A molecular interpretation of stress corrosion in silica. *Nature* **295**, 511–12.

Parks, G. A. 1984. Surface and interfacial free energy of quartz. *J. Geophys. Res.* **89**, 3997–4008.

Paterson, M. S. 1978. *Experimental rock deformation: the brittle field*. Berlin: Springer.

Rice, J. R. 1978. Thermodynamics of the quasi-static growth of Griffith cracks. *J. Mech. Phys. Solids* **26**, 61–78.

Robinson, P. C. 1983. Connectivity of fracture systems, a percolation theory approach. *J. Phys. A: Math. Gen.* **16**, 605–14.

Rudnicki, J. W. 1980. Fracture mechanics applied to the Earth's crust. *Ann. Rev. Earth Planet. Sci.* **8**, 489–525.

Sano, O., I. Ito & M. Terada 1981. Influence of strain rate on dilatancy and strength of Oshima granite under uniaxial compression. *J. Geophys. Res.* **86**, 9299–311.

Sen, P. N., S. Peng, B. I. Halperin & M. F. Thorpe 1985. Elastic properties of depleted networks and continua. In *Physics of finely divided matter*. N. Boccara & M. Daoud (eds), 171–9. Berlin: Springer.

Stauffer, D. 1985. *Introduction to percolation theory*. London: Taylor and Francis.

Swanson, P. L. 1984. Subcritical growth and other time- and environment-dependent behaviour in crustal rocks. *J. Geophys. Res.* **89**, 4137–52.

Trantina, G. G. 1977. Stress analysis of the double torsion specimen. *J. Am. Ceram. Soc.* **60**(7–8), 338–41.

Walsh, J. B. 1965. The effect of cracks on the compressibility of rocks. *J. Geophys. Res.* **70**, 381–9.

Wiederhorn, S. M., E. R. Fuller & R. Thomson 1980. Micromechanisms of crack growth in ceramics and glasses in corrosive environments. *Metal Sci.* Aug.–Sept., 450–8.

CHAPTER THREE

Fracture of polycrystalline ceramics

Stephen W. Freiman & Peter L. Swanson

3.1 Introduction

Catastrophic failure in polycrystalline ceramics results from stressed cracks growing to critical dimensions which can span a range of size scales. As critical flaw dimensions increase in size from a scale less than characteristic microstructure dimensions to a size which encompasses many grain diameters, the resistance to fracture increases, in certain polycrystals, by a factor of 5–10. This increase represents the difference between the fracture resistance of the polycrystal and that of its individual constituent single crystals. In this chapter, we: (1) show how relatively small variations in grain size and shape affect the fracture toughness – crack size relationship (*R*-curve behaviour); (2) briefly review several microstructural mechanisms suggested to be responsible for both the high fracture energy of polycrystals and the rising resistance to fracture with crack extension; (3) present the results of *in-situ* microscopy observations of subcritically propagating cracks which lend support to crack-interface traction as an important fracture resistance mechanism; and (4) examine the complicating influence that the traction mechanism has on prediction of time-dependent failure from flaws propagating under the influence of stress-enhanced chemical reactions.

3.2 Flaw size/grain size effects

Quantitative methods of fracture characterization have been developed for use in predicting both the strength and lifetime of materials under variable applied-stress conditions. This linear elastic, continuum-mechanics approach to failure prediction requires specification of (1) the magnitude of the crack-tip stress field (K_I, the stress intensity factor for Mode I, or tensile cracking), (2) the kinetics of environmentally assisted slow crack growth rates (the crack velocity v–K_I relations), and (3) a failure criterion (i.e. $K_I = K_{Ic}$, the fracture toughness). When the material-breakdown processes are constrained to occur in a sufficiently small volume about the crack tip such that the small-scale inelastic deformation assumption of linear elastic fracture mechanics applies, K_I is an adequate measure of the intensity of the stress field responsible for fracture and K_{Ic} is considered to be a material property.

One of the primary difficulties in applying linear elastic fracture mechanics to the failure analysis of polycrystalline ceramics is the similarities in the sizes of critical flaws and grains. When flaws are completely embedded within individual grains at the point of failure criticality, or when they are very large in comparison to the dominant scale of the microstructure, a continuum description of the elastic and fracture properties of the cracked solid is appropriate. Under these conditions, meaningful fracture mechanics parameters (e.g. K_I) may be estimated from measurements of crack length and applied load. As reported by Rice *et al.* (1980), and shown schematically in Figure 3.1, the transition in critical failure conditions from failure governed by fractures within individual grains to macroscopic fractures encompassing many grains is accompanied by an increase in fracture surface energy, Γ (Γ is defined as $K_{Ic}^2/2E$, where E is Young's modulus). Factors that influence the range of flaw-size to grain-size ratios over which the transition occurs include preferred orientation and number of easy cleavage planes in the particular crystallographic system. The transition from single-crystal to polycrystalline fracture energy values may not occur until the flaw radius is up to 10 or more times the average grain size (Rice *et al.* 1980).

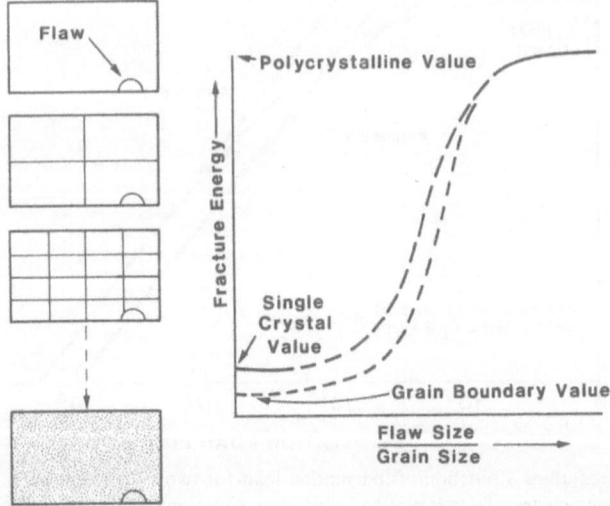

Figure 3.1 Schematic of the typical variation of fracture energy with flaw to grain size ratio, a/G, in polycrystalline ceramics (after Rice *et al.* 1980).

3.2.1 Testing procedure

As noted earlier, in many practical applications of structural ceramic compo-nents, failure results from flaws which become critical in a range where their size is of the same order, or slightly larger than, the scale of the microstruc-ture. A significant advancement in testing procedures which has enabled us to better characterize the effects of microstructure in this transition region has been the development of the indentation-fracture test (Chantikul *et al.* 1981). In this test a standard microhardness testing machine equipped with a Vickers (or Knoop) diamond indenter is used to introduce a well characterized flaw into the surface of a flexural-test specimen. By varying the indenter load, a range of controlled crack sizes, *a*, can be introduced which span the transition region. Specimens are subsequently fractured in either uniaxial or biaxial flex-ure. The resulting fracture strength, σ, is given in equation (3.1) in terms of the indentation load, P, Young's modulus, E, hardness, H, and the critical fracture toughness observed at very long crack lengths, K_{Ic}:

$$\sigma = \eta_v^R K_{Ic}^{4/3} P^{-1/3} (E/H)^{-1/6} \tag{3.1}$$

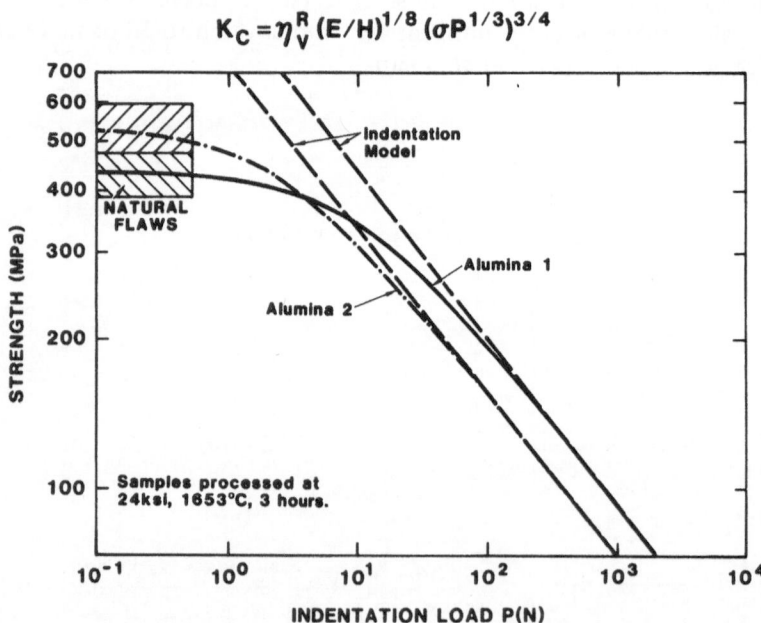

$$K_C = \eta_v^R (E/H)^{1/8} (\sigma P^{1/3})^{3/4}$$

Figure 3.2 Strength as a function of indentation load for two 94 per cent aluminas made from different starting powders. Data is plotted based upon equation (3.1). Deviation of the data from the indentation model at small indentation loads is indicative of direct microstructural effects (after Hellmann *et al.* 1986).

where η_v^R is an empirically determined constant (Chantikul *et al.* 1981). The E/H term reflects the effectiveness with which the indentation damage produces a residual tensile stress field at the specimen surface. The residual field is partially relaxed by the formation of the indentation–crack system which then remains under a state of residual tension. If we assume that K_{Ic}, H, and E are material constants, a logarithmic plot of strength versus indentation load yields a straight line having a slope of $-1/3$. In an isotropic, homogeneous material such as glass, or a very fine-grained ceramic, (3.1) is obeyed over a wide range of indentation loads. Deviations of indentation fracture data from the predicted $-1/3$ slope at small values of the indentation load mark the onset of transition from equilibrium polycrystalline fracture toughness (that which remains constant for increasing crack size) to a region where the strength is relatively insensitive to the initial flaw size. The plateau in the data shown in Figure 3.2 indicates that certain microstructure-related toughening mechanisms develop in this flaw size range which resist the propagation of critical flaws.

3.3 Toughening mechanisms

What makes polycrystalline ceramics more resistant to fracture than single crystals of the same material? This question has been discussed in some detail by Wiederhorn (1984) and Swain & Rose (1984) in terms of the various possible toughening mechanisms. The more generally accepted toughening mechanisms, namely *crack deflection*, *microcracking*, and *phase transformations*, are briefly summarized. We then discuss a recently observed mechanism referred to here as *crack-surface tractions*, the effects of which can account for many of the hitherto unexplained observations of toughness behaviour in polycrystalline ceramics.

3.3.1 Crack deflection

Grain boundaries and cleavage planes represent paths of lower fracture energy. As a result, cracks are locally deflected along these paths to produce a surface with a roughness that scales with the microstructure dimensions. From a stress analysis point of view, independent of fracture energy considerations, the deflected-crack geometry provides a reduction in the local K_I driving force compared to a straight crack of equal projected length. If we consider the condition for local crack extension to be the achievement of a local critical value of K_{Ic}, then we must apply a larger load or applied K_I to the deflected crack to achieve this local failure condition. Macroscopically this appears to represent a 'toughening' of the material. Microscopically, however, the specific fracture surface energy associated with single crystals or grain boundaries has not actually changed; the total fracture surface area has only increased due to the surface roughness.

Faber & Evans (1983a,b) derived a model which predicts the increase in toughness due to both twist and tilt of the local crack plane. The model suggests that elongate grains should have a more significant effect on K_{Ic} than equi-axed ones, in agreement with intuition and experimental observations. However, the model cannot explain the relatively high toughness values observed for many polycrystalline ceramics, nor the long crack-extension distances required to reach equilibrium toughness values.

3.3.2 Microcracking

A dispersion, or cloud, of isolated microcracks surrounding a macroscopic crack tip is another postulated mechanism for increasing fracture energy and producing R-curve behaviour (see also Meredith, this volume). Microcracks are suggested to form in the elevated stress field near a macrocrack tip wherever a local critical-stress microcracking criterion is satisfied (Hoagland et al. 1975). Once formed, suitably positioned and oriented microcracks shield the macrocrack tip from high levels of stress and necessitate an increase in applied load, or equivalent applied K_I, to further advance the macrocrack. According to recent refinements of the theory (Evans & Faber 1984), a wake of residually stressed material is left behind the crack tip along the crack flanks, and constitutes an additional source of absorbed fracture energy which produces R-curve behaviour. However, even for a substantial volume fraction of microcracks, this particular model predicts toughness increases with crack extension of only 10%, much smaller than the 300–500 per cent increases observed experimentally (Steinbrech et al. 1983, Wieninger et al. 1986).

Rice & Freiman (1981) developed a fracture model based on the energy absorption of microcracking which predicts a maxima in the crack-growth resistance as a function of grain size. Although the model does not attempt to treat R-curve behaviour, the predictions of failure from short cracks are in excellent agreement with many experimental observations recorded in the published literature.

3.3.3 Phase transformations

The primary ceramic material in which phase transformation toughening is observed is partially stabilized zirconia, i.e. zirconia in which the tetragonal phase has been stabilized by the addition of other oxides. Details of the phase transformation toughening process are described elsewhere (Evans et al. 1981). Briefly, the tetragonal zirconia transforms to the monoclinic form in the stress field of the crack tip. This transformation is associated with shear strain and volume changes which induce compressive stresses on the crack surfaces behind the crack tip. Toughness values as large as 8 MPa m$^{1/2}$ have been reported. There has also been extensive work on composites of partially stabilized zirconia in which zirconia is combined with an alumina matrix to yield large values of fracture toughness (e.g. Wieninger *et al.* 1986). The concept of phase transformation toughening has also been extended to other materials (Freiman *et al.* 1986).

3.3.4 Crack-surface tractions

Recent real-time, *in-situ* microscopy observations of fracture have provided direct documentation of fracture resistance mechanisms operating in polycrystalline ceramics (Swanson *et al.* 1987, Swanson 1988) and rocks (Swanson 1987). No crack-tip stress-singularity-related distributed microcracking surrounding macroscopic fracture tips was observed in any of the materials using optical reflection/transmission microscopy techniques. We caution that this null result cannot be used as evidence of the non-existence of diffuse-zone microcracking; however, a different mechanism was observed which provides a relatively simple explanation for several different observations associated with macroscopic fracture, including non-linear applied-force/crack-opening displacement relations, R-curve behaviour, specimen-geometry dependent toughness, and inconsistencies in subcritical crack growth data. This mechanism is one of tractions (Fig. 3.3) which act across the nascent fracture

(i) FRICTIONAL INTERLOCKING

Figure 3.3 Schematic of crack-surface tractions which can occur in ceramics, thereby increasing fracture toughness.

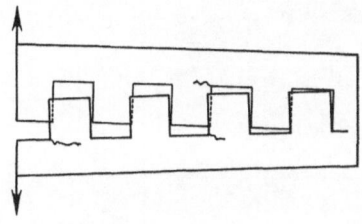

(ii) LIGAMENTARY BRIDGING

(INTACT–MATERIAL 'ISLANDS' LEFT
BEHIND ADVANCING FRACTURE FRONT)

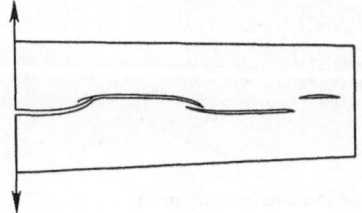

77

surfaces. Two sources of traction have been identified: (1) geometrical or frictional interlocking of the rough fracture surfaces; and (2) ligamentary bridging by remnant islands of unbroken material left behind the advancing fracture front. Each source of traction serves to shield the macrocrack tip from high levels of stress: the K_I field experienced at the crack tip is less than the applied K_I by an amount proportional to the restraining-force magnitude. Each traction site is also associated with microcracking localized along the crack interface *behind* the macrocrack tip as increasing crack-opening displacement

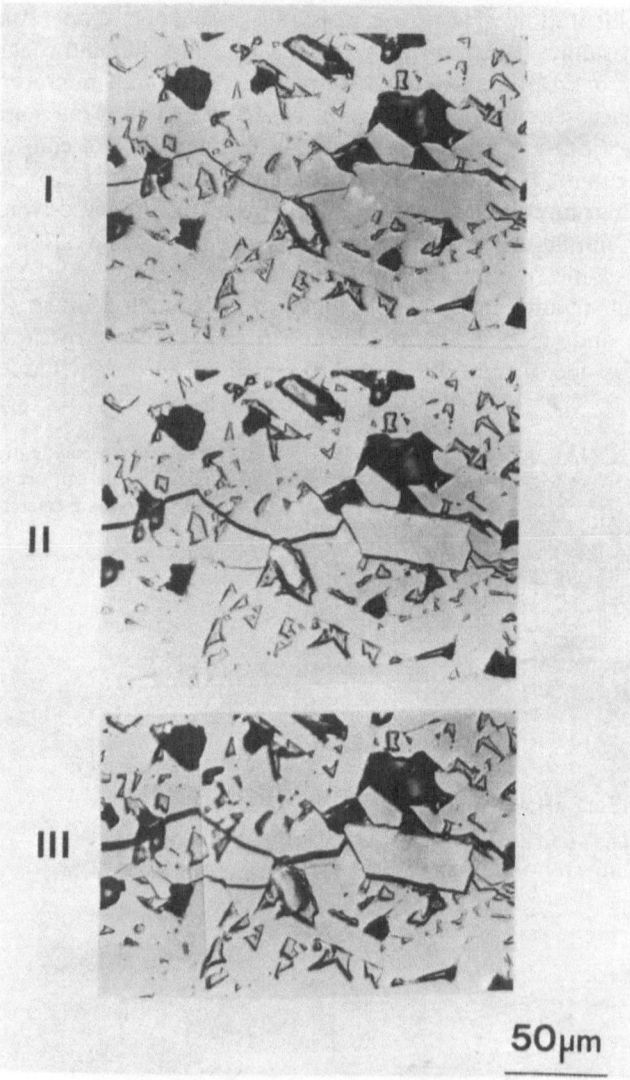

Figure 3.4 Evolution of a traction site in Alumina 1.

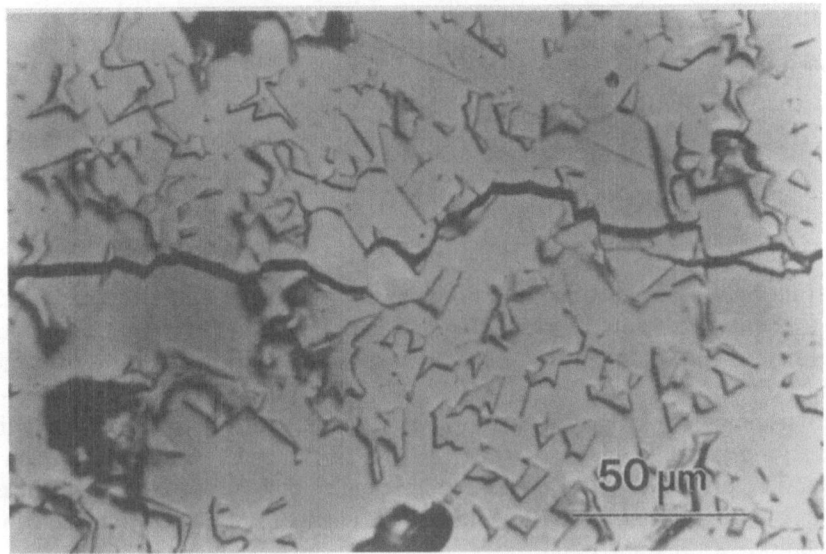

Figure 3.5 Typical traction site in Alumina 2.

reduces the restraining-force magnitude. Several fracture mechanics models have recently been developed to describe this behaviour by incorporating the crack-interface traction force–separation relationship into a tension-softening constitutive law (Shah 1985, Li & Liang 1986, Mai & Lawn 1987).

In this study, we have used the *in-situ* observation technique to examine fracture in two different 94% aluminium oxides (referred to as alumina 1 and alumina 2). Both of these materials were made by standard pressing and sintering procedures (Hellmann *et al.* 1986). The average grain size for both aluminas was 9–11 μm. However, the aspect ratio of the grains in alumina 1 was greater than that in alumina 2 (1.9 compared to 1.6). Both aluminas exhibited R-curve behaviour, as interpreted from indentation-strength data (Fig. 3.2), but the extent of such behaviour was much greater for alumina 1 than for alumina 2. Both sources of interface traction, i.e. ligamentary bridging and frictional interlocking, were observed.

The evolution of a single traction site in alumina 1 is illustrated in Figure 3.4. Alumina 1 exhibited approximately twice as many traction sites as alumina 2 (Fig. 3.5). In addition, the amount of interface-localized microcracking associated with each traction site was twice as much in alumina 1 as in alumina 2. Thus, alumina 1 exhibited approximately four times as much interface-localized microcracking (as observed on the sample's surface) as alumina 2. This observation is qualitatively in good agreement with the indentation-strength data and provides direct observational support for crack-interface tractions as the fracture-resistance mechanism responsible for R-curve behaviour. Although not directly demonstrated in these observations, the causal relationship between crack-interface tractions and increasing fracture energy with crack extension has previously been demonstrated in experiments in which the crack interface is removed by saw cutting (Knehans & Steinbrech 1982).

3.4 Environmental effects

It is well known that glasses and ceramics can fail after periods of time under constant load in a phenomenon known as *static fatigue*. Static fatigue arises due to the slow growth of flaws in the material because of the combined effects of moisture and stress. When a flaw reaches some critical size, catastrophic failure results.

Environmentally enhanced crack growth occurs at ambient temperatures in almost every ceramic material, ranging from silicate glass to nitrides and carbides. The primary environmental constituent giving rise to crack growth is water, although other molecules can also be agents for fracture enhancement (Michalske & Freiman 1984).

Fundamental crack-growth data can best be obtained by optically measuring crack velocities as a function of crack-tip stress intensity, K_I, using a variety of experimental fracture-mechanics procedures. In the case of opaque materials such as polycrystalline ceramics, such direct measurements of crack growth become more difficult. In such materials the relevant crack-growth parameters are best determined by a method known as dynamic fatigue. This procedure involves measuring the strength of the material (preferably after indentation) as a function of the rate of loading. If we assume that the crack velocity can be expressed as a power law function of stress intensity, namely:

$$v = v_0 (K_I/K_{Ic})^n \tag{3.2}$$

where v_0 and n are empirical constants which depend on the material and test environment (Evans 1972). Equation (3.2) can be integrated to yield the following expression for the strength of an indented specimen in a dynamic fatigue test:

$$\sigma = (\lambda' \dot\sigma)^{(1/n' + 1)} \tag{3.3}$$

where $\dot\sigma$ is the stressing rate, λ' is a constant, and $n' = \frac{4}{3}n - \frac{2}{3}$, for normal point-like flaws (Fuller *et al.* 1983). A logarithmic plot of strength versus stressing rate should yield a straight line of slope n' (Fig. 3.6). This value of n', together with a measure of the initial flaw severity determined from the strength of the material under inert conditions, can be used to predict the lifetime of a structure under a specified load.

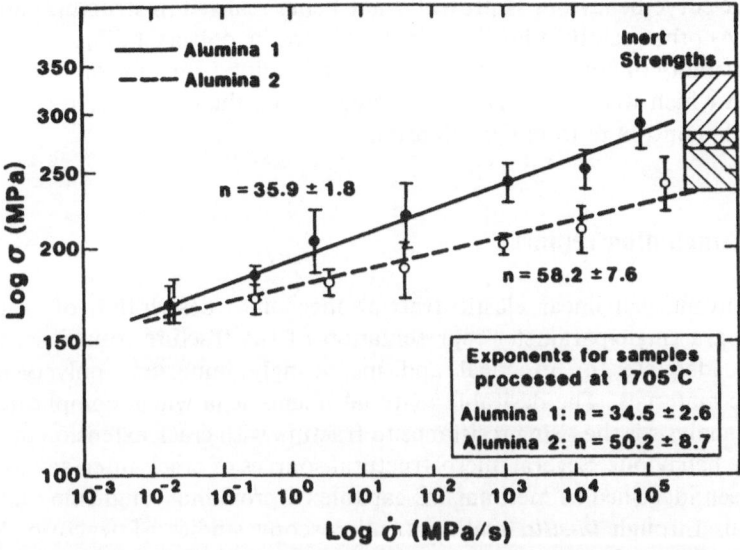

Figure 3.6 Dynamic fatigue data for Aluminas 1 and 2. Note in box that different processing conditions did not change crack growth susceptibility (after Hellmann *et al.* 1980).

3.5 Implications of crack-interface tractions on failure predictions

In materials exhibiting a crack-length-independent toughness, such as glass, single crystals, and certain very-fine-grained ceramics, the strength under known loading rates or the time to failure under known static loads can be predicted with useful accuracy. This is achieved by measuring both the fracture toughness and the subcritical crack growth (v–K_I) curves, integrating the v–K_I data, and calculating the strength or failure time for critical flaw conditions (Davidge *et al.* 1973; Meredith, this volume). The existence of R-curve behaviour gives this fracture-mechanics-based failure prediction logic a significant additional complication; the magnitude of the crack-interface restraining force depends upon the crack history (crack length and crack-opening displacement history). In order to accurately determine the driving force experienced by the crack tip, one cannot merely measure the crack length and applied load as in the familiar $K_I \propto \sigma\sqrt{a}$ relation. Instead, one must also know the crack history to accurately evaluate the hysteretic traction force, the effect of which must be subtracted from the applied-K_I field in the failure prediction calculations. Neglect of the traction-shielding contribution to K_I (tip) results in large uncertainties in crack-growth rates and hence gross errors in failure predictions for these and similar materials (Swanson 1987). Present strategies used to compensate for these uncertainties call for extremely conservative design; K_{Ic} is assumed to be constant and equal to values observed for short cracks. Consequently, the potential performance of ceramic materials with significantly

81

rising R-curve behaviour is presently not being realized in structural applications. Another ramification of microstructure in polycrystalline ceramics is the possibility of crack growth during cyclic loading from tension to compression, in which wedging stresses generated during the compressive portion of the cycle contribute to crack extension.

3.6 Concluding remarks

The conventional linear elastic fracture-mechanics description of fracture, based on a single-parameter representation of the fracture resistance, is not always adequate for practical and increasingly important polycrystalline ceramic materials. The desirable material phenomena which complicates the failure analysis is the rising resistance to fracture with crack extension or rising R-curve behaviour. Several microstructural sources of crack-interface traction have been identified as mechanisms capable of providing toughening in these materials through *in-situ*, real-time, microscopy studies of fracture. While toughened ceramics are not necessarily accompanied by significant strength increases, a practical result of the R-curve behaviour is to increase the likelihood of sustaining larger stable flaws before reaching critical failure conditions. This offers a tremendous benefit to flaw detection efforts in these brittle materials which fail from extremely small cracks. Present efforts are directed at enhancing the interfacial restraining effect through microstructural modification using controlled ceramic processing techniques.

References

Chantikul, P., G. R. Anstis, B. R. Lawn, & D. B. Marshall 1981. A critical analysis of indentation techniques for measuring fracture toughness: II, strength method. *J. Am. Ceram. Soc.* **64**, 539–43.

Davidge, R. W., J. W. McLaren, & G. Tappin 1973. Strength–probability–time (STP) relationships in ceramics. *J. Mat. Sci.* **8**, 1699–705.

Evans, A. G. 1972. A method for evaluating the time-dependent failure characteristics of brittle materials and its application to polycrystalline alumina. *J. Mat. Sci.* **7**, 1137–46.

Evans, A. G. & K. T. Faber 1984. Crack growth resistance of microcracking materials. *J. Am. Ceram. Soc.* **67**, 255–60.

Evans, A. G., D. B. Marshall & N. H. Burlingame 1981. Transformation toughening in ceramics. In *Advances in ceramics*, vol. 3, A. H. Heuer & L. W. Hobbs (eds), 202–16. Columbus, Ohio: American Ceramic Society.

Faber, K. T. & A. G. Evans 1983a. Crack deflection processes: I, theory. *Acta Metall.* **31**, 565–76.

Faber, K. T. & A. G. Evans 1983b. Crack deflection processes: II, experiment. *Acta. Metall.* **31**, 577–84.

Freiman, S. W., L. Chuck, J. J. Mecholsky, D. L. Shelleman & L. J. Storz 1986. Fracture mechanisms in lead zirconate titanate ceramics. In *Fracture mechanics of ceramics*, vol. 8, R. C. Bradt, A. G. Evans, D. P. H. Hasselman & F. F. Lange (eds), 175–85. New York: Plenum.

Fuller, E. R., B. R. Lawn & R. F. Cook 1983. Theory of fatigue for brittle flaws originating from residual stress concentrations. *J. Am. Ceram. Soc.* **66**, 314–21.

Hellmann, J. R., J. Matsko, S. W. Freiman & T. L. Baker 1986. Microstructure-mechanical property relationships in 94% alumina ceramics. In *Tailoring multiphase and composite ceramics*, R. E. Tressler, G. L. Messina, C. G. Pantano & R. E. Newnham (eds), 367–79. New York: Plenum Press.

Hoagland, R. G., J. D. Embury & D. J. Green 1975. On the density of microcracks formed during the fracture of ceramics. *Scripta Metall.* **9**, 907–9.

Knehans, R. & R. Steinbrech 1982. Memory effect of crack resistance during slow crack growth in notched Al_2O_3 bend specimens. *J. Mat. Sci. Lett.* **1**, 327–9.

Li, V. C. & E. Liang 1986. Fracture processes in concrete and fiber reinforced cementitious composites. *J. Engng. Mech.* **112**, 566–86.

Mai, Y.-W. & B. R. Lawn 1987. Crack-interface grain bridging as a fracture resistance mechanism in ceramics: II. Theoretical fracture mechanics model. *J. Am. Ceram. Soc.* **70**, 289–94.

Meredith, P. G. 1989. Fracture and failure of brittle polycrystals: an overview. This volume, 5–47.

Michalske, T. A. & S. W. Freiman 1983. A molecular mechanism for stress corrosion in vitreous silica. *J. Am. Ceram. Soc.* **66**, 284–8.

Rice, R. W. & S. W. Freiman 1981. Grain size dependence of fracture energy in ceramics, II: a model for non-cubic materials. *J. Am. Ceram. Soc.* **64**, 350–4.

Rice, R. W., S. W. Freiman & J. J. Mecholsky 1980. The dependence of strength controlling fracture energy on the flaw size to grain size ratio. *J. Am. Ceram. Soc.* **63**, 129–36.

Shah, S. P. (ed.) 1985. *Applications of fracture mechanics to cementitious composites*. Boston: Martinus Nijhoff.

Steinbrech, R., R. Knehans & W. Schaarwachter 1983. Increase of crack resistance during slow crack growth in Al_2O_3 bend specimens. *J. Mat. Sci.* **18**, 265–70.

Swain, M. V. & L. R. F. Rose 1984. Toughening of ceramics. *Proc. Sixth Int. Conf. on Fracture* **1**, 473–94.

Swanson, P. L. 1987. Tensile fracture resistance mechanisms in brittle polycrystals: An ultrasonics and *in-situ* microscopy investigation. *J. Geophys. Res.* **92**, 8015–36.

Swanson, P. L. 1988. Crack-interface traction: a fracture-resistance mechanism in brittle polycrystals. In *Advances in ceramics*, **22**, 135–55. Columbus, Ohio: Am. Ceram. Soc.

Swanson, P. L., C. J. Fairbanks, B. R. Lawn, Y.-W. Mai, & B. J. Hockey 1987. Crack-interface grain bridging as a fracture resistance mechanism in ceramics: I. Experimental study on alumina. *J. Am. Ceram. Soc.* **70**, 279–89.

Wiederhorn, S. M. 1984. Brittle fracture and toughening mechanisms in ceramics. *Ann. Rev. Mat. Sci.* **14**, 373–403.

Wieninger, H., K. Kromp & R. F. Pabst 1986. Crack resistance curves of alumina and zirconia at room temperature. *J. Mat. Sci.* **21**, 411–18.

CHAPTER FOUR

Compressive brittle fracture and the construction of multi-axial failure maps

Sheila D. Hallam & Michael F. Ashby

4.1 Introduction

Cracks, holes, and inclusions in an elastic solid can interact with a compressive stress field in a way which causes new cracks to grow from them. If these cracks extend to the sample surface, or if they interact with each other so that they grow in an unstable manner, then a macroscopic failure may follow. The initiation and growth of cracks from pores has been considered by Sammis & Ashby (1986): that from small angled cracks is analysed by Nemat-Nasser & Horii (1982), Cooksley (1984), and Ashby & Hallam (1986). In this chapter we consider the growth of cracks in compressive stress states and how they interact to cause a macroscopic failure.

Crystalline ceramics, rocks, and minerals often contain a distribution of fine cracks with a size about equal to the grain size: they are caused by thermal or elastic stress during earlier thermal or mechanical loading. Acoustic and dilatometric studies show that they start to propagate or extend when the axial stress reaches about one half of the ultimate failure stress, and microscopic observations show that they extend parallel to the compression axis (see, for example, Paterson 1978).

Their subsequent behaviour depends on the confining pressure, as shown in Figure 4.1. Simple axial or radial compression, shown in Figures 4.1a and 4.1d, causes one or a few cracks to propagate and combine to give failure on planes parallel to the maximum, compressive stress ('slabbing'). A modest confining pressure prevents this unlimited crack growth; failure then occurs by the interaction of cracks to give the macroscopic shear failure shown in Figure 4.1b. Larger confining pressures limit the growth of individual cracks even further, and the sample deforms in a pseudoductile way with large-scale deformation taken up by many short, homogeneously distributed micro-cracks, shown in Figure 4.1c.

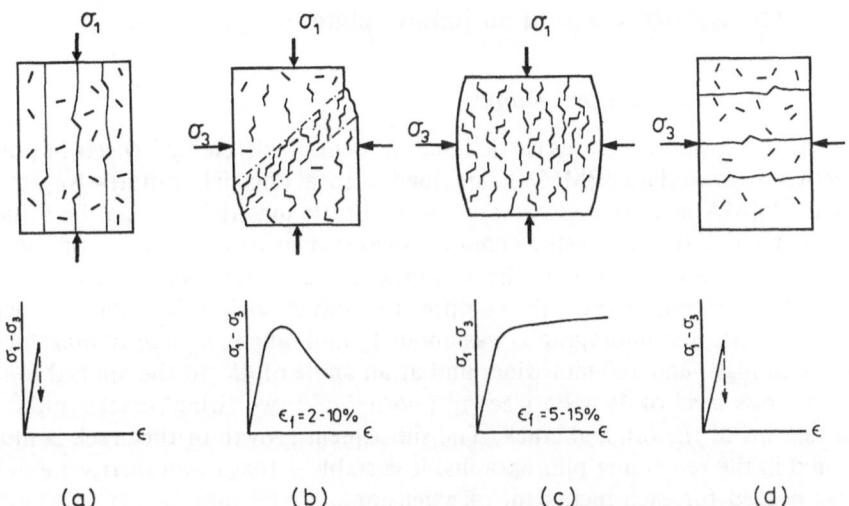

Figure 4.1 Failure modes in compression: (a, d) slabbing; (b) linking of cracks to form a shear zone; and (c) near-homogeneous deformation by distributed microcracking. (Reprinted with permission from *Acta Metallurgica* **34**, Hallam & Ashby, Copyright 1986, Pergamon Journals Ltd.)

The tensile fracture of a brittle solid is relatively simple and well understood: a single flaw (the longest or most favourably oriented) propagates in an unstable manner when the stress intensity at its crack tip exceeds the critical stress intensity (or fracture toughness), K_{IC}, of the solid. The mechanics of compressive brittle fracture is more complex and less well understood. Early attempts considered a critical shear stress criterion for failure (Coulomb 1773). Subsequent work generally starts from the stress field of an elliptical flaw and calculates the conditions under which a crack first initiates at some point on its surface (e.g. Griffiths 1924; McClintock & Walsh 1962; Murrell 1963, 1965; Murrell & Digby 1970a,b): this initiation stress was often assumed to correspond to failure. But the propagation of the microcracks in a compressive stress field is, initially, stable, meaning that a progressively higher stress is needed for each further increment of crack length: the initiation process does not correspond to final failure (e.g. Hoek & Bieniawski 1965, Kobayashi 1971, Cooksley 1984, Nemat-Nasser & Horii 1982, Ashby & Hallam 1986).

In this chapter we summarize recent calculations of Cooksley (1984) and Ashby & Hallam (1986), concentrating on the main results rather than the detailed derivations (which can be found in the source references). The basic problems described include the initiation and stable growth from a single isolated crack in a compressive stress field, and the interaction process leading to final failure. The results allow the construction of a *failure surface* in stress space, which shows the combination of stresses required for failure. This is combined with yield and creep failure surfaces to show how changes of failure mechanism occur as the confining pressure or the temperature are changed.

4.2 The isolated crack in an infinite plate

4.2.1 Experimental observations

A series of experiments was carried out on single cracks cut into sheets of poly-methyl-methacrylate (PMMA) and loaded in a uniaxial compressive stress field. PMMA is a homogeneous isotropic brittle material that can be drilled and cut with a fretsaw to study compressive crack growth. The series of photographs in Figure 4.2 shows the response of one such isolated crack to a steadily increasing stress – the compression axis is vertical. The starter crack, shown in the first photograph, was about 14 mm long in a plate 10 mm thick, 170 mm high, and 100 mm wide, and at an angle of 45° to the applied load. At a stress level of 39 MPa (second photograph) two 'wing' cracks initiated at the tips of the original crack. The subsequent growth of the crack is illustrated in the remaining photographs: it is stable – that is, an increasing load was needed for each increment of extension.

Experiments were carried out on cracks at various angles, ψ, to the compressive stress axis and on plates of various widths. Typical data for single-crack growth are plotted in Figures 4.3 and 4.4 using normalized axes. The length of the wing cracks l is plotted in units of a, the half-length of the initial crack; and the compressive stress, σ, is plotted in units of $K_{IC}/\sqrt{(\pi a)}$, where K_{IC} is the fracture toughness of PMMA ($K_{IC} = 1.0$ MPa m$^{1/2}$ for slow crack

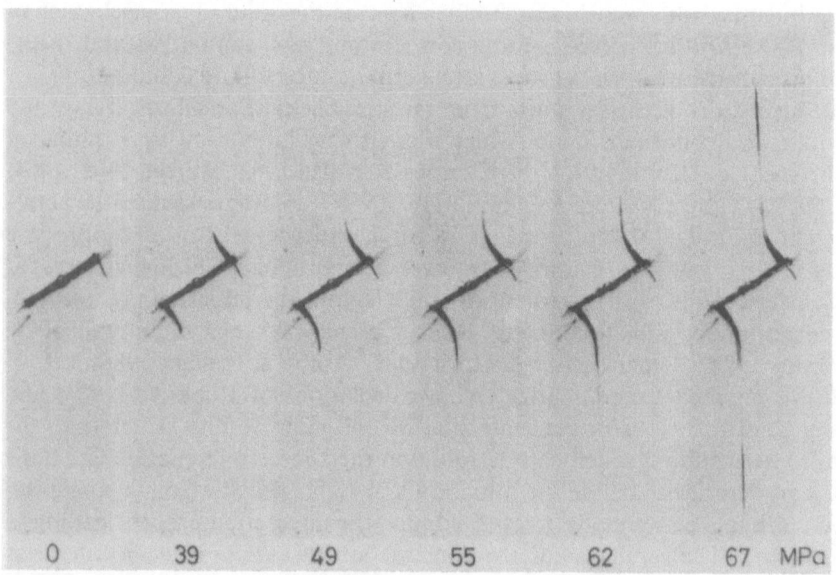

Figure 4.2 A sequence of photographs of wing-crack initiation and growth under progressively higher compressive stress. (Reprinted with permission from *Acta Metallurgica* **34**, Hallam & Ashby, Copyright 1986, Pergamon Journals Ltd.)

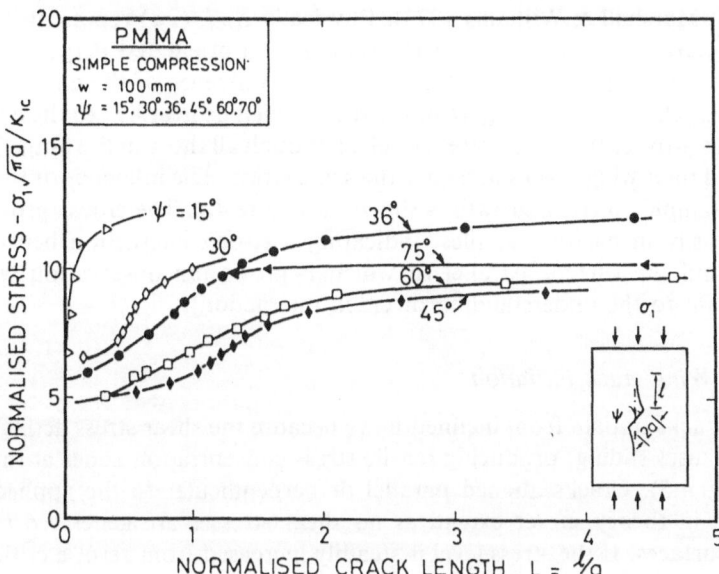

Figure 4.3 Wing-crack growth in PMMA in simple compression, showing the influence of the initial crack angle. (Reprinted with permission from *Acta Metallurgica* **34**, Hallam & Ashby, Copyright 1986, Pergamon Journals Ltd.)

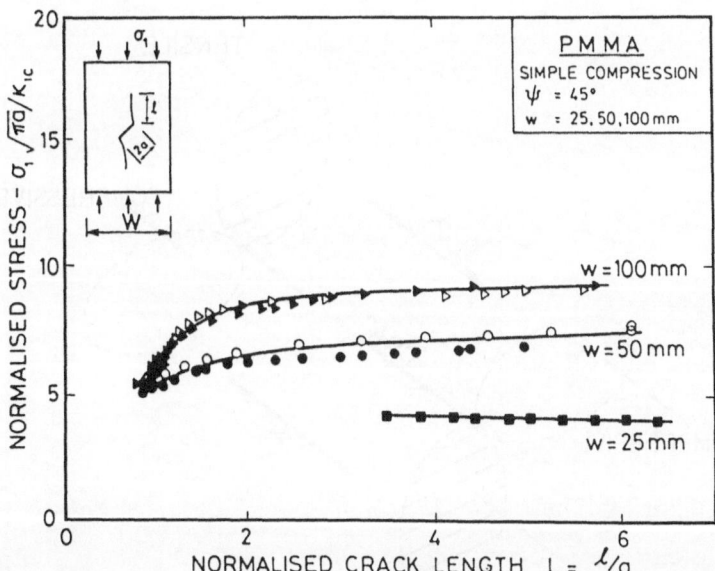

Figure 4.4 Wing-crack growth in simple compression, showing the influence of specimen width. (Reprinted with permission from *Acta Metallurgica* **34**, Hallam & Ashby, Copyright 1986, Pergamon Journals Ltd.)

growth; Marshall & Williams 1973). Physically $K_{IC}/\sqrt{(\pi a)}$ represents the tensile fracture stress of a crack of the same length orientated perpendicular to a tensile stress field in the same material. The influence of the angle ψ of the starter crack is shown in Figure 4.3. Cracks which lie near $45°$ to the compression axis grow at the lowest stress levels, although all those in the range $30-60°$ initiated their wing cracks at nearly the same stress. The influence of the width of the sample on crack growth is shown in Figure 4.4. The cracks grow much more easily in narrow samples, indicating a strong interaction between the crack and the surface parallel to which it grows (an observation which is important to the understanding of crack interaction).

4.2.2 Wing-crack initiation

Wing cracks initiate from inclined flaws because the shear stress acting on the crack causes sliding, producing tensile stress concentration zones at the crack tip (Fig. 4.5). Cracks aligned parallel or perpendicular to the applied stress should, in theory, never extend as no shear stresses are generated on their crack surfaces. If the stress level is steadily increased from zero, a critical tensile stress is eventually reached at which wing cracks initiate and grow into the tensile zone.

Consider an infinite elastic plate containing an initial crack of length $2a$, at an angle ψ to the compression axis, and subjected to principal stresses σ_1 and σ_3 (Figure 4.6). The stresses are treated as positive when tensile, and negative

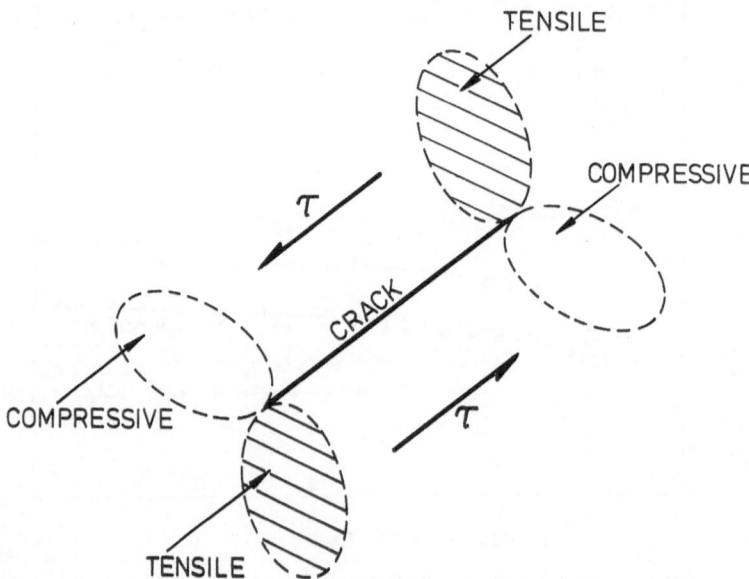

Figure 4.5 Shear stresses acting on angled cracks generate zones of concentrated tensile stress at the crack tip.

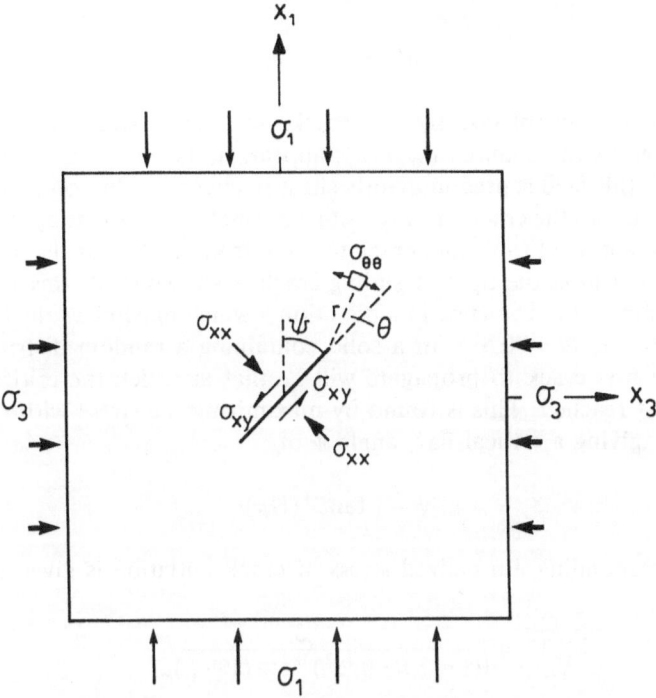

Figure 4.6 The stress fields involved in analysing wing-crack initiation. (Reprinted with permission from *Acta Metallurgica* **34**, Hallam & Ashby, Copyright 1986, Pergamon Journals Ltd.)

when compressive. σ_1 is the most compressive stress and σ_3 the least principal stress. The remote stress field generates shear (σ_{xy}) and normal (σ_{xx}) stresses on the crack plane, where:

$$\sigma_{xy} = \frac{\sigma_3 - \sigma_1}{2} \sin 2\psi \qquad (4.1a)$$

and

$$\sigma_{xx} = \frac{\sigma_3 + \sigma_1}{2} + \frac{\sigma_3 - \sigma_1}{2} \cos 2\psi \qquad (4.1b)$$

(It is implicit in the sign convention that σ_{xy} is always positive and σ_{xx} is always negative.) If the crack surfaces are in contact, the normal stress is transmitted directly across the crack and does not produce any crack-tip stress concentrations. The shear stress tends to make the crack surfaces slide; but because the cracks are closed, a frictional stress, $\mu\sigma_{xx}$, opposes the sliding (μ is the coefficient of friction between the crack surfaces). Then the effective sliding stress is

$$\sigma'_{xy} = \sigma_{xy} + \mu\sigma_{xx} \qquad (4.2)$$

This stress is intensified by the crack, so that a singular stress field, characterized by the quantity $\sigma'_{xy}\sqrt{(\pi a)}$, appears at its tips. In the plane of the initial crack this field is predominantly shear in character, but on planes which lie at an angle θ to the crack tip (Fig. 4.6) a normal tensile stress appears, tending to cause a mode I (opening) crack to grow from the tips of the initial flaw.

The stress field at the tip of a sliding crack is known (Williams 1957). The crack is assumed to propagate in a direction θ which maximizes the tangential tensile stress $\sigma_{\theta\theta}$ ($\theta = 70.5°$). In a solid containing a random distribution of cracks, the first crack to propagate will be that at which the critical tensile stress is first reached. This is found by maximizing the stress with respect to flaw angle, giving a critical flaw angle ψ of

$$\psi = \tfrac{1}{2}\tan^{-1}(1/\mu) \qquad (4.3)$$

The corresponding normalized stress of crack initiation is given by:

$$\frac{\sigma_1\sqrt{\pi a}}{K_{IC}} = \frac{-\sqrt{3}}{[(1-\lambda)(1+\mu^2)^{1/2} - (1+\lambda)\mu]} \qquad (4.4)$$

where λ is the ratio of the principal stresses ($\lambda = \sigma_3/\sigma_1$).

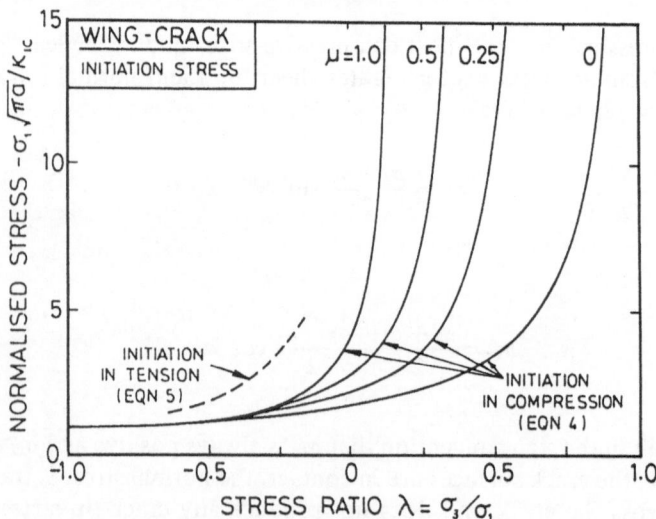

Figure 4.7 Wing-crack initiation in an infinite plate as a function of friction coefficient. (Reprinted with permission from *Acta Metallurgica* **34**, Hallam & Ashby, Copyright 1986, Pergamon Journals Ltd.)

When σ_3 is sufficiently positive, a tensile fracture will occur from cracks where $\theta = \psi = 0$ and the initiation condition becomes

$$\sigma_3\sqrt{(\pi a)} = K_{IC} \qquad \text{or} \qquad \frac{\sigma_1\sqrt{(\pi a)}}{K_{IC}} = -\frac{1}{\lambda} \qquad (4.5)$$

This second criterion truncates the first. The initiation conditions are plotted in Figure 4.7.

Strictly, these equations are only valid when σ_{xx} is compressive. When it is tensile the crack opens and the frictional force disappears: in addition, the normal stress is now concentrated by the crack, giving a new term in the equation for the stress intensity. This calculation is covered in detail by Cooksley (1984) and briefly by Ashby & Hallam (1986). It provides a smooth transition from the closed-crack solution to the tensile truncation. For most purposes, it is adequate to approximate crack initiation by equation (4.4) truncated by (4.5).

4.2.3 Wing-crack growth

Once wing cracks have initiated, further sliding of the main crack wedges them open (Fig. 4.8), causing them to grow further. The relationship between axial stress and crack length is approximated by (Ashby & Hallam 1986):

$$\frac{\sigma_1\sqrt{\pi a}}{K_{IC}} = \frac{-(1+L)^{3/2}}{[\{1 - \lambda - \mu(1+\lambda) - 4.3\lambda L\}\{0.23L + 1/(\sqrt{3}(1+L)^{1/2})\}]} \qquad (4.6)$$

where $L = l/a$.

Physically, this result incorporates three régimes: when L is zero, the stress intensity is determined by shear of the initial crack and (4.6) reduces to (4.4), the initiation condition. When L is about unity (although it depends on λ), the stress intensity is dominated by the wedging open of the wing cracks caused by the sliding of the initial angled crack, and is not much influenced by the confining stress. Finally, when L is large the stress intensity is greatly influenced by the confining stress (because it acts on the entire crack length, $2(l + a)$, whereas σ_1 can be thought of as acting only on the angled segment of the crack of length $2a$). When λ is zero (simple compression) the crack grows in a stable manner such that, for large L,

$$\frac{\sigma_1\sqrt{\pi a}}{K_{IC}} = \frac{-\sqrt{L}}{0.23(1 - \mu)} \qquad (4.7)$$

Equation (4.6) is plotted as full lines in Figures 4.9, 10 & 11 for three values of the coefficient of friction. Where possible, the results have been compared with those of Nemat-Nasser & Horii (1982), who used a Green's function method (broken lines). Figure 4.9 is for zero friction: it shows a plot of

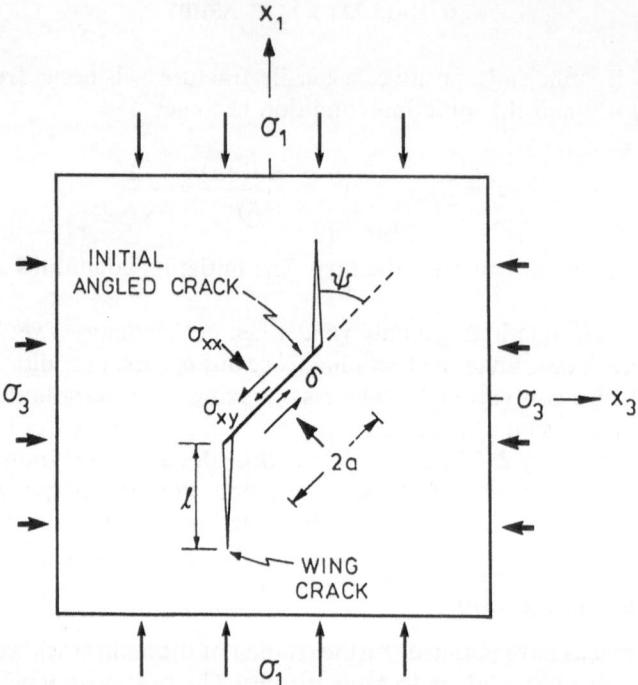

Figure 4.8 Schematic of wing-crack growth from an angled crack in a compressive stress field. (Reprinted with permission from *Acta Metallurgica* **34**, Hallam & Ashby, Copyright 1986, Pergamon Journals Ltd.)

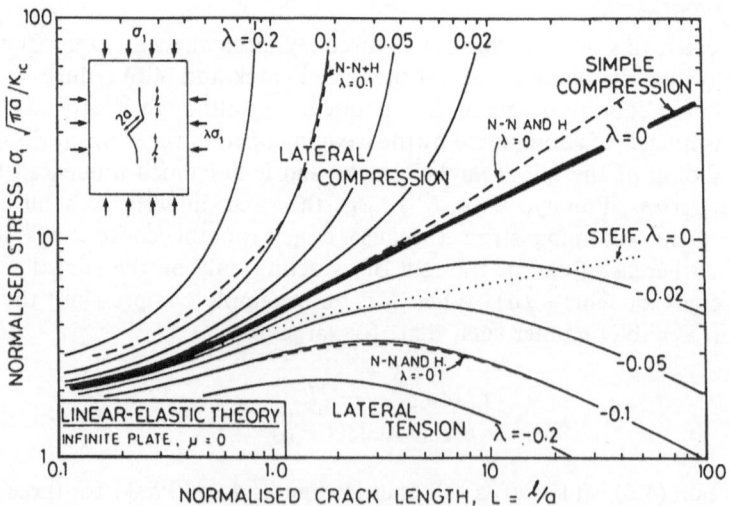

Figure 4.9 Wing-crack growth in an infinite plate with zero friction and comparison with the results of Nemat-Nasser & Horii (1982). (Reprinted with permission from *Acta Metallurgica* **34**, Hallam & Ashby, Copyright 1986, Pergamon Journals Ltd.)

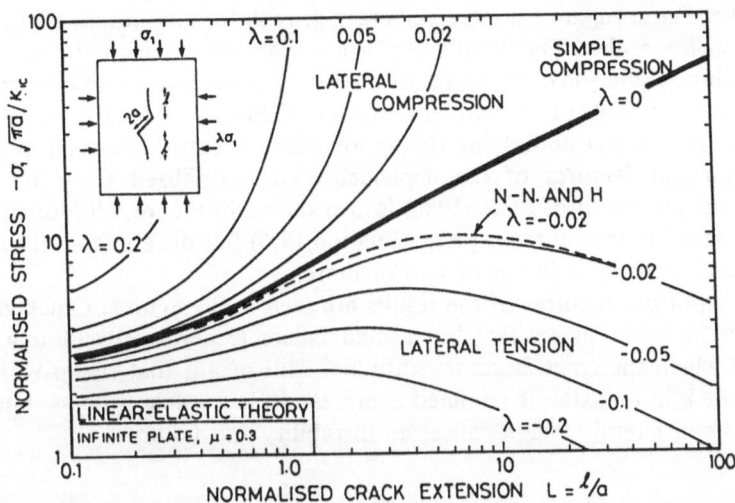

Figure 4.10 Wing-crack growth in an infinite plate with a friction coefficient of 0.3. (Reprinted with permission from *Acta Metallurgica* **34**, Hallam & Ashby, Copyright 1986, Pergamon Journals Ltd.)

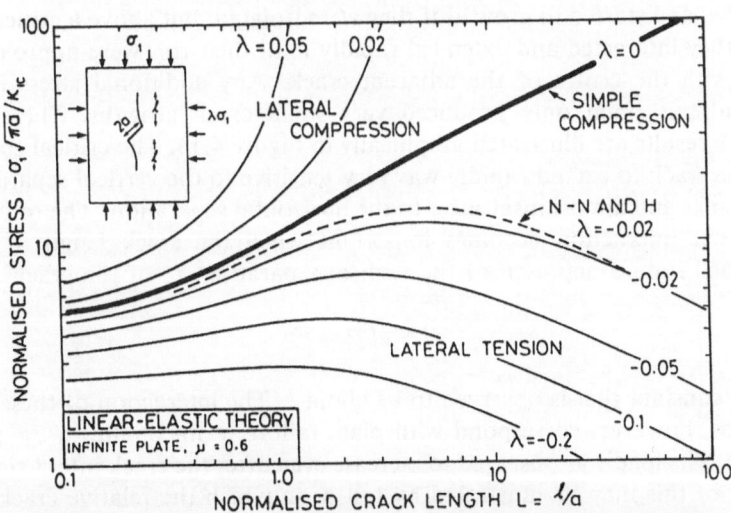

Figure 4.11 Wing-crack growth in an infinite plate with a coefficient of friction of 0.6. (Reprinted with permission from *Acta Metallurgica* **34**, Hallam & Ashby, Copyright 1986, Pergamon Journals Ltd.)

normalized stress against normalized crack growth for nine values of the stress ratio, on log scales. The results are in good agreement with the numerical results throughout the entire range of interest. Similar plots, and comparisons for other coefficients of friction, are shown in Figures 4.10 & 11. It is clear that the analogue model underlying the approximate equation properly describes the important features of the dependence of normalized stress on crack extension. The result of Steif (1984) is also shown in Figure 4.9 (dotted line): it gives roughly the same shape as equation (4.6) but differs from it and the numerical result by a factor of two or more.

The important features of the results are seen in the figures. Crack growth is stable if λ is zero or positive, but a small tension (λ negative) leads to a peak, beyond which the crack becomes unstable. This means that the growth of a single crack in uniaxial or confined compression is a stable process, but even a very small lateral tension causes an instability and failure.

4.3   Crack interaction

4.3.1   Experimental observations

Experiments were carried out on arrays of three cracks machined into PMMA sheets, loaded in simple compression. The experiments carried out to investigate the effects of the crack vertical and horizontal separation are summarized in Figure 4.12. Experiments k, l, m, and n all showed similar results, and a series of photographs taken for test m is shown in Figure 4.13. The cracks initiated and started to grow as if they were isolated; but above a critical stress level they interacted and extended rapidly until their tips were approximately level with the centre of the adjacent crack. Any additional stress increase beyond this point only produced very small crack growths. These crack-growth results are illustrated graphically in Figure 4.14. The critical condition for the crack to extend rapidly was very sensitive to the vertical separation of the cracks and almost insensitive to the horizontal separation. The results suggest that interaction becomes important when the crack length ($l + a/\sqrt{2}$) exceeded some fraction F of the vertical separation, v, of the cracks:

$$l + a/\sqrt{2} = Fv \tag{4.8}$$

F is a constant that is observed to be about $\frac{1}{4}$. The interaction of these cracks did not, however, correspond with plate failure (a maximum in the applied stress). The load was observed to increase even after the crack interaction. The origin of this may be illustrated by test p, in which the relative crack orientations were altered (Fig. 4.15). This test started in a similar manner to the previous tests. Wing cracks initiated on the starter cracks, and showed a critical stress when they rapidly extended to the adjacent crack. Then followed

94

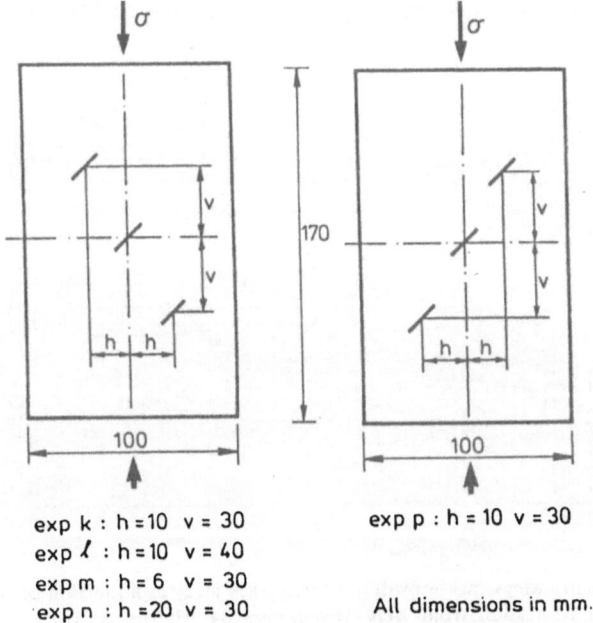

exp k : h = 10   v = 30
exp l : h = 10   v = 40
exp m : h = 6   v = 30
exp n : h = 20   v = 30

exp p : h = 10   v = 30

All dimensions in mm.

Figure 4.12 A summary of crack interaction tests.

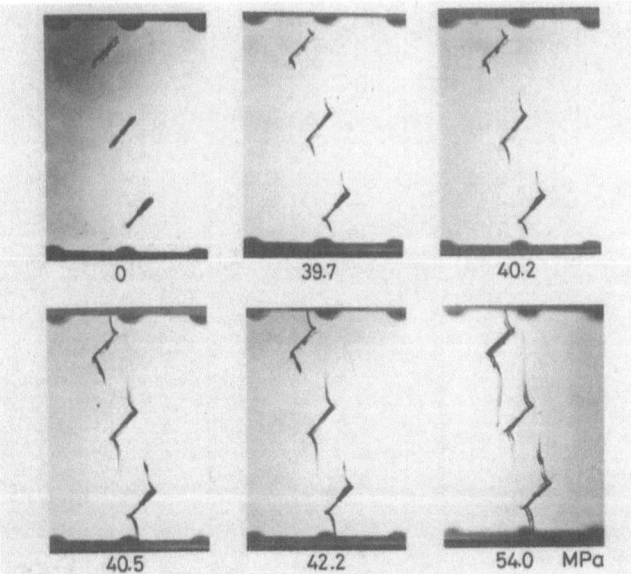

Figure 4.13 A sequence of photographs of crack interaction under progressively higher compressive stresses. The upper and lower cracks do not reach the edge of the plate.

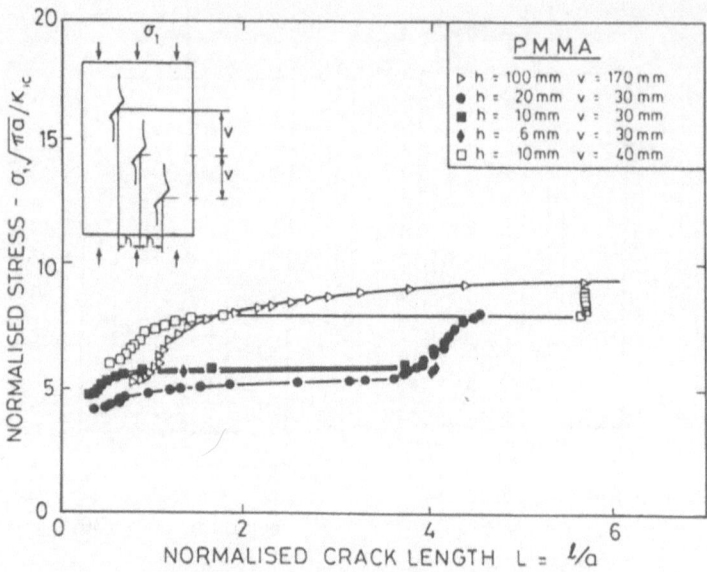

Figure 4.14 Centre wing-crack growth in a three-crack array as a function of crack separation. (Reprinted with permission from *Acta Metallurgica* **34**, Hallam & Ashby, Copyright 1986, Pergamon Journals Ltd.)

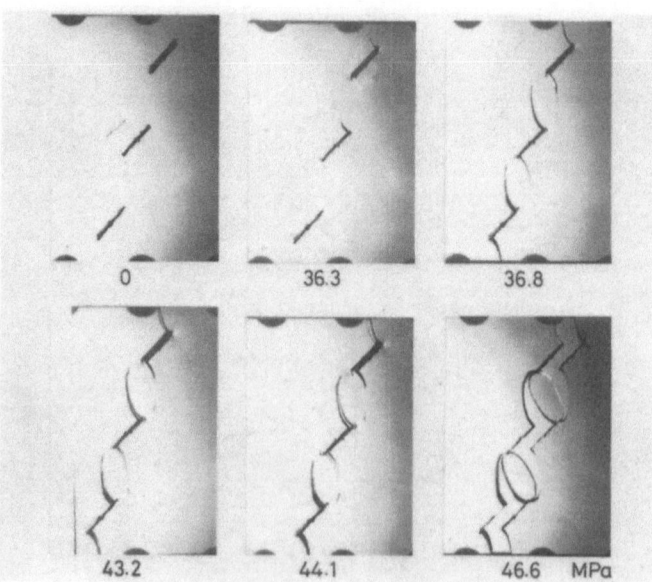

Figure 4.15 A sequence of photographs of crack interaction under progressively higher compressive stresses. The upper and lower cracks do reach the edge of the plate.

a period of relative inactivity when the stress was increased with very little increase in crack extension (up to about 44 MPa). At this point something really very different started to happen (final photograph). Large shear displacements occurred, with associated crack opening. The important occurrence in this test, which did not happen in the others, was that the wing cracks from the upper and lower flaws linked with the upper and lower edge of the PMMA plate. This provided a failure mechanism: the plate was free to shear along the damaged plane, whereas in the previous cases the failure band had been contained by surrounding material. This leads to the conclusion that, in a material containing distributed microcracks throughout its bulk, failure is likely to be associated with the crack interaction process (4.8). Where microcracks are not distributed throughout the bulk it is necessary to form a failure mechanism to allow fault formation to occur. In the test samples, which contained the few central cracks, macroscopic failure could only occur when the extreme fractures reached the plate boundary. Crack interaction and shear is the process that is believed to happen in the type of failure described in Figure 4.1b. In Figure 4.1c the confining stress is sufficiently high to prevent the wing cracks growing to a sufficient length for the crack interaction process to take place. The strength would then increase until another failure mode occurred.

4.3.2 Fracture criterion

The previously described observations of interacting cracks lead to a model for crack interaction in a solid containing many cracks.

If we assume that failure is governed by the crack interaction criterion described by a critical normalized crack extension, L^*, then the resulting failure criterion, derived from (4.6) becomes:

$$\sigma_3 = \frac{\sigma_1(1-\mu)}{1+\mu+\frac{\sqrt{3L^*}}{0.4}} - \frac{\sigma_T}{A\left(1+\mu+\frac{\sqrt{3L^*}}{0.4}\right)} \tag{4.9}$$

where

$$A = \frac{1}{\sqrt{3}(1+L^*)^{3/2}}\{0.4L^* + 1/\sqrt{1+L^*}\}$$

$\sigma_T = K_{IC}/\sqrt{\pi a}$, and σ_1 and σ_3 are the most compressive and least compressive principal stresses, respectively. When normalized by the simple compressive strength, σ_c, the failure criterion becomes

$$\frac{\sigma_3}{\sigma_c} = \frac{(1-\mu)}{\left(1+\mu+\frac{\sqrt{3L^*}}{0.4}\right)}\left\{\frac{\sigma_1}{\sigma_c}+1\right\} \tag{4.10}$$

97

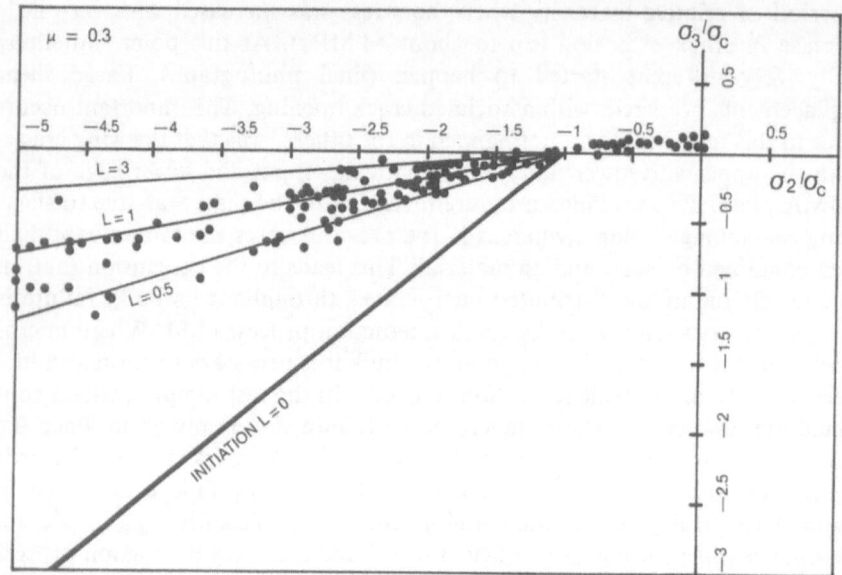

Figure 4.16 Fracture data for various brittle rocks compared to fracture initiation and failure surfaces for friction coefficient of 0.3.

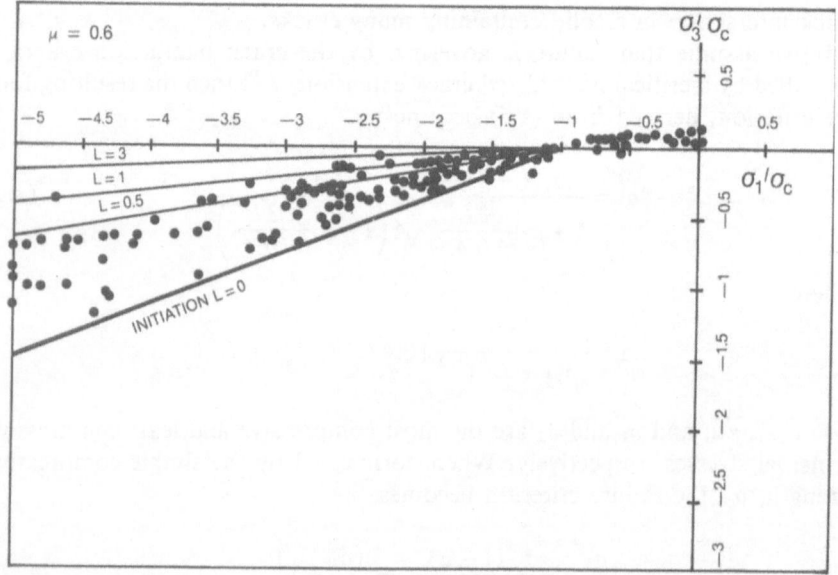

Figure 4.17 Fracture data for various brittle rocks compared to fracture initiation and failure surfaces for a friction coefficient of 0.6.

For a constant value of L these criteria show a linear increase in strength with confining pressure, the same form as the original Coulomb criterion. The solution is independent of the intermediate principal stress.

Hoek & Bieniawski (1965) present a large amount of data for many rocks normalized by the compressive strength for axisymmetric loading. Their data are reproduced, with failure surfaces for various values of L^*, in Figure 4.16 for $\mu = 0.3$ and in Figure 4.17 for $\mu = 0.6$. The failure of the rocks can indeed be identified in terms of a critical crack growth. The scatter in the results would be expected, as different rocks would have different values of the coefficient of friction and initial number of cracks per unit area. For a large volume of data a value of $L^* = 0.5$ described the best failure surface.

4.4 Multi-axial failure maps

Solids which are brittle in tension become plastic under sufficiently large confining pressures. Yield and creep criteria can be combined with the current work on brittle fracture to construct surfaces in stress space to indicate the interaction of these mechanisms. Data have been assembled in Cooksley (1984) for a number of potentially brittle materials (granite, sandstone, concrete, marble, cast iron, rock salt, and ice), and diagrams have been constructed which match the data. The general methodology is described here, together with the results of the above studies. The reader is directed to the source for more detailed information.

4.4.1 Yielding

If fracture is suppressed by applying a sufficiently large hydrostatic pressure, materials yield plastically when the von Mises equivalent stress becomes equal to the yield strength:

$$\bar{\sigma} = \sigma_y$$

where

$$2\bar{\sigma}^2 = (\sigma_1 - \sigma_2)^2 + (\sigma_2 - \sigma_3)^2 + (\sigma_3 - \sigma_1)^2 \qquad \text{(von Mises)}$$

or

$$\bar{\sigma} = \sigma_1 - \sigma_3 \qquad \text{(Tresca)}$$

where σ_y is the yield stress in simple tension. Both criteria are adequate approximations to the yield surface, but the von Mises criterion has the additional advantage of being a smooth function over all stress space, and is the more convenient of the two for use in the mapping process.

Large hydrostatic pressures, p, can raise the yield strength significantly. Studies on steels show a linear increase in the material property σ_y with pressure:

$$\sigma_y/\sigma_y^0 = 1 + \alpha p/K^0$$

where

$$\alpha = \frac{K^0}{\sigma_y^0} \frac{d\sigma_y}{dp}$$

σ_y^0 is the yield strength, and K^0 the bulk modulus at atmospheric temperature and pressure. The value of α is about 10 for metals, and about 5 for minerals and silicates. Pressure influences fracture far more strongly than yielding, so it is normal to find a transition from brittle to ductile behaviour with increasing pressure.

Temperature also influences the yield strength, and its effects can be allowed for in a similar way:

$$\frac{\sigma_y}{\sigma_y^0} = 1 + \beta \frac{(T - 300)}{T_M}$$

where

$$\beta = \frac{T_M}{\sigma_y^0} \frac{d\sigma_y}{dT}$$

and T_M is the melting-point temperature. The value of β is roughly 4 for metals and ceramics near room temperature.

4.4.2   Creep deformation

Solids can deform plastically at an equivalent stress less than the yield strength if the homologous temperature is sufficiently high. The mechanism, creep, is a diffusion-controlled process, and is not usually important unless T/T_m is greater than 0.5 for a ceramic or greater than 0.3 for a metal. The creep behaviour of materials is usually studied under constant stress, and three régimes are identified. Initially, the plastic strain rate, $\dot{\varepsilon}$, decreases with time, t, in the régime known as transient or primary creep. This is followed by a stage lasting perhaps some 90% of the lifetime, in which $\dot{\varepsilon}$ is proportional to t, referred to as steady state or secondary creep. In the final or tertiary stage a rapid increase in strain rate occurs. In the steady state creep régime, the strain rate is related to the stress by an Arrhenius law:

$$\bar{\varepsilon} = A\sigma^{-n} \exp -Q/RT$$

where $\bar{\varepsilon}$ is the equivalent strain rate defined by:

$$\tfrac{9}{2}\bar{\varepsilon}^2 = (\dot{\varepsilon}_1 - \dot{\varepsilon}_2)^2 + (\dot{\varepsilon}_2 - \dot{\varepsilon}_3)^2 + (\dot{\varepsilon}_3 - \dot{\varepsilon}_1)^2$$

The exponential, n, is unity for diffusional flow and in the range 3–10 for power-law creep; Q is the activation energy of the process, R is the universal gas constant, and A is a material constant.

In many cases, the situation of interest is not the application of a steady stress but of an enforced deformation rate. For ice it has been shown that the constant stress producing a steady state creep rate in a creep test corresponds almost exactly with the peak stress or strength in a constant strain rate test at the steady state creep rate. This means that the strength of such a material can be treated as a rate-sensitive yield stress. For a given temperature and strain rate, a simple von Mises yield surface is obtained. If either temperature or strain rate is fixed, a series of 'creep-yield' surfaces can be generated, corresponding to a variation in the other.

In rocks and ceramics, however, creep due to stress-corrosion-enhanced subcritical crack growth is generally more important than the power-law creep described here, especially at low homologous temperatures. Stress-corrosion creep may be modelled in a similar manner incorporating the appropriate dependence of deformation rate on stress (Costin 1983, 1985). This mechanism is not considered here since the deformation rates involved are not normally considered by engineers. The reader is referred to Atkinson & Meredith (1987), Guguen *et al.* (this volume), and Meredith (this volume) for a review and discussion of stress corrosion.

4.4.3 *Map construction*

The failure criteria that have so far been discussed are essentially three-dimensional in nature. The failure surfaces can be represented in two dimensions by selecting a stress state such as plane stress, plane strain, or axisymmetry. This reduces the number of independent variables (stresses) to two. In addition, it is assumed that in constructing the diagrams all loading is proportional and monotonic, and thus a fixed trajectory in stress space is followed to fracture.

The maps presented here, by way of example, are for axisymmetric loading ($\sigma_2 = \sigma_3$), as this is generally the most useful. The equivalent shear stress and shear strain then reduces to:

$$\bar{\sigma} = \pm (\sigma_1 - \sigma_3), \qquad \bar{\varepsilon} = \dot{\varepsilon}_1$$

and

$$p = -\tfrac{1}{3}(\sigma_1 + 2\sigma_3).$$

The axisymmetric map for granite is shown in Figure 4.18. The granite fails in a brittle mode at all confining pressures at engineering rates of strain, and shows a linear increase of strength with confining pressure. A similar surface for many concretes, normalized by the compressive strength is shown in Figure 4.19.

The failure map for Carrara marble is shown in Figure 4.20. The marble is brittle in uniaxial tension and compression, but shows a brittle-to-ductile transition with increasing confining pressure.

Cast iron (Fig. 4.21) is brittle in tension, and very close to the brittle-to-ductile transition in uniaxial compression. With increasing confining pressure independent yielding would be expected.

Ice and rock salt are materials which deform by power-law creep under some conditions of temperature and strain rate, and are brittle under others. In Figures 4.22 & 23 is shown a family of creep yield surfaces for these materials at a fixed temperature and a variety of strain rates. In Figure 4.24 it is shown how the mapping ideas can be used to predict the deformation behaviour at other temperatures if the thermal activation energy, Q, is known. To interpret these diagrams, consider Figure 4.22. At very low strain rates, less than about $10^{-6}\,\mathrm{s}^{-1}$, pure polycrystalline ice is ductile in tension, compression, and confined compression. At a higher rate of loading in the range 10^{-6}–$10^{-3}\,\mathrm{s}^{-1}$

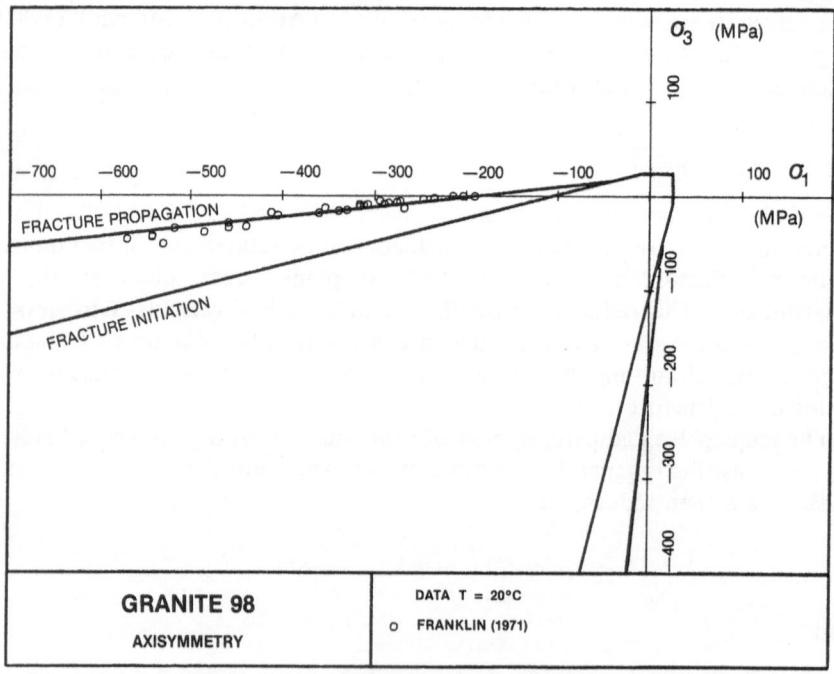

Figure 4.18 Multi-axial failure map for granite.

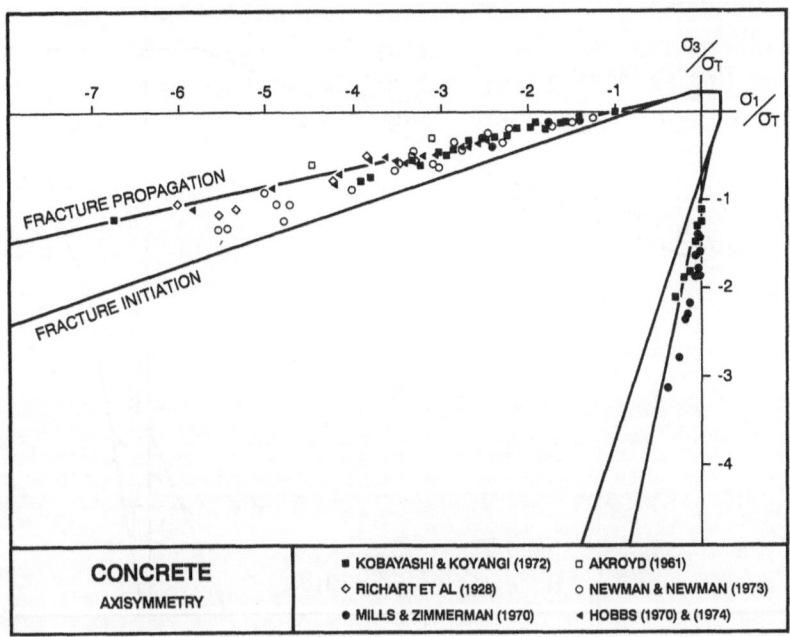

Figure 4.19 Multi-axial failure map for concretes. All data is normalized by the uniaxial compressive strength.

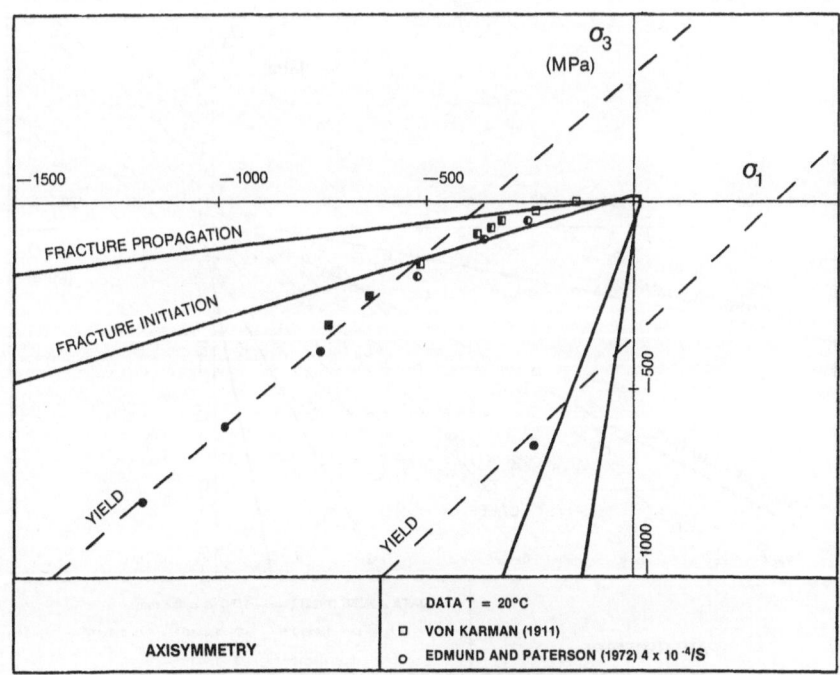

Figure 4.20 Multi-axial failure map for Carrara marble.

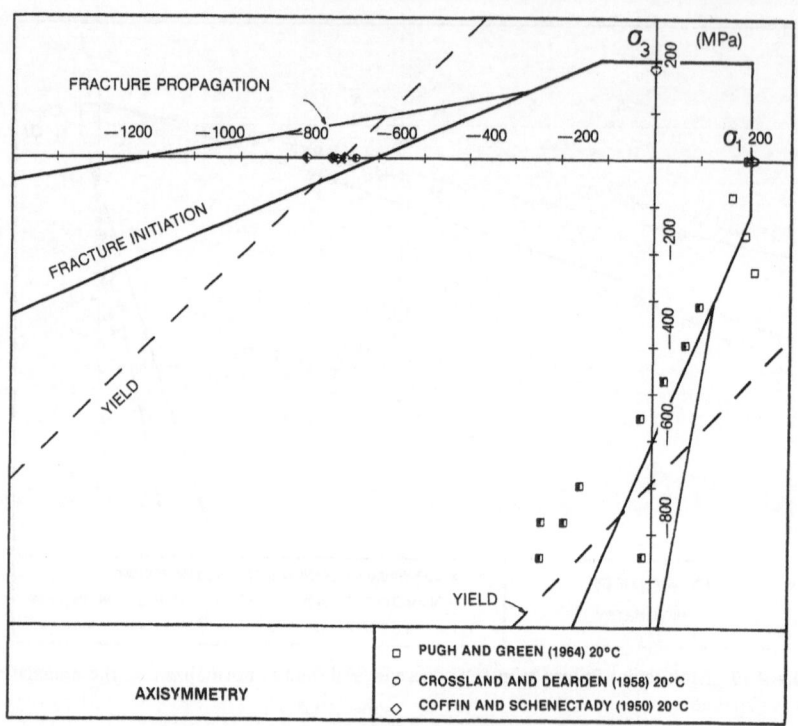

Figure 4.21 Multi-axial failure map for cast iron.

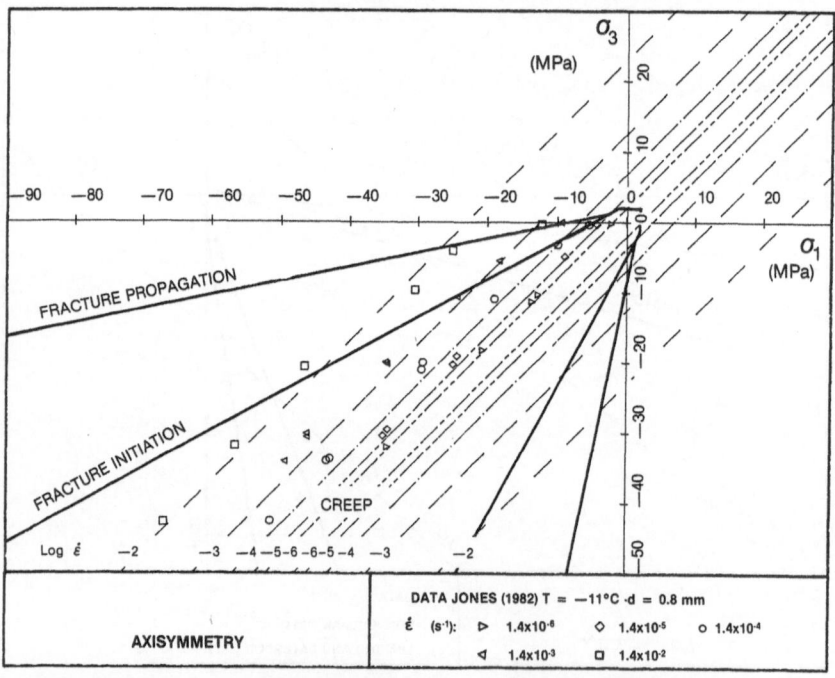

Figure 4.22 Multi-axial failure map for pure polycrystalline ice at $-11°$C.

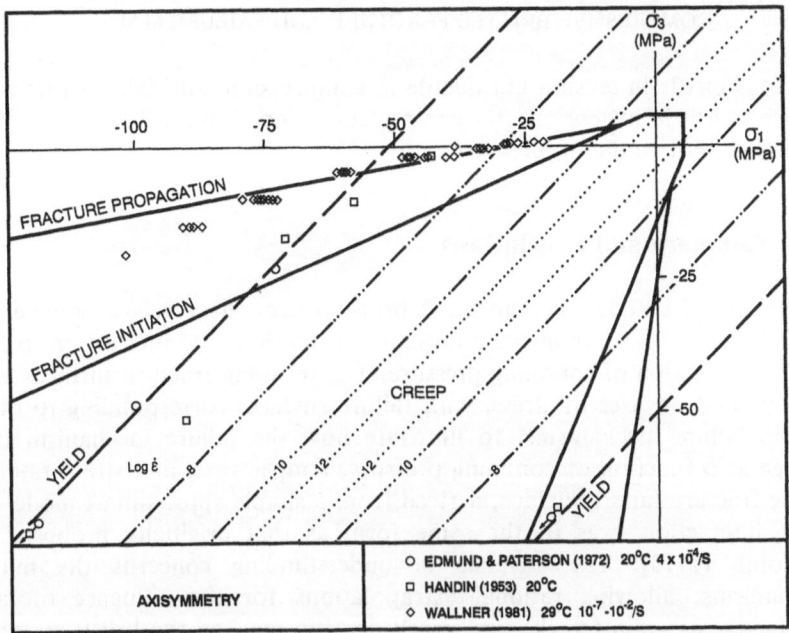

Figure 4.23 Multi-axial failure map for rock salt at approximately 20°C.

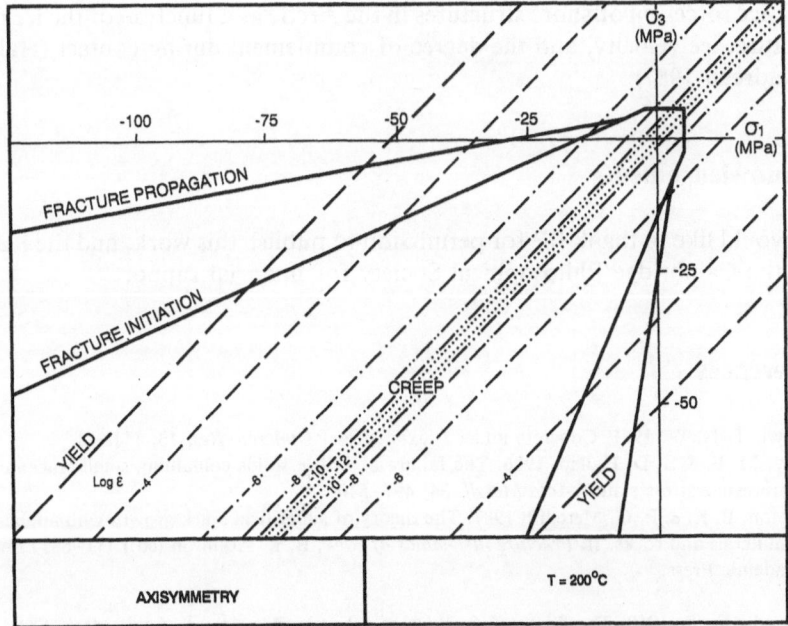

Figure 4.24 Derived multi-axial failure map for rock salt at approximately 200°C.

the ice is brittle in tension but ductile in compression. At higher strain rates the ice is brittle in tension and compression, and shows a brittle-to-ductile transition with confining pressure.

4.5 Summary and conclusions

The results of a study on compressive brittle fracture have been described and modelled in terms of wing-crack initiation, extension, and interaction to failure, as a function of confining pressure. The resulting fracture surfaces have been plotted on axes of stress, with failure surfaces corresponding to other ductile failure mechanisms, to illustrate how the failure mechanism may change as a function of confining pressure, temperature, and strain rate.

The fracture failure surface, derived from a simple approximate model for crack interaction, was of the same form as that originally proposed by Coulomb (1773). The advance in understanding concerns the micromechanisms, allowing rational extrapolations for the influence of such parameters as grain (crack) size, crack density, etc., on the brittle strength. The application of damage mechanics (current work) may improve the approach to the crack interaction problem further.

Multi-axial failure maps provide a useful tool in the understanding of the failure strength and deformation mode for materials which undergo a brittle-to-ductile transition. One such area of application is in the calculation of pack-ice forces on offshore structures in the Arctic as a function of the ice temperature, ice velocity, and the degree of confinement during contact (Hallam & Nadreau 1987).

Acknowledgements

We would like to thank BP for permission to publish this work, and the SERC and the Cambridge Philosophical Society for financial support.

References

Akroyd, T. N. W. 1961. Concrete under triaxial stress. *Concrete Res.* **13**, 111–18.

Ashby, M. F. & S. D. Hallam 1986. The failure of brittle solids containing small cracks under compressive stress states. *Acta Metall.* **34**, 497–510.

Atkinson, B. K. & P. G. Meredith 1987. The theory of subcritical crack growth with application to minerals and rocks. In *Fracture mechanics of rock*, B. K. Atkinson (ed.) 111–66. London: Academic Press.

Coffin, L. F., Jr 1950. The flow and fracture of a brittle material. *J. Appl. Mech.* Sept., **17**, 233–48.

Cooksley, S. D. 1984. *Yield and fracture surfaces of brittle solids under multi-axial loading*. PhD thesis. Engineering Department, University of Cambridge.

Costin, L. S. 1983. A microcrack model for the deformation and failure of brittle rock. *J. Geophys. Res.*, **88**, 9485–92.

Costin, L. S. 1985. Damage mechanics in the post-failure regime. *Mech. Mat.*, **4**, 149–60.

Coulomb, C. A. 1773. Sur une application des règles de maximis et minimis à quelques problèmes de statique relatifs à l'architecture. *Acad. Roy. des Sciences Mémoires de Math. et de Physique par Divers Savants* **7**, 343–82.

Crossland, B. & W. H. Dearden 1958. Plastic flow and fracture of a 'brittle' material (grey cast iron) with particular reference to the effect of fluid pressure. *Proc. Instn Mech. Engrs* **172**, 805–20.

Edmund, J. M. & M. S. Paterson 1972. Volume changes during the deformation of rocks at high pressures. *Int. J. Rock Mech. Min. Sci.* **9**, 161–82.

Franklin, J. A. 1971. Triaxial strength of rock materials. *Rock Mech.* **3**, 86–98.

Griffiths, A. A. 1924. Theory of rupture. In *Proc. 1st Int. Conf. Appl. Mech.*, Delft: Waltman, 55–63.

Gueguen, Y., T. Reuschlé and M. Darot 1989. Single-crack behaviour and crack statistics. This volume, 48–71.

Hallam, S. D. & J. P. Nadreau 1987. Failure maps for ice. In *Proc. 9th Int. Conf. Port Ocean Engng under Arctic Conditions*, Fairbanks, Vol III, 45–56.

Handin, H. 1953. An application of high pressure in geophysics; experimental rock deformation. *Trans Am. Soc. Mech. Engrs* **75**, 315–24.

Hobbs, D. W. 1970. *Strength and deformation properties of plain concrete subject to combined stress, part 1: strength results obtained on one concrete*. Technical Report 42.451. Cement and Concrete Association.

Hobbs, D. W. 1974. *Strength and deformation properties of plain concrete subject to combined stress, part 3: results obtained on a range of flint gravel aggregate concretes*. Technical Report 42.497. Cement and Concrete Association.

Hoek, E. & L. T. Bieniawski 1965. Brittle fracture propagation in rock under compression. *Int. J. Frac. Mech.* **1**, 135–55.

Jones, S. J. 1982. The confined compressive strength of polycrystalline ice. *J. Glaciol.* **28**, 171–7.

Kobayashi, S. 1971. Initiation and propagation of brittle fracture in rock-like materials under compression. *J. Soc. Mat. Sci.* **20**, 164–73.

Kobayashi, S. & W. Koyangi 1972. Fracture criteria of cement paste, mortar and concrete subjected to multi-axial compressive stresses. *RILEM Int. Symp.*, Cannes, October, 131–48.

Marshall, G. P. & J. G. Williams 1973. The correlation of fracture data for PMMA. *J. Mat. Sci.* **8**, 138–40.

McClintock, F. A. & J. B. Walsh 1962. Friction on Griffith's cracks in rocks under pressure. In *Proc. Fourth US Nat. Congr. Appl. Mech.* II, New York, 1015–21.

Meredith, P. G. 1989. Fracture and failure of polycrystals: an overview. This volume, 5–47.

Mills, L. L. & R. M. Zimmerman 1970. Compressive strength of plain concrete under multi-axial loading conditions. *J. Am. Concrete Inst.* Title 67-47, 802–7.

Murrell, S. A. F. 1963. A criterion for brittle fracture of rocks and concrete under triaxial stress, and the effect of pore pressure on the criterion. In *Proc. Fifth Symp. Rock Mech.* 563–77. New York: Pergamon.

Murrell, S. A. F. 1965. The effect of triaxial stress systems on the strength of rocks at atmospheric temperatures. *Geophys. J. R. Astron. Soc.* **10**, 231–81.

Murrell, S. A. F. and P. J. Digby 1970a. The theory of brittle fracture initiation under triaxial stress conditions – I. *Geophys. J. R. Astron. Soc.* **19**, 309–34.

Murrell, S. A. F. & P. J. Digby 1970b. The theory of brittle fracture initiation under triaxial stress conditions – II. *Geophys. J. R. Astron. Soc.* **19**, 499–512.

Nemat-Nasser, S. & H. Horii 1982. Compression induced non planar crack extension with application to splitting, exfoliation and rockburst. *J. Geophys. Res.* **87**, 6805–21.

Newman. K. & J. B. Newman 1973. Design criteria for concrete under combined states of stress. In *Criteria of concrete strength*, Report 1, CIRIA contract, Imperial College, London.

Paterson, M. S. 1978. *Experimental rock deformation – the brittle field*. Berlin: Springer.

Pugh, H. L. L. D. & D. Green 1964. The effect of hydrostatic pressure on the plastic flow of metals. *Proc. Inst. Mech. Engrs* **179**, 415–38.

Richart, F. E., A. Brandtzaeg & R. L. Brown 1928. A study of the failure of concrete under combined stress. *Bull. Univ. Ill. Engng Expt Stat.* **185**, 104.

Sammis, C. G. & M. F. Ashby 1986. The failure of porous solids under compressive stress states. *Acta Metall.* **34**, 511.

Steif, P. 1984. Crack extension under compressive loading. *Engng Fract. Mech.* **20**, 463.

von Kármán, Th. 1911. Festigkeitsversuche unter allseitgim druck *Z. Ver. dt. Ing.* **55**, 1749–57.

Wallner, M., C. Caninenburg & H. Gonther 1979. Ermittlung zeit- und temperaturabhängiger mechanischer Kennwerte von Salzgesteinen. In *Proc. 4th Int. Cong. on Rock Mechanics*, Montreux, **1**, 313–18.

Williams, M. L. 1957. On the stress distribution at the base of a stationary crack. *J. Appl. Mech. Trans* **24**, 109–14.

CHAPTER FIVE

Brittle-to-ductile transitions in polycrystalline non-metallic materials

Stanley A. F. Murrell

5.1 Introduction

5.1.1 Isomechanical solid behaviour (plasticity, creep, and cleavage): deformation-mechanism maps

At room temperatures and pressures minerals, ceramics, and rocks have only limited ductility, and this distinguishes them from most metals. The characteristic strength and brittle or ductile deformation behaviour of crystalline solids is determined by the nature of their atomic bonding and crystal structure. Thus Frost & Ashby (1982) have recognized 21 isomechanical groups in their study of deformation mechanism maps to represent the plasticity and creep behaviour of metals and ceramics. The behaviour of the solids belonging to any particular isomechanical group is closely similar and clearly different from that of solids belonging to the other groups. Their study covered the cubic, tetragonal, hexagonal, and trigonal crystal systems, together with the orthorhombic olivines, in particular the forsterite end member. These materials may be broadly classified into seven groups (of which three are the three classes of metals with cubic crystal structures – f.c.c., b.c.c, and h.c.p.), showing a progression of plastic flow properties (Frost & Ashby 1982, Tables 18.3 & 18.4) and of cleavage properties (Kelly, Tyson & Cottrell 1967; Gandhi & Ashby 1979, Table 4 & Fig. 28). These are illustrated in Table 5.1.

5.1.2 Factors governing ductility and flow stress

An important factor governing ductility is the multiplicity of low-strength slip systems available for polycrystal plasticity without the production of voids (cracks). The von Mises criterion requires five independent systems to maintain material continuity at grain boundaries. This is discussed in detail by Kelly

Table 5.1 Examples of isomechanical groups.[*]

Group	Crystal structure	Examples	Number of independent slip systems (easy glide)
metals	f.c.c.	(Pb, Cu, Ni)	5
	b.c.c. transition	(α-Fe, W)	5
	h.c.p	(Zn, Mg)	2
ionic solids	rock salt cubic (alkali halides)	(NaCl)	2
	rock salt cubic (simple oxides)	(MgO)	2
ionic-covalent solids	rhombohedral (oxides)	(Al_2O_3)	2
	orthorhombic (silicates)	(Mg_2SiO_4)	2
covalent solids	diamond cubic (elements)	(Si, Ge, C)	5
	hexagonal (compounds)	(SiC, Si_3N_4)	2
hydrogen-bonded solids	hexagonal	(H_2O)	2

[*] See Tables 5.5 and 5.8 for examples of mechanical properties.

(1966) (see also Groves & Kelly 1963, who give a simple test for the number of independent systems). The multiplicity of slip systems (and hence the ductility) depends both on the symmetry of the crystal system to which a solid belongs (cubic crystals having greater possibilities of achieving five independent slip systems), and also on the bonding (ionic crystals typically have fewer than five independent slip systems; see Table 5.1).

However, ductility also requires that the available independent slip systems be operative at comparable critical resolved shear stresses, and that the slip be flexible in the sense that cross-slip and slip-band interpenetration readily occur. The behaviour of magnesium oxide (MgO) in this respect exemplifies that of many non-metallic materials (see Table 5.2).

The behaviour of the diamond cubic elements (C, Si, Ge) and of b.c.c. iron (α-Fe) shows, however, that even when there are five independent slip systems operating, the nature and behaviour of the slip dislocations (in particular their cores), which are strongly affected by the bonding character, have a marked influence on ductility and strength. The important factor is the shear stress, τ_p (the Peierls stress), required to move a straight dislocation. Kelly (1966)

Table 5.2 Plastic behaviour of MgO.

Temperature	Behaviour
$<350°C$	slip on $\{110\}\langle110\rangle$; only two independent slip systems; single crystals and polycrystals fully brittle
$>350°C$	slip on $\{110\}\langle110\rangle$ and on $\{001\}\langle110\rangle$ (dislocations cross-slip from $\{110\}$ planes); this gives five independent slip systems
(a) close to 350°C	CRSS* for $\{001\}\langle110\rangle$ dislocations $\approx 10 \times$ CRSS for $\{110\}\langle110\rangle$ dislocations; cracks occur where $\{110\}$ slip bands are impeded by other slip bands or grain boundaries; polycrystal ductility limited (≈ 0.01 strain at failure)
(b) at 1200°C	ratio of CRSS values falls to about 3
(c) at 1500°C	CRSS values equal; slip lines wavy but slip bands still cannot interpenetrate sufficiently
(d) at 1700°C	slip bands interpenetrate fully, and polycrystals are fully ductile

*CRSS is the critical resolved shear stress.

gives an extensive discussion of this. Cottrell (1953) shows that:

$$\tau_p = \{2G/(1-\nu)\} \exp(-4\pi\zeta/b) \tag{5.1}$$

where ζ is the width of the dislocation, given by

$$\zeta = Gb/2\pi(1-\nu)\tau_m \tag{5.2}$$

and τ_m is the theoretical shear strength of a perfect crystal, given by Frenkel's formula (Kelly 1966)

$$\tau_m = Gb/2\pi h \tag{5.3}$$

where G is the shear modulus, b is the magnitude of the Burgers vector, ν is the Poisson ratio, and h is the interatomic spacing perpendicular to the slip plane. Thus the higher the value of τ_m/G the larger the value of τ_p/G. Values of τ_m/G calculated by Kelly range from 0.034 (for zinc) to 0.24 (for diamond) (see Table 5.3). ζ also depends on the ratio, R, of central to non-central forces between the atoms of a crystal. When R and G/τ_m are both small then dislocations are narrow (small ζ), and difficult to move because the Peierls stress τ_p is large. τ_p is smaller for close-packed planes of atoms. Recently calculated values of the Peierls stress are compared with experimental estimates in Table 5.4.

In f.c.c. metals and in alkali halides, the Peierls stress is small and the flow stress is limited by elastic interaction between freely moving dislocations.

Table 5.3 Theoretical shear (τ_m) and cleavage (σ_m) strengths of perfect solids (Kelly 1966, Lawn & Wilshaw 1975).

Material	τ_m/G	τ_m (GPa)	σ_m/E	σ_m (GPa)	τ_m/σ_m	ν
f.c.c. metal						
Cu (10K)	0.039	1.29				
(20°C)	0.039	1.2	0.20	3.9	0.31	0.34
b.c.c. metal						
α-Fe	0.11	0.66	0.2	4	0.2	0.28
W	0.11	16.5	0.22	8.6	1.92	0.29
h.c.p. metal						
Zn	0.034	2.3	0.11	0.38	6	
alkali halide						
NaCl	0.12	2.84	0.10	0.43	6.6	
simple oxide						
MgO	0.12	1.6	0.15	3.7	0.43	0.18
covalent solid						
diamond	0.24	121	0.17	205	5.9	0.10[*]
quartz	0.15	4.4	0.22	16	0.28	0.17
Si	0.24	13.7	0.17	32	0.43	0.27

ν is the Poisson ratio; G is the shear modulus; E is the Young's modulus;
[*] S_{12}/S_{11} (Kelly 1966, Appendix C).

However, in covalent crystals the flow stress is limited by the Peierls stress (i.e. by interactions between dislocations and the atomic bonds). A dislocation in the midglide position in a covalent crystal has an unpaired electron associated with it, which is therefore excited from the valence band to the conduction band. Gilman (1973) therefore suggests that flow stresses will be strongly correlated with the homopolar component of bonding.

In ionic crystals both point defects and dislocations are electrostatically charged, and this has important consequences for the plastic and cleavage properties of ionic and covalent compounds (Poirier 1985; the usefulness of the concept of ionicity in modelling defects in solids is discussed by Catlow & Stoneham 1983).

Kelly (1966) points out that low values of the Poisson ratio ($\nu < 0.25$; see Table 5.3) and high ratios of indentation hardness to Young's modulus (ratios > 0.01) also correlate with high Peierls stresses. He defines as 'inherently strong' those solids in which dislocations can not normally be moved at stresses much below the theoretical shear strength at room temperature. They have marked directional bonding. For such solids dislocation motion is controlled by thermal vibrations of the atoms in the dislocation cores, so the onset of plasticity occurs only at high homologous temperatures ($T/T_m > 0.45-0.65$; see Kelly 1966, Table 3.1).

Gilman (1973) discusses the indentation hardness of different classes of

Table 5.4   Peierls stress (Puls 1981).

Material	ν	Calculated Peierls stress $(10^{-3}G)$	Experimental $(10^{-3}G)$
α-Fe	0.28	4–69	7[*]
W	0.29	18–26	—
NaCl	0.3	0.75–9.33	0.5[*]
MgO	0.18	0.92–3.81	1.0

[*] Measured at temperature of 4K.

solids. The Vickers indentation hardness, H_v, is proportional to the yield stress σ_y, but the ratio between them differs for different isomechanical classes. Thus, for cubic metals, $H_v/\sigma_y \approx 3$, but for cubic ionic solids (such as NaCl, MgO and TiC), $H_v/\sigma_y \approx 35$. In crystals of low symmetry the lower multiplicity of slip systems causes plastic anisotropy to become a factor in the yield of polycrystals (Gilman 1973). Hardness testing has revealed important correlations between (a) yield stress and elastic modulus in f.c.c. metals and covalent crystals, (b) yield stress, glide activation energy, and the band gaps of electron energy levels in covalent crystals, and (c) yield stress and bond length in III–V covalent compounds (such as GaP, InP and InSb) and in isomorphous CaF_2 structure compounds (Mg_2Si, Mg_2Pb, etc.).

The metal carbides (e.g. TiC) reveal another interesting feature: the bonding character, and hence the isomechanical class, can change with temperature. At low temperatures these compounds are hard and have a covalent character, but at high temperatures they become soft and metallic in character (Atkins 1973; Gilman 1973; Frost & Ashby 1982, Ch. 11). At homologous temperatures, $T/T_m < 0.2$, TiC is several times harder than Si and Ge, and ≈ 50 times harder than Cu, but at $T/T_m \approx 0.5$, while Si, Ge, and Cu retain their hardness, the hardness of TiC drops to a value similar to that of Cu (Atkins 1973).

The physical characteristics of dislocations (and their effect on plasticity) are affected not only by the character of the chemical bonding in the crystals concerned but also by the steric or packing characteristics of the crystallographic structure. The latter causes the splitting of dislocations into partial dislocations separated by stacking faults. Twinning is another example of the importance of steric effects. This factor is particularly important in solids, such as silicates, which have complex structures.

5.1.3   Basis for isomechanical classification

The empirical basis for the recognition of isomechanical groups by Frost & Ashby (1982, Ch. 18) was the systematic study by Brown (1980) and Ashby & Brown (1981) of the Vickers hardness of single crystals of 25 elements and compounds, forming seven distinct groups, over a wide range of temperatures.

They found that the best correlation uses temperatures normalized by melting temperature (T/T_m) (this is closely related to the normalization by $G\Omega/k$ suggested by plasticity models (Kocks *et al.* 1975) where Ω is the molar volume, and k is Boltzmann's constant). The best normalization for flow stress or hardness uses the cohesive energy $\Delta H_c/\Omega$, but this is hard to measure and the data base is poor, so Frost & Ashby (1982) chose to normalize by shear modulus G because dislocation line energies are proportional to Gb. An alternative normalization uses Young's modulus E, and in their fracture mechanism maps Gandhi & Ashby (1979) normalize the tensile stress by E.

Problems that occur in developing deformation mechanism maps for minerals, ceramics, and rocks are that in some cases these materials undergo phase changes (such as dehydration in serpentine minerals, or polymorphism, e.g. calcite–aragonite, or order–disorder transitions, e.g. in albite) and that they also may not have a sharply defined melting temperature. In the latter case the solidus temperature is chosen (see Brown & Ashby 1980). It might be thought that the preparation of maps for multiphase materials (e.g. metallic alloys or rocks) would also be difficult, or might not be sensible. However Frost & Ashby (1982) and Gandhi & Ashby (1979) successfully constructed maps for metallic alloys, and point out that these are the ones which are most likely to find practical application.

5.1.4 Brittle behaviour

So far we have discussed only the flow stress, measured as yield stress or hardness, of isomechanical groups. Frost & Ashby (1982, Table 18.4) show how, between major isomechanical groups, there is a natural progression of the material flow properties τ_f/G (at 0 K), $\Delta F/kT_m$, and Q_v/RT_m, where τ_f is the shear stress to overcome obstacles to dislocation motion (either other dislocations or the Peierls force), ΔF is the activation energy for glide, and Q_v is the activation energy for lattice diffusion (see Table 5.5). However, hardness and flow stress do not in themselves explain the origin of brittle behaviour and notch-sensitivity.

Table 5.5 Material properties by isomechanical class.[*]

Property[†]	f.c.c. metals	b.c.c. metals	h.c.p. metals	alkali halides	simple oxides	covalent solids[‡]	ice
$\tau_f/G\,(\times 10^2)$	$<10^{-3}$	1	0.7	2.7	3.7	7	9
$\Delta F/kT$	$<10^{-5}$	3	3	7.5	10	35	36
Q_v/RT_m	18	18	17	23	23	34	26
$\sigma_y^0/K\,(\times 10^{-2})$	0.3	0.9	1.7	2.6	4.0	7	4.7

[*] Data from Frost & Ashby (1982) and Gandhi & Ashby (1979).
[†] See text.
[‡] Including covalent and ionic-covalent solids.

Table 5.6 Classification of brittle behaviour (after Lawn & Wilshaw 1975, Tables 2.2, 4.1 & 6.1).

Classification	Crack-initiation mechanism	Materials	Work to form unit area of fracture surface, Γ (J m^{-2})	Surface energy, γ (J m^{-2})	Fracture energy class
highly brittle	bond rupture	covalent or some ionic-covalent (diamond and ZnS structures, silicates, alumina, mica, W, carbides)	10^{-1}–10	diamond 30 quartz 2.4 silicon 6.2	$\Gamma \approx \gamma$
semi-brittle	bond rupture, dislocation mobility	some ionic-covalent, ionic, some monocrystal b.c.c. and h.c.p. metals	1–10^2	W 22 NaCl 2.4 MgO 10.0 α-Fe 10.0 Zn 6.4	$\Gamma > \gamma$
		polycrystal b.c.c. and h.c.p. metals	10^2–10^4		$\Gamma \gg \gamma$
non-brittle	dislocation mobility	f.c.c. metals, some b.c.c. metals	$> 10^4$		$\Gamma \gg \gamma$

Since the work of Griffith (1920, 1924) it has been recognized that the origin of brittle behaviour lies in microcracks. What physical features cause a solid to develop microcracks, and what physical processes cause the cracks to grow in a brittle fashion, so that elastic deformation is followed by brittle behaviour and fracture rather than by plastic or viscous flow?

Lawn & Wilshaw (1975) give an excellent review of the fracture of brittle solids; see also Meredith (this volume). They recognize several categories of solids, classified according to their degree of brittleness (see Table 5.6).

5.1.5 Strength of perfect solids

Kelly (1966) gives the expression (following Orowan) for the theoretical tensile breaking stress (or cleavage stress) of a perfect solid as

$$\sigma_m = (E\gamma/b)^{1/2} \qquad (5.4)$$

where γ is the specific surface energy, and b is the equilibrium atomic spacing. From more exact calculations it is found that this expression is, for covalent and ionic solids, too large by a factor of ≈ 2. An expression for the theoretical shear strength (τ_m) is given above in equation (5.3). Examples of calculated strengths of perfect solids are given in Table 5.3. Kelly shows that strong solids, having high values of both σ_m and τ_m are those having the highest density of strong directional bonds. These solids have a combination of high elastic moduli, low density, and high melting point, and generally contain one of the elements Be, B, C, N, O, Al, or Si.

5.1.6 Effect of defects on strength

The measured strength of real solids is controlled by defects: mobile dislocations in plastic solids, and surface steps, cracks, and notches in brittle solids.

Griffith (1920) recognized the crucial rôle of microcracks in brittle behaviour and he derived an expression for the tensile strength of a cracked solid. On the basis of energy considerations (Murrell 1964, Kelly 1966) the strength is

$$\sigma_T = \{2E\Gamma/\pi(1 - \nu^2)c\}^{1/2} \tag{5.5}$$

for a crack of length $2c$ in plane strain. By considering the stress concentration at the crack tip we arrive at an alternative expression, which includes the crack-tip radius ρ and the interatomic spacing b (Murrell 1964):

$$\sigma_T = (1.09E\Gamma\rho/4cb)^{1/2} \tag{5.6}$$

For brittle solids $\rho \approx b$ and the two expressions are numerically very close. In very brittle solids Γ is the specific surface energy γ, but where cleavage is accompanied by surface damage (by microplasticity or cracking) the value of Γ may be much larger than γ (see Table 5.6). This was first recognized by Irwin (1948) and Orowan (1949). Γ is now known as the fracture surface energy (Lawn & Wilshaw 1975; Meredith, this volume), and is closely related to the parameter known as fracture toughness.

The role of dislocations in the plastic behaviour of crystalline solids was recognized later (1934) by Taylor, Orowan, and Polanyi (see Cottrell 1953). The critical shear stress, τ_p, required to move a dislocation (the Peierls stress) is given by equation (5.1) above.

5.1.7 Brittle and ductile behaviour of crystals

Kelly, Tyson & Cottrell (1967) showed that the division of crystals into brittle and ductile types could be made on the basis of the ratio σ_m/τ_m. A high ratio gives ductile behaviour. This is relevant to the brittle or ductile behaviour of microcracks, which is discussed by Digby & Murrell (1975). Rice & Thomson (1974) have calculated the stability of a sharp cleavage crack against blunting due to emission of a dislocation, and they arrived at a classification which is essentially similar to that of Kelly, Tyson & Cottrell (1967). The important factors are the ratio γ/Gb and the dislocation width ζ. Small values of these factors (in particular $\gamma/Gb < 0.1$), which also correspond to small values of σ_m/τ_m, give rise to brittle behaviour.

5.1.8 Fracture and flow

A solid subjected to compressive loading will flow if it is ductile (e.g. by plastic flow, by creep, or by cataclastic flow – see below), but it may fracture (by splitting or by shear faulting) if it is in a brittle condition (e.g. rocks at room temperature and pressure).

Solids subjected to tensile loading, with the exception of most f.c.c. metals and their alloys, fail by cleavage at sufficiently low temperatures. Otherwise they fail by ductile fracture through void growth brought about by plastic flow, or at higher temperatures by a transgranular or intergranular creep process, or by diffusional grain-boundary flow; but if no void growth mechanism exists the stretched solid eventually becomes mechanically unstable and ruptures by highly localized deformation (see Gandhi & Ashby 1979). Tensile creep fracture in ceramic polycrystals has recently been reviewed by Porter, Blumenthal & Evans (1981) and Hsueh & Evans (1981). Cavities appear to nucleate with a spheroidal shape at grain junctions, but rapidly grow into a cylindrical shape. These are called equilibrium cavities. Under stress these eventually grow into crack-like cavities, which in due course coalesce to form fractures. In multiphase ceramics at high temperatures a liquid phase may be present at grain boundaries and may have a dominant effect on creep fracture (Tsai & Raj 1982).

The temperatures at which transitions occur between these various failure modes depend on the strain rate, which strongly influences plasticity and creep mechanisms, and on the applied stress system (i.e. on the magnitude and sign – tensile or compressive – of the hydrostatic component of the stress tensor, and on the magnitude of the deviatoric component), since the latter strongly affects the nature of fracture.

The purpose of this chapter is to discuss these transitions, and to present methods of mapping the fields of operation of the deformation and fracture mechanisms, with allowance for the influence of the applied stress system.

5.2 Effect of temperature (T) and strain rate ($\dot{\varepsilon}$) on plastic flow and brittle cleavage

Brittle cleavage characteristically occurs at low temperatures in solids for which the plastic flow stress is higher than the fracture stress. While cleavage is only weakly dependent on T and $\dot{\varepsilon}$ (which do not strongly influence the elastic moduli or specific surface energy; see equation (5.4) above), the plastic flow and creep stresses are strongly dependent on the strain rate, and also depend very strongly on temperature, especially at high temperatures, since these latter deformation processes are thermally activated. Associated with the wide variation of plastic strength there is a range of different mechanisms of flow, and correspondingly a range of fracture mechanisms in which plasticity interacts with the crack nucleation or propagation processes. These are shown schematically on deformation and fracture mechanism maps of the type developed by Ashby and his co-workers (Gandhi & Ashby 1979, Frost & Ashby 1982) in Figures 5.1 and 5.2, based on the maps for MgO with a grain size of 10 μm. Important deformation and fracture mechanisms are more fully described below.

The characteristics of these deformation mechanism maps change systematically between the different isomechanical groups, and are also affected by grain size. The low-temperature normalized plastic-flow stress increases in the progression from the f.c.c. metal group through to the covalent solids and to ice, and the area of the map occupied by the diffusional flow mechanism also increases in this progression. On the other hand, the area occupied by diffusional flow increases at the expense of the area occupied by power-law creep as the grain size decreases, while at the same time the lines of constant strain rate move to lower homologous temperatures (see Table 5.7). In fracture mechanism maps the area occupied by cleavage 1/brittle intergranular fracture (BIF) 1 (due to pre-existing cracks) increases from zero for the f.c.c. metal group to reach its maximum extent for covalent solids and for ice, while at the

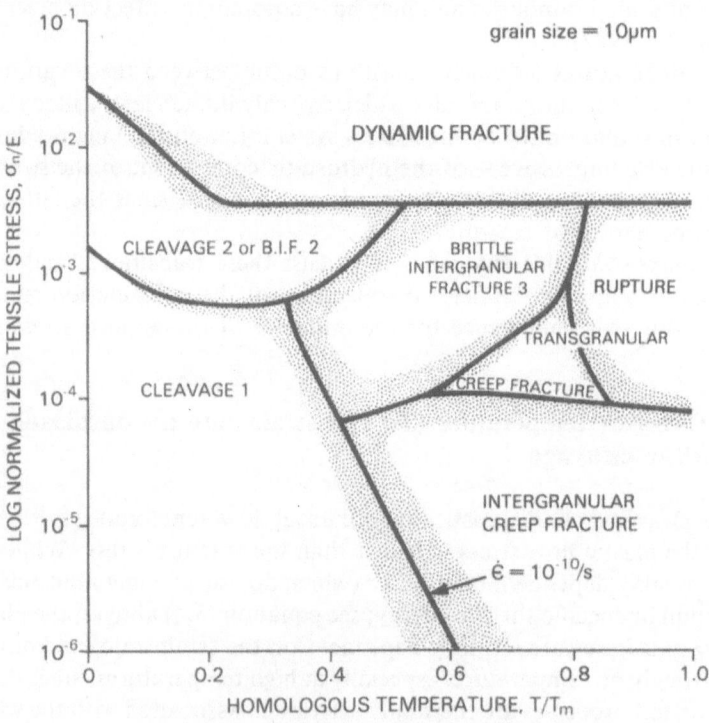

Figure 5.1 Schematic fracture-mechanism map illustrating seven different fracture modes, based on a map for MgO with 10 μm grain size (after Gandhi & Ashby 1979, Fig. 29).

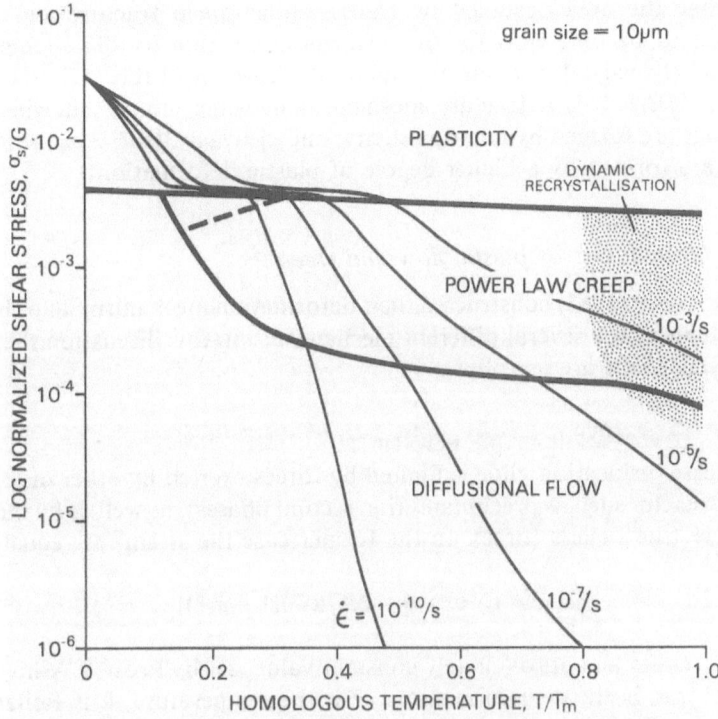

Figure 5.2 Schematic deformation-mechanism map illustrating four different plastic flow modes, based on a map for MgO with 10 μm grain size (after Frost & Ashby 1982, Fig. 16).

Table 5.7 Homologous temperature for MgO of three different grain sizes at which the flow stress is 1 MPa at a strain rate of $10^{-8}\ s^{-1}$.

	1 μm	10 μm	100 μm
T/T_m	0.39	0.54	0.9

Table 5.8 High-temperature bound for cleavage 1 in various solids (after Gandhi & Ashby 1979).[*]

Solid	T/T_m
W	0.2–0.66
Mg	0.3–0.55
NaCl	0.4–0.8
MgO	0.35–0.6
SiC	0.2–0.66
H_2O (ice)	0.6–0.9

[*] Cleavage 1 is caused by pre-existing cracks. The larger the crack size the lower the strength and the higher the temperature up to which brittle cleavage 1 fracture occurs. The high-temperature boundary is marked by creep at a strain rate of $10^{-10}\ s^{-1}$. At smaller crack sizes and higher stresses there is a transition to cleavage 2 fracture.

same time the area occupied by transgranular creep fracture and rupture decreases to become zero for the covalent solids and ice. The temperature range of cleavage 1 for various solids is shown in Table 5.8. (Note that cleavage 2/BIF 2 is a fracture mechanism in which cracks showing brittle behaviour are formed by microplasticity, and cleavage 3/BIF 3 is one in which cracks are formed by a higher degree of plastic deformation.)

5.2.1 Mechanisms of plastic flow and creep

Frost & Ashby (1982) constructed their deformation mechanism maps by using rate equations for several different mechanisms. In this discussion the important mechanisms are as follows.

5.2.1.1 LOW-TEMPERATURE PLASTICITY
In this case dislocation glide is limited by forces exerted by other dislocations or by obstacles such as precipitates (i.e. second phases), as well as by the lattice resistance (the Peierls force). In the former case the strain-rate equation is

$$\dot{\varepsilon} = C_1 \exp \left[(-\Delta F/kT)(1 - \sigma_s \hat{\tau}) \right] \tag{5.7}$$

where C_1 takes an approximately constant value, set by Frost & Ashby (1982) at $10^6 \, \text{s}^{-1}$, σ_s is the deviatoric stress, T is the temperature, k is Boltzmann's constant, ΔF is an activation energy (large in this case) to overcome the obstacle, and $\hat{\tau}$ is the athermal shear strength (the flow strength at 0 K). For obstacles consisting of other dislocations, or small or weak second-phase particles, $\Delta F \approx 0.5 Gb^3$ and $\hat{\tau} \approx Gb/l$, where G is the shear modulus, b is the length of the Burgers vector, and l is the spacing of the obstacles.

In cases where the lattice resistance is important

$$\dot{\varepsilon} = C_2 (\sigma_s/G)^2 \exp \left[(-\Delta F_p/kT)\{ 1 - (\sigma_s/\hat{\tau}_p)^{3/4} \}^{4/3} \right] \tag{5.8}$$

where C_2 has a value set by Frost & Ashby (1982) at $10^{11} \, \text{s}^{-1}$, and the activation energy ΔF_p (the free energy to form a dislocation kink pair) is in this case small (typically $0.1 \, Gb^3$) and $\hat{\tau}_p$ is approximately the flow stress at 0 K (typically $10^{-2} G$) for the hardest slip system.

The multiplication of and increasing interaction between dislocations during plastic deformation causes strain hardening, and this is the dominant process in low-temperature plasticity.

5.2.1.2 HIGH-TEMPERATURE PLASTICITY
At temperatures above $\approx 0.3 T_m$ for pure metals, $\approx 0.4 T_m$ for ionically bonded ceramics, and $\approx 0.6 T_m$ for covalently bonded ceramics, the plastic flow stress becomes much more strongly dependent on the strain rate.

Equations (5.7) and (5.8) show that the low-temperature flow stress is linearly dependent on temperature, and depends only logarithmically on the strain rate. If ΔF is small the exponential term in (5.7) approaches a linear dependence on σ_s, and this couples with a dependence on σ_s^2 of the C_1 term to give a glide-controlled power-law creep relationship,

$$\dot{\varepsilon} \propto (\sigma_s/G)^3 \tag{5.9}$$

which is thought to be applicable to ice, some ceramics, and to some metals, at temperatures below $0.5T_m$. The cross-slip of screw dislocations is an important aspect of this form of creep at temperatures above $0.2T_m$ (Murrell & Chakravarty 1973, Poirier 1985). At still higher temperatures the climb of edge dislocations by a diffusive mechanism becomes the rate-controlling process in plastic flow, in competition with the strain-hardening process.

Cross-slip and dislocation climb are key processes in the recovery of solids from strain hardening. When the rate of recovery equals the rate of hardening the solid undergoes the process of steady-state creep, which is maintained at constant stress. Prior to the establishment of this steady-state deformation is still dominated by strain hardening and it follows a primary or transient creep law, under which the creep rate decreases with time (Murrell & Chakravarty 1973). However, Frost & Ashby (1982) state that present theoretical models for creep are unsatisfactory (see also Poirier 1985), since they are unable to explain the wide range of power-law exponents for the stress dependence, or the high values found experimentally for the strain rate. Their best model gives the following rate equation:

$$\dot{\varepsilon} = C_3(D^*Gb/kT)(\sigma_s/G)^n \tag{5.10}$$

where D^* is the effective diffusion coefficient, the exponent n ranges in value between 3 and 10, and C_3 is a dimensionless constant (with values ranging up to 10^{15}).

At temperatures above $0.6T_m$ dislocation climb is generally controlled by lattice diffusion, and if the climb velocity is proportional to the applied stress, σ_s, the rate equation has $n = 3$, with $D^* = D_v$ (where D_v is the lattice-diffusion coefficient), and $C_3 \approx 1$. In some exceptional cases power-law creep of this form is observed (see equation (5.9) above).

However, at lower temperatures the activation energy is often less than that for lattice diffusion. In addition, both at lower and at higher temperatures the power law exponent n often exceeds 3. Two factors are suggested by Frost & Ashby (1982) to explain these observations: one is that the density and velocity of mobile, climbing dislocations perhaps depends on stress σ_s to higher powers than the simple theory assumes, and the second is that at lower temperatures and higher stresses dislocation core diffusion becomes important. A fuller discussion of these questions is given by Poirier (1985). Allowing for both

121

lattice and core diffusion the effective diffusion coefficient is

$$D^* = D_v[1 + \alpha(\sigma_s/G)^2 D_c/D_v] \tag{5.11}$$

where D_c is the core diffusion coefficient, and $\alpha = 10a_c/b^2$ (where a_c is the fast diffusion cross section of a dislocation core). Thus at high temperatures and low stresses lattice diffusion becomes dominant, whereas at low temperatures and high stresses core diffusion becomes dominant and the power-law exponent increases to $n + 2$ (as D^* tends to $\alpha\sigma_s^2 D_c$; see equation (5.11)).

At very low stresses the power-law exponent reduces to one and the dislocation density remains constant (although low). This form of climb-controlled dislocation creep is called Harper–Dorn creep. In this case $n = 1$, $D^* = D_v$, and $C_3 \approx 10^{-10}$–10^{-9} (e.g. for Al and Pb).

At high temperatures ($> 0.6T_m$) and high strain rates (i.e. high stress), plastic flow in relatively pure solids is often accompanied by repeated waves of recrystallization (dynamic recrystallization). This occurs in metals, ceramics, and ice. However, it has not yet been satisfactorily modelled. It tends to be suppressed by second-phase particles. The formation and growth of new strain-free grains is driven by the internal energy of strain hardening associated with the development of tangled networks of dislocations. At lower temperatures the process involves the formation of small grains with a new orientation due to constrained, localized plastic deformation. At higher temperatures grain boundary migration plays a part, assisted by diffusional processes. Dynamic recrystallization is a process that occurs in bursts, unlike normal recovery processes. Consequently it causes strain softening in constant-strain-rate tests, and episodes of accelerating and decelerating strain in constant-stress tests (see Poirier 1985 for a fuller discussion).

5.2.1.3 DIFFUSIONAL FLOW

In solids at high temperatures and subject to a deviatoric stress field there is a diffusive flux of matter across and along bounding surfaces and grain boundaries. If this is coupled with grain-boundary sliding the diffusive flux results in strain.

The strain-rate equation proposed by Frost & Ashby (1982) for diffusional flow is

$$\dot{\varepsilon} = C_4 \sigma_s \Omega D^* / \mathbf{k} T d^2 \tag{5.12}$$

where Ω is the atomic or ionic volume, d is the grain diameter, and C_4 is a numerical constant (assigned the value 42). In this case the effective diffusion coefficient is given by

$$D^* = D_v[1 + \pi\delta D_b/dD_v] \tag{5.13}$$

where D_b is the grain-boundary diffusion coefficient, and δ is the effective thickness of the grain boundary.

At the highest temperatures the rate of flow is controlled by lattice diffusion, and is determined primarily by D_v/d^2, whereas at lower temperatures it is controlled by grain-boundary diffusion, and is determined by D_b/d^3. At high temperatures and low strain rates superplasticity (giving rise to very large strains but without the formation of a strain texture) may occur (Poirier 1985).

5.2.2 Temperature dependence of plastic flow and creep

Low-temperature plasticity shows a linear relationship between flow stress and temperature and a logarithmic relationship between flow stress and strain rate (see equations (5.7) & (5.8)). However, high-temperature plasticity (equations (5.10) & (5.11)) and diffusional flow (equation (5.12)) both show a much stronger temperature and strain-rate effect because of their dependence on diffusional processes with effective diffusion coefficients which have an exponential Arrhenius dependence on temperature, given by

$$D^* = \alpha^* \exp(-\Delta G/kT) \tag{5.14}$$

where α^* is a frequency factor, and ΔG is an activation energy, and both depend on pressure. In this case the flow stress depends on $(\dot{\varepsilon}/D^*)^{1/n}$, where $n \geqslant 1$, which gives rise to a strong dependence on temperature and strain rate.

5.2.3 Transition from brittle cleavage to plastic flow and creep

The marked difference between the degrees to which the brittle cleavage strength and the plastic flow and creep strength depend on temperature and strain rate would be expected to lead to a clear transition in deformation behaviour at a critical temperature, this temperature being strain-rate dependent (see Orowan 1952, Pugh 1971). At any given strain rate, strength will remain approximately constant at temperatures below the critical temperature, and the transition from low-temperature brittle behaviour to high-temperature plasticity or creep will be indicated by a knee in the curve of strength versus temperature, and a marked strength reduction at temperatures above the critical temperature (see Fig. 5.3). The strain to fracture would also show a transition from low values (≈ 0.01) for brittle cleavage to larger values (≈ 0.1) at and above the critical temperature. However, as temperature is increased there would typically be a transition first from brittle cleavage 1 (or brittle intergranular failure BIF 1) behaviour, caused by pre-existing cracks, to cleavage 3 (or BIF 3) which is preceded by 1–10 per cent of plastic strain (that tends to cause crack blunting and raises the fracture surface energy Γ; see equation (5.5) above), and then a further transition from cleavage 3 to plastic flow. In the case of less brittle solids, such as NaCl, Mg, or W, the intermediate stage

123

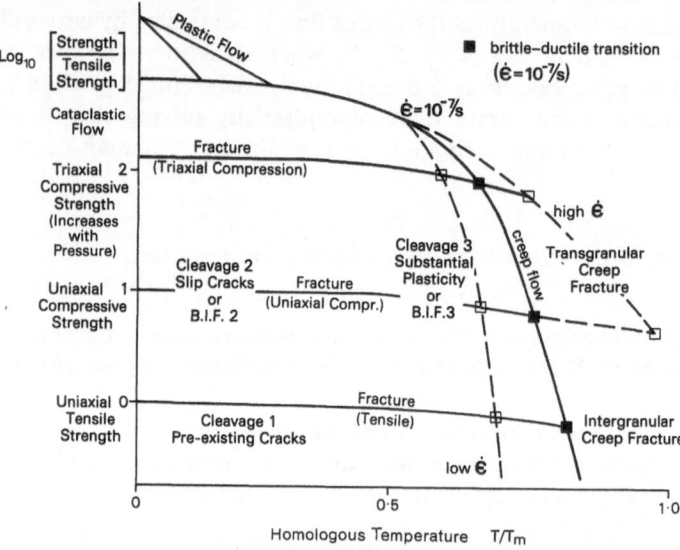

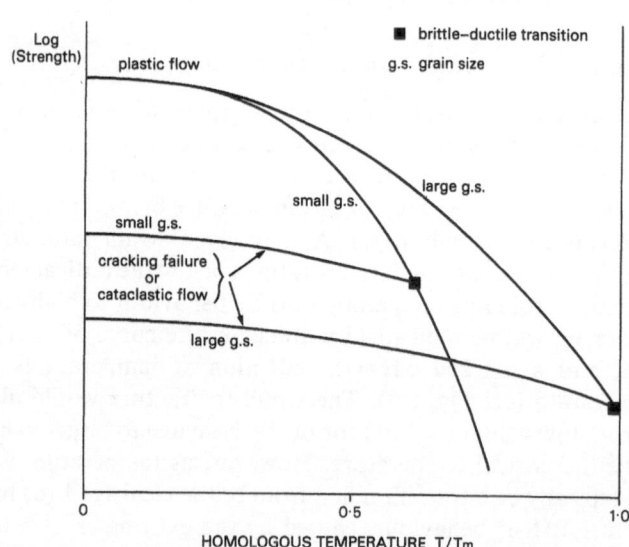

Figure 5.3 Schematic deformation and fracture-mechanism maps illustrating fracture–flow transitions, showing transitions from brittle cleavage fracture to creep fracture and then to creep flow/plasticity. (a) The transition temperature is lower at low strain rates and for greater degrees of compression. (b) The transition temperature is lower for smaller grain size.

Table 5.9 Estimates of transition temperature (T_1) from cleavage 1 to cleavage 3 or creep fracture, and of transition temperature (T_2) from fracture to creep flow at a strain rate of $10^{-7}\,s^{-1}$. Data are for a range of solids containing cracks of length $2c = 100\,\mu m$, with grain diameters $d = 100\,\mu m$, under uniaxial tension.

Solid		T_m (K)	$\Gamma\;(J\,m^{-2})$	ν	σ_T/E (10^{-4}), $T = 0K$	σ_T/E (10^{-4}), $T = T_m$	T_1/T_m	T_2/T_m
W		3683	4.2	0.29	3.7	4.7	0.30	0.37
Mg		923	(1.0)	(0.30)	5.2	6.5	0.36	0.49
NaCl	†	1073	0.28	(0.20)	3.9	6.3	0.44	0.50
MgO	‡	3073	1.2	0.18	2.4	3.8	0.33	0.46
Al$_2$O$_3$	‡	2323	1.0	0.20	1.8	2.3	0.50	0.77
	§	—	—	—	5.7	7.1	—	0.66
Mg$_2$SiO$_4$‖		2140	(1.0)	(0.20)	9.7	11.9	0.50	0.64
	¶	—	—	—	0.97	1.2	—	0.80
H$_2$O (ice)		273	0.1	(0.20)	3.7	4.1	0.63	0.88

* Figures in brackets are estimated. Data for Γ from Lawn & Wilshaw (1975), and for E from Gandhi & Ashby (1979). σ_T is the tensile strength calculated from Griffith's equation (5.5) in the text. Creep and fracture data from Frost & Ashby (1982) and Gandhi & Ashby (1979).
† Estimated using flow stress curve for $d = 10\,\mu m$, taking $2c = 50\,\mu m$.
‡ Estimated using fracture data for $d = 10\,\mu m$.
§ Estimated using flow stress curve for $d = 10\,\mu m$, taking $2c = 10\,\mu m$.
‖ Estimated using flow stress curve for $d = 10\,\mu m$ and fracture data for Al$_2$O$_3$.
¶ Estimated using flow stress curve for $d = 1$ mm.

consists of creep fracture (transgranular or intergranular). The temperatures of the two transitions are given in Table 5.9 for a range of solids.

5.3 Effect of tension, compression, and confining pressure (stress triaxiality) on plastic flow and brittle cleavage

For any given microstructural state, plastic flow takes place in response to a fixed non-zero state of shear and deviatoric stress – the flow stress (σ_s). This shear-stress dependence arises from the critical resolved shear stress law for plastic slip and dislocation glide. At low temperatures, where strain hardening dominates over recovery processes, the flow stress depends on the strain and temperature history. Hydrostatic confining pressures less than $0.001K$ (where K is the bulk modulus) cause negligible change in the plastic flow stress, σ_s or the creep rate, $\dot{\epsilon}$, although σ_s is increased and $\dot{\epsilon}$ is decreased by higher pressures.

In terms of the principal stresses σ_1, σ_2, and σ_3, we find that

$$\sigma_s = [(\sigma_1 - \sigma_2)^2 + (\sigma_2 - \sigma_3)^2 + (\sigma_3 - \sigma_1)^2] \tag{5.15}$$

(this is the von Mises octahedral shear stress law).

In a Mohr diagram, therefore, the flow stress is represented by a straight line

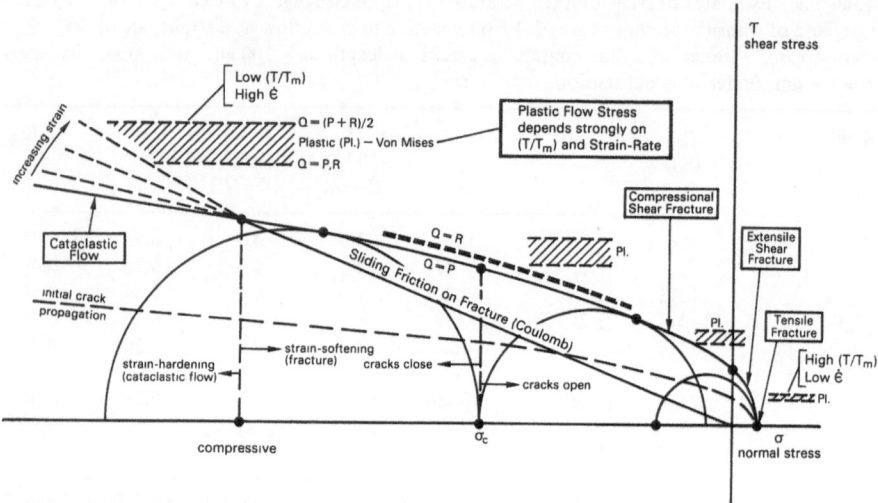

Figure 5.4 Mohr diagram illustrating the dependence of fracture stress and plastic flow stress on the stress system. T/T_m = homologous temperature; $\dot{e}$ = strain rate; $P = \sigma_1$, $Q = \sigma_2$, $R = \sigma_3$, are principal stresses at failure; σ_c = crack closure stress. The condition for initial crack propagation is given by equation (5.17) when cracks are open and equation (5.18) when cracks are closed.

parallel to the normal stress axis (see Fig. 5.4). Under steady-state conditions the location of this line is fixed (i.e. σ_s = constant). σ_s is unaffected by the signs of the principal stresses. For example, the magnitude of the flow stress is the same in uniaxial compression as in uniaxial tension.

By contrast, brittle cleavage and fracture are in general strongly affected by hydrostatic confining pressure and by the sign of the components of normal stress.

In uniaxial tension fracture occurs at a low tensile stress across surfaces normal to the stress (Murrell 1964). This is unaffected by additional confining pressures up to a limiting value of

$$\sigma_3 = \sigma_2 = \sigma_T(A^2 - 1) \tag{5.16}$$

where A is a numerical factor (≈ 1.5) depending on crack shape and on the Poisson ratio, and σ_T is the tensile strength (Murrell 1958, 1965; Brace 1964; Murrell & Digby 1970). In this case, therefore, fracture is determined by the tensile stress only, and is unaffected by the deviatoric stress value, which increases as the confining pressure is increased. The tensile fracture stress coincides with the stress required to initiate crack propagation. In a fracture-mechanism map outside of the region in which cleavage 1 is dominant (in which cracks are pre-existent), tensile fracture must be preceded by a process of crack nucleation and subcritical crack growth.

In uniaxial compression truly brittle fracture occurs at a deviatoric stress

vhich is an order of magnitude higher than the uniaxial tensile strength; either
y a splitting mechanism (Ashby & Hallam 1986) caused by buckling stresses,
r by a shearing mechanism (Griffith 1924; Murrell 1958, 1964, 1965, 1967;
Murrell & Digby 1970; Ashby & Hallam 1986; Horii & Nemat-Nasser 1986).
Jnder stress states intermediate between uniaxial tension and fully compressional states, fracture occurs by an extensional shearing mechanism (Murrell
965, 1976). In all of these states the deviatoric fracture stress increases as the
lydrostatic stress component (or confining pressure) increases. However, the
fracture process involves out-of-plane crack propagation which is stable,
ollowed by crack linkage which eventually becomes unstable. A general
heoretical treatment of the initiation of crack propagation in triaxially
tressed solids is given by Murrell & Digby (1970). This places a lower limit
n the increase of fracture strength with confining pressure. The initiation
tress for penny-shaped cracks is given by:

$$\sigma_S = (\sigma_1 - \sigma_3)/2 = \{ \sigma_T (1 - \nu/2)^2 [\sigma_T \nu (4 - \nu) - 2(\sigma_1 + \sigma_3)] \}^{1/2} \quad (5.17)$$

vhile cracks remain open, and

$$\tau_S = (\sigma_1 - \sigma_3)/2 = \{ \sigma_T (2 - \nu)(1 - \sigma_c/\sigma_T)^{1/2} - \mu_f [(\sigma_1 + \sigma_3)/2 - \sigma_c]/(1 + \mu_f^2)^{1/2} \}$$
$$(5.18)$$

vhen cracks are closed by a pressure σ_c, where μ_f is the coefficient of friction.

Attempts have recently been made by Ashby & Hallam (1986) and Horii &
Nemat-Nasser (1986) (see also Hallam & Ashby, this volume) to model the
crack linkage and instability processes in fracture.

In fully compressional stress states at lower confining pressures, shear
fracture is associated with a stress drop (i.e. strain softening), the magnitude
of which is given by the difference between the fracture load and the load
supported by friction on the shear fault. This stress drop depends on confining
pressure and decreases to zero at a sufficiently high confining pressure (Murrell
1965, Ismail 1974). At still higher confining pressures deformation is entirely
distributed, accompanied by strain hardening, and involves densely disseminated microcracking. This is described as cataclastic flow. The rate of strain
hardening in cataclastic flow depends on confining pressure (see Murrell 1965,
Edmond & Murrell 1973, Paterson 1978 for further details).

The strength at high confining pressure can be more than an order of
magnitude higher than the uniaxial compressive strength. Therefore the low-
temperature fracture strength varies by more than two orders of magnitude
between the tensile cleavage strength and the cataclastic flow stress.

Cataclastic flow has the macroscopic appearance of ductile behaviour, but
the flow mechanism involves multiple widely disseminated and interlocked
microcracks, and is therefore only weakly dependent on temperature and
strain rate, as in the case of tensile cleavage.

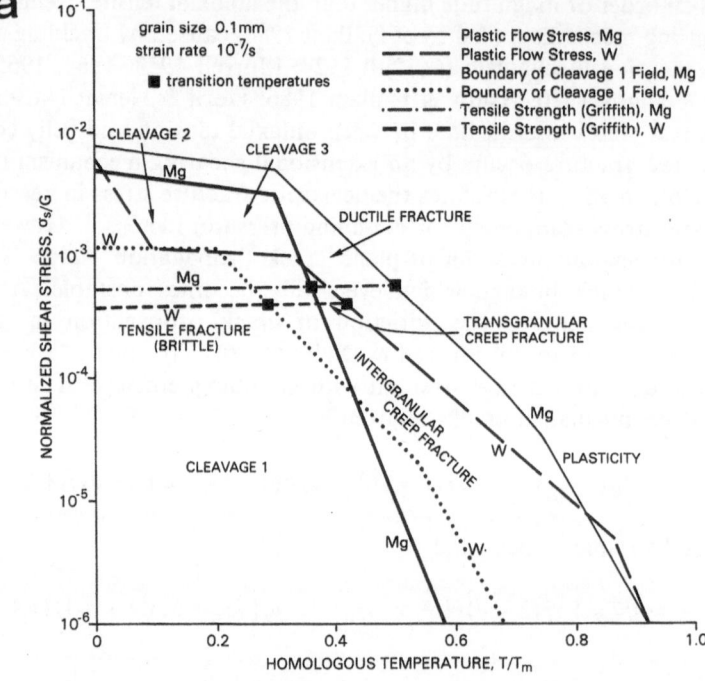

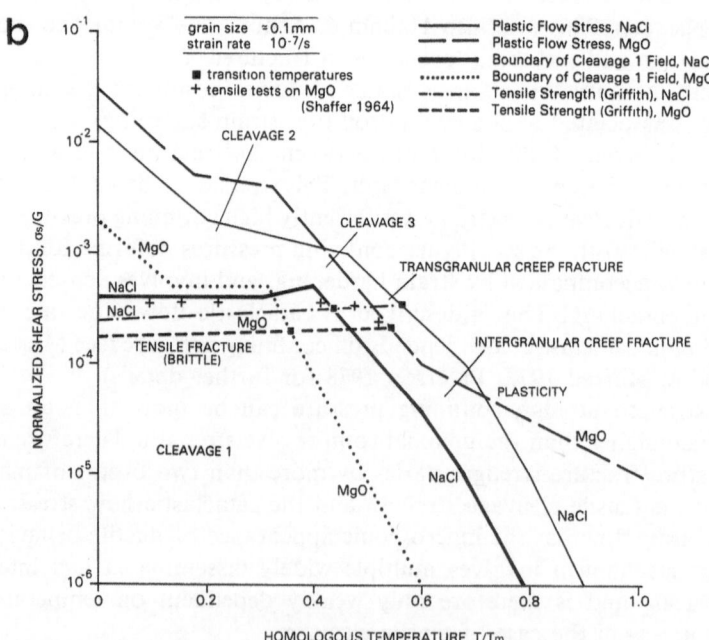

5.4 Transition from fracture to plastic flow

The expected transitions are shown in typical deformation mechanism maps in Figures 5.5a–c, and in a Mohr diagram in Figure 5.4.

In tension, or in compression with low confining pressures, as the temperature is increased there is a fracture-mechanism transition from cleavage 1 (or BIF 1) to cleavage 3 (or BIF 3), or to creep fracture (at lower stresses). This transition occurs at a temperature ranging from $\approx 0.3T_m$ for W up to $\approx 0.63T_m$

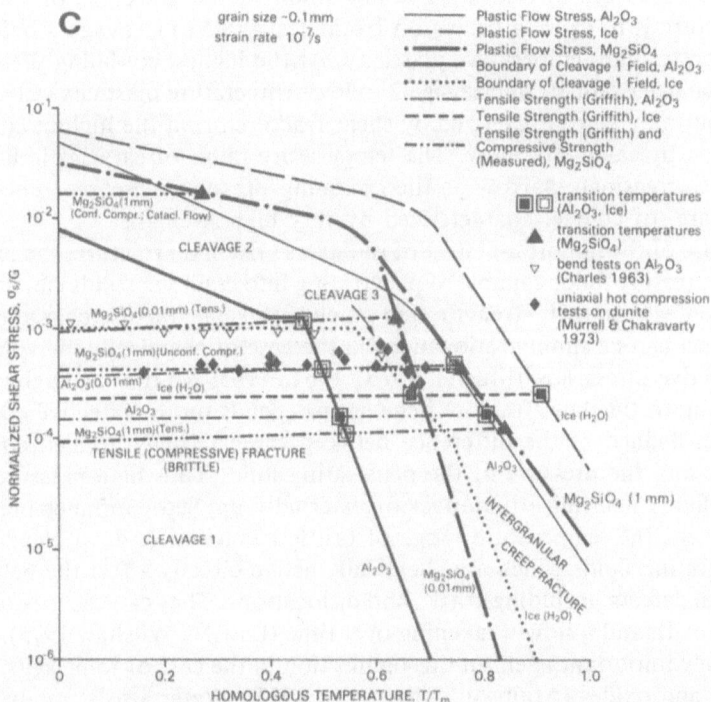

Figure 5.5 Schematic deformation and fracture-mechanism maps for materials belonging to different isomechanical classes, showing the temperature and stress fields in which brittle cleavage fracture and low- and high-temperature plasticity operate, and illustrating the sharp change from fracture to flow. The figure is based on data from Gandhi & Ashby (1979) and Frost & Ashby (1982) for materials with a grain size mainly of 0.1 mm and deformed at a strain rate of 10^{-7} s^{-1}. The brittle tensile fracture stress is calculated from equation (5.5) (see Table 5.9). (a) Map for the semi-brittle metals tungsten and magnesium. (b) Map for the brittle ionic solids rock salt and magnesium oxide. The fracture–flow transition for MgO can be seen in tensile test data points (+) due to Shaffer (1964) extracted from Gandhi & Ashby's (1979) paper. (c) Map for the highly brittle ionic-covalent solids alumina and forsterite and for H$_2$O ice. Indications of the fracture–flow transition for Al$_2$O$_3$ can be seen in bend test data points ($\triangledown$) due to Charles (1963) extracted from Gandhi & Ashby's (1979) paper. Hot compression data points for dunite ($\blacklozenge$) (near Mg$_2$SiO$_4$ composition) due to Murrell & Chakravarty (1973) show fracture strength nearly independent of temperature.

129

for H_2O-ice, and is $\approx 0.5T_m$ for alumina and olivine. Only at still higher temperatures ($0.37T_m$ to $0.88T_m$) does the fracture – plastic flow transition occur in this case, so the transition is preceded by plastic flow or creep.

At higher confining pressures the brittle–ductile transition is preceded by a fracture-mechanism transition from cleavage 1 to cleavage 2 as pressure is increased and the fracture strength increases correspondingly. This is preceded by a phase of closure of any pre-existing cracks by the confining pressure (see Murrell 1976). In the cleavage 2 régime cracks are nucleated by small-scale slip or twinning (general microplasticity). At a given confining pressure in the range characterized by cleavage 2 at low temperatures, the effect of increasing temperature is to cause a transition from cleavage 2 to cleavage 3 before the transition to high-temperature plasticity. At the highest confining pressures a transition can occur from cleavage 2 to low-temperature plasticity. Cleavage 2 behaviour may be characterized by shear fracture or, at the highest confining pressures, by cataclastic flow. The temperature range of cataclastic flow will become increasingly narrow as the confining pressure increases. These conditions are, of course, characterized by very high strengths.

Because of the important effect of the stress system on fracture, the presence of high-pressure fluids capable of penetrating into cracks in solids can have significant effects on their strength, and on cleavage and fracture characteristics. This effect can be simply rationalized in the case of chemically inert fluids by the effective stress law (Murrell 1963, 1964, 1965; Murrell & Digby 1970). According to this law, fracture behaviour depends on the effective confining pressure, defined as the difference between the externally applied confining pressure and the pressure of the penetrating fluid. Thus high-pressure penetrating fluids neutralize the effect of externally applied confining pressures. This widens the temperature range of brittle behaviour (e.g. of cleavage 1).

Penetrating fluids which are chemically active directly affect the behaviour of crystal defects, including cracks and dislocations. They can cause subcritical crack growth and a slow weakening over time (Lawn & Wilshaw 1975). Water is a widely important agent in this connection in the case of ionic salts and of silicates and oxides (Atkinson & Meredith 1987). Water also strongly affects plastic flow in some cases (e.g. in rock salt; see Urai *et al.* 1986). There is insufficient space here for a more extensive discussion of this topic, but the influence of water on subcritical crack growth in silicates and oxides is discussed in more detail by Gueguen *et al.* (this volume) and Meredith (this volume).

5.5 Discussion and conclusions

Typical fracture and flow behaviour of a wide range of solids has been described above in some detail as a function of temperature and of the applied stress system.

Tensile stresses and the confining pressure have significant effects on fracture, to the extent that the strength (measured by the deviatoric stress) at a given temperature and strain rate can vary by several orders of magnitude, depending on the stress system. The lowest strengths occur under tensile stress conditions, and then fracture persists up to the highest temperatures (up to the melting point, at the higher strain rates). However, at a temperature below the temperature at which the transition from fracture to flow occurs there is another transition from highly brittle cleavage fracture (cleavage 1, involving pre-existing cracks) to creep fracture or brittle intergranular fracture, which involve some plasticity.

Fracture can also occur under uniaxial or triaxial compressive stresses, when the strength is much higher than the tensile strength. At very high confining pressures there is a transition from shear fracture to cataclastic flow which, nevertheless, is still a deformation process that is fundamentally dependent on cracking. In this case, cracking could be nucleated by microplasticity (cleavage 2). The temperature range of cataclastic flow or shear fracture is smaller than that for tensile fracture, and there may be a direct transition from brittle deformation (involving cleavage 2) to low-temperature plasticity in this case.

For a given stress system the temperature at which the transition from fracture to flow occurs depends on the strain rate and the grain size, and is higher for higher strain rates and larger grain sizes. At lower temperatures than this, if the stress is less than the fracture stress small amounts of creep can occur, this being mostly transient creep at the lower temperatures (e.g. see Murrell & Chakravarty 1973).

Although there is a large amount of data available on the tensile strength of solids, both at low temperatures when brittle cleavage fracture occurs, and at high temperatures when creep occurs, together with data on the indentation hardness of such solids (which is a measure of compressive strength under confining pressure), there appears to be little systematic data available concerning the transition between fracture and flow. Data for MgO and for Al_2O_3 extracted from the paper by Gandhi & Ashby (1979) is show in Figures 5.5b & c respectively. Hot compression tests by Murrell & Chakravarty (1976) on three silicate rocks (microgranodiorite, dolerite, and dunite) show that the normalized uniaxial strength remains almost constant up to $\approx 0.8T/T_m$, when flow becomes strongly evident, in the case of the first two of these rocks (although this behaviour may have been complicated by partial melting associated with 'impurities' in the form of hydrated minerals). This is close to the figure estimated for forsterite (see Fig. 5.5c), but the experiments on dunite (mineralogically similar to forsterite) extended only to homologous temperatures of $\approx 0.69T/T_m$, so that a more direct confirmation of the fracture–flow transition is not available. The study of this transition requires constant-strain-rate experiments over a range of temperatures spanning the transition and using a variety of different stress systems (uniaxial tension and compression, and triaxial compression with a range of confining pressures).

At very high confining pressures an interesting transition is expected to occur between cataclastic flow (involving brittle cleavage 2 processes, and characterized by strain hardening with a pressure-dependent strain-hardening rate) and plastic flow (involving low-temperature plasticity and also characterized by strain hardening, which in this case is temperature dependent).

In the above discussion attention has been concentrated particularly on the temperature and strain rate at which the brittle–ductile transition occurs. However, because fracture is so strongly affected by the character of the applied stresses, confining pressure also influences the transition (and in particular the transition temperature). This has been clearly demonstrated by Pugh (1971) in the case of the tensile fracture of zinc and molybdenum. Unfortunately, neither the tensile strength, nor the grain size or the strain rate used in the tests are given. For zinc samples the transition temperature was lowered from $0.48T_m$ ($58°C$) at zero pressure to $0.39T_m$ ($0°C$) at a pressure of 113 MPa (corresponding to an increase in the normalized shear stress of 1.15×10^{-3}), and for molybdenum samples it was lowered from $0.11T_m$ ($50°C$) at zero pressure to $0.095T_m$ ($2°C$) at a pressure of 132 MPa (corresponding to an increase in the normalized shear stress of 4.9×10^{-4}). Comparison of these figures with the data in Figure 5.5a for magnesium and tungsten (which belong to the same isomechanical classes as zinc and molybdenum respectively) suggests that cleavage 1 behaviour occurred in the zinc samples, but that cleavage 2 behaviour occurred in the molybdenum samples, and also shows that the rate of change of the homologous transition temperature with the normalized pressure was of the expected order of magnitude.

The general effect of an increase in pressure is a reduction in the brittle–ductile transition temperature, as the applied shear stress increases. In Pugh's (1971) experiments it seems likely that the condition of equation (5.16) applied, so that although the tensile strength probably remained constant the maximum shear stress in the samples increased linearly with the increase of confining pressure. At higher confining pressures, under which (5.16) no longer applies, the fracture strength becomes dependent on the confining pressure (in a non-linear manner) discussed above, and it is the strength which determines the maximum shear stress.

For tests on a material of a given grain size at some given temperature and strain rate, a sufficiently large increase of confining pressure which caused the fracture stress or the cataclastic flow stress to rise to a level exceeding the plastic flow stress (at the given temperature and strain rate) would produce a transition from brittle to ductile behaviour. In fact, however, the author is not aware of any data which has clearly demonstrated this in non-metallic materials. For example, at atmospheric temperatures Ismail (1974) showed that brittle behaviour in a wide variety of silicate rocks, in the form of cataclastic flow, persisted at confining pressures up to 750 MPa, when strengths as high as 2000 MPa were achieved. This clearly showed that a com-

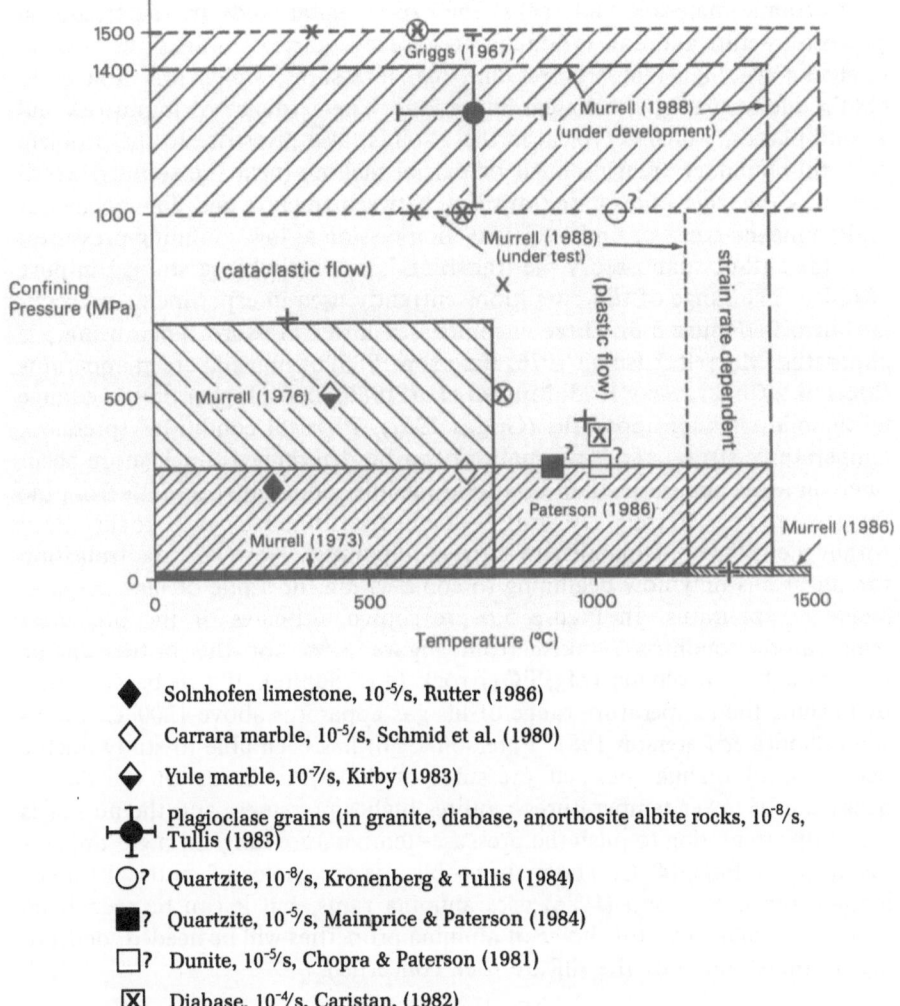

♦ Solnhofen limestone, 10^{-5}/s, Rutter (1986)

◇ Carrara marble, 10^{-5}/s, Schmid et al. (1980)

⬟ Yule marble, 10^{-7}/s, Kirby (1983)

⊕ Plagioclase grains (in granite, diabase, anorthosite albite rocks, 10^{-8}/s, Tullis (1983)

○? Quartzite, 10^{-8}/s, Kronenberg & Tullis (1984)

■? Quartzite, 10^{-5}/s, Mainprice & Paterson (1984)

□? Dunite, 10^{-5}/s, Chopra & Paterson (1981)

⊠ Diabase, 10^{-4}/s, Caristan, (1982)

⊗ Diabase, 10^{-6}/s, Caristan (1982)

X Granite, 10^{-6}/s, Tullis & Yund (1977)

+ Forsterite, 10^{-7}/s, based on Fig. 5c

Figure 5.6 The range of operating conditions for rock deformation apparatus (see text) and observations on the brittle–ductile transition for several rocks and rock-forming minerals. This illustrates the need to extend the range of operating temperatures and pressures in order that the transition can be effectively studied in silicates.

133

bination of high temperatures, low strain rates, and high pressures is necessary to achieve ductile behaviour in silicates.

In ceramic materials and rocks, the experimental study of the transition between fracture and flow is difficult because it requires combinations of accurately known high temperatures and high pressures, which are difficult to obtain due to material technical limitations. The existence of impurities and second phases, which is typical in these solids, also gives rise to the problem of grain-boundary embrittlement or partial melting (near the solidus) which can mask the true solid-phase transition between fracture and flow behaviour under uniaxial stress or under triaxial compression at low confining pressures.

In the author's laboratory the transition is currently being studied in pure H_2O-ice. The range of test conditions currently used in experiments on rocks is shown in Figure 5.6. These encompass a lower pressure range using gas apparatus (Murrell & Ismail 1976, Paterson 1986) or uniaxial creep apparatus (Murrell & Chakravarty 1973, Murrell et al. 1986), and a higher pressure range using solid-medium apparatus (Griggs 1967). Physical conditions (pressure, temperature, stress, and deformation) can be determined much more accurately in a gas apparatus than in a solid-medium apparatus, so data from the former is to be preferred. The brittle–ductile transition for calcite rocks is well within the technical capability of current apparatus. However, the transition for silicates is only now beginning to come within the scope of new developments in apparatus. In Figure 5.6 are shown estimates of the pressure–temperature conditions, taken from Figure 5.5c, for the brittle–ductile transition for an olivine (Mg_2SiO_4) rock (e.g. dunite). It can be seen that by pushing the temperature range of his gas apparatus above 1200°C, Paterson (Chopra & Paterson 1981, Paterson 1986) has been able to study ductile behaviour of olivine rocks at pressures of 300 MPa. Study of the ductile behaviour at lower temperatures requires higher pressures, and the author is currently attempting to push the pressure–temperature range of a gas apparatus up to 1 GPa/1200°C. The main problem is the choice of material for the loading rams. Paterson (1986) uses alumina rams, but it can be seen from Figure 5.5c that a careful choice of alumina properties will be needed, depending on the strength of the silicate rock concerned.

References

Ashby, M. F. & A. M. Brown 1981. In *2nd Riso Conf. Mat. Sci.* N. Hanson, A. Horswell, T. Leffers & H. Lilholt (eds), 1.

Ashby, M. F. & S. D. Hallam 1986. The failure of brittle solids containing small cracks under compressive stress states. *Acta Metall.* **34**, 497–510.

Atkins, A. G. 1973. High-temperature hardness and creep. In *The science of hardness testing and its research applications*, J. H. Westbrook & H. Conrad (eds), 223–40. Metals Park, Ohio: American Society for Metals.

Atkinson, B. K. & P. G. Meredith 1987. The theory of subcritical crack growth with applications to minerals and rocks. In *Fracture mechanics of rock*, B. K. Atkinson (ed.), 111–66. London: Academic Press.

Brace, W. F. 1964. Brittle fracture of rocks. In *State of stress in the Earth's crust*, W. R. Judd (ed.), 111–74. New York: Elsevier.

Brown, A. M. 1980. *The temperature dependence of the Vickers hardness of isostructural compounds*. PhD thesis, University of Cambridge.

Brown, A. M. & M. F. Ashby 1980. Correlations for diffusion constants. *Acta Metall.* **28**, 1085–101.

Caristan, Y. 1982. The transition from high temperature creep to fracture in Maryland diabase. *J. Geophys. Res.* **87**, 6781–90.

Catlow, C. R. A. & A. M. Stoneham 1983. Ionicity in solids. *J. Phys. C: Solid State Phys.* **16**, 4321–38.

Charles, R. J. 1963. Studies of the brittle behaviour of ceramic materials. ASD-TR-61-628, 467.

Chopra, P. N. & M. S. Paterson 1981. The experimental deformation of dunite. *Tectonophysics* **78**, 453–73.

Cottrell, A. H. 1953. *Dislocations and plastic flow in crystals*. Oxford: Clarendon Press.

Digby, P. J. & S. A. F. Murrell 1975. The role of shear stresses in brittle fracture. *J. Mech. Phys. Solids* **23**, 185–96.

Edmond, O. & S. A. F. Murrell 1973. Experimental observations on rock fracture at pressure up to 7 kbar and the implications for earthquake faulting. *Tectonophysics* **16**, 71–87.

Frost, H. J. & M. F. Ashby 1982. *Deformation-mechanism maps: the plasticity and creep of metals and ceramics*. Oxford: Pergamon Press.

Gandhi, C. & M. F. Ashby 1979. Fracture-mechanism maps for materials which cleave: f.c.c., b.c.c. and h.c.p. metals and ceramics. *Acta Metall.* **27**, 1565–602.

Gilman, J. J. 1973. Hardness – a strength microprobe. In *The science of hardness testing and its research applications*, J. H. Westbrook & H. Conrad (eds), 51–74. Metals Park, Ohio: American Society for Metals.

Griffith, A. A. 1920. The phenomena of rupture and flow in solids. *Phil Trans R. Soc. Lond. A* **221**, 163–98.

Griffith, A. A. 1924. The theory of rupture. In *Proc. 1st Int. Congr. Appl. Mech.*, C. B. Biezeno & J. M. Burgers (eds), 54–63. Delft: Tech. Boekhandel en Drukkerij, J. Waltman Jr.

Griggs, D. T. 1967. Hydrolytic weakening of quartz and other silicates. *Geophys. J. R. Astron. Soc.* **14**, 19–31.

Groves, G. W. & A. Kelly 1963. Independent slip systems in crystals. *Phil Mag.* **8**, 877–87.

Gueguen, Y., T. Reuschlé & M. Darot 1989. Single-crack behaviour and crack statistics. This volume, 48–71.

Hallam, S. D. & M. F. Ashby 1989. Compressive brittle fracture and the construction of multiaxial failure maps. This volume, 84–108.

Horii, H. & S. Nemat-Nasser 1986. Brittle failure in compression: splitting, faulting and the brittle–ductile transition. *Phil Trans R. Soc. Lond. A* **319**, 337–74.

Hsueh, C. H. & A. G. Evans 1981. Creep fracture in polycrystals – II. Effects of inhomogeneity on creep rupture. *Acta Metall.* **29**, 1907–17.

Irwin, G. R. 1948. Fracture dynamics. In *Fracturing of metals*, 147–66. Metals Park, Ohio: American Society for Metals.

Ismail, I. A. H. 1974. *Experimental studies of mechanical instabilities in rocks, with particular reference to earthquake focal mechanisms.* PhD thesis, University of London.

Kelly, A. 1966. *Strong solids.* Oxford: Clarendon Press.
Kelly, A., W. R. Tyson & A. H. Cottrell 1967. Ductile and brittle crystals. *Phil Mag.* **15**, 567.
Kirby, S. H. 1983. Rheology of the lithosphere. *Rev. Geophys. Space Phys.* **21**, 1458–87.
Kocks, U. F., A. S. Argon & M. F. Ashby 1975. Thermodynamics and kinetics of slip. *Prog. Mat. Sci.* **19**, 1.
Kronenberg, A. K. & J. A. Tullis 1984. Flow strength of quartz aggregates: grain size and pressure effects due to hydrolytic weakening. *J. Geophys. Res.* **89**, 4281–97.

Lawn, B. R. & T. R. Wilshaw 1975. *Fracture of brittle solids.* Cambridge: Cambridge University Press.

Mainprice, D. H. & M. S. Paterson 1984. Experimental studies of the role of water in the plasticity of quartzites. *J. Geophys. Res.* **89**, 4257–69.
Meredith, P. G. 1989. Fracture and failure of brittle polycrystals: an overview. This volume, 5–47.
Murrell, S. A. F. 1958. The strength of coal under triaxial compression. In *Mechanical properties of non-metallic brittle metals,* W. H. Walton (ed.), 123–45. London: Butterworth.
Murrell, S. A. F. 1963. A criterion for brittle fracture of rocks and concrete under triaxial stress, and the effect of pore pressure on the criterion. In *Rock mechanics,* C. Fairhurst (ed.), Proc. 5th Symp. Rock Mech., 563–577. New York: Pergamon.
Murrell, S. A. F. 1964. The theory of the propagation of elliptical Griffith cracks under various conditions of plane strain or plane stress. Part I. *Br. J. Appl. Phys.* **15**, 1195–210; Parts II and III, ibid., 1211–23.
Murrell, S. A. F. 1965. The effect of triaxial stress systems on the strength of rocks at atmospheric temperatures. *Geophys. J. R. Astron. Soc.* **10**, 231–81.
Murrell, S. A. F. 1967. The effect of triaxial stress systems on brittle fracture and on the brittle–ductile transition. In *Conference on Physical Basis of Yield and Fracture, Oxford, 1966,* 225–34. Institute of Physics and The Physical Society (London), Conf. Series No. 1.
Murrell, S. A. F. 1975. Natural faulting and the mechanics of brittle shear failure. *J. Geol Soc. (Lond.)* **133**, 175–89.
Murrell, S. A. F. & S. Chakravarty 1973. Some new rheological experiments on igneous rocks at temperatures up to 1120°C. *Geophys. J. R. Astron. Soc.* **34**, 211–50.
Murrell, S. A. F. & P. J. Digby 1970. The theory of brittle fracture initiation under triaxial stress conditions: I. *Geophys. J. R. Astron. Soc.* **19**, 309–34; II, ibid., 499–512.
Murrell, S. A. F. & I. A. H. Ismail 1976. The effect of decomposition of hydrous minerals on the mechanical properties of rocks at high pressures and temperatures. *Tectonophysics* **31**, 207–58.
Murrell, S. A. F., P. G. Meredith, G. D. Price, A. Wall & K. Wright 1990. Uniaxial creep apparatus for temperatures up to 1700°C. Work in progress.

Orowan, E. 1949. Fracture and strength of solids. *Rep. Progr. Phys.* **12**, 185–232.
Orowan, E. 1952. Fundamentals of brittle behaviour in metals. In *Fatigue and fracture of metals,* W. M. Murray (ed.), 139–67. New York: Wiley.

Paterson, M. S. 1978. *Experimental rock deformation: the brittle field.* Berlin: Springer.
Paterson, M. S. 1986. Personal communication.
Poirier, J.-P. 1985. *Creep of crystals.* Cambridge: Cambridge University Press.
Porter, J. R., W. Blumenthal & A. G. Evans 1981. Creep fracture in ceramic polycrystals – I. Creep cavitation effects in polycrystalline alumina. *Acta Metall.* **29**, 1899–906.

Pugh, H. Ll. D. 1971. Data from a lecture, quoted by M. Brandes, Mechanical properties of materials under hydrostatic pressure. In *Mechanical behaviour of materials under pressure*, H. Ll. D. Pugh (ed.), 236–98. London: Elsevier Applied Science.

Puls, M. P. 1981. Atomic models of single dislocations. In *Dislocation modelling of physical systems*. Proc. Int. Conf., Gainsville, Florida, USA, June 1980, M. F. Ashby, R. Bullough, C. S. Hartley & J. P. Hirth (eds), 249–68. Oxford: Pergamon.

Rice, J. R. & R. Thomson 1974. Ductile versus brittle behaviour of crystals. *Phil Mag.* **29**, 73.

Rutter, E. H. 1986. On the nomenclature of mode of failure transitions in rocks. *Tectonophysics* **122**, 381–7.

Schmid, S. M., M. S. Paterson & J. N. Boland 1980. High temperature flow and dynamic recrystallization in Carrara marble. *Tectonophysics* **65**, 245–80.

Tsai, R. L. & R. Raj 1982. Creep fracture in ceramics containing small amounts of a liquid phase. *Acta Metall.* **30**, 1043–58.

Tullis, J. 1983. Deformation of feldspars. In *Feldspar mineralogy, reviews in mineralogy*, vol 2, P. H. Ribbe (ed.), 297–324. Washington, DC: Mineralogical Society of America.

Tullis, J. & R. A. Yund 1977. Experimental deformation of dry Westerly granite. *J. Geophys. Res.* **82**, 5705–18.

Urai, J. L., C. J. Spiers, H. J. Zwart & G. S. Lister 1986. Weakening of rock salt by water during long-term creep. *Nature* **324**, 554–7.

CHAPTER SIX
Régimes of plastic deformation – processes and microstructures: an overview

David J. Barber

6.1 Introduction

The preceding chapters are broadly concerned with brittle régimes of deformation and fracture, a research field that has rightfully assumed growing importance in recent years, but that is still poorly covered in textbooks. The contributions that follow are concerned with aspects of plastic flow. To many readers, the territorial background to the plastic deformation of solids will be more familiar than the background to theories of brittle behaviour, given in the overview by Meredith (this volume). As will be apparent from the earlier contributions, deformation involving microcracks, brittle fracturing, and other cataclastic effects leads to marked discontinuities in strain. Ductile behaviour in the deformation of materials, on the other hand, is essentially characterized by smoothly varying distributions of strain. The experimental signatures of ductile and brittle régimes and the transition between the two opposed behaviours under experimental conditions are well known (see, for example, Hobbs *et al.*, 1976, p. 62). Factors bearing on the transition have already been discussed in the chapter by Murrell (this volume).

Although ductile deformation can be achieved by the action of several different mechanisms, these generally require changes of shape of the grains within the material, and so we find that intragranular slip, twinning, and diffusion-assisted processes are commonly involved. The potential for competition between brittle and ductile mechanisms is always inherent, but is strongly influenced by temperature, pressure, grain size and, at high temperatures, by strain rate. It is well known by Earth scientists that microcracking and failure by fracture are inhibited by increasing the hydrostatic pressure, promoting processes of ductile flow. Similarly, increasing the temperature or decreasing the strain rate assist the thermally activated processes such as slip, diffusion, and recrystallization. In this overview, it will be assumed that readers will be familiar with the accepted ideas concerning slip systems in crystals, the rôle of defects in the viscous flow of crystals, diffusion mechanisms, etc.

Progress in understanding the processes of plastic or ductile flow has been evident for many years, but much of the theory has been addressed to rather simple systems, materials with high crystal symmetry, metals and their alloys, etc. The ideas and results from this research, although encouraging and often pointing ways forward, are seldom directly applicable to the more complex systems and environments encountered by ceramists and Earth scientists. Moreover, theories of plasticity are generally phenomological in nature, interpreting stress–strain or strain–time relationships on the basis of assumptions about insensitivity to other factors (e.g. typically strain rate in the former case). As experimental data has been accumulated, we have become increasingly aware that the flow of solids, even when the complicating effects of cataclastic processes are absent, is responsive to several factors, including the obvious variables of temperature, T, and strain rate, $\dot{\varepsilon}$, but also pressure, P, strain, ε, grain size d and strain history, chemical environment and, additionally, the effects of possible phase transformations. The microstructural complexities of rocks and ores, coupled with the adverse environments in which they are frequently found or employed, often makes it difficult to discount effects arising from these additional parameters. Thus, not only is it problematic to devise meaningful laboratory tests, but the interpretation of the deformation microstructures of a particular geological system is consequent upon knowing what factors were pertinent. This type of problem is seldom faced by those outside the Earth sciences: modern ceramics are becoming very complex, but it is usually clear in what environments they will operate.

The contributions in the second half of this monograph indicate some of the directions in which progress is being made in studies of complex non-metallic systems. As in the first part, most authors provide a review element relevant to the topics. This material demonstrates the enthusiasm and success with which Earth scientists, especially, have applied ideas about, and engaged in methods for, the study of micromechanisms of flow in crystalline solids, particularly to minerals and essentially monomineralic rocks. Inevitably, much of the progress has been in the interpretation of the results of laboratory testing of single crystals, ceramics, mineral aggregates and rock samples, but this is already paying off in the analysis of the quite complicated natural features such as shear zones. Moreover, we now see geologists extending their work into the study of deformation effects in even more complex rock systems, and into situations in which chemical and metamorphic effects are important or rate-determining.

In this overview it is clearly not feasible to give a full review of our current state of knowledge about ductility in non-metals. However, the task is simplified by the fact that the ε, $\dot{\varepsilon}$, and T conditions frequently relevant to both ceramics and rocks decree that creep is the most important phenomenon for both. During creep a material can be said to flow 'at its own pace' (Poirier 1985). Creep tests are often carried out under constant load, but clearly this does not guarantee constant stress, even though in creep the deformation processes tend to act to maintain a constant average level of stress. Rate-dependent plasticity, or creep, is the most important form of plastic deformation for rocks, since it pertains to conditions existing in the Earth's mantle. Earth scientists are sometimes uncertain about what constitutes creep and how other processes relate to it, probably because so much of the literature about creep has been written from the viewpoint of metallurgists and engineers. To define creep precisely is not easy, since it embraces both differing test conditions and various régimes of deformation. An attempt to clarify the breadth of application of the term as used by the different groups of materials scientists, for both natural and applied materials, is given in Figure 6.1.

An understanding of creep is very important for engineering applications of ceramics, which generally aim to exploit their superior long-term strength to metals at high temperatures. However, it is worth emphasizing here that creep tests on ceramics (and metals) differ in one important aspect from those carried out on rocks and minerals. The latter almost invariably use a confining pressure to inhibit cataclasis, and thus better to represent subsurface geological conditions. This monograph has no chapter that specifically covers creep, and so this overview is largely a summary of its present-day treatment, with emphasis on results from experimentally deformed specimens.

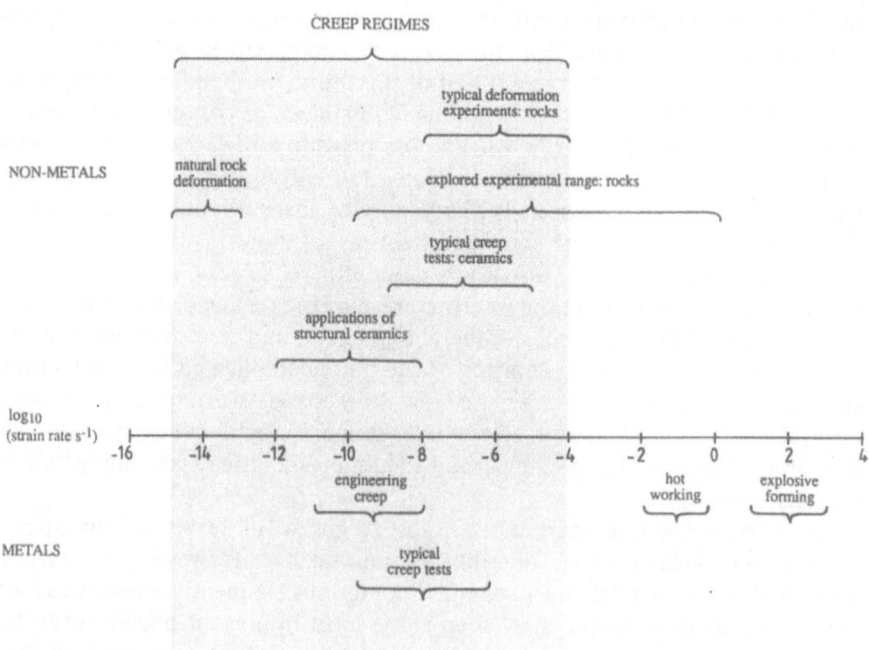

Figure 6.1 Ranges of strain rates applying to various experimental, natural, and commercial cases of plastic deformation.

6.2 The phenomenology of creep

Many deforming materials exhibit linear relationships between strain and time, over a restricted range of strain, for some particular combinations of temperature and applied stress. This régime of constant strain rate is known as *steady-state creep*, and corresponds to the region of constant gradient in the solid curve illustrated schematically in Figure 6.2. The broken curves indicate qualitatively the effects of increasing the testing temperature and increasing the applied stress.

As will become clear in the following section, creep can arise from several different micromechanisms of deformation, which are unified by the mathematical treatment of the resulting flow laws. Creep phenomena and the microprocesses responsible for them have been described and quantified in the books by Gittus (1975), Frost & Ashby (1982), and Poirier (1985), the latter written with the geologist and geophysicist particularly in mind. A more general view of plastic deformation mechanisms in rocks is provided by Nicolas & Poirier (1976), who also provide details of the mid-1970s state of knowledge about deformation in some individual minerals that are major rock constituents. Davidge (1980) provides an introduction to the deformation behaviour of ceramic materials. Other useful reviews concerning creep, and the microprocesses contributing to it, are those of Weertman (1970), Stocker & Ashby (1973), Murrell & Chakravarty (1973), Ashby & Verrall (1978), Frost & Ashby

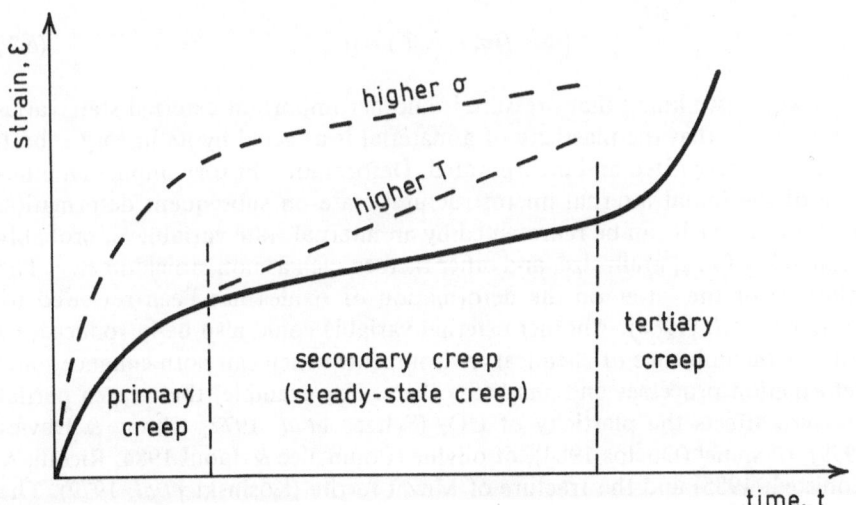

Figure 6.2 Typical form of a strain versus time creep curve (solid line) for a material at a sufficiently high temperature ($T \gtrsim 0.5T_m$) for there to be no net hardening, showing the transient (primary) creep stage, the steady-state-creep régime, and the beginning of accelerating (tertiary) creep. The broken curves show the effects of increased stress and increased temperature.

(1982), and Langdon (1985). To understand the microprocesses, it is important to have an appreciation of the microstructures of materials. Hobbs *et al.* (1976) and Wenk (1976) contain good introductions to microstructures in rock-forming minerals. Later sections of this overview (Sections 6.3.2.2 and 6.6) discuss microstructures further.

Creep tests conducted at constant applied load may produce behaviour in which an approximately constant strain rate, $\dot{\varepsilon}$, occurs. Restricting our considerations to these régimes of constant $\dot{\varepsilon}$ implies creep in which there is no *net* strain hardening or work hardening; this is usually achieved through the countervailing effects of recovery or recrystallization processes. Such behaviour is analogous to the viscous flow of a fluid and, for crystalline solids, applies to creep occurring at moderate to high temperatures. This type of behaviour goes under the broad heading of 'power-law creep', for which $\dot{\varepsilon} \propto \sigma^n$ and, speaking very approximately, this corresponds to temperatures above $\sim 0.3 T_m$ for pure metals, and above $\sim 0.4 T_m$ for ceramics (which may be thought to include rocks!). However, as subsequent discussion will indicate, perhaps rather higher temperatures are needed for true power-law behaviour, and the ranges of temperature over which various laws operate vary considerably with the nature of the materials. The temperature must be such that diffusive mechanisms can overcome any tendency for stress concentrations to build up (these are manifest as strain hardening). Thus we are dealing with the balancing of microprocesses of strain hardening and recovery, both of which are time-dependent processes. Hart (1970, 1976) has argued that a constitutive equation of state for a flowing solid is

$$f(\sigma, \varepsilon, \dot{\varepsilon}, T) = 0 \qquad (6.1)$$

However, we know that pressure is another important external state variable, and also that the plasticity of a material is affected by its history – both need to be recognized and incorporated. Deformation history implies an influence of the initial internal microstructural state on subsequent deformation behaviour, which can be represented by an internal state variable Ξ, probably a function of ε, $\dot{\varepsilon}$, grain size, and other factors such as non-stoichiometry. The influence of the latter on the deformation of oxides has been reviewed by Castaing *et al.* (1984). Another external variable could also be introduced, to indicate the influence of chemical environment, which can both enhance some deformation processes and generate others. For example, the oxygen partial pressure affects the plasticity of UO_2 (Seltzer *et al.* 1972, Chung & Davies 1979), of spinel (Duclos 1984), of olivine (Poumellec & Jaoul 1984, Ricoult & Kohlstedt 1985) and the fracture of MnZn ferrite (Kosinski *et al.* 1979). The effects of hydroxyl ions on the strength of quartz (e.g. Griggs & Blacic 1965, Griggs 1967, Hobbs *et al.* 1972) and amethyst (Kekulawala *et al.* 1978) are well known, while water-induced effects have also been found for albite, $NaAlSi_3O_8$ (Tullis *et al.* 1979) and for diopside, $CaMgSi_2O_6$ (Boland & Tullis

1986). The possible scope and interpretation of such chemical effects have been considered by Hobbs (1981) and others, and are being pursued (Ord & Hobbs 1986), but as yet are not well developed. For simplicity, we now exclude them and, following Hart (1976), on the basis of equation (6.1) we take as a suitable, fairly general, empirical constitutive equation

$$f(\sigma, \dot{\varepsilon}, \Xi[\varepsilon, \dot{\varepsilon}], P, T) = 0 \qquad (6.2)$$

so that steady-state creep can be represented by an equation of form

$$\dot{\varepsilon} = \dot{\varepsilon}(\sigma, \Xi, P, T) \qquad (6.3)$$

These variables are likely to have wide ranges of values, so that it is common to normalize them in some fashion, thereby making it possible to correlate data from a group of materials (e.g. oxide ceramics) in terms of a single curve (e.g. see Ashby & Brown 1981 for more details). The normalization of the temperature of a substance to its melting point, T_m, in degrees Kelvin, giving the homologous temperature, is one example that will be familiar to most readers, and is used in this overview and elsewhere in this monograph.

The variation of creep rate with σ and T within the quasi-steady-state régime is often employed as the equation of state for creep. Parameters used to express the variation are examined in terms of a thermodynamic treatment of deformation, which allows comparison to be made between measured parameters and the values anticipated if particular physical processes predominantly control the creep deformation. Evans & Wilshire (1986) have expressed doubts that creep ever really conforms to true steady state, and have suggested that quoted constant strain rates in constant stress tests probably only represent minima. They have based a theoretical treatment of creep on the idea that there is a balance struck between decaying primary and accelerating tertiary stages.

It is usually found that creep rate increases exponentially with T, so that it is convenient to make plots of the logarithmic values of $\dot{\varepsilon}$ against $1/T$, assuming an Arrhenius-type relationship of the form

$$\dot{\varepsilon} \propto \exp(-Q/RT) \qquad (6.4)$$

where Q is an apparent activation energy which can be associated with the rate-controlling process, and is proportional to the gradient of the plotted curve. R is the universal gas constant.

How the creep rate will depend on the applied stress is less clear, because the predominant micromechanisms are likely to alter as the stress changes, and very different mechanisms can be involved. This innate difficulty is frequently overcome by a log–log plot of $\dot{\varepsilon}$ against σ, a device that will often yield a straight-line relationship for a restricted range of stresses, thereby implying that there is a power law relating creep rate to stress:

$$\dot{\varepsilon} \propto \sigma^n \tag{6.5}$$

where n is a constant called the stress exponent. In practice n tends to increase from a value of unity at very low stresses to higher values at very high stresses.

The influence of pressure on creep rate is complex, since it varies with the processes that facilitate the deformation. Pressure effects are therefore not discussed here, and readers should refer to Sherby *et al.* (1970), Brown & Ashby (1980a), and Sammis *et al.* (1981) for an introduction.

It is clearly possible to combine the relationships (6.4) and (6.5) to produce an empirical equation of state:

$$\dot{\varepsilon} = \dot{\varepsilon}_0 \sigma^n \exp(-Q/RT) \tag{6.6}$$

In this equation $\dot{\varepsilon}_0$ is not entirely temperature independent, since it involves the rigidity modulus, G. The apparent activation energy, Q, can be shown to be equal to the activation enthalpy, ΔH, and so from (6.6) we have

$$\Delta H = -R \left. \frac{\partial \ln(\dot{\varepsilon}/\dot{\varepsilon}_0)}{\partial (1/T)} \right]_\sigma \tag{6.7}$$

Since R is the gas constant, ΔH is consequently the enthalpy per mole. In practice, ΔH can be determined during a creep test by making 'jumps' between two temperatures, T_1 and T_2, and measuring the corresponding quasi-steady-state strain rates $\dot{\varepsilon}_1$ and $\dot{\varepsilon}_2$, whereupon, from (6.7),

$$\Delta H = R \frac{\ln(\dot{\varepsilon}_2/\dot{\varepsilon}_1)}{1/T_2 - 1/T_1} \tag{6.8}$$

Values of ΔH and Q determined by such means for the high-temperature creep of simple solids (especially metals) are found to be close to the activation energy for self diffusion (see Section 6.3.3).

Heard (1985) gives a concise review of modern experimental methods for determining mechanical properties, and the chapter by Tullis (this volume) illustrates the effectiveness of such methods in studies of quartzo-feldspathic rocks. Poirier, Sotin & Beauchesne (this volume) present a new method whereby rheological laws for the high-temperature creep of minerals and rocks can be reliably derived from experimental data. Instead of separately fitting straight lines to Arrhenius plots of strain rate versus stress, the best overall fitting surface is applied to the experimental data by means of a global non-linear inversion procedure. Poirier *et al.* discuss the advantages of this new method over the traditional approach, which include the facilitating of objective comparisons between several possible laws. An example of the practical application of the method is given for the creep of single crystals of barium titanate. These have the perovskite structure, and studies of their rheological properties are relevant to the flow behaviour of the deep mantle.

In this overview, only passing mention has so far been made of the dependence of the creep rate on previous deformation history, embodied in the microstructural parameter Ξ of equations (6.2) and (6.3). The parameter should be taken to represent both intrinsic physical properties (e.g. elastic constants) and materials variables such as microstructural state (grain size and phase distribution). This is a murky area about which insufficient is known. Microstructural studies show that a creep specimen usually develops a characteristic microstructure, which changes continuously and yet retains its essential properties (see also Section 6.3.2.2). (More pronounced examples of such behaviour are to be found when materials deform superplastically: see Section 6.3.4.2.) Thus, starting with a set of test specimens that have comparable microstructures leads to similar quasi-steady-state microstructures. However, it has been noted that, even with single-phase materials, the dislocation microstructures that develop during creep are markedly inhomogeneous (Nix & Ilschner 1980). Thus the quasi-steady state apparently does not indicate a uniform local balance between hardening and softening processes, but instead the presence of regions that are hard and regions that are soft, with continual interchanges in these states (e.g. Ecob & Bilde-Sørensen 1987). These findings imply that even initially quite different microstructures in single-phase materials can result in comparable creep microstructures, but this may not apply for all creep régimes. Some types of microstructure are found to be very resistant to removal (by annealing, etc.), and can at least be expected to have marked effects on the early course of deformation, which may possibly persist beyond the transient (primary) stage of creep (indicated in Fig. 6.2).

6.3 Creep mechanisms

As has already been mentioned, the quasi-steady-state creep of a polycrystalline aggregate does not necessarily imply a single mechanism of deformation. Two or more may act simultaneously, but they can be modelled individually, and may sometimes act independently of each other. The possible mechanisms are often classified in terms of intragranular (or lattice) processes, and intergranular (or boundary) processes, and treated separately (e.g. see Langdon 1985). For simplicity and brevity, this practice is not adopted here, and only a summary is given, although mechanisms not involving diffusion are included.

6.3.1 Creep solely by dislocation glide

The glide of dislocations almost inevitably gives rise to interactions with obstacles in the slip planes (pinned dislocations, precipitates, grain boundaries,

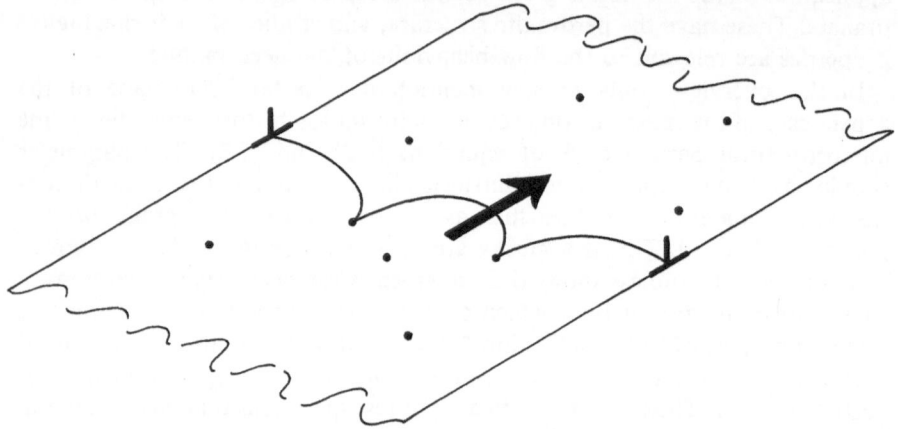

Figure 6.3 Glide of dislocations impeded and controlled by discrete obstacles in the slip plane.

etc.), as indicated in Figure 6.3. The strain rate is then limited by the rate at which such obstacles can be overcome (e.g. by bowing out and pinching-off loops around the obstacles or by the thermally activated cross slip of dislocations). The absence of climb to assist in freeing dislocations from obstacles implies creep at low strain rates and of limited extent, although able to give behaviour resembling power-law creep. Weertman (1957) has examined the more universal case in which glide is controlled by the Peierls stress. He proposed that this be overcome by the motion of double kinks, and showed that this mechanism gives a stress exponent, n, of 2.5. In general, however, stress exponents arising from the fitting of low-temperature creep to a power law have no physical meaning, as pointed out by Poirier (1985). Glide-controlled creep may be important in ice (Weertman 1983), in covalently bonded solids, and possibly in the low-temperature creep of halite.

6.3.2 Dislocation (recovery-controlled) creep

6.3.2.1 GENERAL

Above $\sim 0.4T_m$, gliding dislocations in metals are still arrested by any physical obstacles in the slip planes, but these may be overcome by climb stimulated by the enhanced lattice diffusion associated with the higher temperatures. In ceramics and minerals, the temperatures at which climb may become effective can differ considerably from those predicted by the above-mentioned metals yardstick since, in a given compound, climb will be controlled by the slowest diffusing atomic species. There is more than one way in which climb can assist a dislocation to circumvent an obstacle, but a simple method is that proposed by Weertman (1968, 1972), illustrated schematically in Figure 6.4. In this model, segments of edge dislocation of opposite sign, trapped in close proximity by

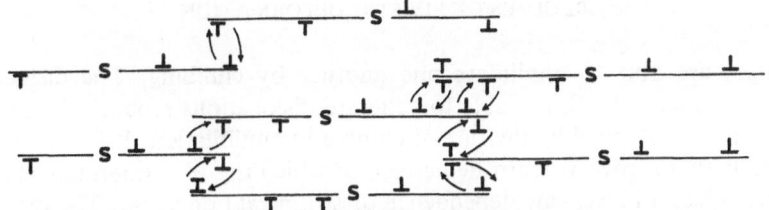

Figure 6.4 The generation of dislocations from sources (S) in the glide planes and the climb of the dislocations, as proposed by Weertman (1968) for power-law creep.

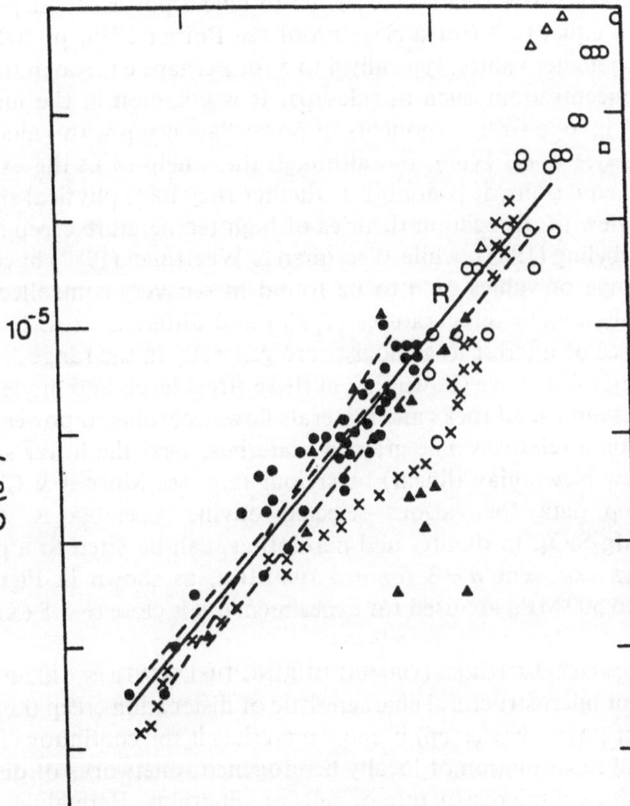

Figure 6.5 Stress – strain rate diagram of selected experimental test data for olivine. The unbroken line gives the fit to the data provided by power-law creep with a stress exponent of 3 (from Ranalli 1982).

pile-ups, are able to annihilate one another by climbing. The dislocation sources, denoted by S, are able to generate dislocations repetitively, and the strain rate is governed by the rate of climb and annihilation. It is found fairly generally that recovery-controlled creep, of which the Weertman model is one variety, gives a power-law dependence of strain rate on stress. The predicted stress exponent, n, is 4.5. The descriptive and specific terminology 'recovery-controlled dislocation creep' is usually shortened to the more ambiguous form 'dislocation creep'. Several authors have shown that any such recovery-controlled creep process can be expected to give a power-law dependence on stress with n equal to 3 (for a clear proof see Poirier 1985, p. 109), but that n may rise to higher values, typically 4 to 5, or perhaps 6 through the existence of stress concentrations such as pile-ups. It is common in the metallurgical literature to invoke stress exponents in power-law creep with values up to 10 (e.g. Mukherjee et al. 1969), but although these help to fit the experimental data to rate equations, it is doubtful whether they have physical significance. A recent review of dislocation theories of high-temperature creep is provided by Nix & Gibeling (1985), while Weertman & Weertman (1987) have discussed the wide range of values of n to be found in recovery-controlled creep for different homologous temperatures (T/T_m) and different stress levels.

The stresses of interest to geologists are generally in the range 1–100 MPa, and laboratory tests have shown that at these stress levels and moderately high temperatures almost all rocks and minerals flow according to power-law creep. However, some relatively fine-grained materials, near the lower stress limit, seem to show Newtonian (linear) behaviour (e.g. see Murrell & Chakravarty 1973). Creep data for various selected olivine assemblages, from pure forsterite, Mg_2SiO_4, to dunites and peridotites, can be fitted to a power law, with a stress exponent $n = 3$ for $\sigma \leqslant 100$ MPa, as shown in Figure 6.5. If stresses up to 500 MPa are used for experiments, n is close to 3.5 (Kirby 1983).

6.3.2.2 MICROSTRUCTURES FORMED DURING DISLOCATION CREEP

An important microstructural characteristic of dislocation creep (i.e. recovery-controlled or power-law creep) in most materials is the continuous formation, dispersal and re-formation of locally heterogeneous networks of dislocations, which constitute a microstructure of cells or subgrains. Here glide, climb and partial annihilation of dislocations all contribute to generate the cell or sub-grain structure, but climb is the rate-controlling process. Such microstructures, consisting of cells separated by tangles of dislocations, or more well defined arrays or nets of dislocations forming subgrain boundaries, have been reported in a variety of metal samples from creep tests and have been widely illustrated in the metallurgical literature. Corresponding microstructures in non-metals are less well documented and so, for reference, dense tangles of dislocations defining evolving cells in experimentally deformed calcite are illustrated in Figure 6.6a, while more regular arrays of dislocations constituting subgrain boundaries in a naturally deformed dolomite are shown in Figure

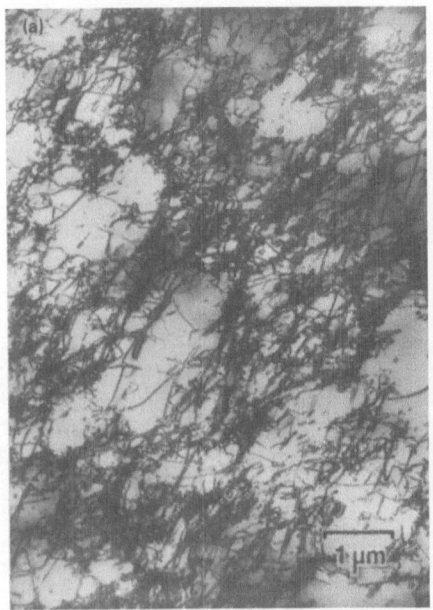

Figure 6.6 (a) Microstructure of cells separated by tangles of dislocations in calcite experimentally deformed at 500°C and 10^{-4} s^{-1}, under 1 GPa confining pressure (1 MV electron micrograph from Barber & Wenk, unpublished). (b) Microstructure of small subgrains separated by arrays and nets of dislocations in a naturally deformed dolomite (1 MV electron micrograph from Barber 1977, Tectonophys. **39**, 193–213).

6.6b. The creation of a well defined deformation microstructure with many of the dislocations in subgrain boundaries has been called creep polygonization (Poirier 1985), as being analogous to the polygonization generated by the static recovery of a deformed material. Such a substructure is indicated diagrammatically in Figure 6.7; although it is not static on a subgrain scale it does not undergo an overall evolution with time, i.e. there is dynamic equilibrium. As noted in Section 6.2, the concept of steady-state creep is only an approximation, but the assumption of a steady state is useful because it admits the simplifying idea of a substructure in dynamic equilibrium. The substructure can then be characterized in terms of parameters for dislocation density, ρ, and subgrain size d'.

The average size of the cells and subgrains in creep microstructures is independent of temperature, but their nature is affected. Most is known about metals, where it is observed that an increase in test temperature and an increase in the stacking fault energy of the material depletes the numbers of dislocations within the cells, while the cell walls are decreased in width as more regular nets and arrays of dislocations form (i.e. 'true' subgrain boundaries, as discussed in many dislocation textbooks). The two microstructural parameters, ρ, and d', are commonly found to have the following relation-

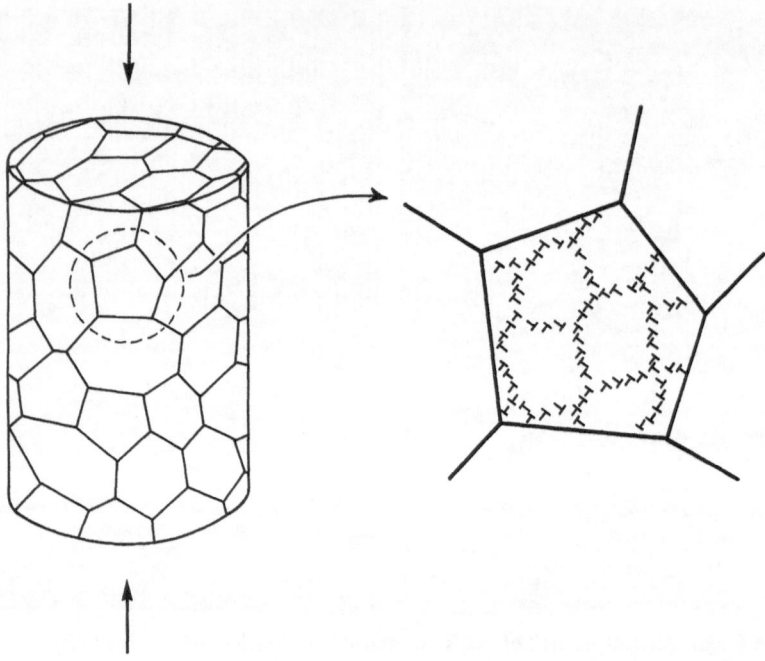

Figure 6.7 Schematic illustration of power-law creep in a polycrystalline aggregate resulting from the climb of dislocations to form a microstructure of subgrains (after Frost & Ashby 1982).

ships to stress, σ:

$$\rho = C(\sigma/Gb)^2 \qquad d' = K_m b(G/\sigma)^m \qquad (6.9)$$

where b is the modulus of the Burgers vectors of the dislocations, $(0.7 \lesssim m \lesssim 1)$, and both C and K_m are constants. K_m is a constant of about 10 for metals, and is much larger (25–80) for ionic crystals and oxides (Takeuchi & Argon 1976). The chapter by Derby (this volume) makes use of these relationships in discussing the effect of dynamic recrystallization on grain size. For some representative creep-related microstructures, see Caillard & Martin (1982), Hobbs (1968), White (1973), and Durham *et al.* (1977).

6.3.2.3 DYNAMIC RECRYSTALLIZATION DURING DISLOCATION CREEP
Dynamic recrystallization may accompany dislocation creep, thus providing softer grains than those in which there is simply a balance between glide and climb, and thereby facilitating deformation. Dynamic recrystallization (sometimes called syntectonic recrystallization by Earth scientists) appears to be particularly important in the deformation of non-metals. (It should be distinguished from *static* recrystallization, which is the migration of grain boundaries and the creation of new grains in the absence of external stress.

150

This is a well known phenomenon described in various books, and is not relevant to creep).

Dynamic recrystallization can occur either by strain-induced grain-boundary migration or by progressive increase in subgrain misorientation (Sellars 1978, Poirier 1985).

Strain-induced grain-boundary migration (SIGB) can be recognized by the appearance of local bulges on high-angle boundaries. First reported by Bailey & Hirsch (1962) in metals, it has since been found in non-metals, and the types of boundaries which may bulge or migrate have been widened to include kink-bands, twins, and deformation bands (e.g. see Etheridge & Hobbs 1974 (mica), McClay & Atkinson 1977 (galena)). Tullis & Yund (1985) report that the process plays a critical rôle in the lower-temperature portion of the dislocation-creep régime. Dynamic recrystallization by SIGB generally does not occur in metals that have rapid rates of dynamic recovery (Roberts 1986), although pure iron appears to be an exception (Glover & Sellars 1973). For many metals it is thus excluded from their creep régimes and occurs principally during hot working. This is not true for silicates, or probably for the majority of non-metals. Pervasive waves of extensive grain growth and dynamic recrystallization by SIGB, as can occur in the creep of NaCl (Guillopé & Poirier 1979) and in some metal alloys at high temperatures, can repeatedly return the deformation to primary creep, thus removing any semblance of steady-state behaviour. It also gives anomalously high strain rates and, according to Frost & Ashby (1982), confuses the high-stress – high-temperature regions of deformation-mechanism maps. Derby (this volume) investigates the link between applied stress and grain size when dynamic recrystallization by SIGB occurs, and suggests that there may be a universal relationship.

Dynamic recrystallization by means of progressive increase of subgrain rotation, without SIGB (i.e. just by increasing the misorientations across subgrain boundaries) has been documented for many non-metals, e.g. olivine (Poirier & Nicolas 1975) and calcite (Schmid *et al*. 1980). In this process, subgrain boundaries that have formed by dynamic recovery progressively absorb more dislocations, so that the orientation relationships between parent grain and daughter subgrains become less and less evident with increasing strain. Subgrains do not form in all materials, e.g. they are absent when creep is controlled by solute drag (see Nix & Ilschner 1980), and thus this mechanism of dynamic recrystallization is denied to such materials. However, it appears to be an important aspect of dislocation creep in many rock-forming minerals, and a good discussion of its rôle has been provided by Urai *et al*. (1986).

The chapter by Derby (this volume) starts from the observation that dynamic recrystallization by means of SIGB leads to a steady-state mean grain size during the deformation of laboratory samples of metals and alloys. For example, in deformation at constant strain rate, the grain size is associated with a steady-state stress. This is caused by cycles of nucleation and recrystallization being initiated at different times at different sites within a deforming aggregate, so that the mean properties tend to constant values. Derby examines whether this interrelationship between grain size and stress can be used as a predictor of stress.

The arguments presented by Derby centre around the accepted idea of Bailey & Hirsch (1962) that the basic mechanism for dynamic recrystallization is essentially the same as that for static recrystallization, and that the small strain-free volumes of recrystallized material that are created when grain boundaries bow out under stress are the nuclei for grain growth. Why this process should lead to a steady-state relationship between stress and grain size is contained in a model for the mechanism proposed by Derby & Ashby (1987). This model suggests that the steady-state grain size is the product of a balance between the grain-boundary migration rate and the nucleation rate for new recrystallization events. It predicts a simple relationship between grain size and the stress exponent of the creep law prior to the onset of recrystallization. A new analysis of data from tests is shown to support a relationship between mean grain size and stress and to be consistent with the model of recrystallizing microstructure. Normalization of the grain size in terms of Burgers vector and normalization of average stress in terms of temperature-compensated shear modulus further implies a relationship that is universal for many materials with different crystal structures. It is suggested that the relationship can be used as an indicator of process stress from measurements of mineral grain size.

6.3.2.4 POWER-LAW BREAKDOWN
Power-law breakdown can occur at high stresses, even well within the temperature régimes where diffusion is easy. This is generally attributed to a return to glide-controlled flow, as shown in Figure 6.8. Various equations have been proposed to fit the $\dot{\varepsilon}-\sigma$ dependences observed experimentally, but are not detailed here (see, for example, Brown & Ashby 1980b, Frost & Ashby 1982).

When discussing the microstructures developed during dislocation creep (Section 6.3.2.2), a subtle distinction was drawn between cells and subgrains on the basis of the dislocation content and orderliness in the boundaries. This may not always be important, but it has been pointed out by Pharr (1981) that the more regular arrays and nets that constitute subgrain boundaries are less resistant to destruction by gliding dislocations than are dislocation tangles. Pharr has inferred that a subgrain microstructure more readily leads to power-law breakdown than one of irregular cells.

6.3.2.5 HARPER–DORN CREEP
A special case observed for some materials with a larger grain size is that deformation at low, constant stress conforms to a linear (Newtonian) law between $\dot{\varepsilon}$ and σ (i.e. $n = 1$). Macroscopically it is therefore similar to Nabarro–Herring creep (see Section 6.3.3), but it differs microscopically, and is therefore known as Harper–Dorn creep (Harper & Dorn 1957). The specimens exhibit a low, constant dislocation density (e.g. $\sim 10^3 \, \text{cm}^{-2}$ in

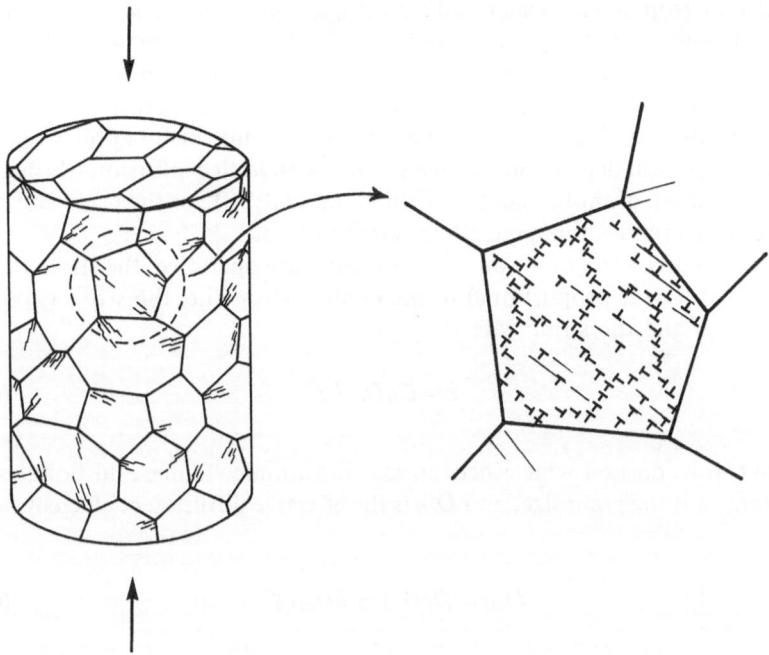

Figure 6.8 Schematic illustration of the breakdown of power-law creep, caused by the competing effects of glide upon the climb/subgrain mechanism (after Frost & Ashby 1982).

Al–5%Mg; Yavari *et al.* 1982). The observed microstructures have led Langdon & Yavari (1982) and others to conclude that Harper–Dorn creep is a special case of creep controlled by the climb of edge dislocations. Lee & Ardell (1986) have attributed Harper–Dorn creep entirely to the coarsening of dislocation networks. The mechanisms for creating and sustaining the dislocations are still unclear. Although this form of creep is mainly known for metallic systems (e.g. Al alloys), it has been reported in CaO (Coath & Wilshire 1977, Dixon-Stubbs & Wilshire 1982) and in the perovskite-structured fluoride, KZnF$_3$ (Poirier *et al.* 1983).

6.3.3 Diffusion creep

Creep at high temperatures and low stresses enables the processes of diffusion and grain-boundary sliding (GBS) to predominate over all others. Deviatoric (i.e. non-hydrostatic) stress fields at grain surfaces produce gradients in chemical potential and, as first suggest by Nabarro (1948), these gradients produce diffusive fluxes of vacancies and atoms which can accommodate changes in grain shapes. Provided that it is also possible for sliding to occur on grain surfaces, macroscopic straining of an aggregate is facilitated without the opening-up of intergranular cracks and the loss of the aggregate's

153

integrity. Creep in which such diffusive migrations govern the deformation of an aggregate or polycrystal is well described by the expression 'creep by diffusional flow', but this is usually abbreviated to 'diffusion creep'.

The flow of a compressed polycrystal by diffusion creep is illustrated schematically in Figure 6.9. The precise régime that applies at high temperatures will depend on the balance between lattice diffusion (through the grain interiors), grain-boundary diffusion, and GBS. Thus the grain size of the material determines the controlling microprocess.

If both lattice and grain-boundary diffusion are operative, the strain rate for diffusional flow is proportional to the applied stress (i.e. the stress exponent, n, is unity) and the strain rate is

$$\dot{\varepsilon} = C\sigma D_{\text{eff}}/d^2 \qquad (6.10)$$

where C is a constant which incorporates the atomic volume and Boltzmann's constant, d is the grain size, and D_{eff} is the effective diffusion coefficient, which is

$$D_{\text{eff}} = D_v(1 + \pi \, \delta D_b/dD_v) \qquad (6.11)$$

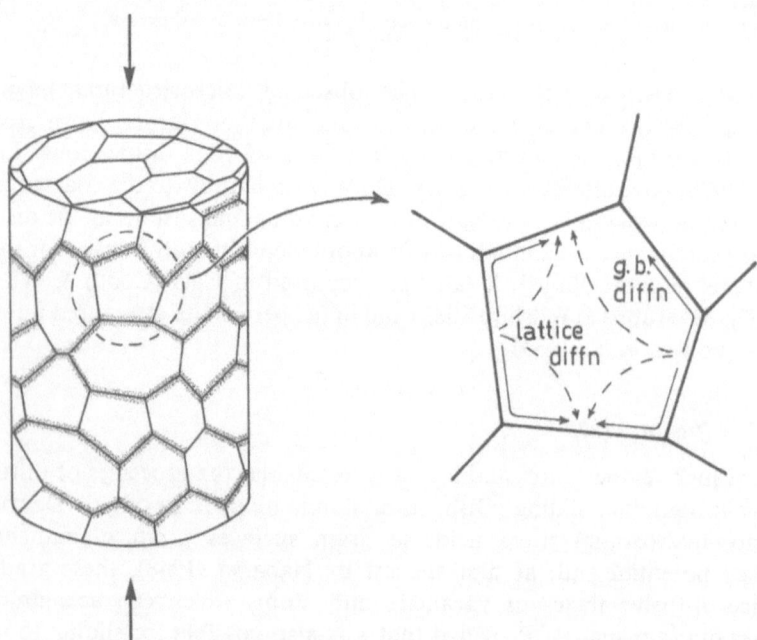

Figure 6.9 Schematic illustration of creep of a polycrystalline aggregate by diffusional flow (after Frost & Ashby 1982).

where D_v and D_b are the volume (lattice) and boundary-diffusion coefficients respectively, and δ is the effective thickness of the grain boundaries.

At the highest temperatures, lattice diffusion is rate controlling and thus the strain rate scales simply, as in (6.10), as D_v/d^2. This régime is known as Nabarro–Herring creep, in recognition of Herring's refinement (1950) of Nabarro's (1948) theory. At lower temperatures, grain-boundary diffusion predominates over lattice diffusion and the strain rate scales as D_b/d^3, giving a régime called Coble creep (after Coble 1963). The constitutive equations for both forms of diffusional creep are derived assuming that grain boundaries are perfect sources and sinks for vacancies. Knowledge gained recently about grain-boundary structure shows that some forms of grain boundaries have high degrees of lattice coincidence (e.g. see the work of Bollmann 1970), while others contain arrays of grain-boundary dislocations (Balluffi et al. 1983). The latter are most likely to act as the sources and sinks of vacancies in grain boundaries, and thus not all boundaries will be equally effective in diffusive creep processes (see also Section 6.5).

The quantity δ in (6.11) is not the average structural thickness of the grain boundaries but the width of the associated regions of high diffusivity. This could be larger than the structural thickness, especially in rocks and ceramics (see Section 6.5). In one oxide, NiO, there is nevertheless good agreement between the quantity δ and the structural widths of grain boundaries (see Atkinson & Taylor 1981, Ricoult & Kohlstedt 1983). However, even the structural thicknesses of grain boundaries in simple polycrystalline metals could not be directly measured until a few years ago, which is why there is even now relatively little information on the structure of grain boundaries and their properties (however, see Chadwick & Smith 1976, Balluffi 1980, Peterson 1983, Yan & Heuer 1983; also see Section 6.5).

It is apparent from (6.10) and (6.11) that the measurement of strain rates for aggregates with well defined grain sizes within the Coble and Nabarro–Herring régimes could yield measures of the grain-boundary diffusivity and lattice diffusivity, respectively. Such approaches have been applied to several ceramic systems, e.g. aluminium oxide (Cannon & Coble 1975) and manganese–zinc ferrite (Nishikawa & Okamoto 1980). However, there are often discrepancies between the diffusivities derived from creep tests on ceramics and from more direct methods, and these may have one of several origins. Similar problems have arisen with minerals, as illustrated by forsterite and olivine. The activation energies for diffusion derived from creep (e.g. Kohlstedt & Ricoult 1984) disagree with values derived for oxygen from diffusion experiments (Jaoul et al. 1980), leading to questions about what is the rate-controlling process. Even more serious problems seem to exist with the data for silicon diffusion in olivine (Condit 1985).

Both diffusion coefficients and microstructural parameters have ranges of values in polycrystals. As yet, few models take account of this fact. This implies that there can be local variations in the predominant microprocess. Raj

& Ghosh (1981) have used a model with distributed parameters to show that they can lead to a transition from dislocation creep to diffusion creep that is spread over several orders of magnitude in strain rate.

6.3.4 Other grain boundary processes

6.3.4.1 GRAIN-BOUNDARY SLIDING

Grain-boundary sliding is seen to be a natural partner to the grain-shape changes occurring in diffusional creep, but there is evidence that it can also occur at low stresses in power-law creep, especially when the grain size is very small. The stress exponent for GBS alone appears to be > 1 (Langdon & Vastava 1982), but not enough is yet known about GBS and its relevance under various conditions and for different materials. However, GBS cannot be considered as a self-sufficient mechanism for generating macroscopic strains since an aggregate deforming only by GBS soon begins to lose integrity because of the effects of the displacements at the grain triple junctions. Therefore GBS must be accompanied by processes for accommodating the excess material and filling the voids that are otherwise generated.

Tvergaard (1988) has studied the influence of GBS on creep and creep rupture by model analysis, finding that some models predict that GBS should have little effect on creep at high stresses, whereas it should accelerate creep at lower stresses. Ruano & Sherby (1988) have suggested that diffusion creep is not always as dominant at high temperatures as the literature indicates, and that GBS or Harper–Dorn creep are more likely for low stresses in fine-grained materials. Clearly, a material's microstructure and, especially, the nature of its grain boundaries will be crucial. It is found that many ceramics undergo strong GBS at high temperatures because intergranular glassy films form during their sintering. An example is silicon nitride, Si_3N_4 (Soma et al. 1984), in which glassy phases form by oxidation when compacts are being sintered.

6.3.4.2 SUPERPLASTICITY

Superplasticity is the stable flow of a fine-grained material under tension, to very large strains, without necking. In metals (to which the term only applies, sensu strictu, as above), superplasticity usually occurs at low stresses and moderate-to-high homologous temperatures (Padmanabhan & Davies 1980). Although superplasticity is rather loosely defined and indeed does not imply a unique mechanism, it is not really an appropriate term to apply to rocks and ceramics, which are very seldom tested for ductility in tension. If they were, none would approach strains of ~5000 per cent as occurs with several metals (e.g. Ahmed & Langdon 1977). However, the term 'superplasticity' is found in the literature of non-metals with the even looser sense of meaning the flow of polycrystalline materials to moderate strains at low stress levels.

Superplasticity can arise in at least two ways; (a) by means of phase transformations, and (b) through the strong involvement of GBS processes.

Behaviour of type (a) is called transformational superplasticity and will not be described here in detail. Vaughan *et al.* (1984), Poirier (1985), and Green (1986) provide introductions appropriate to the Earth sciences. The process may have relevance to tectonically active regions of the Earth's crust and mantle, where cycling of materials through different P and T conditions may occur (e.g. see Sammis & Dein 1974). In ceramics, phase transformations are mainly used to promote high strength rather than softening, the prime example being the striking improvements possible with partially stabilized zirconias such as MgO–PSZ, all such developments stemming from the paper by Garvie *et al.* (1975).

The second form of superplasticity, (b), is usually called 'structural superplasticity' or 'micrograin superplasticity', since it is dependent on a very

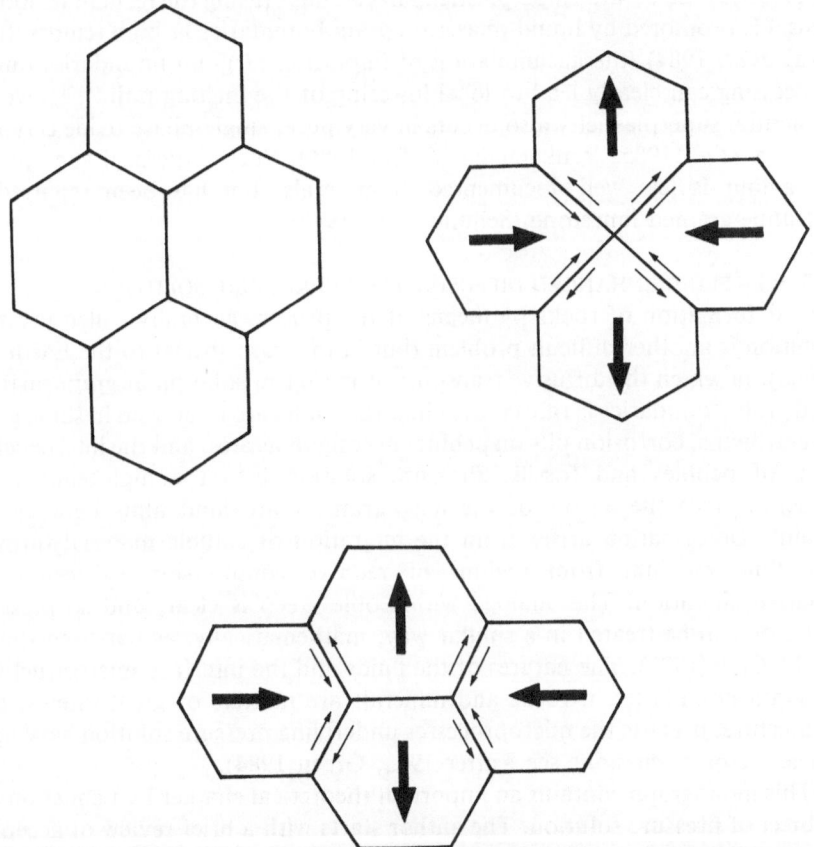

Figure 6.10 Adjacent-grain switching process in a superplastic régime. The aggregate undergoes a strain of more than 50 per cent without overall deformation of the grains. There is some local deformation at the intermediate stage to accommodate the grain-boundary sliding (after Ashby & Verrall 1973).

fine grain size. When this occurs, it is observed that grains remain equi-axed, even after large strains, which is only possible if GBS occurs. Ashby & Verrall (1973) proposed GBS in a new 'grain-switching' guise (see Fig. 6.10), in an attempt to model this aspect of superplasticity, incorporating diffusive mechanisms for grain-shape accommodation. Although the model produces a constitutive equation that can give $\dot{\varepsilon} \propto \sigma^2$, as observed for some materials, it has serious limitations, as do most of the microscopic models (see Gifkins & Langdon 1978 for a review and criticisms). The literature on this subject reveals a range of behaviour, with dynamic recrystallization playing a role in some cases (as modelled by Ghosh & Ghandi 1986), while in others some larger grains are found to elongate considerably. On one point there is fairly general agreement, however, and that is that GBS typically makes more than a 50 per cent contribution to the total strain (e.g. Chokshi & Langdon 1985, Kashyap et al. 1985). Structural superplasticity in ceramics (using the term in its loosest sense) is promoted by liquid phases on grain boundaries at high temperatures (Raj et al. 1984) (the accumulation of impurities at grain boundaries during processing can clearly lead to local lowering of the melting point). However, structural superplasticity also occurs in very pure, single-phase oxide ceramics (Panda et al. 1985, Venkatachari & Raj 1986). Such structural superplastic behaviour is less well documented in minerals, but has been reported in ultrafine-grained limestone (Schmid et al. 1977).

6.3.4.3 FLUID-ENHANCED DIFFUSIVE FLOW: PRESSURE SOLUTION

The deformation of rocks by means of the process generally called pressure solution is another difficult problem (but in this case special to the Earth sciences), in which the diffusive transport of matter in solution at grain surfaces and grain boundaries is rate controlling. Its results are to be seen in some grain overgrowths, corrosion pits on pebbles in conglomerates, and the interpenetration of pebbles and fossils. Pressure solution is not a high-temperature process, since the agents of the mass transfer are fluid films between the grains. Deformation arises from the migration of soluble material through the fluid medium from regions of relative compression to regions of relative dilatation. The analogy with Coble creep is clear, and so pressure solution can be treated in a similar way, mathematically, as has been shown by McClay (1977). The natures of the fluids and the interface microstructures at grain boundaries in rocks and minerals are matters of great interest and conjecture, because the microprocesses underlying pressure solution have been unclear (for discussions see Rutter 1983, Green 1984).

This monograph contains an important theoretical chapter by Lehner on the subject of pressure solution. The author starts with a brief review of accepted facts and previous relevant work, and then proceeds to postulate models for the micromechanisms involved, from which he is ultimately able to derive constitutive relations, in the form of transport equations and rate equations for the deformation of an aggregate by means of pressure solution.

The development of macroscale constitutive theories for the behaviour of materials during deformation has tended to lag behind progress in documenting and understanding microscale mechanisms. This is clearly seen in the case of geological materials and is especially true when solid deformation and liquid-phase transport are coupled with phase changes or chemical reactions. Lehner selects the special problem of pressure solution in order to expound and discuss a general and systematic approach to the development of constitutive theories. The aim of Lehner's analysis is to establish a framework for theoretical studies of large deformations in porous rocks which exhibit stress-enhanced solubility in their aqueous pore fluid. Few scientists encountering his approach for the first time will find it easy to assimilate, but the chapter is an outstanding contribution to the difficult task of understanding pressure-solution phenomena. Reading it with sufficient tenacity to understand at least the basic principles of Lehner's approach will repay the effort. This task is facilitated by the author's summary of the main steps in the theoretical development and the results thus derived.

The basic entity addressed by Lehner is a representative elementary volume of a granular rock in which grain boundaries, possessing an island or asperity structure, are invaded by the pore-fluid solvent phase during pressure solution. By using averaging techniques and a three-scale description, a macroscale Gibbs equation is derived for the aggregate sample. Grain boundary displacements resulting from pressure solution and their conjugate thermodynamic forces are incorporated into this Gibbs equation, using a formalism that involves internal variables. Phenomenological relations which link the phase-boundary displacements to the thermodynamic forces are discussed for the limiting cases of diffusion-rate control and interface-reaction-rate control. The general form of a rate-type constitutive description for rock deformation by pressure solution is derived. It is suggested that this can form a basis for future quantitative studies of pressure solution.

The chapter by Spiers & Schutjens (this volume, immediately following Lehner's chapter), continues with the theme of pressure solution, but with a mixture of theory and experiment. The experimental data come from compaction creep tests on rock salt, NaCl, which the physicist and ceramist tend to view as a model solid, but one lacking importance as a practical structural material. Unfortunately, perhaps, the practising Earth scientist cannot take such an esoteric stance, since halite forms natural enclosures for hydrocarbon reservoirs, etc. Moreover, halite formations have recently been considered for the storage of radioactive waste products, so that the rheological and dilatant properties of rock salt and its marked susceptibility to pressure solution assume even greater significance. This is the stimulus underlying the work described by the Utrecht researchers. It has led to a piece of fundamental research that advances theoretical modelling of fluid-enhanced deformation, and usefully illustrates and illuminates the nature of grain-contact microstructures in one particular type of aggregate.

Spiers & Schutjens develop a theoretical model for the densification of porous polycrystalline aggregates by fluid-phase diffusional creep, using an approach based on considerations of thermodynamic dissipation. In this approach, grain contacts are assumed to possess a time-statistically stable island structure. An expression for the creep rate is obtained by assuming that all work performed on a sample is dissipated in driving the mechanism responsible for the fluid-phase-assisted creep. When grain-boundary diffusion is rate controlling, the theory leads to an equation for the rate of densification creep at a given temperature that is proportional to the effective pressure and varies inversely with the cube of the grain size. The predictions of the constitutive model are tested against the results of compaction creep experiments on sieved NaCl powder immersed in saturated brine at room temperature. For effective pressures less than 2.2 MPa, the experimental results are in good agreement with the model, with all the samples exhibiting classical pressure-solution microstructures, together with clear evidence that grain boundaries contained a fluid-filled network of channels during deformation.

6.3.4.4 EFFECTS OF CHEMICAL REACTIONS AND METAMORPHISM ON DEFORMATION BEHAVIOUR

Research into fluid films is also given an impetus by findings which indicate that the presence of fluids can give significant oxygen exchange between some silicates at low temperatures (Giletti *et al.* 1978). The whole question of the interplay between chemical reactivity and deformation has become the focus of much thought amongst Earth scientists, and papers in this area, plus many references, are contained in Thompson & Rubie (1985). Rubie (this volume) provides a more specific review and contribution on the subject of reaction-enhanced deformability and the effects of metamorphism. Metamorphism embraces all processes whereby rocks are changed under the actions of pressure, or heat, or a combination of both. Rocks which have been metamorphosed under similar P and T conditions are said to belong to the same metamorphic facies and normally will have attained the same metamorphic grade. Metamorphic reactions in themselves can be very complex, since several minerals in close proximity may be involved. Additionally, there are reactions possible as a result of pore fluids under pressure, retrograde metamorphism (the result of holding a high-metamorphic-grade rock for a very long time at a temperature below that at which it achieved that grade), etc. Some of the effects of such complexities (e.g. pressure solution) lie outside the usual experience of those not working in the Earth sciences. Others do not, since one result of retrograde metamorphism may be the activation of an atomic ordering mechanism which can only occur at a low temperature. Systems that exhibit such behaviour are well known to ceramists and metallurgists since they can, for example, preclude the use of slow cooling to minimize built-in elastic strains in structural members.

The data reviewed by Rubie shows that, in many (but not all) cases, the deformability of rocks can be greatly enhanced by concomitant metamorphism, and that mineral reactions can result in a large increase in strain rate at constant stress, or a large decrease in strength at constant strain rate. In experimental studies, the mechanisms responsible for this behaviour can be evaluated from a combination of mechanical properties data and microstructural observations. In studies of naturally deformed rocks, however, because only microstructural data are available, it is much more difficult to identify unequivocally the deformation mechanisms that operate.

The existing data are interpreted as suggesting that an important process of reaction-enhanced deformability involves the formation of transiently fine-grained reaction products. Consequently, the deformation mechanism may change from dislocation creep to a grain-size-sensitive process such as diffusion creep, possibly with a large contribution from grain-boundary sliding (i.e. diffusion-accommodated grain-boundary sliding). Evidence is presented which indicates that this process can operate during dehydration, solid–solid, and hydration reactions, both in laboratory experiments and during natural rock deformation. The limited amount of pertinent data suggests that strain rates may increase by a factor of up to 10^6–10^8 at constant stress.

Rubie's chapter continues by discussing and making an assessment of the relative importance of other reaction-augmented deformation processes. These include enhancement of the rate of diffusion creep by fluids released during dehydration reactions, cataclastic flow as the result of a reduction in the effective confining pressure during devolatilization reactions, incongruent pressure solution, and transformation plasticity. It is concluded that some of these mechanisms may only have a small effect on rheological behaviour, but current difficulties in confidently evaluating their relevance emphasize the need for further carefully designed experimental studies.

6.4 The classification of deformation régimes and mechanical properties

In the past decade or so, the construction of deformation-mechanism maps to indicate the operational fields of the various deformation régimes has become established (Ashby 1972, Stocker & Ashby 1973, Frost & Ashby 1982). Examples are presented and used in chapters within this volume (Murrell, Tullis), which show that such maps are a useful way of classifying the deformation behaviour of different materials and making comparisons between them. Another way of bringing out similarities in deformation properties is by the use of isomechanical groups (substances with similar structures and bonding that have analogous normalized mechanical properties, e.g. see Ashby & Brown 1981). This approach has been discussed by Murrell (this volume).

6.5 The structure of grain boundaries and interfaces between phases, and the nature of intergranular films

Electron microscopy and related methods can play important rôles in studies of the fundamental structures of grain boundaries and other types of interfaces. It is also possible to investigate whether very thin intergranular films exist, and to quantify the thicknesses and compositions of more substantial asperities and second phases on surfaces and at interfaces. If we can also characterize zones of enhanced chemical reactivity, it will be possible to constrain our models and thus enable their predictions to be examined more critically.

The structures of grain boundaries and interfaces between phases are clearly most important, but the once widely held idea that grain boundaries were amorphous zones is not long out of favour, and so our present knowledge is still inadequate. Considerations of grain-boundary structures essentially began with the work of Bollmann (1970), and considerable progress has been made with metals (Balluffi 1980). Rather less is known about grain boundaries in ceramics (Clarke 1987), and the information that is available largely stems from the work of Balluffi et al. (1981, 1983), and Sass, Carter, and their colleagues (e.g. Carter et al. 1979, 1980; Carter & Sass 1981). Yan & Heuer (1983) also contains some representative papers. Much less still is known about the structures of grain boundaries in minerals, and very few studies have yet been undertaken (e.g. White & White 1981, Ricoult & Kohlstedt 1983, Hay & Evans 1988). Some knowledge exists about interfaces between phases in rocks and minerals because of the use of TEM methods (including lattice imaging) in various studies of phase transformations (e.g. Van Landuyt et al. 1987), of the contacts between hosts and exsolved phases (e.g. Williame et al. 1976), and of alteration processes (e.g. Eggleton & Banfield 1985, Ahn et al. 1985). However, the properties of such interfaces are not necessarily directly relevant to interfaces formed by igneous processes or by sintering.

The importance of grain boundaries in diffusion creep (Section 6.3.3) has led to attempts to determine how effective various types of grain boundary are as sources and sinks of point defects. Jaeger & Gleiter (1978) and Gleiter (1979) have concluded that the efficiency of grain boundaries as vacancy sources depends on their atomic structures. High-angle boundaries are better sources than those with small angles of misorientation, and grain-boundary dislocations play a significant rôle. (Grain-boundary dislocations, essential to the structures of low-angle ($\leqslant 10°$) boundaries, also occur in most high-angle boundaries, where they accommodate any small deviations from the nearest coincidence geometry of the two abutting lattices (see Balluffi et al. 1983).)

A study by Ricoult & Kohlstedt (1983) is particularly interesting because it finds that the structural thicknesses of low-angle grain boundaries in olivine are ~7 nm, which is about one-third of the thickness of similar boundaries in metals. Ricoult & Kohlstedt employed the TEM methods pioneered by Carter *et al.* (1979), which have also been used to study grain boundaries in Al_2O_3 (Carter *et al.* 1980). (Essentially, the lengths of the streaks of diffracted intensity from a grain boundary are inversely proportional to the structural thickness of the boundary.) The results are surprising at first sight, because the structures of grain boundaries in ceramics and minerals are generally expected to be complicated by the electrostatic interactions between the two or more charged ion species forming the compounds. As yet we do not know whether the structural widths obtained for olivine are representative of silicates in general, but in some oxides the widths appear to be even less (see the discussion in Ricoult & Kohlsted 1983). However, as already mentioned in Section 6.3.3, the structural width of grain interfaces is not necessarily the most important parameter for grain-boundary processes, and what matters is the thickness of any boundary zone having a chemical potential differing from that of the grain interiors. Modern techniques such as SIMS and AES are capable of providing this type of data, but so far they have been exploited almost entirely for materials with commercial importance, especially semiconductors.

6.6 Microstructures, preferred orientation, and models of crystalline plasticity

6.6.1 Microstructures in rocks and ceramics

Rocks constitute most of the Earth (at least, most of the crust and mantle) and also most of the inner planets of the solar system. All terrestrial rocks that have been molten and many that are of interest from the viewpoint of deformation processes possessed melt-crystallization microstructures at the onset of deformation. Man-made ceramics, unlike rocks and man-made metals, are seldom formed by crystallizing a melt. Most ceramics are currently produced by the sintering together of compactions of particles at comparatively high temperatures (typically $>0.75T_m$) to consolidate the compact and eliminate most porosity. Newer forms of ceramics have different fabrication methods, which include the crystallization of glass (therefore quite akin to a melt-derived microstructure), sputter-deposition, etc. Despite the differences in the processing of ceramic materials, the need to equilibrate them at high temperatures tends to imbue them with microstructures that superficially resemble those of rocks of high metamorphic grade, especially when considering monomineralic examples of each category.

Minerals are, of course, the basic crystal components of rocks, classified in terms of crystal structure and chemical constituents. Minerals are products of nature, *senso strictu*. They tend to incorporate trace impurities that are a signature of their formation environments and therefore have a degree of uniqueness. Nevertheless, the crystalline phases to be found in ceramics largely overlap the mineral phases to be found in rocks, exhibiting identical symmetries and crystal structures, and often only minor differences in chemical composition. It is thus convenient to adopt the view that minerals are the basic crystalline constituents of both rocks and ceramics.

The term 'microstructure' has been widely used in the foregoing sections. While many readers may be clear about its meaning, one should stress that the term has tended to expand in scope as modern techniques achieve ever greater resolution. Most materials scientists (and here we count rocks as materials) now take the term 'microstructure' to mean all forms of structural inhomogeneity extending from that visible with low-power optical magnification down to features separated by distances as small as 1 nm. Indeed, transmission electron microscopy (TEM; resolution limit $\simeq 0.15$ nm) has effectively produced an overlap between the terms 'microstructure' and 'crystal structure'. Microstructures are very diverse in form as well as in scale and there is no agreed way of describing them, which introduces an element of confusion. Hornbogen (1986) has suggested a scheme for classifying microstructures, but as yet it seems to have won few adherents.

The chapter by Tullis (this volume) is an excellent illustration of what can be achieved through a research programme that allows comparison of the microstructures generated by deformation of polycrystalline materials under both laboratory and natural, geological conditions. Knowledge of the P and T conditions applying in the laboratory is used to interpret the natural deformation conditions.

Tullis's experimental deformation studies have been remarkably successful in producing microstructures in monomineralic and polymineralic quartz-feldspar aggregates which are very similar to those observed in naturally deformed aggregates from a range of metamorphic grades. This indicates that the same processes have been operative. Monomineralic aggregates of quartz and of feldspar show many similarities but also display important differences in deformation behaviour. Experiments on quartzites show that with increasing P and T there is a broad régime of faulting, a narrow régime of cataclastic flow, and then a broad régime of climb-accommodated dislocation creep. Up until now, the experimental studies have been unsuccessful in producing diffusion creep or pressure solution, In feldspar aggregates the faulting régime is narrower, the cataclastic régime is broader, the experimentally accessible portion of the dislocation creep régime is accommodated by recrystallization rather than by climb, and fluid-assisted grain-boundary-diffusion creep occurs in very fine-grained aggregates. Observations of naturally deformed feldspathic rocks indicate a higher temperature régime of climb-accommodated dislocation creep. Tullis notes that quartz and feldspar violate the simple metallurgical rule that ductility is proportional to fractional melting temperature, since quartz has a much higher T_m than feldspar but a lower temperature for the onset of dislocation creep.

The experimental deformation studies of Tullis on polymineralic aggregates

164

are limited in direct application to relatively low metamorphic grades, because of the onset of partial melting at the high temperatures necessitated by the relatively fast experimental strain rates employed (10^{-4}–$10^{-6}\,s^{-1}$). Within the experimentally accessible window, studies of granite and aplite demonstrate a switch in the relative strengths of the two components and a corresponding change in textural evolution. At the lower temperatures, where dislocation activity is very limited, dispersed quartz grains remain undeformed while the feldspar matrix undergoes ductile cataclastic flow. At higher temperatures in the dislocation creep régime, portions of original feldspar grains remain only slightly deformed and become dispersed, while quartz and recrystallized feldspars form a weak, continuous matrix. In rocks in which feldspars form the stress-supporting framework, the difference in strength between original and dynamically recrystallized grains at low-to-intermediate metamorphic grades tends to favour ductile shear zones on all scales, whereas deformation is more homogeneous at high grades where climb-accomodated creep becomes possible.

Following the Tullis chapter is one by Knipe, reporting work that uses similar microstructural characterization methods to those employed by Tullis and her colleagues, but which serves to demonstrate the ways in which micro-scale observations of deformation processes can be used to build up a picture of macroscale faulting and the emplacement of thrust sheets, while incident-ally providing compelling evidence of the complexity of tectonic systems. The author reports the results of his studies of the Assynt region of the famous thrust fault of north-west Scotland, the Moine Thrust.

Analysis of the deformation features preserved in quartzites involved in the evolution of the Moine Thrust Zone is used to illustrate how microstructural studies can aid the assessment of the deforming mechanisms, kinematics, and dynamics that were operative in tectonic events. The microstructures preserved in fault rocks developed along the major thrusts, together with the features characteristic of deformation within each of the thrust sheets of the Assynt region, are reported. Each of the fault zones has been active over a different range of metamorphic conditions and preserves evidence of a different combi-nation of deformation mechanisms experienced during different pressure–temperature–time paths. The microstructures are used to infer the deformation mechanism path of each thrust fault and integrated into the likely P–T–t paths.

Knipe's study reveals changes in the deformation mechanisms, the strain path, and the type of strain-rate cycles associated with the emplacement of the different thrust sheets. Deformation along the thrust faults which have experienced the highest temperatures is dominated by dislocation creep pro-cesses. However, Knipe finds that the later stages of emplacement along the Moine Thrust saw an increase in the importance of fracturing and involved a complex history of alternating thrust sheet thinning and shearing. The last stages of deformation in the fault zone rocks were accompanied by vertical shortening by diffusive mass transfer, indicating a decrease in the differential stress levels that induced deformation. Cataclastic flow dominated the emplacement of the lower-level thrust sheets, which took place over lower temperatures. Plastic deformation by dislocation movement is found to be an important subsidiary deformation mechanism in the intermediate level Assynt Thrust sheet but is more pervasive in the higher structural levels of this sheet. Evidence of a change to vertical shortening by diffusive mass transfer after emplacement of the Assynt Thrust is preserved in the fault rocks; this deformation pre-dates deformation and rotation of the Assynt Sheet by the underlying Sole Thrust system. Although most faults in the lowest thrust sheet, the Sole Thrust system, are dominated by cataclastic rocks, a fault tip preserves evidence of slow-strain-rate deformation by diffusive mass transfer, and probably represents the final decline of displacement activity on the whole thrust system.

Where possible, Knipe uses the deformation microstructures to speculate on the magnitude of the deformation and displacement rates associated with the evolution of the Moine Thrust Zone. In addition, the stability of the microstructures generated in the different thrust sheets is related to the different $P-T-t$ paths of each thrust sheet. The variation in microstructures of cataclastic gouge zones preserved in the different thrust sheets is interpreted as indicating that these zones are particularly susceptible to annealing during cooling.

6.6.2   Preferred orientation–deformation textures

One of the most important effects produced by the deformation of a crystalline aggregate is the changes in the crystallographic orientations of the individual crystals. These changes, most obvious in materials deformed to strains $\geqslant 10$ per cent, result mainly from slip within individual crystals, leading to their crystallographic rotation. As a consequence, initially isotropic aggregates develop a pronounced crystallographic preferred orientation or 'texture', which can be measured by X-ray, optical, or electron beam methods. Materials deforming by brittle mechanisms generally do not develop preferred grain orientations, but do exhibit preferred orientations for cracks.

Earth scientists also employ the term 'fabric' in describing the appearance of rocks, a term which is rather all-embracing and tends to de-emphasize the heterogeneities to be found on the macroscale. With the increased importance of microstructural studies to Earth scientists, the word 'microfabric' has come into use, embracing both microstructure and preferred orientation. The geological significance of microfabrics is discussed by Hobbs (1985). The word 'texture' as a synonym for preferred orientation is widely used by metallurgists, and has growing acceptance in the wider materials and Earth science communities.

The pattern of preferred orientation of a deformed aggregate or rock is typically represented by a map showing how the distribution of a crystallographic property varies in space. This is obtained by means of a projection into two dimensions, resulting in a pole figure, or fabric diagram. The information in these pole figures must correlate with microstructures, since both are a consequence of the activity of particular slip systems (GBS and cracking may also be involved). Introductions to the rôle of slip systems are to be found in the texts of Barrett & Massalski (1981) and Hobbs et al. (1976). Other papers (e.g. Van Houtte & Wagner 1985) give more detailed arguments. A deformation texture is a record of the last major deformation process to affect a fabricated material or a natural rock. In the former case, the process parameters will normally be well known, but for the rock they will not and must be deduced, if possible. The question for geologists is whether or to what extent it is possible to use observed preferred orientations to interpret recent deformation history. In attempting to answer this question by studying the change in orientation of grains during deformation, and in simulations combining the effects on all such grains in a polycrystalline assemblage, it is generally assumed that slip and deformation twinning are the important mechanisms (Gil Sevillano et al. 1980). In this review, space is too limited to describe the often more minor rules of twinning, kinking, etc. (Barber (1985) gives a brief introduction and useful references to these mechanisms, while Van Houtte & Wagner (1985) review the rôle of twinning in texture development and how it can be modelled.)

Recent activity in the development of texture analysis and its use in interpreting the processes underlying the deformation of materials is well illustrated in the text by Wenk (1985), which contains references to many papers reporting recent research activity in this field. Mention was made in Section 6.2 of the influence of initial microstructure on $\sigma-\varepsilon$ relationships. This aspect has also been engaged by those working in the field of texture (e.g. Lister & Hobbs 1980, Bunge 1987), and there have been attempts to include changes in texture and microstructure in constitutive equations (e.g. Kocks 1976). Tullis (this volume) briefly discusses some recent developments in texture studies that are especially relevant to rock deformation.

6.6.3 Modelling of crystalline plasticity

Theories of polycrystalline plasticity that take into account the deformational mechanisms active in individual grains tend to be very complex, but all start from simplifying assumptions about the relationships between macroscopic stress, or strain, and the microscopic distributions of stress, or strain. A prime requirement for a polycrystalline aggregate to deform to significant plastic strain is that parting should not occur at grain boundaries, which implies that strain continuity must be maintained. Taylor (1938) met this point by making the hypothesis that all the grains in an aggregate undergo the same homogeneous strain as experienced by the body as a whole. In general this can only be achieved if at least five slip systems are active, enabling each grain to undergo a general shape change and thus maintain contact with its neighbours (von Mises 1928). However, consideration of the conditions of stress equilibrium implies that, in general, both stress and strain will be heterogeneous within the grains. Nevertheless, Taylor theory often approximates to the way that materials of high crystal symmetry behave during deformation. An introduction to modified forms of Taylor theory and other approaches to the simulation of texture development are to be found in Van Houtte & Wagner (1985).

There are several examples of the successful use of Taylor theory to model plastic deformation in monomineralic rocks. Simulated textures for quartzites and limestones are found to be in reasonable agreement with those observed in experimentally deformed samples (e.g. Lister & Hobbs 1980, Wenk *et al.* 1986). The main problems with Taylor theory stem from the basic assumption of homogeneous strain and the requirement that all grains in an aggregate deform at the same rate and, accordingly, undergo the same shape change as the whole. In general, minerals tend to have low symmetry, with the consequence that their slip systems are asymmetrically disposed and often have large differences in critical resolved shear stresses. This causes the single-crystal yield surface to be strongly anisotropic. For such low-symmetry minerals, some crystal orientations are much stronger than others, and thus there is a spread in strain rates between the grains of their polycrystals. The situation becomes even worse if a material fails to meet the von Mises criterion (von Mises 1928) and lacks five independent slip systems. The yield surface is then not closed, so that crystals cannot deform homogeneously. It is clear that in such cases Taylor theory may only be applied if some of its rigorous constraints are relaxed. Others argue that Taylor theory is so inappropriate to strongly anisotropic materials that other types of theory must be developed and applied.

The chapter by Takeshita, Wenk, Canova & Molinari (this volume) begins with a brief review of research pertinent to rocks in regard to Taylor theory, and then proceeds to the particular case of olivine. The authors examine modelling of the development of texture and plastic anisotropy by applying both relaxed-constraint Taylor theory and a viscoplastic, self-consistent theory, a comparison that has recently been made for halite (Wenk *et al.* 1989). It is shown that the two approaches give similar result for olivine, corresponding well with what is observed from experiments and with natural fabrics. On the basis of this model, it appears that dislocation climb is rate controlling, and that simple shear is the most economical strain mode, at least at high temperature. It is also shown that (010) and (100) pole figures can be used to determine the sense of shear if the shear plane can be identified.

Turning to a different type of anisotropy, fabrics that result from the dynamic recrystallization of rocks from the upper mantle, dunite and lherzolite, have been used to explain the anisotropy of seismic-wave velocities in the upper mantle (Carter *et al.* 1972). Past arguments, based on experimental creep data, as to whether ⟨100⟩ or [011] slip predominates in the peridotites (essentially olivines) of the upper mantle, may be irrelevant if dynamic recrystallization predominates (for more details see Lliboutry 1987). Clearly, the modelling of upper mantle flow by laboratory-derived steady-state-creep laws is open to doubt, and texture analysis is demonstrably relevant to some important problems.

It will be clear from this overview that any understanding or modelling of plastic deformation is dependent on knowledge of the basic microprocesses that are available for relieving stress in a given substance, how the critical resolved shear stresses for slip systems vary with temperature, etc. In recent years the development of new ceramic materials for structural applications in hostile environments (e.g. nuclear reactors or gas turbines) has necessitated a great deal of research into their deformation mechanisms, especially at high temperatures. Many oxide ceramics are notable for being able to retain the same basic crystal structure over a range of stoichiometries. Non-stoichiometry not only changes the kinetics of diffusion (which may therefore modify climb behaviour) but may also affect deformation mechanisms directly (e.g. by changing the height of Peierls barriers). The chapter by Heuer, Keller & Mitchell (this volume) describes research to find which slip systems operate as a function of temperature, composition, and crystal orientation in uranium dioxide, UO_2. This material is mostly used as a nuclear fuel, and its mechanical properties are important because of the large stresses suffered by fuel rods during use. It crystallizes with the fluorite structure (as does zirconia in its cubic, fully stabilized, form). Uranium dioxide compositions can embrace a wide range of oxygen contents at high temperatures while still maintaining the fluorite structure.

The chapter by Heuer et al. illustrates the investigation of deformation mechanisms by the use of single crystals. It is well known that UO_{2+x} shows a tendency to soften with increasing excess oxygen content (x) when deformed at high temperatures (Ronchi & Blank 1970). However, a more precise description of its behaviour is still needed. It is shown that, at low temperatures, $\{111\}\langle 0\bar{1}1\rangle$ is the preferred slip system, and that the same slip system also operates at high temperatures for some orientations of non-stoichiometric crystals. It is also shown that additional mechanisms are, for various combinations of x and crystal orientation, either $\{100\}\langle 0\bar{1}1\rangle$ slip or non-crystallographic slip, combining activity on $\{100\}$ and $\{111\}$ planes, or on $\{110\}$ and $\{111\}$ planes. The findings are then discussed in terms of the structure of dislocations in the fluorite structure, and the chapter concludes by listing a number of questions that still remain unresolved.

An understanding of slip mechanisms and factors that influence them may often entail an even more detailed study of the properties of the dislocations involved. A sublime example of this situation is presented by Hennig-Michaeli & Couderc for the ore mineral chalcopyrite, $CuFeS_2$. Its tetragonal structure can be thought of as an 'ordered' sphalerite (ZnS) structure, but a further complication is that the strongly covalent bonding represents an effective ionic state between $Cu^+Fe^{3+}S_2^{2-}$ and $Cu^+Fe^{2+}S_2^{2-}$. The plastic deformation behaviour of chalcopyrite is more complex than that of sphalerite, because a multiplicity of glide mechanisms contribute. The study focuses on dislocation reactions occurring in experimentally deformed crystals, including the dissociation of perfect dislocations into partials, cross slip of extended dislocations, and interactions between dislocations associated with different glide modes. The single-crystal specimens were tested by compression along several low index directions, under a confining pressure, at two different temperatures.

It is shown that, at $200°C$, the main slip modes are $\{112\}\langle 3\bar{1}\bar{1}\rangle$ and $\{112\}\langle \bar{3}11\rangle$, producing high-density bands of straight dislocations with Burgers vector, $\mathbf{b} = \frac{1}{2}\langle 3\bar{1}\bar{1}\rangle$, which are dissociated into four non-collinear partials. Other observed Burgers vectors in the $\{112\}$ plane are $\mathbf{b} = \frac{1}{2}\langle \bar{1}\bar{1}1\rangle$, $\mathbf{b} = \langle 1\bar{1}0\rangle$ and $\mathbf{b} = \langle \bar{2}01\rangle$. All of these also undergo dissociation. Twinning is found to proceed by the motion of partial dislocations with $\mathbf{b} = \frac{1}{6}\langle 11\bar{1}\rangle$. It is found that dislocations in crystals suitably orientated to slip on $(001)\langle 1\bar{1}0\rangle$ are homogeneously distributed, and they dissociate into two collinear partials with $\mathbf{b} = \frac{1}{2}\langle 1\bar{1}0\rangle$ which are observed to cross slip readily onto $\{112\}$ planes. Dislocations with $\mathbf{b} = \langle 010\rangle$ introduced by glide on $\{100\}\langle 010\rangle$ are split into two non-collinear partials, $\mathbf{b} = \frac{1}{4}\langle 021\rangle$.

The main deformation modes at $400°C$ are slip on $(001)\langle 110\rangle$ and $\{112\}\langle \bar{1}\bar{1}1\rangle$. Dislocations with $\mathbf{b} = \frac{1}{2}\langle \bar{1}\bar{1}1\rangle$ are homogeneously distributed and cross slip from $\{112\}$ into $\{1\bar{1}0\}$. The interaction of dislocations with $\mathbf{b} = \frac{1}{2}\langle \bar{1}\bar{1}1\rangle$ gliding in different $\{112\}$ planes is found to generate dislocation segments with $\mathbf{b} = \langle 110\rangle$.

6.7 Concluding remarks

It will be apparent from the foregoing sections that much of the progress in understanding the flow properties and creep of materials has come about through research on metals, for which there is now a huge amount of data. Unfortunately, far less information is available for non-metals and, although it is clear that many of the metallurgical findings have relevance, they cannot simply be directly applied to ceramics, or to rocks – as demonstrated by the findings of Tullis (this volume). Important factors are the very different nature of the bonding in non-metals, the generally greater complexity of their atomic structures and their lower symmetries and, often, the very different environments under which they are deformed. Contributory factors to the relative dearth of knowledge about deformation processes in ceramics and minerals are the much greater commercial importance of metals, and practical impediments such as the greater difficulty in studying microstructures in non-metals by TEM prior to the widespread adoption of the ion-thinning method (Paulus & Reverchon 1961) in the 1970s (Barber 1970).

Lack of basic data is still a hindrance to progress in the elucidation of the microprocesses of flow, and perhaps nowhere is this more true than in the case of diffusion data for non-metals. In particular, very little is known about diffusion mechanisms in minerals. Once again, the complexity of obtaining data for compounds where the relevant temperatures are quite high and several atomic species may be mobile largely explains the lack of data for many oxides and most silicates prior to 1960. However, the recognition that diffusion between phases could be sufficiently fast to play a role in metamorphic processes (Fyfe *et al.* 1958, Spry 1969), the charting by EPMA of compositional zoning in metamorphic minerals such as garnets, and interest in geothermometry, have all stimulated the need for diffusion data. The need is widely recognized, but there are relatively few groups engaged in this type of work, which is difficult, not well suited to the present-day dictates of rapid and guaranteed results, and probably not given sufficient encouragement. For an introduction to diffusion, see Manning (1974): for summaries of data on oxides and silicates, see Freer (1980) and Freer (1981) respectively.

Acknowledgements

I am grateful to David Price, Stan Murrell, an anonymous reviewer and, especially, to Jan Tullis for comments on drafts of this overview, which led to considerable improvements.

References

Ahmed, M. M. I. & T. G. Langdon 1977. Exceptional ductility in the superplastic Pb-62 pct Sn eutectic. *Metall. Trans* **8A**, 1832–3.

Ahn, Jung Ho, D. R. Peacor & E. J. Essene 1985. Co-existing paragonite–phengite in blueschist eclogite: a TEM study. *Am. Mineral.* **70**, 1193–204.

Atkinson, A. & R. I. Taylor 1981. The diffusion of ^{63}Ni along grain boundaries. *Phil Mag.* **A43**, 979–98.

Ashby, M. F. 1972. A first report of deformation-mechanism maps. *Acta Metall.* **20**, 887–97.

Ashby, M. F. & L. M. Brown 1981. Flow in polycrystals and the scaling of mechanical properties. In *Deformation of polycrystals: mechanisms and microstructures*. N. Hansen, A. Horsewell, T. Leffers & H. Lilholt (eds), 1–13. Proc. 2nd Risø Symp. Metall. Mat. Sci. Risø National Laboratory, Roskilde. Roskilde, Denmark: Risø National Laboratory.

Ashby, M. F. & R. A. Verrall 1973. Diffusion accommodated flow and superplasticity. *Acta Metall.* **21**, 149–63.

Bailey, J. E. & P. B. Hirsch 1962. The recrystallization process in some polycrystalline metals. *Proc. R. Soc. Lond.* **A267**, 11–30.

Balluffi, R. W. (ed.) 1980. *Grain-boundary structure and kinetics*. Proceedings of a conference at Milwaukee, Wisconsin, Sept. 1979. Metals Park, Ohio: American Society of Metals.

Balluffi, R. W., P. D. Bristowe & C. P. Sun 1981. Structure of high-angle grain boundaries in metals and ceramic oxides. *J. Am. Ceram. Soc.* **64**, 23–34.

Balluffi, R. W., A. Brokman & A. H. King 1982. CSL/DSC lattice model for general crystal–crystal boundaries and their line defects. *Acta Metall.* **30**, 1453–70.

Balluffi, R. W., P. D. Bristowe & A. Brokman 1983. Character of grain boundaries. *Adv. Ceram.* **6**, 15–35.

Barber, D. J. 1970. Thin foils of non-metals made for electron microscopy by sputter-etching. *J. Mat. Sci.* **5**, 1–8.

Barber, D. J. 1977. Defect microstructures in deformed and recovered dolomites. *Tectonophysics* **39**, 193–213.

Barber, D. J. 1985. Dislocations and microstructures. In *Preferred orientation in deformed metals and rocks: an introduction to modern texture analysis*, H.-R. Wenk (ed.), 149–82. New York: Academic Press.

Barrett, C. S. & T. B. Massalski 1980. *Structure of metals*, 3rd edn. New York: McGraw-Hill.

Blacic, J. D. 1975. Plastic deformation mechanisms in quartz: the effect of water. *Tectonophysics* **27**, 271–94.

Boland, J. N. & T. E. Tullis 1986. Deformation behaviour of wet and dry clinopyroxenite in the brittle to ductile transition region. In *Mineral and rock deformation: laboratory studies*, B. E. Hobbs & H. C. Heard (eds), 35–50. Geophysical Monograph **36**. Washington, DC: American Geophysical Union.

Bollmann, W. 1970. *Crystal defects and crystalline interfaces*, Berlin: Springer.

Brown, A. M. & M. F. Ashby 1980a. Correlations for diffusion constants. *Acta Metall.* **28**, 1085–101.

Brown, A. M. & M. F. Ashby 1980b. On the power law creep equation. *Scripta Metall.* **14**, 1297–302.

Bunge, B. J. 1987. The influence of texture on the mechanical properties of polycrystalline materials. In *Constitutive relations and their physical basis*, S. I. Anderson, J. B. Bilde-Sørensen, N. Hansen, T. Leffers, H. Lilholt, O. B. Pederson & B. Ralph (eds), 55–66. Proc. 8th Int. Symp. Metall. Mat. Sci. Roskilde. Roskilde, Denmark: Risø National Laboratory.

Caillard, D. & J. L. Martin 1982. Microstructure of aluminium during creep at intermediate temperatures. I. Dislocation between networks after creep. *Acta Metall.* **30**, 437–45.

Cannon, R. M. & R. L. Coble 1975. Review of diffusional creep of Al_2O_3. In *Deformation of ceramics*, R. C. Bradt & R. E. Tressler (eds), 61–100. New York: Plenum.

Carter, C. B. & S. L. Sass 1981. Electron diffraction and microscopy techniques for studying grain-boundary structure. *J. Am. Ceram. Soc.* **64**, 335–45.

Carter, C. B., A. M. Donald & S. L. Sass 1979. Diffraction effects and images from inclined boundaries in polycrystalline thin foils. *Phil Mag.* **A39**, 533–49.

Carter, C. B., D. L. Kohlstedt & S. L. Sass 1980. Electron diffraction and microscopy studies of the structure of grain boundaries in Al_2O_3. *J. Am. Ceram. Soc.* **63**, 623–7.

Carter, N. L., D. W. Baker & R. P. George, Jr 1972. Seismic anisotropy, flow and constitution of the upper mantle. In *Flow and fracture of rocks*, H. C. Heard, I. Y. Borg, N. C. Carter & C. B. Raleigh (eds), 167–90. Geophysical Monograph **16**. Washington, DC: American Geophysical Union.

Castaing, J., A. Dominguez-Rodriguez & C. Monty 1984. The effects of non-stoichiometry on the deformation of oxides. In *Deformation of ceramics II*, R. C. Bradt & R. E. Tressler (eds), 141–58. New York: Plenum.

Chadwick, G. A. & D. A. Smith (eds) 1976. *Grain boundary structure and properties*. New York: Academic Press.

Chokshi, A. H. & T. G. Langdon 1985. The role of interfaces in superplastic deformation. In *Superplasticity*, B. Baudelet & M. Suery (eds), 2.1–15. Paris: Centre Nationale de la Recherche Scientifique.

Chung, T. E. & T. J. Davies 1979. The low stress creep of fine-grain uranium dioxide. *Acta Metall.* **27**, 627–35.

Clarke, D. R. 1987. Grain boundaries in polycrystalline ceramics. *Ann. Rev. Mat. Sci.* **17**, 57–74.

Coath, J. A. & B. Wilshire 1977. Deformation processes during high temperature creep of lime, magnesia and doloma. *Ceramurgia Int.* **3**, 103–8.

Coble, R. L. 1963. A model for boundary-diffusion controlled creep in poycrystalline materials. *J. Appl. Phys.* **34**, 1679–82.

Condit, R. H. 1985. An approach to analyzing diffusion in olivine. In *Point defects in minerals*, R. N. Shock (ed.), 106–15. Geophysical Monograph **31**. Washington, DC: American Geophysical Union.

Davidge, R. W. 1980. *Mechanical behaviour of ceramics*. Cambridge: Cambridge University Press.

Derby, B. 1990. Dynamic recrystallization and grain size. This volume, 354–64.

Derby, B. & M. F. Ashby 1987. On dynamic recrystallization. *Scripta Metall.* **21**, 832–7.

Dixon-Stubbs, P. J. & B. Wilshire 1982. Deformation processes during creep of single and polycrystalline CaO. *Phil Mag.* **A45**, 519–29.

Duclos, R. 1984. High temperature deformation mechanisms in $MgO.nAl_2O_3$ spinels. In *Deformation of ceramic materials II*, R. C. Bradt & R. E. Tressler (eds), pp. 159–175. New York: Plenum Press.

Durham, W. B., C. Goetze & B. Blake 1977. Plastic flow of oriented single crystals of olivine. 2. Observations and interpretations of the dislocation structure. *J. Geophys. Res.* **82**, 5755–70.

Ecob, R. C. & J. B. Bilde-Sørensen 1987. The structural basis for creep relations. In *Constitutive relations and their physical basis*, S. I. Anderson, J. B. Bilde-Sørensen, N. Hansen, T. Leffers, H. Lilholt, O. B. Pederson & B. Ralph (eds), 67–82. Proc. 8th Int. Symp. Metall. Mat. Sci., Roskilde. Roskilde, Denmark: Risø National Laboratory.

Eggleton, R. A. & J. F. Banfield 1985. The alteration of granitic biotite to chlorite. *Am. Mineral.* **70**, 902–10.

Etheridge, M. A. & B. E. Hobbs 1974. Chemical and deformational controls on recrystallization of mica. *Contrib. Mineral. Petrol.* **43**, 111–24.

Evans, R. W. & B. Wilshire 1986. A new theoretical approach to creep and creep fracture. In *Strength of metals and alloys (ICSMA 7)*, vol. 3, H. J. McQueen, J.-P. Bailon, J. I. Dickson, J. J. Jonas & M. G. Akben (eds), 1807–30. Oxford: Pergamon.

Freer, R. 1980. Self-diffusion and impurity diffusion in oxides. *J. Mat. Sci.* **15**, 803–24.

Freer, R. 1981. Diffusion in silicate minerals and glasses: a data digest and guide to the literature. *Contrib. Mineral. Petrol.* **76**, 440–54.

Frost, H. J. & M. F. Ashby 1982. *Deformation mechanism maps*. Oxford: Pergamon.

Fyfe, W. S., F. J. Turner & J. Verhoogen 1958. Metamorphic reactions and metamorphic facies. *Geol Soc. Am. Mem.* **73**, 60–6.

Garvie, R. C., R. H. Hannink & R. T. Pascoe 1975. 'Ceramic steel'. *Nature* **258**, 703–4.

Ghosh, A. K. & C. Ghandi 1986. Superplasticity in Al–Li alloys. In *Strength of metals and alloys (ICSMA 7)*, vol. 3, H. J. McQueen, J.-P. Bailon, J. I. Dickson, J. J. Jones & M. G. Akben (eds) 2065–72. Oxford: Pergamon.

Gifkins, R. C. & T. G. Langdon 1978. Comments on theories of superplasticity. *Mat. Sci. Engng* **36**, 27–33.

Gil Sevillano, J., P. Van Houtte & E. Arnouldt 1980. Large strain work hardening and textures. *Progr. Mat. Sci.* **25**, 69–412.

Giletti, B. J., M. P. Semet & R. A. Yund 1978. Studies in diffusion III: oxygen in felspars: an ion microprobe determination. *Geochim. Cosmochim. Acta* **42**, 45–57.

Gittus, J. 1975. *Creep, viscoelasticity and creep fracture in solids*. New York: Wiley.

Gleiter, H. 1979. Grain boundaries as point defect sources or sinks – diffusional creep. *Acta Metall.* **27**, 187–92.

Glover, G. & C. M. Sellars 1973. Recovery and recrystallization during high temperature deformation of α-iron. *Metall. Trans*, **4**, 765–75.

Green, H. W. 1984. 'Pressure solution' creep: some causes and mechanisms. *J. Geophys. Res.* **89**, 4313–18.

Green, H. W. 1986. Phase transformation under stress and volume transfer creep. In *Mineral and rock deformation: laboratory studies*, B. E. Hobbs & H. C. Heard (eds), 201–11. Geophysical Monograph **36**. Washington, DC: American Geophysical Union.

Griggs, D. 1967. Hydrolytic weakening of quartz and other silicates. *Geophys. J. R. Astron. Soc.* **14**, 19–31.

Griggs, D. T. & J. D. Blacic 1965. Quartz: anomalous weakness of single crystals. *Science* **147**, 292–5.

Guillopé, M. & J.-P. Poirier 1979. Dynamic recrystallization during creep of single-crystalline halite. *J. Geophys. Res.* **84**, 5557–67.

Harper, J. G. & J. E. Dorn 1957. Viscous creep of aluminium near its melting temperature. *Acta Metall.* **5**, 654–5.

Hart, E. W. 1970. A phenomological theory for plastic deformation of polycrystalline metals. *Acta Metall.* **18**, 599–610.

Hart, E. W. 1976. Constitutive relations for the nonelastic deformation of metals. *J. Engng Mat. Technol.* **98**, 193–202.

Hay, R. S. & B. Evans 1988. Intergranular distribution of pore fluid and the nature of high-angle grain boundaries in limestone and marble. *J. Geophys. Res.* **B93**, 8959–74.

Heard, H. C. 1985. Experimental determination of mechanical properties. In *Preferred orientation in deformed metals and rocks: an introduction to modern texture analysis*, H.-R. Wenk (ed.), 485–506. New York: Academic Press.

Hennig-Michaeli, C. & J.-J. Couderc 1990 A TEM study of dislocation reactions in experimentally deformed chalcopyrite single crystals. This volume, 391–414.

Herring, C. 1950. Diffusional viscosity of a polycrystalline solid. *J. Appl. Phys.* **21**, 437–45.

Heuer, A. H., R. J. Keller & T. E. Mitchell 1990. On the slip systems in uranium dioxide. This volume, 377–90.

Hobbs, B. E. 1968. Recrystallization of single crystals of quartz. *Tectonophysics* **6**, 353–401.

Hobbs, B. E. 1981. The influence of metamorphic environment upon the deformation of minerals. *Tectonophysics* **78**, 335–83.

Hobbs, B. E. 1985. The geological significance of microfabric analysis. In *Preferred orientation in deformed metals and rocks: an introduction to modern texture analysis*, H.-R. Wenk (ed.), 463–84. New York: Academic Press.

Hobbs, B. E., A. C. McLaren & M. S. Paterson 1972. Plasticity of single crystals of quartz. In *Flow and fracture of rocks*, H. C. Heard, I. Y. Borg, N. L. Carter & C. B. Raleigh, (eds), 29–53, Geophysical Monograph **16**. Washington, DC: American Geophysical Union.

Hobbs, B. E., W. D. Means & P. F. Williams 1976. *An outline of structural geology*. New York: Wiley.

Hornbogen, E. 1986. Review: a systematic description of microstructure. *J. Mat. Sci.* **21**, 3737–47.

Jaeger, W. & H. Gleiter 1978. Grain boundaries as vacancy sources in diffusional creep. *Scripta Metall.* **12**, 675–8.

Jaoul, O., C. Froidevaux, W. B. Durham & M. Michaut 1980. Oxygen self-diffusion in forsterite. *Earth Planet. Sci. Lett.* **47**, 391–7.

Kashyap, B. P., A. Arieli & A. K. Mukherjee 1985. Review: microstructural aspects of superplasticity. *J. Mat. Sci.* **20**, 2661–86.

Kekulalawa, K. R. S. S., M. S. Paterson & J. N. Boland 1981. An experimental study of the role of water in quartz deformation. In *Mechanical behaviour of crystal rocks*, N. L. Carter, M.

Friedman, J. M. Logan & D. W. Stearns, (eds), 49–60. Geophysical Monograph **24**. Washington, DC: American Geophysical Union.

Kirby, S. H. 1983. Rheology of the lithosphere. *Rev. Geophysics Space Phys.* **21**, 1458–87.

Knipe, R. J. 1990. Microstructural analysis and tectonic evolution in thrust systems: examples from the Assynt region of the Moine Thrust Zone, north-west Scotland. This volume, 228–61.

Kocks, U. F. 1976. Laws for work-hardening and low temperature creep. *J. Eng Mat. Technol.* **98**, 76–85.

Kohlstedt, D. L. & D. L. Ricoult 1984. High-temperature creep of silicate olivines. In *Deformation of ceramics II*, R. C. Bradt & R. E. Tressler, (eds), 251–80. New York: Plenum.

Kosinski, S. G., S. Vaidya, D. W. Johnson & R. E. Tressler 1979. Oxygen stoichiometry effects on the fracture of manganese–zinc ferrites. *Am. Ceram. Soc. Bull.* **58**, 616.

Langdon, T. G. 1981. Deformation of polycrystalline materials at high temperatures. In *Deformation of polycrystals: mechanisms and microstructures*, N. Hansen, A. Horsewell, T. Leffers & H. Lilholt, (eds), 45–54. Proc. 2nd Risø Symp. Metall. Mat. Sci., Roskilde. Roskilde, Denmark: Risø National Laboratory.

Langdon, T. G. 1985. Regimes of plastic deformation. In *Preferred orientation in deformed metals and rocks: an introduction to modern texture analysis*, H.-R. Wenk, (ed.), 219–232. Orlando, Florida: Academic Press.

Langdon, T. G. & R. B. Vastava 1982. An evaluation of deformation models for grain boundary sliding. In *Mechanical testing for deformation model development*, R. W. Rohde & J. C. Swearengen, (eds), 435–51. Philadelphia: American Society for the Testing of Materials.

Langdon, T. G. & P. Yavari 1982. An investigation of Harper–Dorn creep. II – the flow process. *Acta Metall.* **30**, 881–7.

Lee, S. & A. J. Ardell 1986. Dislocation link length distributions during Harper–Dorn creep of monocrystalline aluminium. In *Strength of metals and alloys (ICSMA 7)*, vol. 1, H. J. McQueen, J.-P. Bailon, J. I. Dickson, J. J. Jones & M. G. Akben (eds), 671–6. Oxford: Pergamon.

Lehner, F. 1990. Thermodynamics of rock deformation by pressure solution. This volume, 296–333.

Lister, G. & B. E. Hobbs 1980. The simulation of fabric development during plastic deformation and its application to quartzite: the influence of deformation history. *J. Struct. Geol.* **2**, 355–70.

Liboutry, L. A. 1987. In *Very slow flows of solids*. Dordrecht: Martinus Nijhoff.

Manning, J. R. 1974. Diffusion kinetics and mechanisms in simple crystals. In *Geochemical transport and kinetics*, A. W. Hofmann, B. J. Giletti, H. S. Yoder Jr. & R. A. Yund (eds) 3–13. Publication 634. Washington, DC: Carnegie Institute.

McClay, K. R. 1977. Pressure solution and Coble creep in rocks and minerals: a review. *J. Geol Soc. Lond.* **134**, 57–70.

McClay, K. R. & B. K. Atkinson 1977. Experimentally induced kinking and annealing of single crystals of galena. *Tectonophysics* **39**, 175–89.

Meredith, P. G. 1990. Fracture and failure of brittle polycrystals. This volume, 5–47.

Mohamed, F. A., K. L. Murty & J. W. Morris Jr 1973. Harper–Dorn creep of metals at high temperatures. In *Rate processes in plastic deformation of materials: proceedings of the John E. Dorn Memorial Symposium*, J. C. M. Li & A. K. Mukherjee (eds), 459–78. Cleveland, Ohio: American Society of Metals.

Mukherjee, A. K., J. E. Bird & J. E. Dorn 1969. Experimental correlations for high-temperature creep. *Trans Am. Soc. Metals* **62**, 155–79.

Murrell, S. A. F. 1990. Brittle-to-ductile transitions in polycrystalline non-metallic materials. This volume, 109–37.

Murrell, S. A. F. & S. Chakravarty 1973. Some new rheological experiments on igneous rocks at temperatures up to $1120°C$. *Geophys. J. R. Astron. Soc.* **34**, 211–50.

Nabarro, F. R. N. 1948. Deformation of crystals by the motion of single ions. In *Report of a conference on strength of solids (Bristol)*, 75–90. London: The Physical Society.

Nicolas, A. & J.-P. Poirier 1976. *Crystalline plasticity and solid state flow in metamorphic rocks*. London: Wiley Interscience.

Nishikawa T. & Y. Okamoto 1980. Creep deformation of polycrystalline Mn–Zn ferrite. In *Ferrites: proceedings of the International Conference*. H. Watanabe, S. Iida & M. Sugimoto (eds), 306–9. Tokyo, Japan: Centre for Academic Publications.

Nix, W. D. & J. C. Gibeling 1985. Mechanisms of time-dependent flow and fracture of metals. In *Flow and fracture at elevated temperatures*, R. Raj (ed.), 1–63. Metals Park, Ohio: American Society for Metals.

Nix, W. D. & B. Ilschner 1980. Mechanisms controlling creep of single phase metals and alloys. In *Strength of metals and alloys (ICSMA 5)*, P. Haasen, V. Gerold & G. Kostorz (eds), 1503–30. Oxford: Pergamon.

Ord, A. & B. E. Hobbs 1986. Experimental control of the water-weakening effect in quartz. In *Mineral and rock deformation: laboratory studies*, B. E. Hobbs & H. C. Heard (eds), 51–71, Geophysical Monograph **36**. Washington, DC: American Geophysical Union.

Padmanabhan, K. A. & J. G. Davies 1980. *Superplasticity*. Berlin: Springer.

Panda, P. C., R. Raj & P. E. D. Morgan 1985. Superplastic deformation in fine-grained $MgO.2Al_2O_3$ spinel. *J. Am. Ceram. Soc.* **68**, 522–9.

Paulus, M. & F. Reverchon 1961. Dispositif de bombardement ionique pour préparations micrographiques. *J. Phys. Radium* **22**, 103–7A.

Peterson, N. L. 1983. Grain boundary diffusion in metals. *Int. Metals Rev.* **28**, 65–91.

Pharr, G. M. 1981. Some observations on the relation between dislocation substructure and power law breakdown in creep. *Scripta Metall.* **15**, 713–17.

Poirier, J.-P. 1982. On transformation plasticity. *J. Geophys. Res.* **87**, 6791–7.

Poirier, J.-P. 1985. *Creep of crystals*. Cambridge: Cambridge University Press.

Poirier, J.-P. & A. Nicolas 1975. Deformation-induced recrystallization due to progressive misorientation of subgrains, with special reference to mantle peridotites. *J. Geol.* **83**, 707–20.

Poirier, J.-P., J. Peyronneau, J. Y. Gesland & G. Brebec 1983. Viscosity and conductivity of the lower mantle: an experimental study on a $MgSiO_3$ analogue: $KZnF_3$. *Phys. Earth Planet. Inter.* **32**, 273–87.

Poirier, J.-P., C. Sotin & S. Beauchesne 1990. Experimental deformation and data processing. This volume, 179–89.

Poumellec, B. & O. Jaoul 1984. Dependence of creep rate of Fe-bearing olivines on pO_2. In *Deformation of ceramics II*, R. C. Bradt & R. E. Tressler (eds), 281–305. New York: Plenum.

Raj, R. & A. K. Ghosh 1981. Micromechanical modelling of creep using distributed parameters. *Acta Metall.* **29**, 283–92.

Raj, R., R. L. Tsai, J. G. Wang & C. K. Chyung 1984. Superplastic flow in ceramics enhanced by a liquid phase. In *Deformation of ceramics II*, R. C. Bradt & R. E. Tressler (eds), 353–78. New York: Plenum.

Ranalli, G. 1982. Deformation maps in grain size-stress space as a tool to investigate mantle rheology. *Phys. Earth Planet. Inter.* **29**, 42–50.

Ricoult, D. L. & D. L. Kohlstedt 1983. Structural width of low-angle grain boundaries in olivine. *Phys. Chem. Minerals* **9**, 133–8.

Ricoult, D. L. & D. L. Kohlstedt 1985. Experimental evidence for the effect of chemical environment on the creep rate of olivine. In *Point defects in minerals*, R. N. Schock (ed.) 171–84. Geophysical Monograph **31**. Washington, DC: American Geophysical Union.

Roberts, W. 1986. Microstructural evolution and flow during hot working. In *Strength of metals and alloys (ICSMA 7)*, vol. 3, H. J. McQueen, J.-P. Bailon, J. I. Dickson, J. J. Jones & M. G. Akben (eds), 1859–91. Oxford: Pergamon.

Ronchi, C. & H. Blank 1970. Lattice defects in the fluorite structure of UO_2 and UO_{2+x}. *Nucl. Metall.* **17**, 174–82.

Ruano, O. A. & O. D. Sherby 1988. On constitutive equations for various diffusion-controlled creep mechanisms. *Rev. Phys. Appl.* **23**, 625–37.

Rubie, D. C. 1990. Mechanisms of reaction-enhanced deformability in minerals and rocks. This volume, 262–95.

Rutter, E. H. 1983. Pressure solution in nature, theory and experiment. *J. Geol Soc. Lond.* **140**, 725–40.

Sammis, C. G. & J. L. Dein 1974. On the possibility of transformational superplasticity in the Earth's mantle. *J. Geophys. Res.* **79**, 2961–5.

Sammis, C. G., J. C. Smith & G. Schubert 1981. A critical assessment of estimation methods for activation volume. *J. Geophys. Res.* **B86**, 10 707–18.

Schmid, S., J. N. Boland & M. S. Paterson 1977. Superplastic flow in finegrained limestone. *Tectonophysics* **43**, 257–91.

Sellars, C. M. 1978. Recrystallization of metals during hot deformation. *Phil Trans R. Soc. Lond.* **A228**, 147–58.

Sherby, O. D., J. L. Robbins & A. Goldberg 1970. Calculation of activation volumes for self-diffusion and creep at high temperatures. *J. Appl. Phys.* **41**, 3961–8.

Soma, T., M. Matsui & I. Oda 1984. Creep phenomena in reaction-bonded and sintered silicon nitrides. In *Deformation of ceramics II*, R. C. Bradt & R. E. Tressler (eds), 379–403. New York: Plenum.

Spiers, C. J. & P. M. T. M. Schutjens 1990. Densification of crystalline aggregates by fluid phase diffusional creep. This volume, 334–53.

Spry, A. 1969. *Metamorphic textures*. Oxford: Pergamon.

Stocker, R. L. & M. F. Ashby 1973. On the rheology of the upper mantle. *Rev. Geophys. Space Phys.* **11**, 391–426.

Takeshita, T., H.-R. Wenk, G. R. Canova & A. Molinari 1990. Simulation of dislocation-assisted plastic deformation in olivine crystals. This volume, 365–76.

Takeuchi, S. & A. S. Argon 1976. Steady state creep of single phase crystalline material at high temperatures. *J. Mat. Sci.* **11**, 1542–66.

Taylor, G. I. 1938. Plastic strain in metals. *J. Inst. Metals* **62**, 307–24.

Thompson, A. B. & D. C. Rubie (eds) 1985. *Metamorphic reactions: kinetics, textures and deformation*. New York: Springer.

Tullis, J. 1990. Experimental studies of deformation mechanisms and microstructures in quartzo-felspathic rocks. This volume, 190–227.

Tullis, J., G. L. Shelton & R. A. Yund 1979. Pressure dependence of rock strength: implications for hydrolytic weakening. *Bull. Mineral.* **102**, 110–14.

Tullis, J. & R. A. Yund 1985. Dynamic recrystallization in felspars: a mechanism for ductile shear formation. *Geology* **13**, 238–41.

Tvergaard, V. 1988. Mechanical models of the effect of grain boundary sliding on creep and creep rupture. *Rév. Phys. Appl.* **23**, 595–604.

Urai, J. L., W. D. Means & G. S. Lister 1986. Dynamic recrystallization of minerals. In *Mineral and rock deformation: laboratory studies*, B. E. Hobbs & H. C. Heard (eds), 161–99. Geophysical Monograph 36. Washington, DC: American Geophysical Union.

Van Houtte, P. & F. Wagner 1985. Development of textures by slip and twinning. In *Preferred orientation in deformed metals and rocks: an introduction to modern texture analysis*, H.-R. Wenk (ed.), 233–58. New York: Academic Press.

Van Landuyt, J., G. Van Tendeloo & S. Amelinckx 1987. Phase transformations as studied by electron microscopy. *Ultramicroscopy* **23**, 371–82.

Vaughan, P. J., H. W. Green & R. S. Coe 1984. Anisotropic growth in the olivine–spinel transformation of Mg_2GeO_4 under non-hydrostatic stress. *Tectonophysics* **108**, 299–322.

von Mises, R. 1928. Mechanik der plastichen Formänderung von Kristallen. *Z. Angew. Math. Mech.* **8**, 161–85.

Weertman, J. 1957. Steady-state creep of crystals. *J. Appl. Phys.* **28**, 1185–9.

Weertman, J. 1970. The creep strength of the Earth's mantle. *Rev. Geophys. Space Phys.* **8**, 145–68.

Weertman, J. 1968. Dislocation climb theory of steady-state creep. *Trans Am. Soc. Metals* **61**, 681–94.

Weertman, J. 1972. High temperature creep produced by dislocation motion. In *Rate processes in plastic deformation of materials: proceedings of the John E. Dorn Memorial Symposium*, J. C. M. Li & A. K. Mukherjee (eds), 315–36. Cleveland, Ohio: American Society of Metals.

Weertman, J. 1983. Creep deformation of ice. In *Annual Review of Earth and Planetary Sciences*, G. Wetherill (ed.), vol. 11, 215–40. Palo Alto, California: Annual Reviews Inc.

Weertman, J. & J. C. Weertman 1987. Constitutive equations and diffusion-dislocation controlled creep. In *Constitutive relations and their physical basis*, S. I. Anderson, J. B. Bilde-Sørensen, N. Hansen, T. Leffers, H. Lilholt, O. B. Pederson & B. Ralph (eds), 191–203. Proc. 8th Risø Int. Symp. Metall. Mat. Sci., Roskilde. Roskilde, Denmark: Risø National Laboratory.

Wenk, H.-R. (ed.) 1976. *Electron microscopy in mineralogy*. Berlin: Springer.

Wenk, H.-R. (ed.) 1985. *Preferred orientation in deformed metals and rocks: an introduction to modern texture analysis*. New York: Academic Press.

Wenk, H.-R., G. Canova, A. Molinari & H. Mecking 1989. Texture development in halite: comparison of Taylor model and self-consistent theory. *Acta Metall.* **37**, 2017–29.

White, S. 1973. Syntectonic recrystallization and texture development in quartz. *Nature* **244**, 276–8.

White, S. & J. C. White 1981. On the structure of grain boundaries in tectonites. *Tectonophysics* **78**, 613–28.

Williame, C., W. L. Brown & M. Gandais 1976. Physical aspects of exsolution in natural alkali felspars. In *Electron microscopy in mineralogy*, H.-R. Wenk (ed.), 248–57. Berlin: Springer.

Yan, M. F. & A. H. Heuer (eds) 1983. *Character of grain boundaries: Advances in ceramics* **6**. Columbus, Ohio: American Ceramic Society.

Yavari, P., D. A. Miller & T. G. Langdon 1982. An investigation of Harper–Dorn creep. I – Mechanical and microstructural characteristics. *Acta Metall.* **30**, 871–9.

CHAPTER SEVEN

Experimental deformation and data processing

Jean-Paul Poirier, Christophe Sotin & Solange Beauchesne

7.1 Introduction

The methods and procedures of experimental deformation of minerals and rocks, now a valued tool of structural geology, tectonics, and geophysics, have been transferred almost in totality from the practice of physical metallurgy. Indeed, even in the newer field of materials science, dealing with ceramics and other materials, the traditional methods for deriving rheological laws from high-temperature creep experiments have been religiously handed down from adviser to student.

The purpose of this chapter is to try and make the Earth scientist reader receptive to the idea that these hallowed methods are not necessarily the best, even though they are used by everybody (including, until very recently, one author of this contribution). The principle of an alternative method will be presented and demonstrated using preliminary results of high-temperature creep experiments on single crystals of barium titanate (a crystal with the perovskite structure).

7.2 The traditional methods

Let us first rapidly – and critically – review the traditional methods used to derive a rheological law from the data obtained in laboratory deformation experiments.

A rheological law makes sense only in the case of steady-state (or quasi-steady-state) deformation, and we will accept here that it is the equation relating all relevant physical and mechanical parameters that fits the experimental data best.

For metals, there are usually four parameters (stress, strain rate, temperature, and grain size), but there can be many more for ceramics and minerals, e.g. oxygen partial pressure, degree of non-stoichiometry and, of course, hydrostatic pressure.

The most usual kinds of experiments are constant-stress creep tests or constant-strain-rate experiments (especially with Griggs-type apparatus). In the former case the steady-state strain rate $\dot{\varepsilon}$ is measured for several values of applied stress σ and temperature T (and possibly other parameters), whereas in the latter case the strain rate is imposed at a given temperature and the measured stress is that corresponding to a plateau of the stress–strain curve. In both cases, the results consist of values of creep rates for couples of values of stress and temperature, which, in most instances, cannot be displayed as a simple matrix, i.e. raw data can seldom be arranged in suites of values of creep rates at various stresses for the same temperature, or creep rates at various temperatures for the same stress. From then on, two main classes of methods have been and still are used:

(a) The experimental values of $\dot{\varepsilon}$ are normalized in temperature using a value of activation energy chosen *a priori* or determined by trial and error. The parameter thus found (the Zener–Hollomon parameter; see Poirier 1985) is logarithmically plotted against the stress normalized by the shear modulus, on a single master curve, thus yielding the stress dependence. This method, popularized by Dorn's school, involves too much *a priori* massaging of the data to be of real value when one wants to find an unknown rheological law. We mention it here for the sake of completeness and will not discuss it further.

(b) The most popular current method for finding a rheological law consists of plotting separately $\ln \dot{\varepsilon}$ versus $1/T$ and $\ln \dot{\varepsilon}$ versus $\ln \sigma$ (Poirier 1985). In each plot, straight lines are fitted to the points corresponding to experiments performed at the same stress or temperature. The slope of the lines in the Arrhenius plot provides the activation enthalpy of creep: if the slope does not depend on stress a power law is acceptable, if not, the activation enthalpy is stress dependent, leading to an exponential stress dependence of the creep rate. The $\ln \dot{\varepsilon}$ versus $\ln \sigma$ plot provides the stress exponent n, at least for the values of σ corresponding to a constant slope. This procedure, although widely used, is far from being satisfactory for many reasons, as listed below:

(i) The rheological law to be found $(\dot{\varepsilon}, \sigma, 1/T, P, d, \text{etc.}) = 0$ can be represented by a surface in the space of the parameters; strain rate $\dot{\varepsilon}$, stress σ, temperature T, pressure P, grain size d, etc. The problem is to find the *best* surface, fitted to the experimental data, represented by points with experimental error bars in that space. Now, in the usual procedure, sections of the surface by planes $\sigma = \text{constant}$, $T = \text{constant}$ etc. are *independently* fitted to the experimental results, yielding parameters such as $d(\ln \dot{\varepsilon})/d(1/T)$ or $d(\ln \dot{\varepsilon})/d \ln \sigma$, which will be identified as the activation enthalpy or the power-law stress exponent. Even though some check of internal

180

consistency can be provided by graphically correlating the Arrhenius and $\ln \dot\varepsilon - \ln \sigma$ plots (Poirier 1985), there is no reason to believe that the best fits (supposing that they are the best, see below) in each of these sections coincide with sections of the best overall fitting surface.

(ii) If, as it often is the case, there are few points, if any, at the same stress or temperature, it is difficult to fit lines to the data in the Arrhenius or $\ln \dot\varepsilon - \ln \sigma$ plots unless one reduces the data to a few common temperatures or stresses, using values of the activation energy or stress exponent derived from few experimental points, a procedure that is at best self-consistent; however, its impact on the results is usually not estimated.

(iii) If the Arrhenius or the $\ln \dot\varepsilon - \ln \sigma$ plots are curved – not an unusual occurrence – it is a matter of arbitrary choice to divide the curves into approximately linear portions and to assume different creep régimes, or to find a law that accounts for the observed temperature and stress dependence (usually an empirical law).

(iv) The technique for fitting a straight line through the experimental points is fraught with often unsuspected difficulties and ambiguities. Most often, straight lines are fitted by linear regression of $\ln \dot\varepsilon$ on $1/T$ or $\ln \sigma$. This means that, due to the usual way of representing the data, the sum of the squares of distances from the experimental points to the line, *parallel to the strain-rate axis*, is minimized. If the correlation coefficient is different from 1, the regression of $\ln \dot\varepsilon$ on $1/T$ will be different from that of $1/T$ on $\ln \dot\varepsilon$ (the correlation coefficient is related to the angle between the two lines), the values found for the activation energy, or the stress exponent, will be different, but there is absolutely no reason why $\ln \dot\varepsilon$ versus $1/T$ (or $\ln \sigma$) should be privileged over $1/T$ (or $\ln \sigma$) versus $\ln \dot\varepsilon$; this is, however, currently done. In a sense, eyeball fitting is more satisfactory. Multilinear regression answers the objections made in (i) and allows the determination of a best fitting surface (instead of separate sections). However, it is the surface $\ln \dot\varepsilon = f(\sigma, 1/T)$ that is determined by minimizing the sum of the squares of the differences in $\ln \varepsilon$ between the experimental points and the surface (e.g. Heard & Kirby 1981). Considering $\sigma = g(\dot\varepsilon, 1/T)$ or $1/T = h(\dot\varepsilon, \sigma, ...)$ would yield different results, equally worthy of consideration.

(v) Whether the fit to scattered experimental points is obtained by eyeball, linear, or multilinear regression, it cannot take into account the experimental errors and uncertainties in the determination of the creep rate and the measurement of stress (especially with the Griggs-type apparatus), temperature, and grain size (usually not unimodal). No statistically meaningful standard deviation can be quoted for the parameters of the rheological law. Even though, in most cases, values of Q or n are quoted with an uncertainty, it has no statistical value and may even be overestimated (for instance, by fitting lines through the extremities of error bars).

(vi)   It follows from these observations that in the case in which different laws can be fitted to the results (e.g. a climb-controlled power-law or a glide-controlled law with a stress-dependent activation energy), it is impossible to decide objectively which law fits the data best – with weighty consequences when the laws are to be extrapolated.

7.3   A global inversion method

We are faced with the problem of extracting a few rheological parameters (activation energy, stress, pressure, grain size, oxygen pressure dependence, etc.) from what is, in the best case, a large number of experimental data values; each experimental point is characterized by the values of the measured (or imposed) strain rate, the imposed (or measured) stress, the temperature, hydrostatic pressure, grain size, oxygen partial pressure, etc. All of these values have experimental errors attached, due to either heterogeneity or fluctuations of the measured quantity, or to the measurement technique.

Similar, even more complicated, problems exist in geophysics, such as, for instance, determining the density and elastic constants of the Earth's material as a function of depth from the arrival times of seismic waves at various stations. These problems are solved by the use of inversion methods.

The method proposed here, to determine rheological laws from creep data, is a generalized non-linear inversion technique originally due to Tarantola & Valette (1984). Sotin & Poirier (1982) proposed use of the method for processing creep data, and it has been presented in detail in Sotin (1986). In what follows, we will summarily present its principle and, by use of an example, illustrate the advantages that it presents over the traditional method.

The method has the great advantage of not privileging any of the data; indeed, it treats all experimental data and unknown parameters on an equal footing, as unknown quantities. The experimental data have a most probable value equal to the measured value and a variance equal to the square of the experimental error, and the creep parameters have a most probable value fixed *a priori* (within reasonable limits) with an infinite or very large variance. Data and parameters are assumed to follow a Gaussian (or lognormal in the case of $\dot{\varepsilon}$) probability law.

We will consider here the simplest case in which the only variables are the temperature and stress, and the creep law to be fitted is a power law of the form:

$$\dot{\varepsilon} = A\sigma^n \exp\left[-(Q/RT)\right]$$

or

$$\ln \dot{\varepsilon} = \ln A + n \ln \sigma - Q/RT$$

The parameters to be determined are $\ln A$, n, and Q. The same method would apply if there were more variables to take into account (P, grain size, etc.) and consequently more parameters to determine (activation volume, grain-size dependence, etc.).

If there are N experiments, for the ith experiment we have a triplet of experimental data:

$$x_i = \begin{pmatrix} \ln \dot{\varepsilon}_i \\ \sigma_i \\ T_i \end{pmatrix} \quad \text{with } i \leqslant N$$

The N x_i's are taken together with the creep parameters ($x_{N+1} = \ln A$, $x_{N+2} = n$, $x_{N+3} = Q$) as the $3N + 3$ components of a single vector X.

The inversion yields the most probable vector X, compatible with the experimental errors and obeying the N, generally non-linear, equations:

$$f_i(X) = \ln \dot{\varepsilon}_i - \ln A - n \ln \sigma_i + Q/RT_i = 0$$

X is calculated by iteration from an algorithm embodying the least-squares minimization procedure in the non-linear case (which involves matrices of partial derivatives of the f_i's with respect to the x_i's; Sotin & Poirier 1984). The iteration is stopped when a test of convergence indicates that $f_i(X) \leqslant 10^{-6}$. Thus, the inversion procedure yields best values of the parameters and recalculated values of the experimental data, which can be compared to the initial values. The *a posteriori* standard deviation can be calculated. Large residuals from some experimental data may warrant re-examination and possibly elimination of the data (if a good reason to eliminate them is found), or division of the data population into subpopulations, which will be treated independently. If the iteration converges too slowly, or not at all, it may be a sign that the law is not appropriate. Thus, this method allows objective comparison between several laws. The best law is the one that leads to the best and quickest inversion with the smallest residuals and standard deviation. Let us also note that statistical tests (chi-squared test) can be performed on the results in order to assess the validity of the assumption of Gaussian behaviour of the errors.

7.4 An example of creep data processing

As an example, we will consider preliminary results of creep experiments on single crystals of barium titanate (with perovskite structure), $BaTiO_3$. No definitive value should be attached to these results, which must be taken here only as a suite of real creep data, to be used in a case study of data processing. The article on creep of barium titanate has been published recently (Beauchesne and Poirier 1989). The creep data considered here consist of the strain-rate values measured in 32 experimental runs at five stresses and various high temperatures. The stresses are applied by putting weights in a pan: it is therefore easy to fix the nominal stresses at predetermined values. However, this is not done as easily for the temperatures, which are measured after stabilization of the furnace temperature. The data are given in Table 7.1 as values of $\ln \dot{\varepsilon}$ and $10^4 T$ for the five applied stresses.

Table 7.1 Data from creep experiments on barium titanate.

σ (MPa)	$10^4/T$	$\ln \dot{\varepsilon}_E$	$\Delta \ln \dot{\varepsilon}$
9	5.75	-12.23	-1.30
	5.75	-12.35	-0.93
	5.75	-13.12	$+0.61$
	5.76	-13.45	$+1.52$
	5.76	-13.34	$+1.22$
	5.91	-14.11	$+0.64$
	5.96	-13.74	-1.28
	6.15	-15.12	-0.81
	6.32	-16.12	-1.12
12	5.75	-11.35	-0.24
	5.76	-10.98	-0.21
	5.76	-11.58	-0.28
	5.96	-13.30	-0.02
	6.15	-14.92	$+1.61$
14	5.74	-10.80	-0.64
	5.94	-12.17	-0.01
	6.16	-13.36	-0.73
	6.61	-16.52	$+0.11$
18	5.96	-11.23	-0.12
	6.04	-12.50	$+2.06$
	6.12	-12.57	$+0.92$
	6.22	-13.43	$+1.33$
	6.37	-14.22	$+0.83$
	6.71	-16.60	$+1.16$
25	6.36	-12.60	$+0.05$
	6.39	-12.41	-1.49
	6.53	-14.33	$+2.49$
	6.53	-13.41	-0.39
	6.76	-15.65	$+1.66$
	6.79	-14.46	-1.97
	6.79	-15.02	-0.58
	7.01	-16.34	-1.93

$\ln \dot{\varepsilon}_E$ are experimental values.
$\Delta \ln \dot{\varepsilon}$ are reduced residuals, expressed in standard deviations.

The rheological law to be fitted to the results is the power law written in the form:

$$\ln \dot{\varepsilon} = \ln A + n \ln \sigma - Q/RT$$

Arrhenius plots of the data for the five stresses (9, 12, 14, 18, and 25 MPa) are shown in Fig. 7.1. In each case the usual linear regression of $\ln \dot{\varepsilon}$ on $1/T$ has been carried out and the line drawn through the data points; the line corresponding to the section of the best fitting surface, determined by global inversion, is also represented.

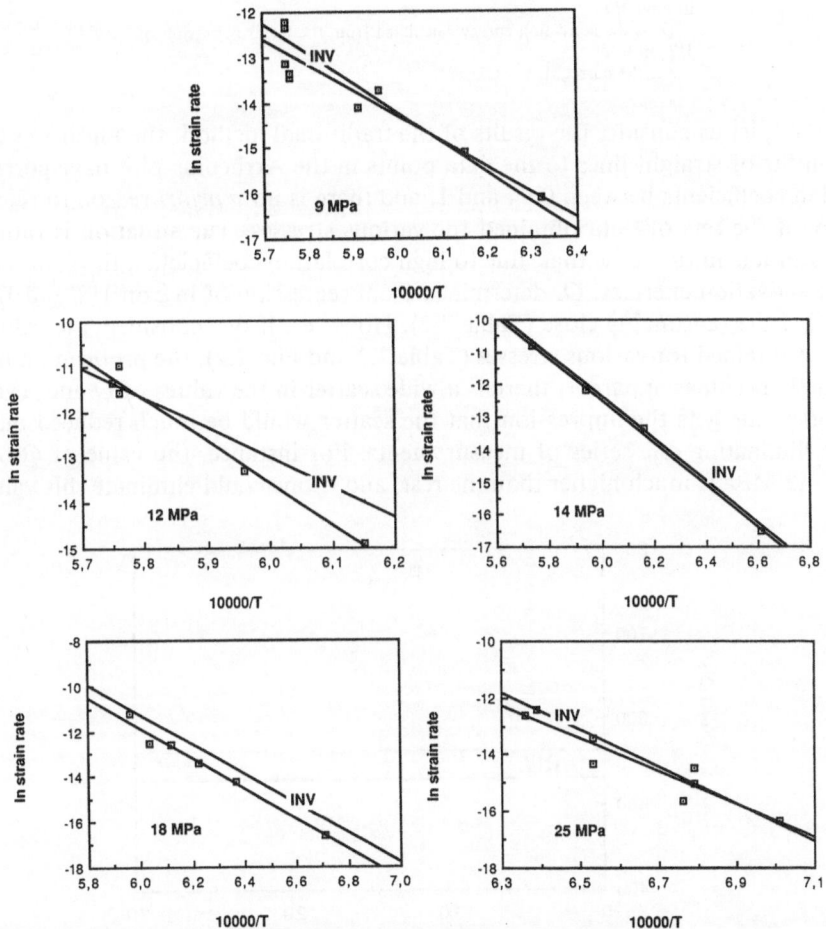

Figure 7.1 Arrhenius plots for high-temperature creep of barium titanate at stresses of 9, 12, 14, 18, and 25 MPa. The unmarked straight line through the data points corresponds to the linear regression fit of $\ln \dot{\varepsilon}$ on $1/T$, and the line marked INV corresponds to the section of the best surface obtained by inversion, by the constant-stress plane.

Table 7.2 Activation energies of creep calculated from the slope of linear regression fits to the data, for runs at various stresses, σ.

σ (MPa)	Q (kJ mol^{-1})	Q' (kJ mol^{-1})	r
9	472	531	0.94
12	775	790	0.99
14	546	547	1.00
18	563	573	0.99
25	484	550	0.94

Q is the activation energy calculated from the linear regression of $\ln \dot{\varepsilon}$ on $1/T$.

Q' is the activation energy calculated from the linear regression of $1/T$ on $\ln \dot{\varepsilon}$.

r is the correlation coefficient.

First, let us consider the results of the traditional method: the linear regression fits of straight lines to the data points in the Arrhenius plot have correlation coefficients between 0.94 and 1, and there is no *a priori* reason to reject any of the sets of data obtained for various stresses. The situation is rather favourable in the sense that, due to high correlation coefficients, the values of the activation energies, Q, determined from regression of $\ln \dot{\varepsilon}$ on $1/T$ and $1/T$ on $\ln \dot{\varepsilon}$ are reasonably close (Table 7.2). However, if one considers the values of Q obtained for various stresses (Table 7.2 and Fig. 7.2), the problem immediately becomes apparent: there is a wide scatter in the values of Q and, even worse, one gets the impression that the scatter would be much reduced only by eliminating one series of measurements. For instance, the value of Q for $\sigma = 12$ MPa is much higher than the rest, and if one could eliminate this value

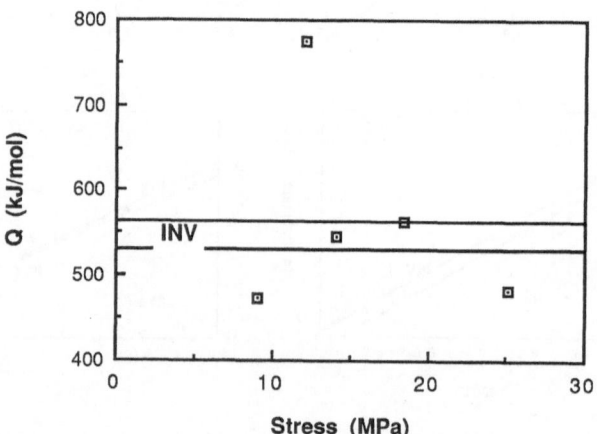

Figure 7.2 Activation energy as a function of stress for high-temperature creep of barium titanate. The best activation energy obtained by inversion is 547.6 ± 16.6 kJ/mole^{-1}, within the band marked INV.

186

Q would appear to be stress independent, as is expected from a power-law creep: conversely, if one could eliminate the value of Q for $\sigma = 9$ MPa, there would appear to be an unmistakable dependence of Q on σ. Indeed, the experiments at 9 MPa were the last to be performed and, until then, we thought that there was a stress dependence of Q.

After carefully screening the experimental procedure and the raw data for the runs at 9 and 12 MPa, it appeared that there was absolutely no good reason to eliminate either one of the runs and that they were quite as good as the others. Even though such quandaries may not occur extremely frequently (or at least are not frequently reported), the situation is rather typical and clearly illustrates the shortcomings of the traditional methods. The same situation obtains for the stress exponent: the $\ln \dot{\varepsilon}$–$\ln \sigma$ plots at various temperatures lead to different values of n between which it is not easy to choose (Fig. 7.3). As we shall presently see, the global inversion technique is free from such inconveniences.

The inversion was carried out using two different rheological laws: a power law with stress-independent activation energy,

$$\ln \dot{\varepsilon} = \ln A + n \ln \sigma - Q/RT$$

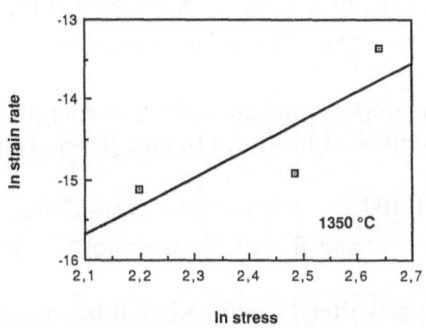

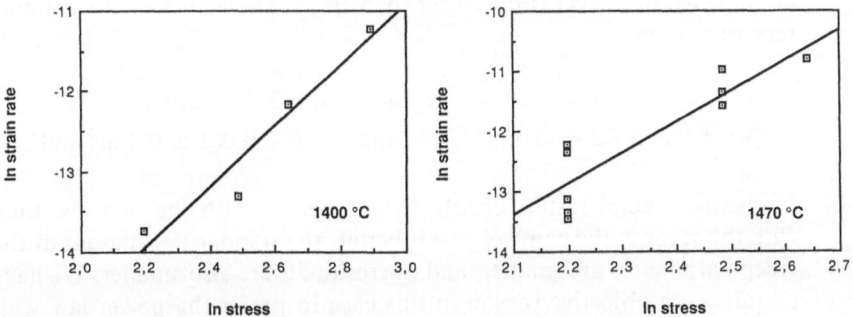

Figure 7.3 Determination of the stress exponent by linear regression of $\ln \dot{\varepsilon}$ on $\ln \sigma$. For $T = 1350°C$, $n = 3.6$; for $T = 1400°C$, $n = 3.8$; and for $T = 1470°C$, $n = 5.1$.

and a law with a stress-dependent activation energy, usually reflecting glide- or cross-slip control,

$$\ln \dot{\varepsilon} = \ln A + n \ln \sigma - Q_0/RT + B\sigma/RT$$

The errors on temperature, stress, and strain rate were respectively taken as

$$\Delta T = \pm 4°C, \qquad \Delta\sigma/\sigma = 0.04, \qquad \text{and} \qquad \Delta \ln \dot{\varepsilon} = \pm \ln 1.2$$

(i) For the power law, the *a priori* values and standard deviations of $\ln A$, n, and Q were respectively taken as

$$\ln A = -50 \pm 100, \qquad n = 5 \pm 5, \qquad \text{and} \qquad Q = 300 \pm 300 \text{ kJ mol}^{-1}$$

After eight iteration steps, convergence to better than 10^{-6} was already obtained, with residuals smaller than 2.5 standard deviations at most (Table 7.1). The chi-squared test showed that the assumption of Gaussian statistics was correct to within a 95 per cent confidence limit. The final values of the parameters were

$$\ln A = -43.46 \pm 2.44, \qquad n = 4.28 \pm 0.19, \qquad \text{and}$$
$$Q = 547.6 \pm 16.6 \text{ kJ mol}^{-1}$$

(ii) For the law with stress-dependent activation energy the *a priori* values and standard deviation of $\ln A$, n, Q_0, and B were respectively taken as

$$\ln A = -50 \pm 100, \qquad n = 5 \pm 5, \qquad Q_0 = 300 \pm 300 \text{ kJ mol}^{-1},$$
$$\text{and } B = 10 \pm 10 \text{ m}^3 \text{mol}^{-1}$$

Iteration was stopped after 18 steps, when it became obvious that there would be no better convergence: successive values of $f_i(X)$ oscillated about zero, but never came closer than 10^{-2}. The values of the parameters were then

$$\ln A = -25.8 \pm 11.06, \qquad n = 3.25 \pm 0.66,$$
$$Q_0 = 571.4 \pm 22.4 \text{ kJ mol}^{-1}, \qquad \text{and} \qquad B = 0.001 \pm 0.4 \text{ m}^3 \text{mol}^{-1}$$

(iii) The experimental results clearly 'invert better' with the first law than with the second: the convergence is better, the standard deviations of the creep parameters are smaller, and the residuals are also smaller. We have therefore an objective reason in this case to prefer the power law with stress-independent activation energy to the law with a stress-dependent activation energy.

The lines representing the sections of the best surface by constant-stress planes are represented on the Arrhenius plot of Figure 7.1. It appears that they can be quite different from the lines obtained by regression. The values of the activation energy obtained by inversion are also reported in Figure 7.2: the inversion procedure leads to a stress-independent activation energy with a reasonably small standard deviation that, despite the appearances, really takes into account the experimental results that, in the traditional approach, led to a value of Q far above the average.

7.5 Conclusion

We have shown by means of one example that global inversion techniques, when used to process experimental creep data, provide much more consistent and reliable results than the traditional method of separately finding the effect of temperature and stress by linear regression on plots at constant stress or temperature. The inversion technique allows an argumented choice between various rheological laws and provides creep parameters with standard deviations that authentically reflect the experimental uncertainties. There is no doubt that it allows safer extrapolation of creep laws to geological conditions.

Acknowledgements

Our thanks are due to David Barber who provided the incentive to write this chapter. This is IPG contribution no. 1102.

References

Beauchesne, S. & J.-P. Poirier 1989. Creep of barium titanate perovskite: a contribution to a systematic approach to the viscosity of the lower mantle. *Phys. Earth. Planet. Inter.* **55**, 187–99.

Heard, H. & S. H. Kirby 1981. Activation volume for steady state creep in polycrystalline CsCl: cesium chloride structure. In *Mechanical behavior of crystal rocks*, N. L. Carter, M. Friedman, J. M. Logan & D. W. Stearns (eds), Geophysical Monograph 24. Washington D.C.: American Geophysical Union.

Poirier, J.-P. 1985. *Creep of crystals*. Cambridge: Cambridge University Press.

Sotin, C. 1986. Contribution á l'Etude de la Structure et de la Dynamique Interne des Planètes. ScD dissertation, University Paris VII.

Sotin, C. & J.-P. Poirier 1984. Analysis of high temperature creep experiments by generalized non-linear inversion. *Mech. Mat.* **3**, 311–17.

Tarantola, A. & B. Valette 1982. Generalized non-linear inverse problems solved in using the least-squares criterion. *Rev. Geophys. Space Phys.* **20**, 219–32.

CHAPTER EIGHT

Experimental studies of deformation mechanisms and microstructures in quartzo-feldspathic rocks

Jan Tullis

8.1 Introduction

Experimental rock deformation studies provide two kinds of information. First, they provide information which will help to *interpret* the preserved micro- and macroscale structures in naturally deformed rocks, in terms of the P and T conditions of deformation, the flow stresses and finite strains, and whether the deformation was brittle or ductile, seismic or aseismic. Second they provide data which can be used in *predictive* modelling of the thermo-mechanical evolution of structures on various scales, ranging from single faults and folds to plate boundaries to whole mantle convection. In all cases the usefulness of the experimental studies depends on a demonstration that, despite large differences in strain rate, the same *processes* have operated in both the experimental and the natural deformations.

In the past there has been considerable scepticism that deformation experiments lasting only a few hours to a few months could possibly involve the same processes as natural deformations lasting millions of years, but a combination of theory, experiments, and observations provides considerable confidence that this is in fact true. For metals, ceramics, and rocks, the current understanding is that there are a relatively small number of distinct grain-scale deformation mechanisms, each of which is dominant at a particular set of P and T conditions, has a particular steady-state flow law relating flow stress to strain rate, and produces a characteristic set of deformation microstructures. Under a differential stress all deformation mechanisms will tend to operate, but at different rates, depending on the P and T conditions; usually one mechanism will contribute most of the strain, and thus be dominant. This is shown conveniently on a deformation mechanism map (Ashby 1972); a map for olivine was calculated by Stocker & Ashby (1973), and numerous examples are compiled by Frost & Ashby (1982). The axes of these maps are commonly normalized flow stress and temperature, and boundaries are drawn where two mechanisms contribute equal strain rates, thereby defining fields in which single mechanisms are dominant. The whole map is contoured for total strain rate. A schematic version of a deformation mechanism map is shown in Fig. 8.1. Although we do not have sufficient fundamental information on diffusivities etc. for most minerals to construct accurate maps, a combination of existing flow law data plus analogies with well studied metals gives us confidence in the overall configuration. Fracture and cataclastic flow occur at low temperature and high stress, dislocation glide at somewhat higher temperature, dislocation creep at still higher temperature and lower stress, and diffusion creep (both grain boundary and volume) at moderate to high temperature and low stress. The locations of the mechanism field boundaries depend on pressure, grain size, presence of fluids, etc. For example, the field of diffusion creep is greatly expanded by smaller grain size (smaller diffusion distances) and the presence of fluids (faster boundary diffusion paths); see the cases calculated by Rutter (1976).

Successful experimental studies of a particular deformation mechanism require that its field of dominance on a deformation mechanism map include both natural and experimental strain-rate contours. Brittle deformation processes have very little strain-rate dependence, and so experimental studies can be carried out at approximately the same pressure and temperature as those of the natural deformation. However, for thermally activated ductile deformation mechanisms, there is a stronger strain-rate dependence and so in order to activate the same process as in nature, experimental studies must utilize unnaturally high temperatures. In Fig. 8.1 it is shown why most experimental

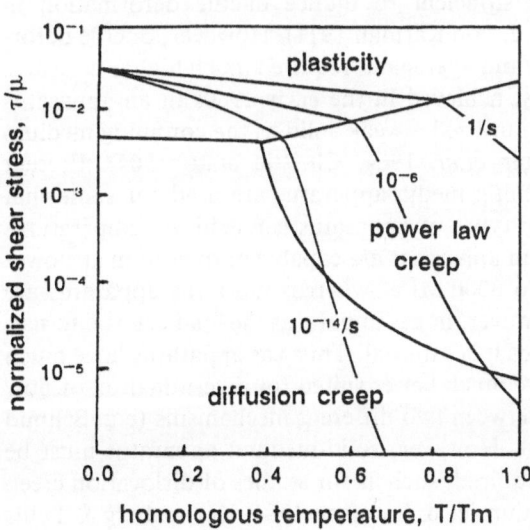

Figure 8.1 Schematic deformation-mechanism map, with axes of shear stress normalized by shear modulus, and temperature normalized by melting temperature, adapted from maps for olivine by Frost & Ashby (1982), for a grain size of ~50 μm. Fields of dominance for plasticity, power-law creep, and diffusion creep are shown, together with an experimental strain-rate contour (10^{-6} s^{-1}) and a natural strain-rate contour (10^{-14} s^{-1}).

studies of ductile deformation in rocks have involved dislocation creep: it is the dominant mechanism for most metamorphic conditions in the crust (and upper mantle), but it is also operative at faster strain rates and higher temperatures, largely because of its dependence on solid-state diffusion (see Poirier 1985). This gives rise to the experimentalist's common rule of trading time for temperatures. In Fig. 8.1 it is also indicated why experimental studies of diffusion creep have been more difficult: the stresses and strain rates are very low. However, recent experience shows that by using very high temperatures and fine grain sizes, experimentally measurable strain rates can be obtained (e.g. Rutter & White 1979, Karato *et al.* 1986).

In this chapter a brief review of some results from experimental deformation studies of quartzo-feldspathic aggregates is presented, with examples of their applicability to naturally deformed rocks. The chapter will focus almost entirely on studies of deformation mechanisms and microstructures, as there has been a recent review of applicability of experimental flow laws by Paterson (1987).

8.2 Experimental techniques

8.2.1 Apparatus and sample assembly

In order to inhibit cracking and allow extensive ductile deformation of aggregates in laboratory times, it is necessary to apply moderate pressure as well as elevated temperatures. Most early apparatus employed gas or liquid confining media and an external furnace; the modest temperatures and pressures achievable in these apparatus were sufficient to induce ductile deformation in materials such as carbonates (e.g. von Kármán 1911). However, ductile deformation of quartz single crystals and aggregates required much higher pressures and temperatures, and was first achieved in the early 1960s in an apparatus designed by D. T. Griggs which utilized a weak solid as the confining medium and an internal furnace (Carter *et al.* 1964, Christie *et al.* 1964). Present versions of gas and solid confining media apparatus are used for somewhat different kinds of studies. Both types of apparatus can achieve temperatures of at least 1200°C. Solid medium apparatus are capable of operation at slower strain rates and pressures up to 3000 MPa, whereas most gas apparatus are constrained to < 600 MPa. However, in gas apparatus the load cell is internal, whereas in solid media apparatus it is external. Thus gas apparatus have much greater stress sensitivity, and are much better suited for determination of flow laws, including the transition between two different mechanisms (e.g. Schmid *et al.* 1977, Karato *et al.* 1986). However, solid-medium apparatus must be used when high pressures are required, such as for studies of dislocation creep of quartz and feldspar (e.g. Mainprice & Paterson 1984, Kronenberg & Tullis

1984), the activation volume for creep of silicates (Green & Borch 1987), and the interaction of deformation and phase changes (e.g. Snow & Yund 1987). In a major innovation, Green & Borch (1989) have recently designed and successfully used a molten salt assembly in a Griggs-type apparatus; this provides a stress sensitivity of ± 10 MPa together with a pressure capability of at least 4 GPa.

The experimental studies described in this chapter have all been carried out using a modified Griggs-type (solid-medium) deformation apparatus; details of the apparatus and sample assembly can be found in Kronenberg & Tullis (1984) and Tullis & Tullis (1986). A few pertinent points may be mentioned here. Early versions of the sample assembly had large longitudinal temperature gradients from the centre to the ends of the sample, resulting in extremely inhomogeneous deformation and complicated deformation histories as material points moved across the gradients (e.g. Fig. 1 in Tullis et al. 1973). Present-day assemblies ensure that 90 per cent of the sample is within $\pm 10°$C of the thermocouple temperature. Earlier versions of the sample assembly utilized talc or alsimag as the confining medium, but the strength of these materials contributed significantly to the externally measured sample strength. At present it is most common to use a thick sleeve of NaCl (Fig. 8.2), or to make the whole assembly of NaCl (Kirby & Kronenberg 1984). Other confining media such as $CaCO_3$ and machinable glass ceramic have been used at temperatures $>900°$C, where NaCl melts (at 1500 MPa), but these materials contribute significantly to the externally measured sample strength due to high friction against the cold top moving piston.

Recently, increasing attention has been paid to controlling the chemical environment of samples during deformation. For over 20 years it has been known that trace amounts of water have a pronounced weakening effect on quartz deformed in the dislocation creep régime; quartz single crystals and aggregates were observed to be stronger when deformed in an anhydrous medium, and weaker (as well as exhibiting 'higher temperature' microstructures) when deformed in a hydrous medium (Griggs & Blacic 1965, Green et al. 1971). More recent experiments, using Pt-jacketed samples with ~ 0.1–0.5 wt% water mechanically sealed inside, show that in a NaCl assembly the added water is retained and induces weakening, but in a $CaCO_3$ assembly it is rapidly lost (e.g. Kronenberg & Tullis 1984). Buffered experiments in gas apparatus on olivine show important effects of f_{O_2} and f_{H_2} (e.g. Kohlstedt & Hornack 1981, Poumellac & Jaoul, 1984). Following predicted effects on other minerals by Hobbs (1985), buffered deformation experiments on quartz and feldspar have been attempted in solid-media apparatus (Ord & Hobbs 1986, Tullis & Yund 1988), although no certain effect has yet been documented.

In contrast to most metallurgical tests, almost all rock deformation apparatus, whether gas or solid confining medium, employ an axial compression geometry, with $\sigma_2 = \sigma_3$. This produces deformation microstructures with

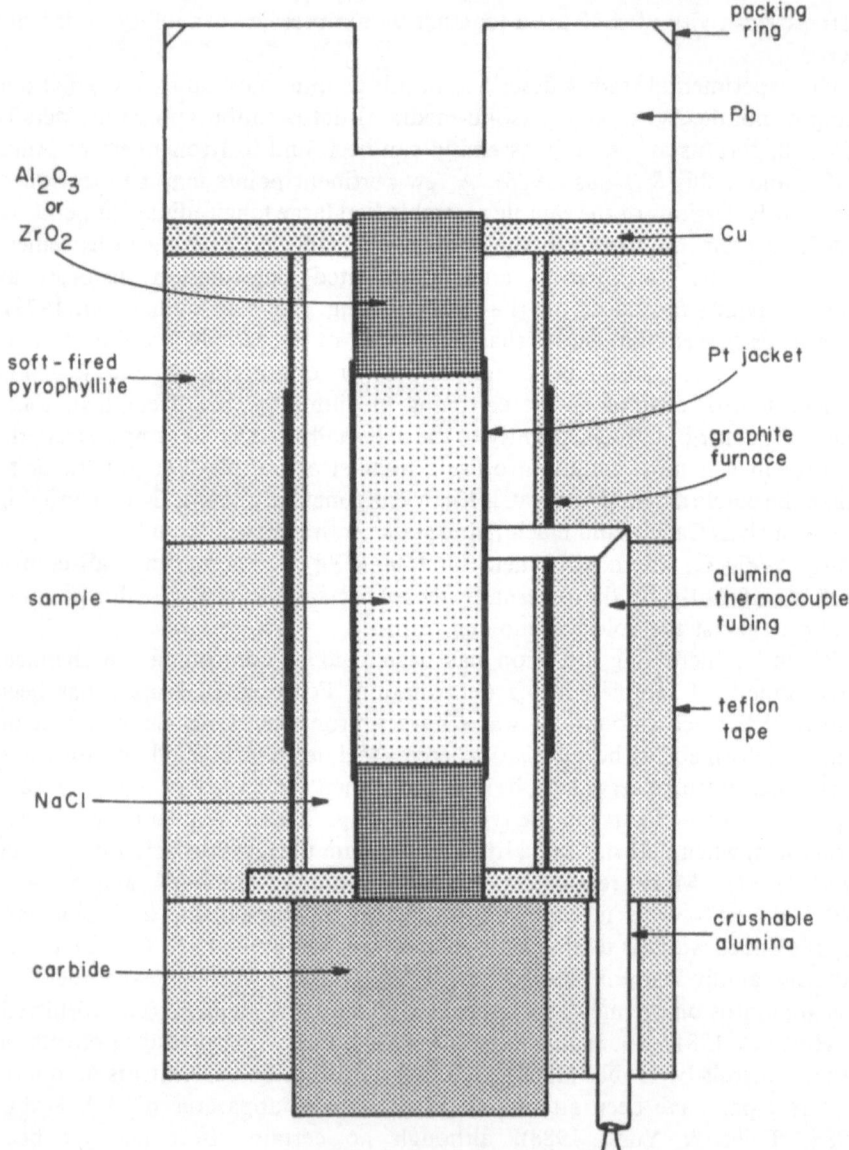

Figure 8.2 Sectional view of sample assembly (diameter 25.4 mm) used in high-temperature and pressure axial compression experiments in a Griggs-type apparatus. The use of a stepped furnace, metal jacket around the sample, and low thermal conductivity end pistons minimizes temperature gradients within the sample. The use of a thick NaCl liner allows high sample strains.

higher symmetry than found in many natural deformations, limiting the usefulness of experimental investigations of preferred orientation development, for example. However, sample assemblies in such apparatus can be modified to produce approximations of plane strain (e.g. Tullis 1977) or simple shear deformation (e.g. Schmid *et al.* 1987; Dell'Angelo & Tullis, 1989). Some rock deformation studies have employed a 'cubic' apparatus with six independently moveable pistons (e.g. Kern & Karl 1969); this apparatus allows more general deformations, but it is somewhat limited in the temperatures and pressures it can achieve, and most studies have been carried out on non-silicates.

Another limitation of experimental studies involves the relatively low total strain. Axial compression experiments in gas apparatus are usually limited to ≤ 20 per cent shortening, due to impingement of the sample on the furnace. In solid-medium apparatus much greater shortening strains (~ 75 per cent) are possible, and it is only at such high strains that quartzo-feldspathic rocks begin to appear 'mylonitic'. Multiple experiments to increasing strains show a substantial evolution of the deformation microstructures, such as original grain flattening, amount of recrystallization, and intensity of preferred orientation. These raise questions about the common practice of assuming steady-state flow and determining flow laws from very low strain experiments.

The level of the applied load in a creep experiment (generally used to determine flow laws), or the flow stress in a constant-displacement-rate test (usually employed to study microstructures), is of some concern. Many rock deformation studies employ very high differential stresses in order to produce measurable strains in laboratory times; 'unnaturally' high confining pressures are then required to prevent significant microcracking. Even if the stress is low enough to avoid cracking, it may be above the power law breakdown stress (e.g. Tsenn & Carter 1987; Blenkinsop & Drury 1988), where dislocation creep changes to dislocation glide. Natural stress magnitudes are difficult to measure, but are generally believed to be < 200 MPa (e.g. Solomon *et al.* 1980). Many experimental studies carried out to date have employed much higher stresses, raising questions about the significance of flow-law extrapolations (e.g. Paterson 1987). In general, the deformation processes characterizing high-grade metamorphic régimes (low stress, high temperature, and slow strain rate) remain less well investigated experimentally than those of lower grade.

As one final caution, it should be remembered that although experimental deformation studies allow one to control the deformation conditions, one must still infer the operative processes from various kinds of observations *after* the deformation. An innovative deformation technique which allows one to make direct optical observations of deformation in progress has been developed recently (Means 1983, Urai 1987). This technique employs analogue materials such as octochlorobenzene or minerals of low melting temperature, in aggregates one grain thick, sheared between glass plates under a microscope. Although the direct applicability of some of these materials to the behaviour of silicates remains to be documented, the ability to trace the evolution of individual recrystallizing grains, for example, has allowed important insights.

8.2.2 Starting materials

Some experimental deformation studies utilize single crystals, in order to investigate fundamental properties such as the yield strength for different slip systems, but microstructural (and flow-law) studies usually employ polycrystalline aggregates. To date most studies have been carried out on natural monomineralic aggregates, such as quartzite (e.g. Figs 8.3A & 8.5A), novaculite, dunite, ortho- and clinopyroxenite, albite rock (e.g. Fig. 8.6A) and anorthosite, with relatively few studies on polyphase aggregates such as aplite (e.g. Fig. 8.7A) and diabase. However, it is not always possible to find a naturally occurring pure, fresh aggregate with a grain size fine enough ($\leq 300 \ \mu$m) to provide a statistically valid number of grains in the small samples required for high-temperature and high-pressure experiments (e.g. 6×12 mm in the Griggs apparatus). Thus increasing use has been made of sintered aggregates, so that the grain size and composition can be controlled. In some cases, carefully sized powders are loaded into Pt capsules and sintered *in situ* at high pressure and temperature, with the deformation experiment performed on that same sample (e.g. Tullis & Yund 1985), and in other cases larger samples are sintered hydrostatically and then cored to provide samples for deformation (e.g. Fuchan & Paterson, 1988). The principal difficulty in the latter case is in achieving theoretical density without melting (e.g. Zeuch & Green 1984). Sintered samples have made it possible to create very pure fine-grained aggregates and thus to achieve diffusion creep (e.g. Karato *et al.* 1986). They have also made it possible to prepare systematically varying amounts of two phases to study the deformation of polyphase aggregates (e.g. Jordan 1987).

8.2.3 Observational techniques

Various kinds of observational techniques are critical to the success of experimental deformation studies. The demonstration of identical microstructures on all scales is the only method we have of inferring that the deformation processes have been identical in experimental and natural deformations. Even in studies designed principally to obtain mechanical data, it is imperative that microstructural observations be made to identify the process(es).

Optical petrographic microscope observations remain of fundamental importance, but new technology in the past 20 years or so has resulted in valuable additional methods of characterizing microstructures. One of the most important developments has been the ion milling technique (Tighe & Christie 1969, Barber 1970), which has allowed silicates to be prepared for transmission electron microscopy (TEM). TEM observations, commonly utilizing magnifications of $\times 20\ 000-80\ 000$, are the only way directly to observe dislocations in most rocks, although dislocations in olivine can be rendered optically visible by oxidation (Kohlstedt *et al.* 1976), providing greatly increased areas that can be viewed. TEM has proved to be of great importance in placing tighter constraints on the operative deformation mechanisms in experimentally deformed samples, and it is the correlation of TEM with optical microstructures between experimentally and naturally deformed rocks which gives the best evidence for operation of the same processes. However, it is important to remember that one cannot assume that the process which produced the optical or especially the TEM microstructures preserved in the rock was necessarily that which contributed most of the finite strain (e.g. Brodie & Rutter 1985).

There has recently been a renaissance of interest in crystallographic preferred orientations, due to new technological developments as well as new geological questions for which this information is useful. The ability to determine complete crystallographic orientations of aggregates using X-ray or neutron diffraction (e.g. Wenk 1985), together with computer models capable of predicting complete patterns for general states of strain and any assumed combination of slip systems (e.g. Lister *et al.* 1978; Etchecopar & Vasseur 1987; Wenk *et al.*, in press) has proved very powerful. Electron channelling techniques have recently been developed to determine the orientations of individual grains (e.g. Lloyd *et al.* 1987). X-ray and universal stage measurements of carefully characterized naturally deformed samples (e.g. Law 1986; Schmid & Casey 1986) as well as experimentally sheared samples (e.g. Schmid *et al.* 1987; Dell'Angelo & Tullis, 1989) have allowed asymmetric preferred orientations to be used as a sense-of-shear indicator. In addition, experimental and modelling studies are being used to interpret seismic anisotropy of the mantle and core in terms of flow mechanism and strain geometry (e.g. Jean-Loz & Wenk 1987, Karato 1988).

Scanning electron microscopy (SEM) is extremely useful for characterizing microcracks, pores, and other surface features at magnifications intermediate between those of the optical and the transmission electron microscopes. It has proved to be especially useful in studies of fault initiation (e.g. Tapponnier & Brace 1976, Wong 1982) and cataclastic flow (e.g. Hirth & Tullis 1989).

8.3 Experimental deformation of quartz aggregates

8.3.1 Overview

Experimental deformation studies of quartz aggregates at strain rates of $10^{-5}-10^{-6}\,\text{s}^{-1}$ show a field of brittle faulting at lower pressures and temperatures (e.g. Heard & Carter 1968); a narrow régime of dislocation glide plus cracking and crushing at intermediate pressures and temperatures (Hirth &

Tullis 1986); and a régime of dislocation creep at high pressures and temperatures (e.g. Green *et al.* 1971, Tullis *et al.* 1973). Quartz has a very high melting temperature ($\sim 1750°$C at 1500 Mpa, dry; $\sim 1680°$C at 300 MPa, dry), and for an experimental pressure and strain rate of 10^{-6}s^{-1} it shows steady-state dislocation creep at $T/T_m > 0.45$, whereas for a natural pressure and strain rate of 10^{-14}s^{-1} this occurs at $T/T_m \sim 0.2$. Trace amounts of water have a large effect on the strength in the dislocation creep régime (e.g. Jaoul *et al.* 1984, Kronenberg & Tullis 1984), and the 'water' defect (see below) appears to be responsible for the restriction of dislocation creep in experimental deformations to 'unnaturally' high pressures ($\geqslant 1000$ MPa; Mainprice & Paterson 1984, Kronenberg & Tullis 1984). To date there have been no reports of high-temperature ('dry') diffusion creep, nor of steady-state pressure solution (although see Rutter & White 1979). Quartzites experimentally deformed at various temperatures within the dislocation creep field develop microstructures which appear to be identical to those observed in naturally deformed rocks from low to high metamorphic grade; the striking similarities have given confidence that experimental studies can be of great use in understanding the deformation of the crust.

8.3.2 Faulting and cataclastic flow

Experimentally deformed non-porous quartzites exhibit a broad régime of brittle faulting, but only a narrow window in P and T space of what might be termed cataclastic flow (Hirth & Tullis 1986). At 10^{-5}s^{-1} new dislocations are first noted at $\sim 300°$C, but quartzites fault in a brittle manner up to 900°C at 300 MPa (Mainprice & Paterson 1984) and up to 600°C at 1500 MPa (Hirth & Tullis 1986). At 1500 MPa and 700–800°C quartzites show macroscopically ductile deformation, with sharp deformation bands, patchy undulatory extinction, and inhomogeneous flattening of original grains (Fig. 8.3a). TEM shows very high densities of tangled dislocations in a roughly cellular arrangement, with cracks and very-fine-grained crush zones developed after about 20 per cent strain (Fig. 8.4a); some grain growth and/or boundary-migration recrystallization of these crush zones occurs with continued strain (Hirth & Tullis 1986).

The presence of porosity has a strong effect on the transition from faulting to cataclastic flow, at least for strains $\leqslant 25$ per cent (Hirth & Tullis 1989).

Figure 8.3 Optical photomicrographs of Heavitree quartzite; all micrographs are the same scale. (A) Starting material: (B–D) samples deformed at 800°C, 1500 MPa, and 10^{-6}s^{-1} to increasing amounts of strain. Compression direction is vertical. (B) 12 per cent shortening: note local grain-boundary migration and deformation lamellae in some grains; horizontal cracks result from unloading. (C) 45 per cent shortening: note grain-boundary recrystallization and sweeping undulatory extinction in many grains. (D) 64 per cent strain: note increased amounts of recrystallization, and deformation lamellae in some grains.

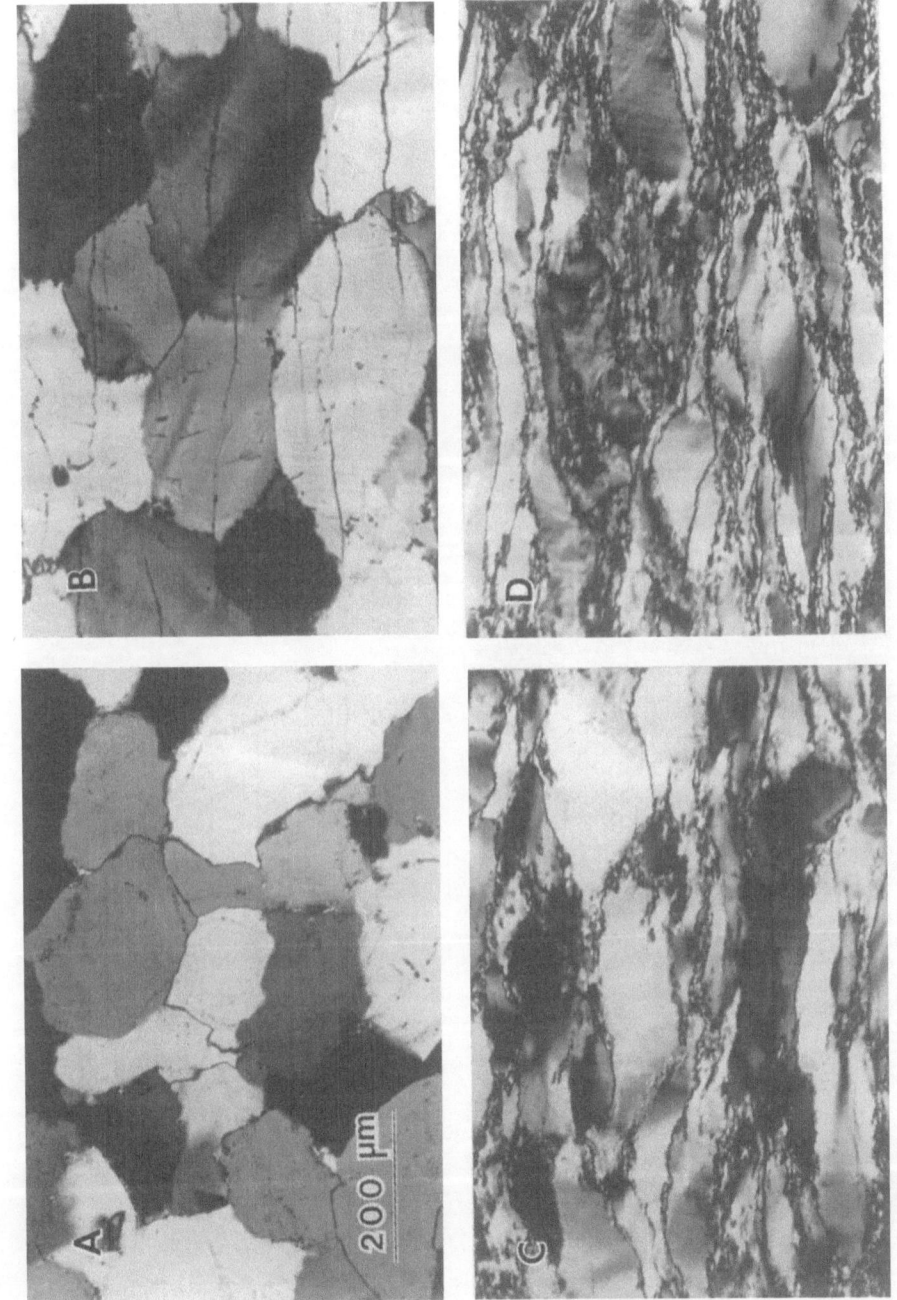

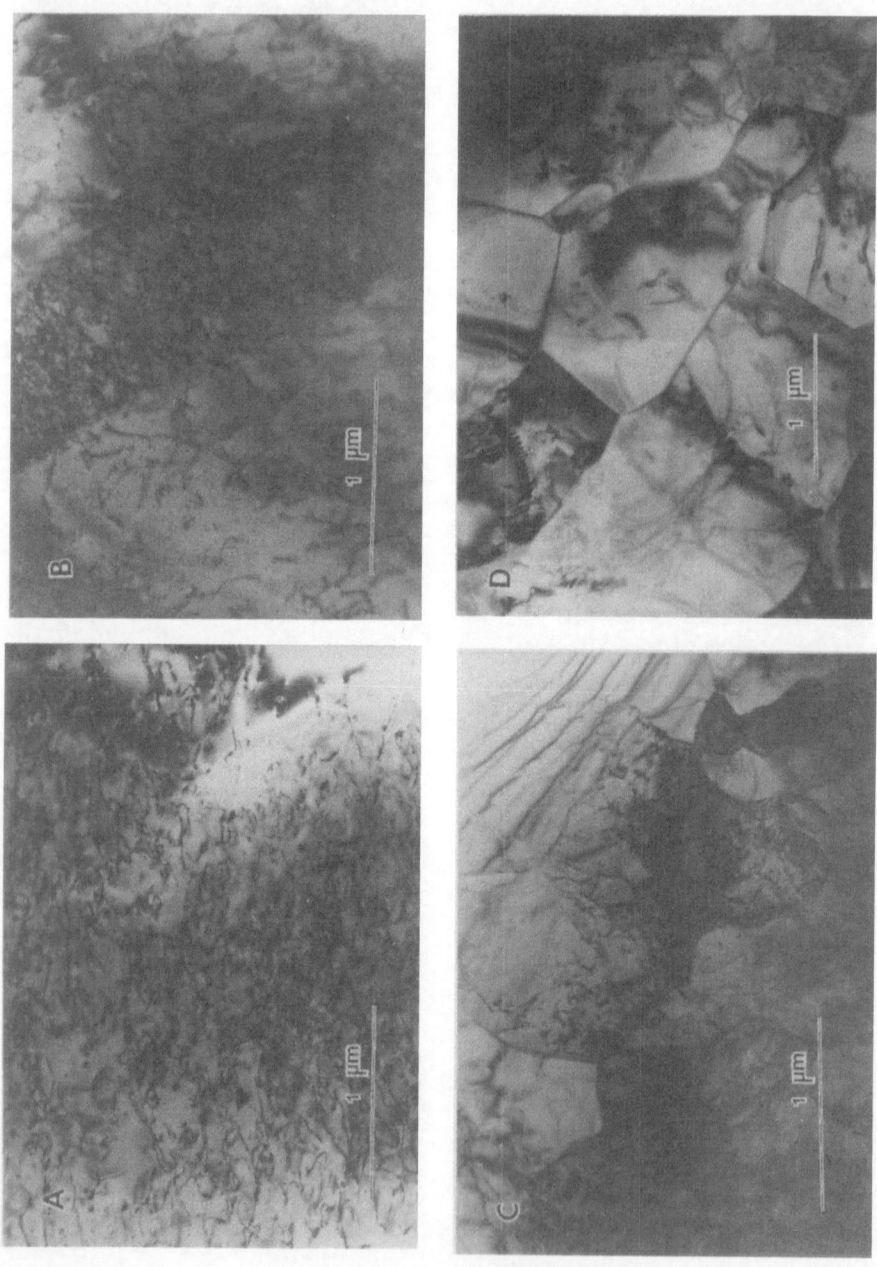

Quartzite with 7 per cent porosity shows a transition from brittle faulting to distributed cataclastic flow at a pressure of ~600 MPa, in low strain tests at room temperature and $10^{-4} s^{-1}$ (Hadizadeh & Rutter 1981), consistent with other studies on porous basalt (Shimada 1986). Cataclastic flow occurs due to gradual pore collapse associated with work hardening. However, this is only a transient behaviour; samples begin to weaken at strains ⩾20 per cent when net dilatancy initiates, and at ~25 per cent strain they undergo brittle faulting (Hirth & Tullis 1989). Thus porous quartzites 'revert' to the behaviour of non-porous quartzites once the porosity is eliminated.

These experiments on dry samples indicate that the low-temperature strength of quartz, with its lack of cleavages, makes distributed cracking (cataclastic flow) impossible in 'flawless' aggregates. Stable distributed cracking occurs only at moderately high confining pressures, when there are high densities of immobile dislocation densities, or pores, to provide local stress concentrations. Trace amounts of chemically active fluids might be expected to slightly enlarge the régime of cataclastic flow by speeding the rate of stable crack propagation (e.g. Atkinson 1984); however, it appears that such fluids enhance dislocation creep even more (e.g. Tullis & Yund 1980).

There are some important applications of these experimental studies of faulting and cataclastic flow to natural deformations. At shallow crustal conditions, non-porous quartzite units should be extremely strong. Porous sandstones and quartzites should initially undergo distributed cataclastic flow, but after a small amount of macroscopically ductile strain they should switch to brittle faulting; such a transition has been described by Aydin & Johnson (1983).

8.3.3 Dislocation creep

8.3.3.1 AXIAL COMPRESSION

Experimental deformation studies of quartz aggregates have been complemented by numerous studies of natural and synthetic single crystals, designed to determine slip systems. The latter have been fraught with some difficulties related to the nature and influence of the 'water' defect in quartz (see below), but there is general agreement between all studies of experimentally and naturally deformed quartz that slip on $(0001) \langle 11\bar{2}0 \rangle$ is easiest at lower temperatures, with slip on $\{10\bar{1}0\}$ [0001] becoming easier at

---

Figure 8.4 Transmission electron micrographs (bright field) of Heavitree quartzite deformed at $800°C$, 1500 MPa, and $10^{-6} s^{-1}$. (A) Sample shortened 12 per cent: dislocation density is quite variable, and this micrograph represents the 'average'; there are few if any subgrain boundaries. (B–D) From sample shortened 60 per cent. (B) Planar zone of very high density of tangled dislocations corresponding to optical-scale deformation lamellae. (C) Subgrains and recrystallized grains derived from them by progressive misorientation. (D) Recrystallized grains which have grown somewhat from original subgrain size; some of them have developed subgrain boundaries.

higher temperatures (e.g. Baeta & Ashbee 1970, Tullis *et al.* 1973). The dislocation creep behaviour of aggregates described in this section pertains to samples deformed at ~ 1500 MPa and $10^{-5}\,s^{-1}$, because quartzites deformed at experimental strain rates are brittle to semi-brittle at confining pressures below ~ 1000 MPa (see section below on effect of water).

The details of the transition from cataclastic flow to dislocation creep remain somewhat sketchy. For a strain rate of $10^{-5}\,s^{-1}$, quartzites deformed at 800°C have a high yield strength (~ 1000 MPa at ~ 20 per cent strain) followed by strain softening (to ~ 400 MPa at 50 per cent strain). Near the peak stress the grains exhibit deformation bands and a patchy undulatory extinction. TEM shows extremely high densities of tangled dislocations, and some very small recrystallized grains both along grain boundaries and within grains. With continued strain, the deformation is inhomogeneous, with some grains highly flattened and others remaining relatively undeformed; recrystallized grains form along most grain boundaries (Fig. 8.5B). TEM shows that in this strain interval the dislocation density is reduced, and subgrains are developed and form recrystallized grains by progressive rotation. The delayed onset of climb and the weakening which accompanies it are not well understood, but have important implications for determining the conditions and rheologies of steady-state flow.

At 900°C and above, quartzites undergo steady-state dislocation creep after strains of only ~ 5 per cent with a flow stress of <400 MPa. In the lower-temperature portions of this régime ($\sim 900-1000$°C), original grains are quite homogeneously flattened with increasing strain (Figs. 8.3B, C & D), and deformation lamellae (primarily sub-basal) are common (Figs. 8.3B, D & 8.4B). Strain-induced grain-boundary migration produces 'suturing' of grain boundaries and some recrystallized grains at low strains (Fig. 8.3B), although at higher strains recrystallized grains result primarily from progressive subgrain misorientation with some subsequent migration (Figs 8.4C & D; Hobbs 1968, Guillopé & Poirier 1979), giving rise to a 'core and mantle' structure (e.g. White 1977). At still higher temperatures (1100–1200°C) there are more (and larger) recrystallized grains even at low strain (Figs 8.5C & D), and these and the original grains contain larger subgrains. Complete recrystallization occurs at moderate strain, and the microstructure maintains a dynamic steady state at all further strains. At these very high temperatures, recovery and recrystallization keep pace with deformation, and the deformation textures are almost indistinguishable from those resulting from annealing.

Figure 8.5 Optical photomicrographs of Black Hills quartzite; all micrographs are the same scale. (A) Starting material: (B–D) samples shortened ~ 50 per cent at 1500 MPa, with increasing temperature and/or decreasing strain rate: compression direction is vertical. (B) 800°C, $10^{-5}\,s^{-1}$: deformation is quite inhomogeneous; some grains are flattened and others remain undeformed; recrystallized grains are very small. (C) 900°C, $10^{-6}\,s^{-1}$: deformation is more uniform, recrystallized grains are larger and more numerous. (D) 1100°C, $10^{-5}\,s^{-1}$: dynamic recrystallization is almost complete, and the recrystallized grain size is quite large.

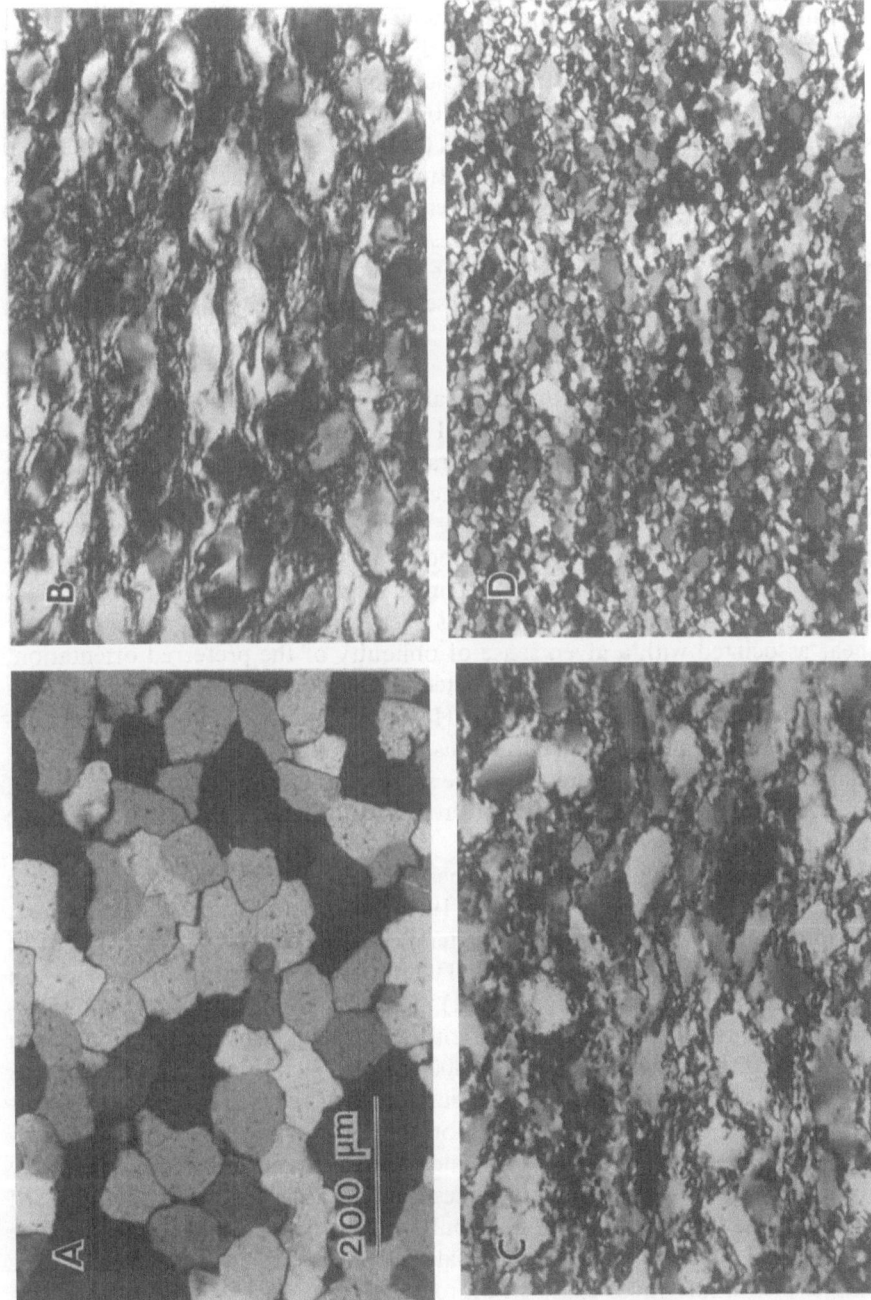

Accompanying the change in dominant slip systems with increasing temperature there is a progressive change in the preferred orientation of deformed original grains (Tullis *et al.* 1973), from a maximum of *c* axes ([0001]) parallel to compression to small circle girdles about compression, the opening angle of which increases with temperature to a maximum of ~40°C. There is no change in pattern at the displacive transition between low or α quartz (trigonal) and high or β quartz (hexagonal), implying no change in the active slip systems (Gleason & Tullis 1989). Flints that have syntectonically recrystallized during deformation at the same conditions show the same patterns of preferred orientation in the α field, but a very different pattern (a *c* maximum) in the β field (Green *et al.* 1971). These results indicate that oriented, anisotropic grain growth during deformation in the β field significantly affects the preferred orientation (e.g. Hobbs 1968, Jessel 1988).

8.3.3.2 SIMPLE SHEAR

The above studies have shown that axial compression can produce 'mylonitic' textures, with highly flattened original grains and necklaces of small recrystallized grains; this helps to demonstrate that the principal requirement for mylonite production is high strain, rather than some particular strain geometry (Tullis *et al.* 1981). However, in nature most mylonites probably involve non-coaxial deformation, as indicated by recrystallized grains and preferred orientations which are oblique to the flattening foliation (e.g. Simpson & Schmid 1983, Lister & Snoke 1984). There has been some debate about the sense of shear associated with a given sense of obliquity of the preferred orientation, because two theoretical models predicted opposite tendencies (Lister & Hobbs 1980, Etchecopar & Vasseur, 1987). However, careful studies of field situations in which there are unambiguous strain markers appear to show that quartz *c* axes are concentrated with the sense of shear (Schmid & Casey 1986), and recent improved models of preferred orientation development predict the same thing (Wenk *et al.* 1989).

Simple shear is difficult to attain experimentally. Rotary shear apparatus are capable of high strains (Handin *et al.* 1960, Tullis & Weeks 1986), but current versions cannot operate at the high temperatures and pressures required to render quartz ductile. Dell'Angelo & Tullis (1989) adapted the inclined-piston technique of Friedman & Higgs (1981) and Schmid *et al.* (1987) to a Griggs-type apparatus. They deformed quartzites in combined shear and compression within the dislocation creep régime (800°C, $10^{-6}s^{-1}$). Use of a TEM grid as a strain marker enabled accurate measurements of the two components; shear strains ranged up to 2.9 and axial shortening to 60 per cent. Optical microstructures in the sheared samples include several features that are commonly observed in naturally sheared quartzites, but are absent in axially compressed samples. The flattening foliation is continuously curved from edges to centre of the thin sample, because thermal and mechanical constraints at the sample/piston interfaces cause strain to be lower there. Although there is no clearly defined planar element in the shearing orientation, there is a faint planar fabric at 35–40° to the shear-zone boundaries, defined by concentrations of recrystallized grains and by impurity segregations. The recrystallized grains are oriented with their long axes distinctly oblique to the original grain flattening foliation and approximately normal to the instantaneous compression direction in samples with a range of shear strains. Thus they may represent a 'steady-state foliation' (Means 1981). Finally, the c axis preferred orientation is distinctly oblique to the original grain flattening foliation; the maximum is displaced with the sense of shear.

8.3.3.3 EFFECT OF CHEMICAL ENVIRONMENT

A considerable effort has been devoted to experimental investigations of the effects of trace amounts of water on the deformation of quartz in the dislocation glide and creep régimes. Many experiments have involved deformation of single crystals of natural or synthetic quartz, and these will not be reviewed (see Blacic & Christie 1984; McLaren et al., 1989). However, there have also been a number of studies demonstrating hydrolytic weakening in quartz aggregates. One of the first was the demonstration that flints, containing 1–2 wt% water, are weaker and undergo considerably more recrystallization than novaculites, containing ~0.1 wt% water (Green et al. 1971). More recent systematic studies have shown that addition of 0.1–0.5 wt% water results in a decrease of strength, activation energy, and stress exponent, for novaculites and quartzites of grain sizes in the range 2–200 μm deformed at 1000–1500 MPa, compared to 'as-received' samples. Similarly, vacuum drying results in an increase in strength and brittleness, as well as in the activation energy and stress exponent (Jaoul et al. 1984, Kronenberg & Tullis 1984). There is a strong effect of confining pressure on the dislocation creep strength for quartzites and novaculites: as-received as well as water-added aggregates become stronger at pressures below ~1000 MPa, and at 300–500 MPa they fail in a brittle manner after only a few per cent strain (Kronenberg & Tullis 1984, Mainprice & Paterson 1984). However, flints deform ductilely even at low pressure (Green et al. 1971, Mainprice 1981).

The mechanism by which water affects the strength of quartz aggregates is not known, but it occurs too fast to be due to volume diffusion of an O-related species. Quartzites (average grain size, $d \sim 100$ μm) deformed with 0.1 wt% water added are significantly weakened after <2 h, which corresponds to a diffusion distance of only ~1–2 μm; this suggests that pipe diffusion along pre-existing dislocations might be important (Tullis & Yund 1985a). However, this would not explain the striking pressure dependence of weakening.

Until recently most experimental rock deformation studies of quartz aggregates have not attempted to control the chemical environment of the sample, other than sometimes to provide a source of water. On the basis of theoretical considerations, Hobbs (1983) suggested that the f_{O_2} and f_{H_2} environment during deformation might have a significant effect on diffusion rates and therefore rock strength, by influencing the types and concentrations of crystal defects present. However, buffered deformation experiments on novaculite (with water added) at 850°C, 10^{-5} s^{-1}, and 1500 MPa showed no measurable effect of f_{H_2} or f_{O_2} on flow strength in the dislocation creep régime (Tullis & Yund 1988).

8.3.3.4 APPLICATIONS

The optical and TEM microstructures and crystallographic preferred orientations produced in quartz aggregates experimentally deformed in the dislocation glide and creep régimes are remarkably similar to those observed in naturally deformed rocks from low to high metamorphic grades (e.g. Wilson 1973; White 1973, 1976; Liddell *et al.* 1985; Price 1985), indicating that it is possible to activate the same processes. Perhaps one area deserving more experimental study is deformation at very high temperatures, where dynamic recrystallization is complete at very low strains. Another area requiring further work is hydrolytic weakening. Recently the first documentation of hydrolytic weakening in natural quartz-bearing rocks has been described (Kronenberg *et al.*, 1990). We need to identify the cause of hydrolytic weakening in experimentally deformed aggregates, the reason for its strong pressure dependence, and the way in which pressure trades off for time; only then can we have confidence in extrapolating experimental results on microstructures, as well as on rheologies, to natural deformations.

8.3.3.5 PRESSURE SOLUTION

Microstructural evidence shows that pressure solution is a common deformation mechanism in quartz aggregates naturally deformed at temperatures of ~ 100–$250°C$ (Kerrich *et al.* 1977, Mitra 1978), but the low stresses and strain rates required by this deformation mechanism mean that laboratory studies have for the most part been unsuccessful. Several studies of compaction of loose sand in the presence of various fluids have succeeded in producing solution at high-stress grain contacts and deposition within pores (e.g. Ernst & Blatt 1963), although local crushing may have contributed. Rutter & White (1979) carried out shearing experiments on saw-cut Tennessee sandstone with artificial gouge and fluid present, and found that the stress exponent changed from ~ 3 to ~ 1 at a strain rate of 10^{-7} s^{-1}, associated with SEM evidence of dissolution and precipitation of faceted microcrystals. More success has been had recently with experimental studies of pressure solution in halite aggregates (e.g. Hickman & Evans 1988; Spiers & Schutjens, this volume). It is possible that very low stress experiments on very-fine-grained aggregates, perhaps using fluids in which quartz is more soluble than it is in water, will allow measurable strain rates without concurrent dislocation motion or fracture. However, it is likely that many advances will have to come from a combination of careful field observations coupled with theory (e.g. Lehner, this volume).

8.4 Experimental deformation of feldspar aggregates

8.4.1 Overview

Experimental deformation studies of feldspar aggregates (mostly fine-grained anorthosites and albite rocks) show many similarities with quartzites, but also important differences. At a strain rate of $10^{-5} s^{-1}$, feldspars show a more restricted field of brittle faulting at low pressures and temperatures (Seifert 1969, Borg & Heard 1970), a wider régime of cataclastic flow at intermediate pressures and temperature (Tullis & Yund 1987a), and a broad transition region to dislocation creep (including mechanical twinning) which only occurs at high pressures and temperatures (Kronenberg & Shelton 1980, Tullis & Yund 1987b). Within the dislocation creep field there appears to be a transition from recrystallization-accommodated creep to climb-accommodated creep at higher temperatures and slower strain rates than are accessible experimentally, with important textural and mechanical effects (Tullis & Yund 1985b). Trace amounts of water have a large effect on the strength in the dislocation creep régime (Tullis & Yund 1980), and as for quartz, seem to cause dislocation creep at laboratory strain rates to be restricted to high pressures (Shelton *et al.* 1981). For very fine grain sizes at high temperatures and low stresses, a régime of grain-boundary diffusion creep has been identified (Ji & Mainprice 1986, Tullis & Yund 1988). Naturally deformed feldspars appear to show evidence of all of these processes, as well as others that are not accessible experimentally due to the onset of melting.

Feldspars have a much lower melting temperature than quartz ($\sim 1200°C$ at 1500 MPa, dry; $\sim 1100°C$ at 300 MPa, dry), but at both experimental and natural strain rates, steady-state dislocation creep of feldspars requires higher temperatures than does that for quartz. For feldspars deformed experimentally at 1500 MPa and $10^{-6} s^{-1}$, steady-state dislocation creep requires $T/T_m > 0.85$, and for natural deformations at 300 MPa and $10^{-4} s^{-1}$ it requires $T/T_m > 0.45$. Thus the behaviour of quartz and feldspar contradicts the simple metallurgical rule that materials with high melting temperatures have higher creep resistance.

8.4.2 Faulting and cataclastic flow

Feldspar aggregates undergo faulting at low temperatures and at low pressures. At a strain rate of $10^{-5} s^{-1}$, anorthosite shows brittle faulting up to at least 1500 MPa at temperatures $< 300°C$, and up to at least $900°C$ at pressures < 500 MPa (Hadizadeh & Tullis 1986). This behaviour is similar to that of non-porous quartzites.

In contrast to quartzites, feldspars have a broad régime of cataclastic flow at intermediate temperatures and pressures. In this régime deformation

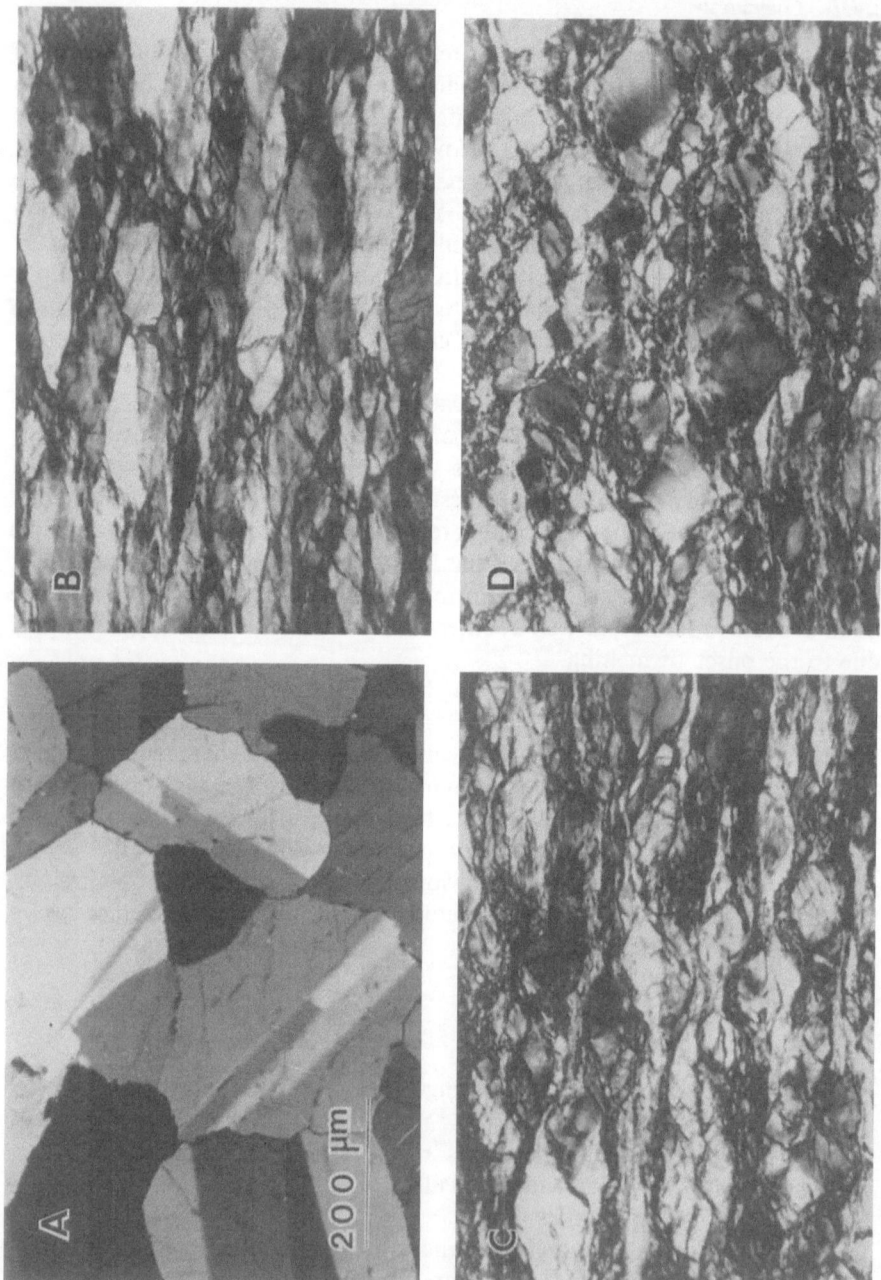

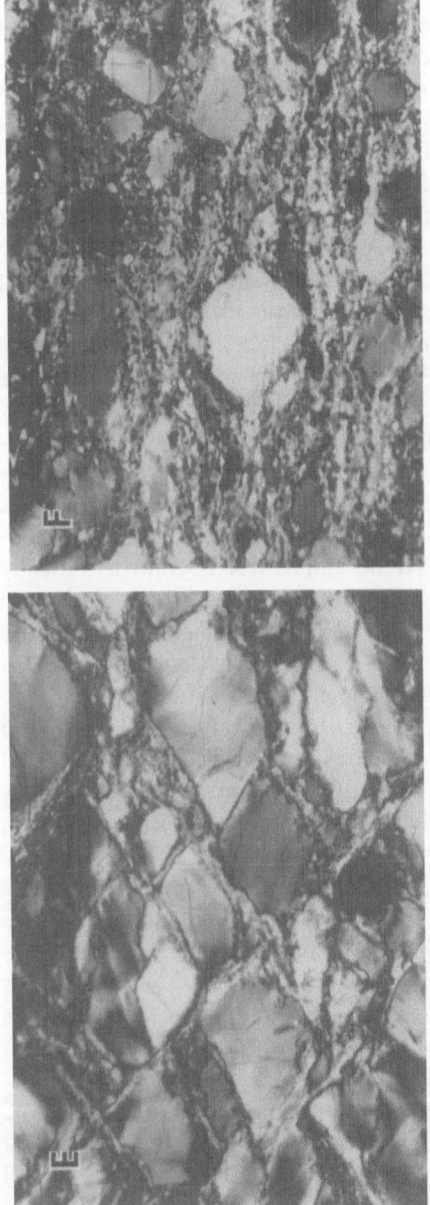

Figure 8.6 Optical photomicrographs of Hale albite rock; all micrographs are the same scale. (A) Starting material: many grains contain growth twins; (B–F) samples deformed at 1500 MPa and 10^{-5} s^{-1}; compression direction is vertical. (B) 60 per cent shortening at 650°C: deformation has occurred by multiple grain-scale faulting and finer-scale microcracking and crushing (cataclastic flow). (C) 60 per cent shortening at 800°C: deformation mostly by cataclastic flow but some of the finely crushed material has recrystallized. (D) 60 per cent shortening at 900°C: deformation by cataclastic flow and recrystallization-accommodated dislocation creep. (E,F) 50 and 86 per cent shortening, respectively, at 1100°C: deformation by recrystallization-accommodated dislocation creep.

involves stable distributed cracking, primarily on the two cleavages, with rotation and frictional sliding of the fragments (Tullis & Yund 1987a). The cracking component is indicated optically by multiple grain scale faults on one or both of the two cleavages, which produces a marked flattening foliation (Fig. 8.6B). A component of strain due to dislocation processes appears to be suggested by optical microstructures such as strong, patchy undulatory extinction, strong crystallographic preferred orientation, and features suggestive of subgrains and recrystallized grains. However, TEM shows only penetrative cracking and development of crush zones; no dislocation multiplication or motion has occurred (Tullis & Yund 1987b). At ~25 per cent strain the largest uncracked region is ~2 μm across, and there are numerous sharply bounded crush zones with fragments down to <0.01 μm in size, which have rotated so as to produce rings in the electron diffraction patterns. This confirms that optical-scale undulatory extinction can be produced by penetrative fine-scale cracking and crushing, as demonstrated earlier by Marshall & McLaren (1977) and Tullis & Yund (1977). The preferred orientation is produced by the multiple faulting on crystallographically controlled (cleavage) planes, which leads to external rotations just as does dislocation glide on a single set of planes (Tullis & Yund 1987b). This same behaviour has been observed in naturally deformed amphiboles (Allison & LaTour 1977).

There has long been a belief that increasing pressure alone serves to stabilize cracks and to accomplish a transition from faulting to cataclastic flow (e.g. Paterson 1978). Most models for this process involve dislocations, either dislocation pile-ups to nucleate cracks which do not travel beyond the locally extensile stress field, or dislocation emission from crack tips, which blunts and thus slows the cracks. Such models closely approximate the behaviour of more ductile minerals, such as calcite (e.g. Heard 1960; Fredrich et al., 1989). Feldspar is the first material for which it has been demonstrated that a transition to distributed cracking occurs in the absence of dislocation activation, and that increases in temperature as well as pressure are required (Tullis & Yund 1987a). The rôle of pressure may be to limit the propagation of cracks away from nucleating flaws such as pores, fluid inclusions, and alteration products, but the rôle of temperature is not yet well understood.

In both albite and anorthosite there is quite a broad transition zone between cataclastic flow and dislocation creep (Tullis & Yund 1987b). Optically, this produces a transition from rather uniformly flattened grains to equant, undeformed grain fragments in a matrix of very fine recrystallized grains (Figs 8.6B–F). For an experimental strain rate of 10^{-5} s^{-1}, mechanical twins (albite and pericline laws) are first noted at 600°C, and high densities of tangled dislocations with a crude cellular arrangement are first noted at 700°C. Most of the strain in the latter samples still appears to occur by grain-scale faulting and crushing. At 800°C the dislocation density is uniformly high, and the absence of subgrains and presence of straight dislocation segments as well as dissociated dislocations indicates that there has been no climb. In these samples there is definite evidence that some of the early formed, very-fine-grained crush zones have undergone grain growth and/or dynamic recrystallization; this may explain the strain weakening shown by these samples. At 900°C there is still a component of grain-scale faulting, but some of the strain occurs by recrystallization-accommodated dislocation creep (Tullis & Yund 1985b).

8.4.3 Dislocation creep

Experimental deformation studies of feldspar aggregates rest on studies of mechanical twinning and slip systems provided by single-crystal deformation experiments, as well as studies of diffusion, ordering and disordering, and exsolution. The latter are well summarized in review volumes (e.g. Brown 1983, Ribbe 1983). Essentially the same slip systems are observed in experimentally and naturally deformed alkali and plagioclase feldspars (Gandais & Willaime 1983, Tullis 1983); the most common system appears to be (010) [001], and the dislocations are frequently dissociated.

Because climb in feldspars is evidently much more difficult than it is in quartz, the accommodation mechanism for creep of feldspars in the lower-temperature portion of the creep régime is different from that for quartz (Tullis & Yund 1985b). At lower temperatures (700–1100°C at $10^{-6} s^{-1}$), at which climb cannot counteract the work hardening associated with intersecting slip systems, original grains rapidly develop a high density of tangled, immobile dislocations. However, along their margins they undergo dynamic recrystallization by a process of strain-induced boundary migration. Small grain-boundary bulges pinch off to form recrystallized grains. These grains are initially strain free and so can undergo an increment of easy glide. When they eventually strain harden they are replaced by new grain-boundary migrations. There is a significant contrast in strength between the very fine ($\sim 1 \mu m$) recrystallized grains and the coarser original grains; thus aggregates show an initial strength peak, followed by strain weakening as an interconnected matrix of fine recrystallized grains develops around the margins of all the original grains.

Optical microstructures in experimentally deformed feldspars clearly reflect this different accommodation mechanism for creep. Whereas quartzites show progressive flattening of original grains, with recrystallized grains developing by grain-boundary migration at the lowest strains but by subgrain rotation thereafter, in feldspar aggregates the original grains remain equant, although they develop a patchy undulatory extinction and sometimes even undergo some grain-scale faulting. Very fine ($\sim 1 \mu m$) recrystallized grains develop along the grain boundaries, grain-scale faults, and deformation bands (Fig. 8.6E). When they form an interconnected matrix they take up almost all of the sample strain; as strain increases, the original grains are gradually reduced in size due to progressive boundary recrystallization, increasing the proportion of fine-grained matrix, but the recrystallized grain size remains essentially the same (Fig. 8.6F).

The contrast in strength between original and recrystallized grains readily leads to strain localization within the latter. For example, in samples of bimodal grain size (half 2–10 μm, half 150–180 μm), almost all of the sample strain occurs within the readily recrystallized fine-grained matrix, and the larger grains remain as undeformed augen (except where impingement occurs) out to very high strain (Tullis & Yund 1985b). Similarly, in samples pre-faulted to very low strain, and subsequently deformed at higher temperature within the dislocation creep régime, the small amount of gouge sinters and recrystallizes and ductile strain is localized in this material (Tullis et al., 1990).

There is some question concerning the deformation processes occurring in the fine, recrystallized grains. Macroscopically these fine-grained regions are superplastic, and so it is tempting to infer that there might be a large component of diffusion creep or grain-boundary sliding, as found by Schmid et al. (1977) in fine-grained calcite aggregates. Although it is almost impossible to prove or disprove dominance of a deformation mechanism just from microstructural examination (e.g. Brodie & Rutter 1985), several features argue against a substantial component of either of these mechanisms, at least in the experimental samples. First, the recrystallized grains contain a variable dislocation density; some contain none, some a moderate density, and some a high density of tangled dislocations, just like the original grains. Second, the recrystallized regions have a strong crystallographic preferred orientation, judging from optical effects. Both of these features are consistent with dislocation creep, but not with grain-boundary sliding or diffusion creep. The most convincing argument is provided by distinct microstructural contrasts with samples which have undergone diffusion creep (see below).

The flow law for steady-state recrystallization-accommodated dislocation creep of feldspars has not yet been determined. However, all experimental determinations of dislocation creep flow laws from coarse-grained aggregates (e.g. Carter et al. 1981, Shelton & Tullis 1981) are for samples not even close to steady state; they reflect the properties of the work-hardened original grains rather than those of the just developing recrystallized grains. The problem in determining an appropriate flow law is that high confining pressures must be used, as for quartz, due apparently to some aspect of the 'water' defect, and this requires solid confining medium apparatus. The development of a liquid salt sample assembly (Green & Borch, 1989) should allow more accurate flow-law determinations.

Experimental studies of albite rock and anorthosite show very similar deformation behaviour, and observations of experimentally deformed aplite and granite indicate that potassium feldspar is also similar (e.g. Tullis & Yund 1977). In all these feldspars there are indications that climb is becoming somewhat easier at the highest experimentally accessible temperatures. However, the relatively low melting temperature of feldspars precludes our ability to document experimentally a transition to climb-accommodated creep. In cases such as this we must depend on careful examination of samples that are naturally deformed at high metamorphic grades to infer a change in mechanism. Another consequence of the high temperatures of experimental deformation studies is the fact that all dislocation creep studies are necessarily carried out on completely disordered feldspars (e.g. Yund & Tullis 1980), whereas in natural deformations at low to moderate grade the feldspars would be ordered.

Observations of naturally deformed feldspars confirm and supplement the experimental results. Feldspars deformed at greenschist to lower amphibolite grade typically occur as relatively undeformed augen with long, fine-grained 'tails', indicating deformation by recrystallization-accommodated dislocation creep (e.g. Simpson 1985). At these conditions, ductile shear zones are common in feldspathic 'basement' rocks (e.g. Simpson 1982), very probably due to the weakening associated with dynamic recrystallization. In contrast, in feldspars deformed at upper amphibolite to granulite grade, dislocation climb apparently becomes sufficiently easy to allow climb-accommodated creep. In these high-grade rocks, feldspars show the same features as exhibited by quartz at lower grade, such as flattened original grains containing subgrains, coarse recrystallized grains, and smooth, sweeping undulatory extinction (e.g. White & Mawer 1986). These higher-grade rocks do not show ductile shear zones, but tend to deform homogeneously (e.g. Voll 1976).

8.4.4 Effect of chemical environment

The addition of trace amounts of water ($\sim 0.1 - 0.2$ wt%) reduces the strength of feldspar aggregates in the dislocation creep régime, as it does in quartz aggregates (Tullis & Yund 1980). Samples deformed with added water at $\sim 1000 - 1500$ MPa have a lower stress exponent and activation energy than as-received samples, whereas vacuum-dried samples have a higher stress exponent and activation energy (Shelton & Tullis 1981). At pressures $\leqslant 1000$ MPa all samples are stronger and fail in a brittle manner, just as found in quartz aggregates (Shelton et al. 1979, Tullis & Yund 1980). The mechanism of the observed hydrolytic weakening is not well understood, although the rates of O and Al/Si diffusion are greatly increased in the presence of water, whereas those of Na and K appear to be unaffected (Yund 1986).

Only recently have deformation experiments under controlled f_{O_2} and f_{H_2} been attempted, motivated by studies of the effect of these variables on diffusion rates. High pressure ($\geqslant 1000$ MPa) and higher f_{H_2} have been shown to increase the rate of Al/Si interchange (disordering) in albite and in potassium feldspar (Goldsmith 1987, 1988), as well as the interdiffusion rate of CaAl/NaSi in peristerites (Yund & Snow, 1989). These results suggest that the concentration of the water-related defect in the crystal structure increases with pressure, but is also influenced by f_{H_2}. This in turn suggests that there should

213

be some effect of f_{H_2} (in addition to the well documented effect of pressure) on flow strength. However, there might not be a strong effect of f_{H_2} on deformation by recrystallization-accommodated dislocation creep, because this mechanism does not involve volume diffusion. An effect might be expected in the higher-temperature climb-accommodated creep régime, but unfortunately the latter appears to be experimentally inaccessible.

8.4.5 Diffusion creep

Diffusion creep has been inferred in very-fine-grained aggregates of albite deformed by creep at high temperatures and atmospheric pressure. For fine-grained synthetic albite, Ji & Mainprice (1986) found a transition in stress exponent from $n \sim 3$ to $n \sim 1.5$ at the temperature at which partial melting began. They inferred that the increased diffusion rate in the melt regions along grain boundaries promoted a transition from dislocation to melt-assisted diffusion creep, similar to the observations at higher pressure of Cooper & Kohlstedt (1984) on olivine basalt and Dell'Angelo et al. (1987) on synthetic granite.

Diffusion creep has also been inferred from experiments on fine-grained ($2-10$ μm) synthetic albite aggregates deformed with added water (~ 1 wt%) at $900-1100°C$ and $10^{-5} s^{-1}$ (Tullis & Yund 1988). The experiments were carried out in a solid confining medium apparatus, and the samples strengths were unmeasurably low (< 100 MPa), so it could not be verified that the stress exponent was close to 1. However, the textural evidence was quite convincing, especially by contrast with similar samples deformed 'dry' by recrystallization-accommodated dislocation creep at the same conditions. Some of the grains in the deformed samples were larger than any in the starting powder, and many had perfect lath shapes with square corners, instead of the polygonal grains seen in the dislocation creep régime. TEM revealed abundant small pores along the grain boundaries, and a very low average dislocation density within the grains. Furthermore, the same behaviour was observed at both 500 and 1500 MPa, in striking contrast to the situation for dislocation creep. Therefore it appears that these samples had undergone some combination of grain-boundary diffusion creep, grain-boundary sliding, and fluid-phase transfer (pressure solution).

8.5   Experimental deformation of quartzo-feldspathic aggregates

8.5.1   Overview

Relatively few experimental deformation studies have been performed on polyphase aggregates of quartz and feldspar, but they are very useful for checking our ability to predict the behaviour of such aggregates from the

214

known behaviour of monomineralic aggregates of the constituents. Several studies have been conducted on fine-grained granite ($d \sim 500 \mu$m) and aplite ($d \sim 200 \mu$m), consisting of roughly 30–35 per cent each of quartz, microcline, and plagioclase, with <5 per cent biotite and accessories, in which the feldspars form the connected matrix phase and the quartz grains are more dispersed. For deformation at a strain rate of $10^{-6} s^{-1}$, these aggregates show a field of brittle faulting and frictional sliding at lower pressures and/or temperatures (Tullis & Yund 1977, Stesky 1978), a field of cataclastic flow at pressures of ~ 600–1500 MPa and intermediate temperatures (Tullis & Yund 1977, Carter *et al.* 1981), and a field of dislocation creep at high pressures and temperatures (Tullis & Yund 1977, 1980; Dell'Angelo & Tullis, 1982). The onset of partial melting affects the strength and may change the dominant deformation mechanism, depending on the grain size, amount of melt, and strain rate (van der Molen & Paterson 1978; Dell'Angelo & Tullis 1988).

8.5.2 Faulting and cataclastic flow

Samples deformed at ⩽ 500 MPa or ⩽ 300°C develop through-going faults. In coarser-grained Westerly granite ($d \sim 500 \mu$m) the faults may be influenced by a given biotite grain or feldspar cleavage; in finer-grained aplites ($d \sim 200 \mu$m) the faults tend to avoid quartz grains and to go through feldspar grains.

At pressures of ~ 700–1500 MPa and temperatures of ~ 300–700°C, aplites undergo macroscopically ductile cataclastic flow, due to distributed cracking and grain-scale faulting in the two feldspars. However, the dispersed quartz grains are stronger (due in part to their lack of cleavages), and they remain almost completely undeformed (Fig. 8.7B). Thus the aplites develop a texture opposite to that of quartzo-feldspathic aggregates at higher metamorphic grade, where feldspar grains remain as relatively undeformed augen in the midst of ductilely deformed quartz and mica. This low-temperature deformation texture of less strained quartz and more strained feldspars has been observed in naturally deformed rocks, in the vicinity of faults (e.g. Evans 1988).

8.5.3 Dislocation creep

At pressures of ~ 1000–1500 MPa, there is a broad transition with increasing temperature from cataclastic flow to dislocation creep, which involves a reversal in the relative strengths of the quartz and feldspar and a profound change in the deformation texture (Dell'Angelo & Tullis 1982). At 800°C and 1500 MPa, the quartz grains deform by climb-accommodated dislocation creep, and the feldspars by a mixture of cataclastic flow and recrystallization-accommodated dislocation creep. The two phases have approximately equal strength, as the grain strains are almost equal to each other and to the sample strain, and the samples show steady-state flow (Fig. 8.7C).

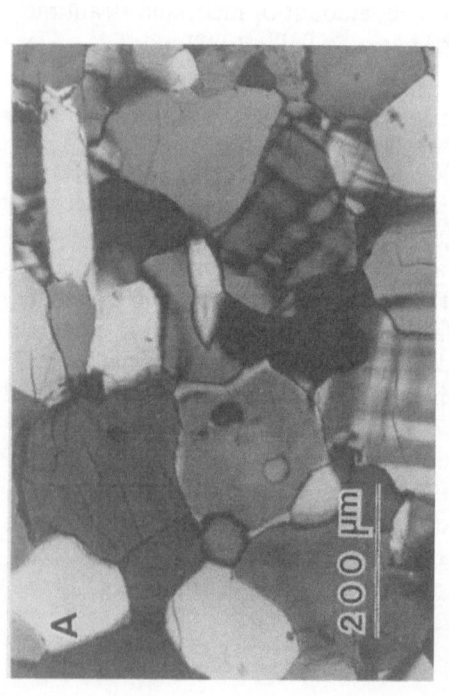

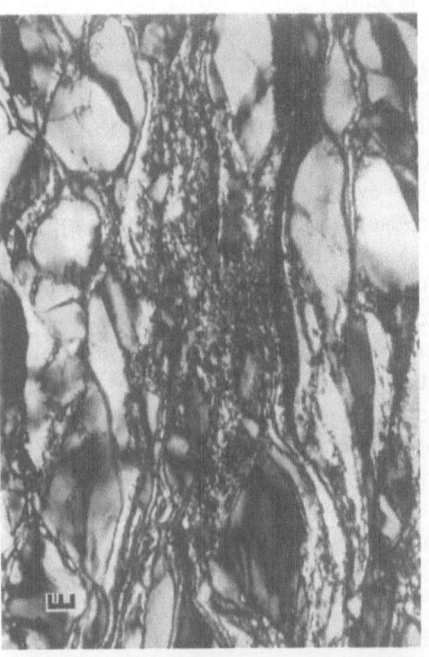

Figure 8.7 Optical photomicrographs of Enfield aplite; all micrographs are the same scale. (A) Starting material: grains with cross-hatch pattern are microcline, grey grains are mostly oligoclase, white grains are mostly quartz, but tabular grain is biotite. (B–D) Samples deformed at 1500 MPa and $10^{-6}\,s^{-1}$, with increasing temperature: compression direction is vertical. (B) 47 per cent shortening at 725°C: dispersed quartz grains (white) remain little deformed, while feldspars have deformed by cataclastic flow. (C) 55 per cent shortening at 800°C: quartz and feldspars have deformed about equally, the former by climb-accommodated dislocation creep, and the latter by a combination of cataclastic flow and recrystallization-accommodated dislocation creep. (D) 55 per cent shortening at 900°C: both quartz and feldspars have deformed by dislocation creep, but the latter are stronger and are becoming less connected, while the former is weaker and is becoming interconnected. (E) 85 per cent shortening, 900°C: portions of relatively undeformed feldspar grains are dispersed, quartz and recrystallized feldspars form interconnected matrix.

At 900°C the behaviour of quartz and feldspar has completed its reversal. Quartz grains undergo quite homogeneous flattening with increasing strain, but their average grain strain exceeds the sample strain (after sample strains of ~40 per cent; Fig. 8.7D). Sub-basal deformation lamellae are quite common. The preferred orientation of the quartz grains appears to be identical in form and strength to that developed in pure quartzites deformed to the same strain at about the same conditions, indicating that the presence of neighbouring feldspar grains has little influence (Dell'Angelo & Tullis 1986). Dynamic recrystallization in the quartz grains is less evident than in quartzites strained an equivalent amount, presumably because there are relatively few quartz/quartz grain boundaries. However, by ~50 per cent sample strain most grains have recrystallized grains ($d = 3-10~\mu m$) at their margins which appear to have been derived by progressive subgrain rotation.

The feldspar grains in the 900°C samples behave very differently. With increasing strain they undergo 'boudinage', and become separated into several lozenge-shaped pieces which do not show significant internal strain (Fig. 8.7D). Connecting the lozenges are thin tails of dynamically recrystallized grains, much smaller in size ($\sim 1~\mu m$) than the recrystallized quartz grains (Fig. 8.7E). The weakness of these recrystallized grains (and the quartz grains) compared to the original feldspar grains means that by strains $\geqslant 60$ per cent the aggregate changes its interconnectedness: fragments of original feldspar grains become dispersed in a matrix of weaker quartz and recrystallized feldspar. Accompanying this change of texture is a progressive strain weakening. Another reflection of the weakness of the recrystallized feldspar grains is the fact that early-formed faults in aplite samples develop highly localized ductile shear during subsequent high-temperature deformation, just as they do in pure feldspar aggregates, whereas pre-faulted quartzites deform homogeneously (Tullis *et al.*, in press). A further consequence of the contrast in strength between original and recrystallized feldspar grains is demonstrated in shearing deformation. Synthetic aggregates of half quartz and half albite (original grain size 100 μm) subjected to combined compression and shear at 900°C and $10^{-6}\,s^{-1}$ develop a C and S fabric (Berthe *et al.* 1979) in which the flattened quartz grains define the S or foliation planes, and feldspar augen with asymmetric tails of recrystallized grains define the shear or C planes (Fig. 8.8A,B; Dell'Angelo & Tullis, in press).

The addition of trace amounts of water (not enough to cause partial melting) to granite samples at pressures of ~1000–1500 MPa reduces their strength and produces 'higher-temperature' microstructures, including more recovery and dynamic recrystallization (Tullis & Yund 1980). Although water is known to speed the rates of stable crack growth (e.g. Atkinson 1984) in addition to causing hydrolytic weakening, the net effect of the added water appears to be to lower the temperature of the transition from cataclastic flow to dislocation creep.

Experimental study of the dislocation creep régime in quartz–feldspar

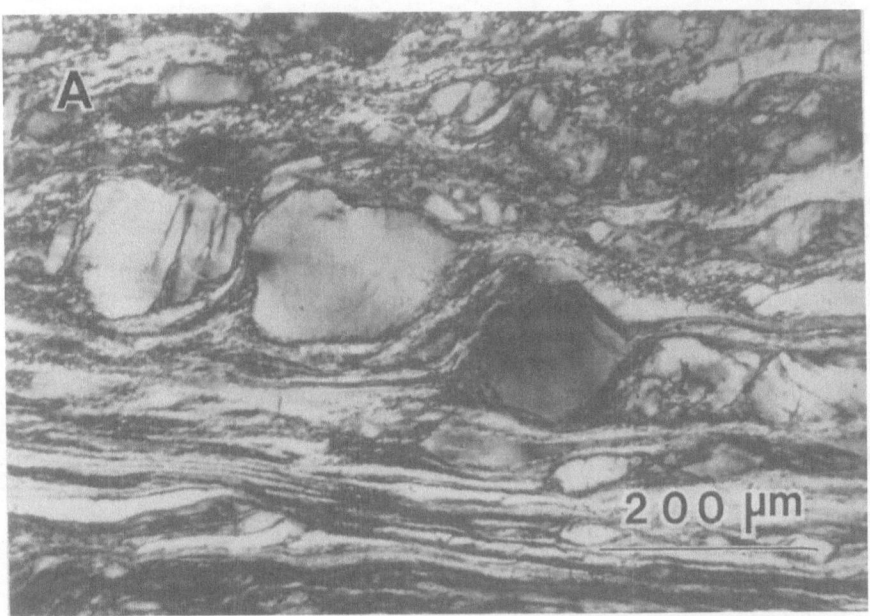

Figure 8.8 Optical photomicrographs of synthetic aggregate of 50 per cent quartz and 50 per cent albite subjected to combined compression plus shear at 900°C, 1500 MPa, and $10^{-6}\,s^{-1}$: (A) taken with crossed polarizers; (B) taken with uncrossed polarizers. Feldspar grains form relatively undeformed 'augen' with finely recrystallized tails, the asymmetry of which indicates the top-to-the-right shear sense. Quartz grains are more homogeneously deformed and more coarsely recrystallized.

aggregates is limited in the same way as that of pure feldspar aggregates, by the relatively low melting temperature. The texture produced experimentally at the highest temperature without melt consists of feldspar augen with recrystallized tails in a matrix of highly deformed and more coarsely recrystallized quartz (and mica), and is identical to that seen in naturally deformed granitic rocks of low to moderate metamorphic grade (e.g. Berthe *et al.* 1979; Simpson 1985). Thus we conclude that there is a range of both natural and experimental conditions in which quartz deforms by climb-accommodated dislocation creep and feldspars deform by recrystallization-accommodated dislocation creep. Feldspars deformed under these conditions typically form ductile shear zones. In contrast, at higher metamorphic grade, both quartz and feldspar appear to deform by climb-accommodated dislocation creep, and to have quite similar strengths and more similar recrystallized grain sizes. At these conditions the rocks are more homogeneously deformed (e.g. Voll 1976).

8.5.4 *Effect of partial melt*

The effect of partial melt on the deformation of granitic aggregates has been investigated at lower pressures and temperatures for which melt-free granite deforms in a brittle manner, as well as at higher temperatures and pressures for which it deforms by dislocation creep. In the former case, the fluid pressure of the melt reduces the stress needed for faulting. Arzi (1978) and van der Molen & Paterson (1979) defined a 'rheologically critical' melt percentage of about 30 per cent at which the grains in the rock lose contact, and the strength is governed primarily by the properties of the melt. At higher temperatures and pressures at which granitic aggregates deform by dislocation creep, experimental studies show distinct effects of grain size, melt amount, and strain rate on the resulting deformation (Dell'Angelo & Tullis 1988). For coarser grain sizes (10–200 μm), relatively small amounts of melt (<5 per cent), and a strain rate of $10^{-6}\,s^{-1}$, the melt is expelled from boundaries normal to σ_1 and collects along boundaries parallel to σ_1, but the dominant deformation mechanism remains dislocation creep. For larger amounts of melt or faster strain rates, cataclastic flow becomes the dominant deformation mechanism. For very-fine-grained aggregates (2–10 μm), if the melt amount is ~ 1 per cent or less the dominant deformation mechanism remains dislocation creep, but if it is 3–5 per cent there is a switch to melt-assisted diffusion creep as the dominant mechanism (Dell'Angelo *et al.* 1987), as noted in pure albite aggregates by Ji & Mainprice (1986).

8.6 Summary

The roughly eight orders of magnitude difference between experimental and natural strain rates might seem to suggest that the operative processes would be quite different. In fact, careful experiments guided by results from materials science and coupled with detailed and 'educated' observations of microstructures in well constrained naturally deformed rocks have been remarkably successful in elucidating details of most of the deformation mechanisms operative in the crust. The results have made it possible to more accurately interpret natural microstructures in terms of the processes that produced them, the P, T, etc., conditions under which they were formed, the orientations and magnitudes of the stresses and strains involved, and the strain-hardening or -softening nature of the deformation. Some mechanisms, such as pressure solution of quartz, may never be amenable to laboratory study, but a combination of experiments on more soluble materials, combined with theory and careful observations of naturally deformed rocks, has already provided valuable information. Other mechanisms that are important in nature, such as climb-accommodated dislocation creep of feldspar rocks and granites, may be partly or entirely cut out from laboratory study by strain-rate-invariant phase transformations or by melting. Again, analogies with other materials and detailed observations of high-grade naturally deformed rocks may provide significant insights. Although there remain many unanswered questions, one can obtain an appreciation for the great amount that has been learned in the past 25 years, since the first successful laboratory ductile deformation of silicates, by re-reading Griggs *et al.* (1960) or Turner & Weiss (1963). A large part of our scientific advances has resulted from technological advances, such as new generations of apparatus and new observational techniques, while some have been stimulated by borrowing from materials science, and others have resulted from detailed study of naturally deformed rocks. Fortunately, many further advances remain to be made.

Acknowledgements

The experimental studies described in this chapter have been generously supported by the NSF and the USGS; their support is gratefully acknowledged. I am particularly indebted to Dick Yund, who has been an equal partner in all of our collaborative projects, and to David Barber for all of his help with this chapter. I would also like to acknowledge stimulating discussions with a large number of colleagues and students, and offer special thanks to Greg Hirth for the TEM prints.

References

Allison, I. S. & T. E. LaTour 1977. Brittle deformation of hornblende in a mylonite: a direct geometrical analogue of ductile deformation by translation gliding. *Can. J. Earth Sci.* **14**, 1953–9.
Arzi, A. A. 1978. Critical phenomena in the rheology of partially melted rocks. *Tectonophysics* **44**, 173–84.

Ashby, M. F. 1972. A first report of deformation-mechanism maps. *Acta Metall.* **20**, 887–97.

Atkinson, B. K. 1984. Subcritical crack growth in geological materials. *J. Geophys. Res.* **89**, 4077–114.

Aydin, A. & A. M. Johnson 1983. Analysis of faulting in porous sandstones. *J. Struct. Geol.* **5**, 19–35.

Baeta, R. D. & K. H. G. Ashbee 1970. Mechanical deformation of quartz, Parts I and II. *Phil Mag.* **22**, 624–35.

Barber, D. J. 1970. Thin foils of non-metals made for electron microscopy by sputter-etching. *J. Mat. Sci.* **5**, 1–8.

Berthe, D., P. Choukroune & D. Gapais 1979. Orthogneiss, mylonite, and non-coaxial deformation of granites: the example of the South Armorican shear zone. *J. Struct. Geol.* **1**, 31–42.

Blenkinsop, T. G. & M. R. Drury 1988. Stress estimates and fault history from quartz microstructures. *J. Struct. Geol.* **10**, 673–84.

Borg, I. & H. C. Heard 1970. Experimental deformation of plagioclases. In *Experimental and natural rock deformation*, P. Paulitsch (ed.), 375–403. New York: Springer.

Brodie, K. H. & E. H. Rutter 1985. On the relationship between deformation and metamorphism, with special reference to the behavior of basic rocks. In *Advances in Physical Geochemistry* **4**, A. B. Thompson & D. C. Rubie (eds), 138–79. New York: Springer.

Brown, W. L. (ed.) 1983. *Feldspars and feldspathoids*. NATO ASI Series C137, 541 pp.

Carter, N. L., J. M. Christie & D. T. Griggs 1964. Experimental deformation and recrystallization of quartz. *J. Geol.* **72**, 687–733.

Carter, N. L., D. A. Anderson, F. D. Hansen & R. L. Kranz 1981. Creep and creep rupture of granitic rocks. In *Mechanical behavior of crustal rocks*, N. L. Carter *et al.* (eds). Geophysical Monograph **24**, 61–102. Washington DC: Amer. Geophys. Union.

Christie, J. M., D. T. Griggs & N. L. Carter 1964. Experimental evidence of basal slip in quartz. *J. Geol.* **72**, 734–56.

Cooper, R. L. & D. L. Kohlstedt 1984. Solution–precipitation enhanced diffusional creep of partially molten olivine-basalt aggregates during hot pressing. *Tectonophysics* **107**, 207–33.

Dell'Angelo, L. N. & J. Tullis 1982. Textural strain softening in experimentally deformed aplite. *Trans Am. Geophys. Union* **63**, 438.

Dell'Angelo, L. N. & J. Tullis 1986. A comparison of quartz *c*-axis preferred orientations in experimentally deformed aplites and quartzites. *J. Struct. Geol.* **8**, 683–92.

Dell'Angelo, L. N. & J. Tullis 1988. Experimental deformation of partially melted granitic aggregates. *J. Met. Geol.* **6**, 495–515.

Dell'Angelo, L. N. & J. Tullis 1989. Fabric development in experimentally sheared quartzite. *Tectonophysics* **169**, 1–21.

Dell'Angelo, L. N., J. Tullis & R. A. Yund 1987. Transition from dislocation creep to melt-enhanced diffusion creep in fine-grained granitic aggregates. *Tectonophysics* **139**, 325–32.

Ernst, G. W. & H. Blatt 1963. Experimental study of quartz overgrowths and synthetic quartzite. *J. Geol.* **72**, 461–9.

Etchecopar, A. & G. Vasseur 1987. A 3-D kinematic model of fabric development in polycrystalline aggregates: comparisons with experimental and natural examples. *J. Struct. Geol.* **9**, 705–18.

Evans, J. P. 1988. Deformation mechanisms in granitic rocks at shallow crustal levels. *J. Struct. Geol.* **10**, 437–44.

Fredrich, J., B. Evans & T.-F. Wong, 1989. Micromechanics of the brittle to plastic transition in Carrara marble. *J. Geophys. Res.* **94**, 4129–45.

Friedman, M. & N. G. Higgs 1981. Calcite fabrics in experimental shear zones. In *Mechanical behavior of crustal rocks*, N. L. Carter *et al.* (eds). Geophysical Monograph **24**, 11–29. Washington DC: Amer. Geophys. Union.

Frost, H. J. & M. F. Ashby 1982. *Deformation mechanism maps*. Oxford: Pergamon.

Gandais, M. & C. Willaime 1983. Mechanical properties of feldspars. In *Feldspars and feldspathoids*, W. L. Brown (ed.), NATO ASI Series C137, 207–46.

Gleason, G. & J. Tullis, 1989. C-axis preferred orientations of quartzites deformed and annealed in the alpha and beta quartz fields. *Trans Am. Geophys. Union*, **70**, 458.

Goldsmith, J. R. 1987. Al/Si interdiffusion in albite: effect of pressure and the role of hydrogen. *Contrib. Mineral. Petrol.* **95**, 311–21.

Goldsmith, J. R. 1988. Enhanced Al/Si diffusion in $KAlSi_3O_8$ at high pressures: the effect of hydrogen. *J. Geol.* **96**, 109–24.

Green, H. W. & R. S. Borch 1987. The pressure dependence of creep. *Acta Metall.* **35**, 1301–5.

Green, H. W. & R. S. Borch, 1989. A new molten salt cell for precision stress measurement at high pressure. *Europ. J. Mineral.* **1**, 213–19.

Green, H. W., D. T. Griggs & J. M. Christie 1971. Syntectonic and annealing recrystallization of fine-grained quartz aggregates. In *Experimental and natural rock deformation*, P. Paulitsch (ed.), 272–335. New York: Springer.

Griggs, D. T. & J. B. Blacic 1965. Quartz: anomalous weakness of synthetic crystals. *Science* **147**, 292–5.

Griggs, D. T., F. J. Turner & H. C. Heard 1960. Deformation of rocks at $500°$ to $800°C$. *Geol Soc. Am. Mem.* **79**, 39–104.

Guillopé, M. & J.-P. Poirier 1979. Dynamic recrystallization during creep of single crystalline halite: an experimental study. *J. Geophys. Res.* **84**, 5557–67.

Hadizadeh, J. & E. H. Rutter 1983. The low temperature brittle–ductile transition in a quartzite and the occurrence of cataclastic flow in nature. *Geol. Rund.* **72**, 493–509.

Hadizadeh, J. & J. Tullis 1986. Transition from brittle faulting to ductile cataclastic flow for anorthosite: both P and T are required. *Trans Am. Geophys. Union* **67**, 372–3.

Handin, J., D. V. Higgs & J. K. O'Brien 1960. Torsion of Yule marble under confining pressure. *Geol Soc. Am. Mem.* **79**, 245–74.

Heard, H. C. 1960. Transition from brittle fracture to ductile flow in Solnhofen limestone as a function of temperature, confining pressure, and interstitial fluid pressure. *Geol Soc. Am. Mem.* **79**, 193–226.

Heard, H. C. & N. L. Carter 1968. Experimentally induced 'natural' intragranular flow in quartz and quartzite. *Am. J. Sci.* **266**, 1–42.

Hickman, S. & B. Evans 1988. Influence of normal stress and contact radius on pressure solution along halite/silica contacts. *Trans Am. Geophys. Union* **69**, 1426.

Hirth, J. G. & J. Tullis 1986. Cataclastic flow of dry non-porous quartzite. *Trans Am. Geophys. Union* **67**, 1186.

Hirth, J. G. & J. Tullis, in press. The effects of pressure and porosity on the micromechanics of the brittle–ductile transition in quartzite. *J. Geophys. Res.*

Hobbs, B. E. 1985. The hydrolytic weakening effect in quartz. In *Point defects in minerals*, R. N. Shock (ed.), Geophysical Monograph **31**, 151–70. Washington, DC: Amer. Geophys. Union.

Hobbs, B. E. 1968. Recrystallization of single crystals of quartz. *Tectonophysics* **6**, 353–401.

Jaoul, O., J. Tullis & A. K. Kronenberg 1984. The effect of varying water contents on the creep behavior of Heavitree quartzite. *J. Geophys. Res.* **89**, 4298–312.

Jean-Loz, R. & H.-R. Wenk 1988. Convection and anisotropy of the inner core. *Geophys. Res. Lett.* **15**, 72–5.

Jessel, M. W. 1988. Simulation of fabric development in recrystallizing aggregates – II. Example model runs. *J. Struct. Geol.* **10**, 779–94.

Ji, S. & D. Mainprice 1986. Transition from power law to Newtonian creep in experimentally deformed dry albite. *Trans Am. Geophys. Union* **67**, 1235.

Jordon, P. G. 1987. The deformational behavior of bimineralic limestone–halite aggregates. *Tectonophysics* **135**, 185–97.

Karato, S.-I. 1988. Seismic anisotropy: mechanisms and tectonic implications. In *Rheology of solids and of the Earth*, S.-I. Karato & M. Toriumi (eds). Oxford: Oxford University Press.

Karato, S.-I., M. S. Paterson & J. D. FitzGerald 1986. Rheology of synthetic olivine aggregates: influence of grain size and water. *J. Geophys. Res.* **91**, 8151–6.

Kern, H. & F. Karl 1969. Eine dreiaxial wirkende Gesteinspresse mit Heizvorrichtung. *Bergbauwissenschaften* **16**, 90–2.

Kerrich, R., R. D. Beckinsdale & J. J. Durham 1977. The transition between deformation regimes dominated by intercrystalline diffusion and introcrystalline creep evaluated by oxygen isotope thermometry. *Tectonophysics* **38**, 241–57.

Kirby, S. H. & A. K. Kronenberg 1984. Deformation of clinopyroxenite: evidence for a transition in flow mechanisms and semibrittle behavior. *J. Geophys. Res.* **89**, 3177–92.

Kohlstedt, D. L. & P. Hornack 1981. Effect of oxygen partial pressure on the creep of olivine. In *Anelasticity in the Earth*, F. D. Stacey, M. S. Paterson & A. Nicolas (eds), 101–7. Geodynamics Series 4. Washington, DC: American Geophysical Union.

Kohlstedt, D., C. Goetz, W. B. Durham & J. B. Vandersande 1976. A new technique for decorating dislocations in olivine. *Science* **19**, 1045–51.

Kronenberg, A. K. & G. L. Shelton 1980. Deformation microstructures in experimentally deformed Maryland diabase. *J. Struct. Geol.* **2**, 341–53.

Kronenberg, A. K. & J. Tullis 1984. Flow strengths of quartz aggregates: grain size and pressure effects due to hydrolytic weakening. *J. Geophys. Res.* **89**, 4281–97.

Kronenberg, A. K., P. Segall & G. H. Wolf, 1990. Hydrolytic weakening and penetrative deformation within a natural shear zone. In *The brittle-ductile transition in rocks*, A. Duba, W. Durham, J. Handin & H. Wang (eds), Geophysical Monograph **56**, Washington DC: Amer. Geophys. Union.

Law, R. D. 1986. Relations between strain and quartz crystallographic fabric in the Roche Maurice quartzites of Plougastel, western Brittany. *J. Struct. Geol.* **8**, 493–516.

Lehner, F. 1990. Thermodynamics of rock deformation by pressure solution. This volume 296–333.

Liddell, N. A., P. P. Phakey & H.-R. Wenk 1976. The microstructure of some naturally deformed quartzites. In *Electron microscopy in mineralogy*, H.-R. Wenk (ed.), 419–27. New York: Springer.

Lister, G. S. & B. E. Hobbs 1980. The simulation of fabric development during plastic deformation: the effects of deformation history. *J. Struct. Geol.* **2**, 355–70.

Lister, G. S. & A. W. Snoke 1984. S–C mylonites. *J. Struct. Geol.* **6**, 617–38.

Lister, G. S., M. S. Paterson & B. E. Hobbs 1978. The simulation of fabric development in plastic deformation and its application to quartzite: the model. *Tectonophysics* **45**, 107–58.

Luan, F. & M. S. Paterson 1988. Deformation of synthetic quartz aggregates. *Trans Am. Geophys. Union* **69**, 1418.

Mainprice, D. H. 1981. *The experimental deformation of quartz polycrystals*. PhD thesis, Australian National University, Canberra.

Mainprice, D. H. & M. S. Paterson 1984. Experimental studies of the role of water in the plasticity of quartzites. *J. Geophys. Res.* **89**, 4257–69.

Marshall, D. B. & A. C. McLaren 1977. Deformation mechanisms in experimentally deformed plagioclase feldspars. *Phys. Chem. Mineral.* **1**, 351–70.

McLaren, A. C., J. D. FitzGerald & J. Gerretsen 1989. Dislocation nucleation and multiplication in synthetic quartz: relevance to water weakening. *Phys. Chem. Mineral.* **16**, 465–82.

Means, W. D. 1981. The concept of steady-state foliation. *Tectonophysics* **78**, 179–99.

Means, W. D. 1983. Microstructure and micromotion in recrystallization flow of octochloro-propane: a first look. *Geol. Rund.* **72**, 511–28.

Mitra, S. 1978. A quantitative study of deformation mechanisms and finite strain in quartzites. *Contrib. Mineral. Petrol.* **59**, 203–26.

Ord, A. & B. E. Hobbs 1986. Experimental control of the water-weakening effect in quartz. In *Mineral and rock deformation: laboratory studies*, B. E. Hobbs & H. C. Heard (eds), 51–72. *Am. Geophys. Monogr.* **36**, Washington, DC: Amer. Geophys. Union.

Paterson, M. S. 1978. *Experimental rock deformation: the brittle field.* New York: Springer.

Paterson, M. S. 1987. Problems in the extrapolation of laboratory rheological data. *Tectonophysics* **133**, 33–43.

Poirier, J.-P. 1985. *Creep of crystals.* Cambridge: Cambridge University Press.

Poirier, J.-P., C. Sotin & S. Beauchesne 1989. Experimental deformation and data processing. This volume.

Poirier, J.-P. & A. Nicolas 1975. Deformation-induced recrystallization due to progressive misorientation of subgrains with special reference to mantle peridotites. *J. Geol.* **83**, 707–20.

Poumellec, B. & O. Jaoul 1984. Influence of pO_2 and pH_2O on the high temperature plasticity of olivine. In *Deformation of ceramics II*, R. C. Bradt & R. E. Tressler (eds), 281–305. New York: Plenum.

Price, G. P. 1985. Preferred orientations in quartzites. In *Preferred orientations in deformed metals and rocks: an introduction to modern texture analysis*, H.-R. Wenk (ed.), 385–406. New York: Academic Press.

Ribbe, P. H. (ed.) 1983. In *Feldspar Mineralogy*, Reviews in Mineralogy, **2**, 2nd edn. Washington DC: Min. Soc. Amer.

Rutter, E. 1976. The kinetics of rock deformation by pressure solution. *Phil Trans R. Soc. Lond.* A **283**, 203–19.

Rutter, E. H. & S. H. White 1979. The microstructures and rheology of fault gouges produced experimentally under wet and dry conditions at temperatures up to 400°C. *Bull. Mineral.* **102**, 101–9.

Schmid, S. M. & M. Casey 1986. Complete fabric analysis of some commonly observed quartz c-axis patterns. *Am. Geophys. Union Monogr.* **36**, 263–86.

Schmid, S. M., J. M. Boland & M. S. Paterson 1977. Superplastic flow in fine-grained limestones. *Tectonophysics* **43**, 257–90.

Schmid, S. M., R. Panozzo & S. Bauer 1987. Simple shear experiments on calcite rocks: rheology and microfabric. *J. Struct. Geol.* **9**, 747–78.

Seifert, K. E. 1969. Strength of Adirondack anorthosite at elevated temperatures and pressures. *Geol. Soc. Am. Bull.* **80**, 2053–60.

Shelton, G. L. & J. Tullis 1981. Experimental flow laws for crustal rocks. *Trans Am. Geophys. Union* **62**, 396.

Shelton, G. L., J. Tullis & R. A. Yund 1979. Pressure dependence of rock strength: implications for hydrolytic weakening. *Bull. Mineral.* **102**, 110–14.

Shimada, M. 1986. Mechanism of deformation in a dry porous basalt at high pressures. *Tectonophysics* **121**, 153–73.

Simpson, C. 1982. Strain and shape fabric variations associated with ductile shear zones. *J. Struct. Geol.* **4**, 61–72.

Simpson, C. 1985. Deformation of granitic rocks across the brittle–ductile transition. *J. Struct. Geol.* **7**, 503–11.

Simpson, C. & S. M. Schmid 1983. An evaluation of criteria to deduce the sense of movement in sheared rocks. *Geol. Soc. Am. Bull.* **94**, 1281–8.

Snow, E. & R. A. Yund 1987. The effect of ductile deformation on the kinetics and mechanisms of the aragonite/calcite transformation. *J. Met. Geol.* **5**, 141–53.

Solomon, S. C., R. M. Richardson & E. A. Bergman 1980. Tectonic stress: models and magnitudes. *J. Geophys. Res.* **85**, 6086–92.

Spiers, C. & P. M. T. M. Schutjens 1990. Densification of crystalline aggregates by fluid-phase diffusional creep. This volume 334–53.

Stesky, R. M. 1978. Mechanisms of high temperature frictional sliding in Westerly granite. *Can. J. Earth Sci.* **15**, 361–75.

Stocker, R. L. & M. F. Ashby 1973. On the rheology of the upper mantle. *Rev. Geophys. Space Phys.* **11**, 391–426.

Tapponnier, P. & W. F. Brace 1976. Development of stress-induced microcracks in Westerly granite. *Int. J. Rock Mech. Min. Sci.* **13**, 103–12.

Tighe, N. J. & J. M. Christie 1969. Deformation structures in quartz rocks. In *Proc. EMSA*, 60–61. Baton Rouge: Claitors Publ. Co.

Tsenn, M. C. & N. L. Carter 1987. Upper limits of power law creep of rocks. *Tectonophysics* **136**, 1–26.

Tullis, T. E. & J. Tullis 1986. Experimental rock deformation techniques. In *Mineral and rock deformation: laboratory studies*, B. E. Hobbs & H. C. Heard (eds), 297–324, Geophysical Monograph **36**, Washington, DC. *Am. Geophys. Union.*

Tullis, J. 1983. Deformation of feldspars. In *Feldspar mineralogy*, 2nd edn, P. H. Ribbe (ed.), *Rev. Mineral.* **2**, 297–322.

Tullis, J. & R. A. Yund 1977. Experimental deformation of dry Westerly granite. *J. Geophys. Res.* **82**, 5707–18.

Tullis, J. & R. A. Yund 1980. Hydrolytic weakening of experimentally deformed Westerly granite and Hale albite rock. *J. Struct. Geol.* **2**, 439–51.

Tullis, J. & R. A. Yund 1985a. Dynamic recrystallization of feldspars: a mechanism for ductile shear zone formation. *Geology* **13**, 238–41.

Tullis, J. & R. A. Yund 1985b. Hydrolytic weakening of quartz aggregates: requirement for rapid water penetration. *Trans Am. Geophys. Union* **66**, 1084.

Tullis, J. & R. A. Yund 1986. Accommodation mechanism for dislocation creep: comparison of quartz and feldspar. *Trans Am. Geophys. Union* **66**, 366.

Tullis, J. & R. A. Yund 1987a. Mechanism of cataclastic flow in anorthosite. *Trans Am. Geophys. Union* **68**, 404.

Tullis, J. & R. A. Yund 1987b. Transition from cataclastic flow to dislocation creep of feldspar: mechanisms and microstructures. *Geology* **15**, 606–9.

Tullis, J. & R. A. Yund 1988. The effects of hydrogen, oxygen, and water fugacities and confining pressure on the strength of quartz aggregates. *Trans Am. Geophys. Union* **69**, 478.

Tullis, J. & R. A. Yund 1988. The effect of hydrogen fugacity and confining pressure on the strength of feldspar aggregates. *Geol Soc. Am. Abstr. Prog.* **20**, A213.

Tullis, J., J. M. Christie & D. T. Griggs 1973. Microstructures and preferred orientations of experimentally deformed quartzites. *Geol Soc. Am. Bull.* **84**, 297–314.

Tullis, J., L. Dell'Angelo & R. A. Yund, 1990. Ductile shear zones from brittle precursors in felspathic rocks. In *The brittle-ductile transition in rocks*, A. Duba, W. Durham, J. Handin & H. Wang (eds), Geophysical Monograph **56**. Washington, DC: Amer. Geophys. Union.

Tullis, J., A. Snoke & V. Todd 1981. Significance and petrogenesis of mylonitic rocks. *Geology* **10**, 227–30.

Tullis, T. E. & J. D. Weeks 1986. Constitutive behavior and stability of frictional sliding of granite. *Pageophage* **124**, 383–414.

Turner, F. J. & L. E. Weiss 1963. *Structural analysis of metamorphic tectonites.* New York: McGraw-Hill.

Urai, J. L. 1987. Development of microstructure during deformation of carnallite and bischofite in transmitted light. *Tectonophysics* **135**, 251–63.

van der Molen, I. & M. S. Paterson 1979. Experimental deformation of partially-melted granite. *Contrib. Mineral. Petrol.* **70**, 299–318.

Voll, G. 1976. Recrystallization of quartz, biotite, and feldspars from Erstfeld to the Leventina-nappe, Swiss Alps, and its geological significance. *Schweiz. Mineral. Petrogr. Mitt.* **56**, 641–7.

von Kármán, 1911. Festigheitversuche unter allseitigem Druch. *Z. Verein. Deutsch. Ingen.* **55**, 1749–57.

Wenk, H.-R. & U. F. Kocks 1987. The representation of orientation distributions. *Metall. Trans.* **18A**, 1083–92.

Wenk, H.-R., G. Canova, A. Molinari & U. F. Kocks 1989. Viscoplastic modeling of texture development in quartzite. *J. Geophys. Res.* **94**, 17895–906.

White, S. 1973. Dislocation structures responsible for optical strain features in deformed quartz crystals. *J. Mat. Sci.* **8**, 490–9.

White, S. 1976. The effects of strain on the microstructures, fabrics, and deformation mechanisms in quartzites. *Phil Trans. R. Soc. Lond. A* **283**, 69–86.

White, S. 1977. Geological significance of recovery and recrystallization processes in quartz. *Tectonophysics* **39**, 143–70.

White, J. C. & C. K. Mawer 1986. Extreme ductility of feldspars from a mylonite, Parry Sound, Canada. *J. Struct. Geol.* **8**, 133–43.

Wong, T.-F. 1982. Micromechanics of faulting in Westerly granite. *Int. J. Rock Mech. Min. Sci.* **19**, 49–64.

Yund, R. A. 1983. Diffusion in feldspars. In *Feldspar Mineralogy*, Reviews in Mineralogy, 2 2nd edn, P. H. Ribbe (ed.), 203–22. Washington DC: Min. Soc. Amer.

Yund, R. A. & E. Snow, 1989. The effects of hydrogen fugacity and confining pressure on the interdiffusion of NaSi–CaAl in peristerites. *J. Geophys. Res.* **94**, 10662–8.

Yund, R. A. & J. Tullis 1980. The effect of water, pressure, and strain on Al/Si order–disorder kinetics in feldspar. *Contrib. Mineral. Petrol.* **72**, 297–302.

Zeuch, D. H. & H. W. Green 1984. Experimental deformation of a synthetic dunite at high temperature and pressure, part I: mechanical behavior, optical microstructure and deformation mechanism, *Tectonophysics* **110**, 233–62.

CHAPTER NINE

Microstructural analysis and tectonic evolution in thrust systems: examples from the Assynt region of the Moine Thrust Zone, Scotland

Robert J. Knipe

9.1 Introduction

Recent microstructural studies of experimentally and naturally deformed rocks have provided important information for the understanding of deformation mechanisms, rates and conditions during tectonic events (Borradaile *et al.* 1982, Schmid 1982, Hobbs & Heard 1986, Tullis 1986, Wang 1986, Atkinson 1987, Schmid *et al.* 1987, Knipe & Rutter 1990). The microstructures characteristic of different deformation mechanisms, which operate under different deformation conditions, have been identified and their preservation in tectonites used to assess natural deformation processes and kinematics (Pfiffner 1982; Simpson & Schmid 1983; Groshong *et al.* 1984; Lister & Snoke 1984; Mitra 1984; Ord *et al.* 1984; Brodie & Rutter 1985; Gibson & Grey 1985; Knipe & Wintsch 1985; Obee & White 1985; Law *et al.* 1986; Rutter *et al.* 1986; Sibson 1986; Twiss 1986; Knipe, 1989). In addition, such studies have helped to construct fault-zone models (Sibson 1982, 1983, 1985; Knipe 1985, 1989; Kuznir & Park 1986; Platt & Behrmann 1986).

Despite these advances a number of problems still exist in the use of preserved microstructures for the assessment of natural deformation mechanisms. For example, the recognition that deformation along fault zones is likely to involve some form of cyclic deformation in which each event incorporates a range of deformation mechanisms (Sibson 1980, 1986; House & Grey 1982; White & White 1983; Knipe 1985, 1989; Platt & Behrmann 1986; Wojtal & Mitra 1986) presents a problem in the interpretation of the finite microstructures preserved (Knipe, 1989). In addition, the superposition of microstructures which arise from the reactivation of fault zones during later tectonic events with different kinematic frameworks also complicates microstructural interpretation. These problems highlight two important requirements for future microstructural studies. First, there is a need to consider the total deformation history experienced by the tectonite, i.e. to assess the deformation mechanism path (Knipe, 1986), and, second, as the generation and stability of many deformation microstructures are temperature- and pressure-dependent, consideration of the pressure–temperature–time path (Thompson & England 1984, Thompson & Ridley 1987) is also necessary (e.g. Platt & Lister 1985). The integration of these aspects leads to a need to consider the pressure (P) – temperature (T) – strain rate – ($\dot{\varepsilon}$) – strain (ε) – time (t) path (i.e. the $PT\dot{\varepsilon}\varepsilon t$ path) when assessing the microstructural development of tectonites. In this chapter the deformation histories and $PT\dot{\varepsilon}\varepsilon t$ paths involved in the evolution of thrust systems are discussed. The aim of the chapter is to illustrate how microstructural analysis can aid the assessment of $PT\dot{\varepsilon}\varepsilon t$ paths and thus help to delineate the tectonic evolution of an area by providing information on: (a) the mechanisms of deformation, the sequence of deformation mechanisms, and the nature of cyclic deformation events or instabilities; (b) the kinematics of deformation; (c) the conditions of deformation; and (d) the dynamics of the deformation.

9.2 Deformation histories in thrust systems

Large-scale thrust systems (Boyer & Elliott 1982; Butler 1982, 1987) are an integral feature of continental collision zones (Coward & Ries 1986) and involve crustal shortening by the movement and stacking of thrust sheets. The deformation history and thus the evolution and preservation of microstructures in rocks incorporated in thrust systems depends upon the geometry and distribution of faults in the thrust system, the displacement pattern during the fault array evolution (Knipe 1985) and the thermal history of the thrust zone (Brewer 1981, Platt 1987, Thompson & Ridley 1987).

In the case of a single master fault bringing material from deep to high levels in the crust, the superposition of deformation at different conditions will produce complex finite microstructures (Sibson 1977) which may be difficult to unravel in detail. A more favourable situation for assessing microstructural development is where a sequential transfer of displacement to different faults takes place as the thrust array evolves. If this sequence involves the transfer of displacement to successively lower faults (i.e. a foreland propagating system), rocks deformed along deeper level faults are carried, piggy-back style, to higher levels; thus each fault in the array may preserve microstructures characteristic of deformation at different crustal conditions. The chances of recognizing deformation mechanisms and characterizing the different types of cyclic deformation events which may occur during faulting at different crustal levels is therefore increased in these latter situations.

The thrust fault system chosen for discussion in this chapter, the Moine Thrust Zone of north-west Scotland, is ideal for the assessment of deformation histories as it is known to have evolved by a piggy-back sequence and contains a range of fault rocks developed under different deformation conditions (Christie 1960; White *et al*. 1982; Law *et al*. 1984, 1986). The large-scale structural evolution of this zone is reviewed in the next section, before the fault-rock microstructures and deformation histories associated with the different faults in the array are discussed in detail.

9.3 The Moine Thrust Zone of north-west Scotland

The Moine Thrust Zone of north-west Scotland forms part of the Caledonian Orogenic Belt and involved the displacement of a series of thrust sheets towards the WNW over a foreland of Precambrian Gneiss and Cambro-Ordovician sediments (Coward 1983, Butler & Coward 1984, Butler, 1986). The Moine Thrust is the most important and highest thrust in the sequence; it probably has the largest displacement and carries Proterozoic metasediments over a deformed foreland succession contained in the lower structural levels of the thrust zone. The Moine Thrust was the earliest of the structures to develop, and was carried into its present position by movement on the lower level thrusts. The total displacement on the thrust system is well in excess of 70 km (Elliott & Johnson 1980) and occurred over the time period between 430 and 408 Ma (Johnson *et al*. 1985).

In places the higher faults have been breached by the lower thrusts and the area has been affected by post-thrust extension, considered by Coward (1982, 1984) and Enfield & Coward (1987) to be related to post-orogenic collapse during the Devonian.

Studies of the deformation processes involved in the emplacement of thrust sheets in the Moine Thrust Zone have taken advantage of the occurrence of the same quartz-rich lithologies (the Cambrian Quartzites) in various deformation states throughout the thrust belt, and have concentrated on the ductile deformation involved in the production of mylonites located in the higher structural levels (Christie 1960, 1963; Weathers *et al*. 1979; White 1979a; Evans & White 1984; Law *et al*. 1984, 1986; Ord & Christie 1984; Knipe & Law 1987; Law & Potts 1987; Law 1987). Only a few studies of the cataclastic rocks present in the lower thrust sheets have been conducted (Blenkinsop & Rutter 1986, Bowler 1987, Lloyd & Knipe 1990). This chapter considers the deformation processes involved in the fault-rock evolution on all the major faults in a traverse across the Moine Thrust Zone at Assynt, and attempts to integrate fault-rock evolution with the larger-scale development of the thrust system.

The Assynt area was chosen because of the recent work on the geometrical evolution, the thermal history, and the deformation timing in this area (Coward 1984, 1985; Butler 1984; Johnson *et al*. 1985). In addition, Assynt

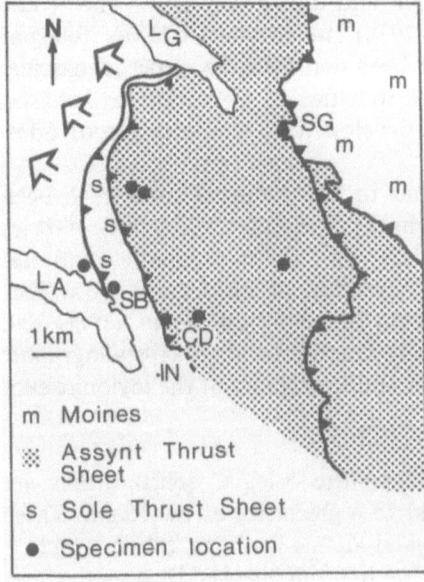

m   Moines

   Assynt Thrust
   Sheet

s   Sole Thrust Sheet

●   Specimen location

Figure 9.1 Sketch map of the Assynt area showing the main thrust sheets (after Coward 1982). The arrows indicate the thrust movement direction. CD, Cnoc Dubh; IN, Inchnadamph; LA, Loch Assynt; LG, Loch Glencoul; SB, Skiag Bridge; SG, Stack of Glencoul.

occupies an important transition area in the thrust belt as it separates a northern sector, within which the Moine Thrust is considered to have developed on a steeper mid-crustal ramp compared with the southern section (Coward 1984). This difference has resulted in a deeper burial and more ductile deformation of Cambro-Ordovician rocks in the north (Coward 1984). The structure of the Assynt area is reviewed in Figure 9.1, which also shows the location of the specimens described in this chapter. The structure can be considered in terms of three thrust units: (a) The Moine Thrust sheet, which occupies the highest structural position; (b) the Assynt Thrust sheet, comprising the Ben More and Glencoul Thrust sheets, which occupy the intermediate structural levels; and (c) the Sole Thrust System, which includes the lowest thrusts exposed. The displacement estimates for these three systems are: ~70 km for the Moine Thrust (Coward *et al.* 1980, Elliott & Johnson 1980); ~28 km for the Ben More thrust (Elliott & Johnson 1980); 20–33 km for the Glencoul Thrust (Coward *et al.* 1980); and 3 km for the Sole Thrust System (Coward 1984). The fault rocks and deformation features present in each of these three units are described in separate sections below.

9.4   The Moine Thrust at the Stack of Glencoul

The rocks exposed along the Moine Thrust at the Stack of Glencoul (Fig. 9.1) are L–S to S–L mylonites derived from the overriding Moine (and Lewisian) rocks and the Cambrian Quartzites of the foot wall. The zone of intense deformation is approximately 100 m thick and dips gently ESE. The Moine Mylonites are composed of quartz (>70%), muscovite (<15%), feldspar <15%), and chlorite (<15%), and have been derived from upper greenschist metasediments retrograded from biotite to chloride assemblages. Johnson et al.(1985) suggest that the mylonites developed at temperatures of 300–350°C.

The quartz mylonites in the foot wall to the Moine Thrust have been extensively studied, particularly in terms of their crystallographic fabrics (Christie 1960, 1963; Law et al. 1986; Law 1987). The mylonitic quartzites have microstructures indicative of the operation of dislocation creep processes. The most recent studies of these mylonites by Law et al. (1986) and Law (1987) involved an integration of microstructural and the crystallographic fabric data into models of the mechanisms and kinematics of the mylonite evolution and revealed the following:

(a) Close to the Moine Thrust (<0.15 m) the original quartz grains are recrystallized [50–100% new grains] to a grain size of 10–15 μm. These new grains are slightly elongated and define a shape fabric which is oblique to the main mylonitic foliation banding marked by domains/lens with slightly different phyllosilicate contents (1–15%) and separated by thin (<1 mm wide) phyllosilicate bands. The c-axes in these rocks define either a single girdle (at 0.5 cm from the Moine Thrust) or a cross-girdle (1–14 cm below the Moine Thrust) which have skeletal outlines and intensity distributions that are asymmetric with respect to the main mylonitic foliation and lineation. Both the microstructures and the crystallographic fabrics are consistent with development during a strain path dominated by non-coaxial deformation associated with WNW-directed thrusting.

(b) At distances greater than 30 cm from the Moine Thrust, the volume fraction of recrystallized grains is 40–75% and the mylonite foliation is defined by flattened relict detrital grains which now have axial ratios in the x–z plane of 50:1 to 100:1. The c-axis fabrics of these rocks are symmetrical with respect to the main foliation and lineation, and, together with the presence of globular grains (equant grains in which the c-axis is subparallel to the shortening direction z, indicate co-axial deformation.

(c) Law et al. (1986) argue that these different deformation paths are synchronous and that large-scale overprinting of a coaxial deformation on a non-coaxial deformation has not occurred. The two kinematic domains recognized at the Stack of Glencoul are also present in adjacent areas of Assynt and Loch Eriboll (Law et al. 1984), and indicate that a component of extensional flow in the direction of movement accompanies the emplacement of the Moine Thrust sheet. Additional field evidence which supports this extensional flow component has also been found by Coward (1982, p. 252) and Butler (1984).

The size of the subgrains (2–4 μm) and the dislocation densities (4×10^8 cm^{-2} to 1×10^9 cm^{-2}), together with the size of the recrystallized grains (10–14 μm) in the mylonites at the Stack of Glencoul have been used by Weathers *et al.* (1979) and Ord & Christie (1984) to estimate the differential stress levels associated with mylonite development. The estimated stress levels ranged from 11.5 to 183 MPa. Such stress levels indicate strain rates in the region of 10^{-12}–10^{-10} s^{-1} (see Rutter 1976, fig. 9). Weathers *et al.* (1979) describe the microstructure of the ribbon mylonites as typical of low-stress, high-temperature, low-strain-rate deformation. In other words, dislocation recovery processes such as cross-slip and climb are indicated by the well organized low-angle boundaries, curved dislocations, and homogeneous dislocation distribution. Examination of these mylonites during the present study reveals more complex and heterogeneous microstructures related to a series of late or post-mylonite events. The features produced by these later deformation events (quartz veins, concentrated shearing in phyllosilicate-rich domains, shear bands, stylolites, minor forethrusts, and backthrusts) are discussed separately below, before their relative chronology and significance is outlined.

9.4.1 Quartz veins

A prolonged history of intermittent quartz veining late in the mylonite evolution close (<0.15 m) to the Moine Thrust is indicated by the presence of quartz veins (up to 600 μm wide) oriented at various angles (110°– 32°) to the main foliation in x–z sections. The veins at low angles (<40°) dip ESE, have been recrystallized to grain size similar to the mylonite matrix, and appear to represent early veins which have been rotated and sheared towards the WNW after their development. In contrast, veins at high angles to the main foliation are coarser grained and exhibit fewer internal deformation features. The veins that appear to be very late in the deformation history have suffered little or no rotation and are composed of large (>15 μm) fibrous subgrains, often aligned perpendicular to the vein margin. They commonly contain dislocation densities 30–40 times those in the adjacent mylonitic fabric (Fig. 9.2), indicating a small amount of late, very localized high-strain-rate deformation. The late, high-dislocation-density features are concentrated within a millimetre of the late vein and are not pervasive in the adjacent mylonite fabric (Fig. 9.2). Many of the veins exhibit extremely irregular margins composed of grains larger (50– 500 μm) than the recrystallized matrix. This microstructure may represent a limited amount of grain-boundary migration at the vein margin (see later discussion).

9.4.2 Concentrated shearing in phyllosilicate-rich domains

Although the late quartz veins are deformed in the quartz-rich mylonite bands, they are more sheared in the phyllosilicate-rich domains of the mylonites. This

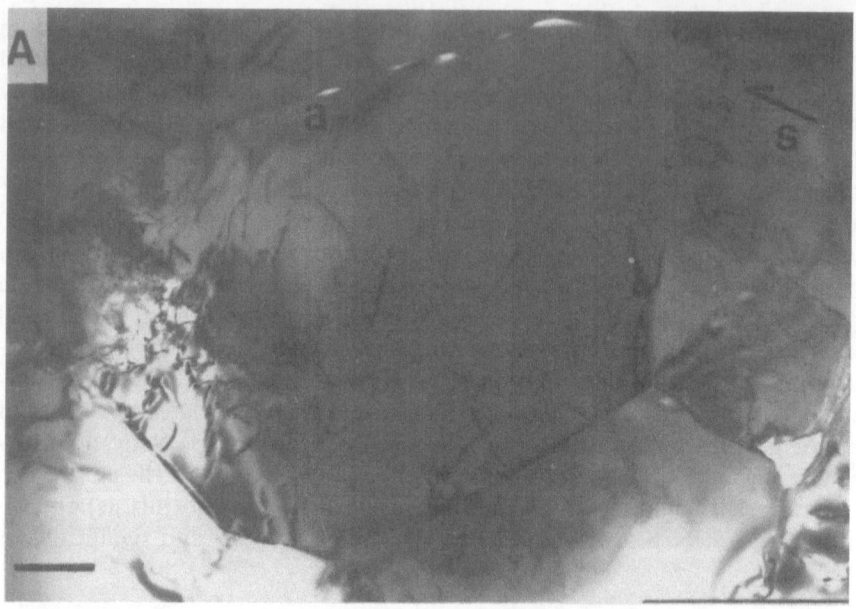

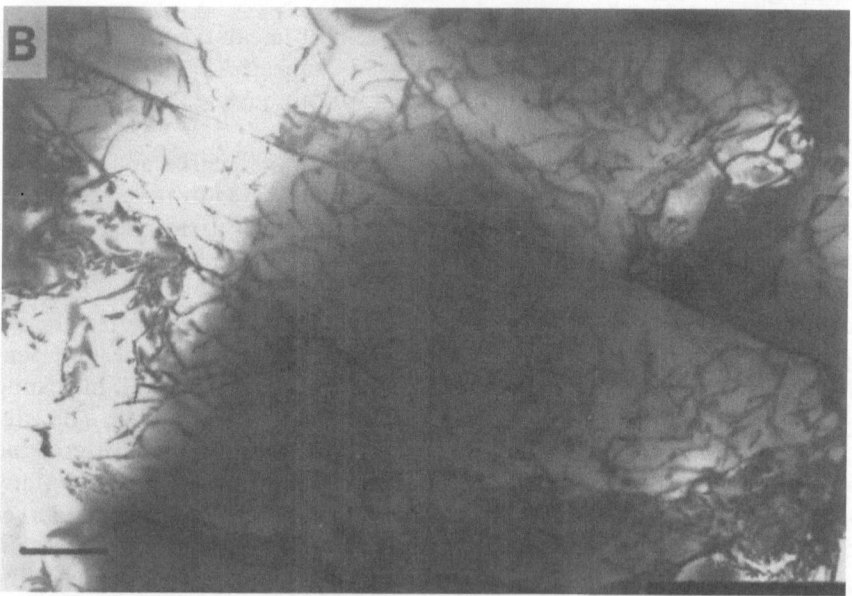

Figure 9.2 TEM micrographs which compare the microstructures of the mylonite (A) with the substructure of a late vein in the mylonites (B). The mylonite is composed of a fine grained recrystallized aggregate with internal dislocation densities of $\sim 4 \times 10^8 \, cm^{-2}$. Note also the bubbles present on grain boundary (a) are located on the boundaries which are extensional during WNW shearing; the shearing sense in this x–z section is indicated. The quartz vein shown in (B) has a substructure of elongate subgrains with high dislocation densities.

Figure 9.3 Optical micrograph of the Moine Thrust plane (MT) separating the phyllosilicate-rich Moines of the hanging wall from the quartz mylonites of the foot wall. Note the early sheared vein at S, and the minor extensional fault E with displacement down to ESE. The vein V truncates this extension but is sheared by deformation in the Moines.

observation indicates that either the deformation associated with shearing to the WNW continued for longer in these phyllosilicate-rich domains or that the shear strain rate was higher in these units after vein formation. One late vein shown in Figure 9.3, which truncates the Moine Thrust, indicates that shearing at this stage of the deformation was concentrated in the phyllosilicate-rich Moine rocks of the hanging wall.

9.4.3 Shear bands and extensional faults

Shear bands oriented at 15–35° to the main foliation are present in the mylonites at the Stack of Glencoul. They are more abundant in selected horizons within the phyllosilicate-rich Moines above the Moine Thrust, but also occur in the quartz mylonites of the foot wall. By far the largest number of these dip to the WNW and indicate WNW shearing. A few examples of extensional features which dip to the ESE are also present. Law *et al.* (1986) have described a small listric normal fault which has a displacement down to the ESE and has associated dynamic recrystallization within the zone of deformation. This particular zone of extension to the ESE predates a late vein which truncates the Moine Thrust but is sheared to the WNW in the Moine rocks. The deformation of this vein and the extensional fault suggests intermittent changes in the strain path and deformation mechanisms from subhorizontal thrust sheet extension (by dislocation creep and veining) to WNW shearing during the late stages of mylonite evolution. The shear bands and the crystallographic fabrics in the mylonites also indicate that the component of extensional flow was not confined to the deformation-producing veins and minor normal faults.

9.4.4 Minor forethrusts and backthrusts

A number of minor forethrusts (overriding to the WNW) and backthrusts (overriding to the ESE) truncate the mylonite fabric at the Stack of Glencoul. The displacements associated with these features range from a few millimetres to ~1 m. The occurrence of Moine-like, phyllosilicate-rich bands below the

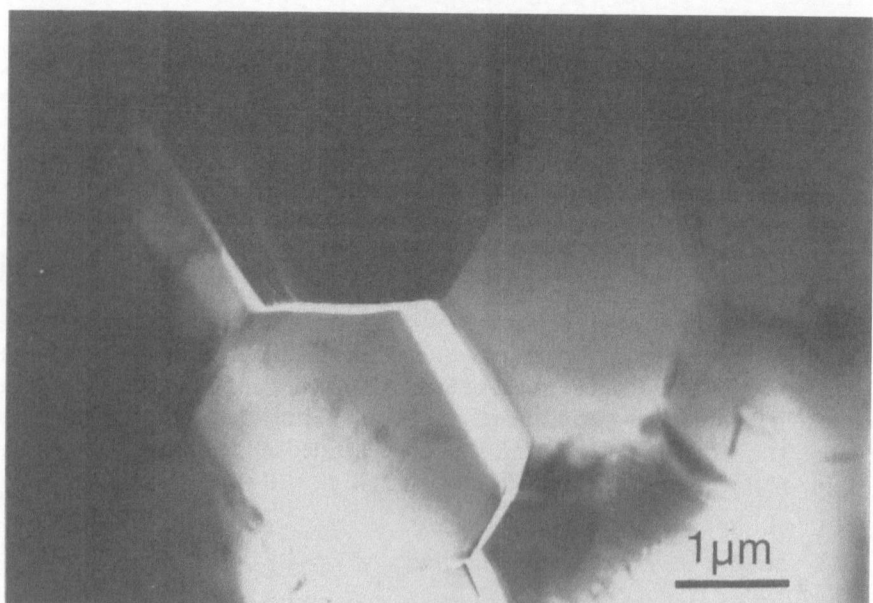

Figure 9.4 TEM micrograph of a cataclastic zone present in the Stack of Glencoul mylonites. Note the hexagonal grains and the lack of internal dislocation microstructures. Scale bar 1 μm.

Moine Thrust has been interpreted by Law *et al.* (1986) as a possible fore-thrust which breaches and thus repeats the Moine Thrust. The quartzite which has overridden the phyllosilicate-rich rocks is brecciated. In thin section both the minor forethrusts and the backthrusts are marked by extremely fine-grained material, which usually lacks any shape fabric and encloses rotated blocks containing the mylonitic fabric; i.e. the microstructures are indicative of cataclastic deformation. TEM of these zones reveals a microstructure composed of occasional large ($>10\,\mu$m) grains with high dislocation densities [$>1 \times 10^9$ cm^{-2}] which represent mylonite fragments with modified substructures, set in a matrix of equant hexagonal grains with a diameter of $5-7\mu$m. The $120°$ grain-boundaries and the extremely low dislocation densities (Fig. 9.4) characteristic of the matrix suggest a high degree of recovery. The significance of this microstructure, which is interpreted as a recovered cataclasite containing fragments of the host rock, is discussed in more detail in a later section. A few extremely fine-grained bands subparallel to the mylonite fabric do not contain any oblique shape fabric and may also represent cataclastic zones.

9.4.5 Stylolites

The veins, backthrusts, and shear bands are all truncated by subhorizontal stylolites, indicating that diffusive mass transfer (DMT) processes (pressure solution) producing vertical shortening were associated with the late stages of the deformation history in the mylonitic rocks (Fig. 9.5). In general, the stylolites represent the last deformation event preserved in the mylonites at the Stack of Glencoul, although occasional quartz veins and late extensional or vertical fractures do truncate these features.

The chronology of these deformation features is reviewed in Figure 9.6, which emphasizes the overlap in time of the development of some features. The deformation history recorded by the mylonites involves changes in the dominant deformation mechanism and the strain path with time. The trend revealed from the microstructural analysis involves the following:

(1) An early phase of mylonite evolution dominated by dislocation creep processes. Close to the Moine Thrust this deformation is dominated by non-coaxial straining, while away from the Moine Thrust a co-axial deformation path was followed (see Law *et al.* 1986 for details).

(2) A late stage of mylonite evolution during shearing to the WNW, where intermittent fracturing produced quartz veins. Subsequent deformation of the quartz veins involving dislocation creep processes has rotated these features. Such rotation is often concentrated into the phyllosilicate-rich bands. In addition, shear-band evolution together with minor extensional faulting indicate a component of extensional flow primarily in the direction of movement, and vertical shortening accompanying the shearing

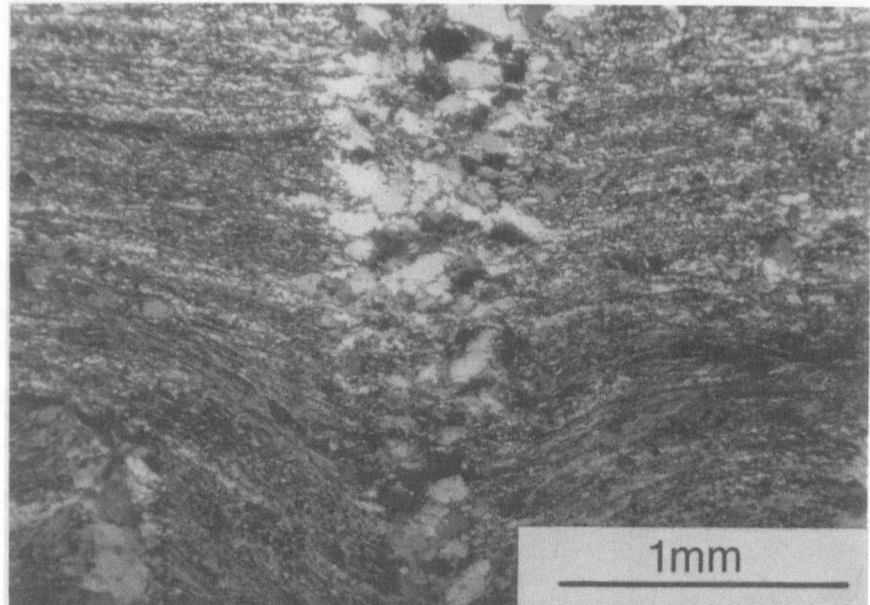

Figure 9.5 Optical micrograph illustrating the indentation of a coarse-grained vein into a phyllosilicate-rich band in the mylonites of the Stack of Glencoul. The early mylonitic fabric in the phyllosilicate-rich band is wrapped around the vein because of the concentration of DMT associated with vertical shortening. Scale bar 1 mm.

towards the WNW. The superposition of these deformation features indicates an alternation of shearing and extensional deformation events.

(3) A post-mylonite phase of forethrusting and backthrusting involving fracturing and development of cataclasites.

(4) A post-cataclastic deformation phase of vertical shortening achieved by diffusive mass transfer. This late deformation produced stylolites and was also postdated by occasional vertical fracturing (and less often by quartz vein development).

The early phase of mylonite evolution represents the largest straining event preserved. The later events, although important to the microstructual evolution, represent small strains.

In general terms, the deformation history in the mylonites represents a transition from ductile deformation by distributed dislocation creep to deformation involving more localized shearing and fracturing. All these events were associated with WNW thrusting which was followed by more ductile deformation (DMT), producing vertical shortening. It should be emphasized that all the deformation phases associated with WNW shearing involve a component of extension in the movement direction. Early in the deformation history extension was achieved by dislocation creep processes, and this is reflected

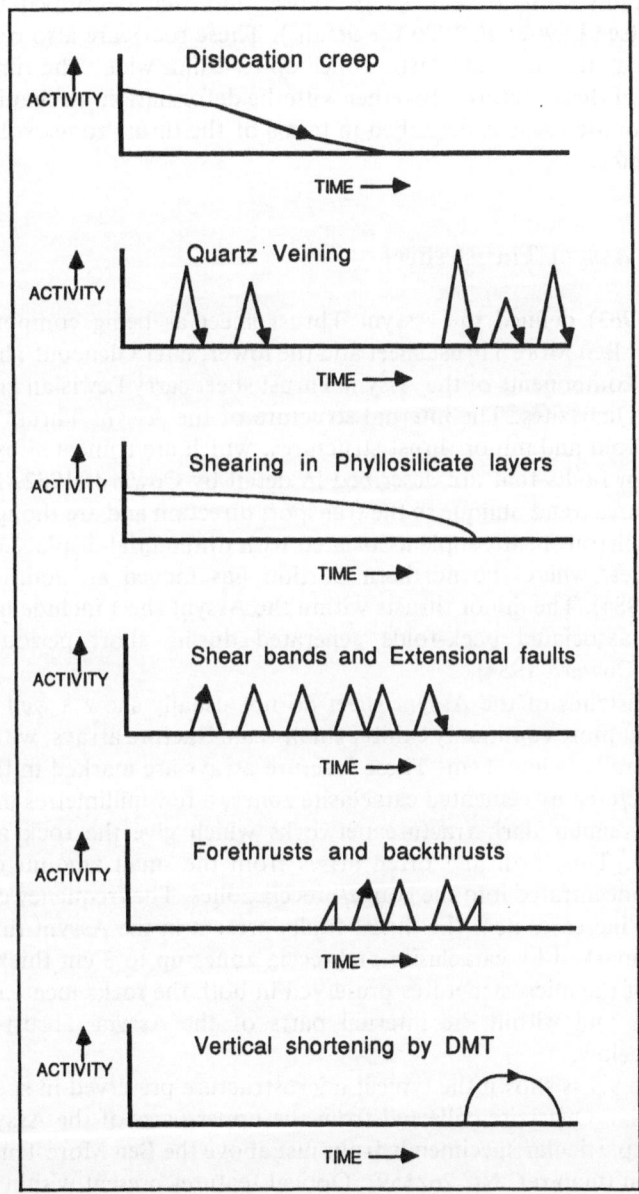

Figure 9.6 Review of the relative chronology of deformation features which develop late during the deformation history of the mylonites at the Stack of Glencoul. See text for details.

by the *c*- and *a*-axis fabrics (Law *et al.* 1986, Law 1987), while later in the deformation this component of strain was produced by either shear-band evolution or extensional veining and faulting.

The quartzites which are more than 40 m below the Moine Thrust contain original grains with a weak shape fabric and well developed undulatory extinction (see Law *et al.* 1986 for details). These rocks are also truncated by transgranular fracture/cataclastic zones up to 2 mm wide. The timing of the formation of these features, together with the deformation-mechanism history of the mylonitic rocks is described in terms of the thrust-zone evolution later in the chapter.

9.5 The Assynt Thrust sheet

Christie (1963) defined the Assynt Thrust sheet as being composed of the higher-level Ben More Thrust sheet and the lower, later Glencoul Thrust sheet. Both these components of the Assynt Thrust sheet carry Lewisian gneisses and Cambrian Quartzites. The internal structure of the Assynt Thrust sheet contains both fold and minor thrust structures, which are truncated by a late set of extension faults that are described in detail by Coward (1982, 1984). The fold structures trend oblique to the transport direction and are thought to arise during a sinistral shear couple associated with differential displacement in the Assynt sheet, where the northern portion has moved an additional 3 km (Coward 1984). The minor thrusts within the Assynt sheet include backthrusts and have associated back-folds generated during short periods of ESE transport (Coward 1984).

The quartzites of the Assynt sheet do not usually show a well developed cleavage but more commonly exhibit small-scale fracture arrays, with displacements generally below 2 cm. These fracture arrays are marked in the field by either white/creamy cemented cataclasite zones, a few millimetres in width, or by more irregular dark fracture networks which give the rock a 'bruised' appearance. This 'bruising' often arises from the small amount of chlorite ($\ll 10\%$) concentrated into the gouge/breccia zones. The frequency of the fracture arrays increases near the thrust faults present in the Assynt thrust sheet, which are marked by cataclasite or breccia zones up to 3 cm thick. Selected examples of the microstructures preserved in both the rocks located along the fault zones and within the internal parts of the Assynt Thrust sheet are described below.

In Figure 9.7 is shown the typical microstructure preserved in specimens of the Cambrian Quartzite collected from the upper parts of the Assynt Thrust sheet. This particular specimen is from just above the Ben More Thrust at Ben na Fhuarain (map ref. NC 262159). Optical features present within the grains include undulatory extinction, deformation bands, and deformation lamellae, all of which are associated with TEM-scale subgrain microstructures. More

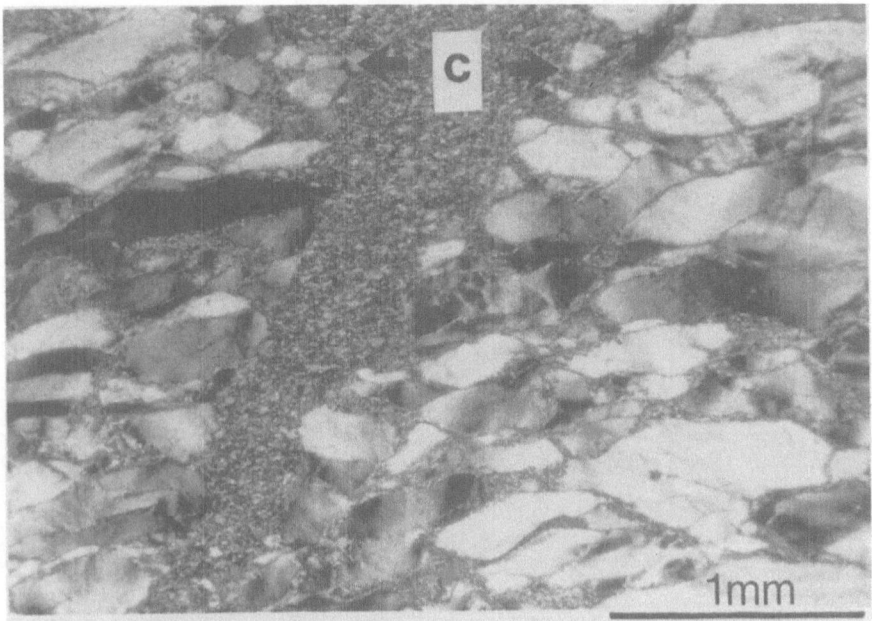

Figure 9.7 Optical micrograph of deformed quartzite from Ben More thrust sheet. Note grain elongation and undulatory extinction together with transgranular cataclastic zone with gouge infill, C. Scale bar 1 mm.

than 40 per cent of the grains have a spread in the *c*-axis orientation of more than $20°$. Recrystallization is confined to grain boundaries and deformation band boundaries, and new grains only account for ~5 per cent of the rock volume. In addition to these microstructures which indicate deformation by dislocation processes, many grains also contain dense arrays of subparallel microfractures.

A three-dimensional network of transgranular fractures is ubiquitous to these specimens and separate blocks which contain the crystal plastic deformation features. The fractures are marked by cemented gouge and microbreccia zones up to 2 mm wide and define blocks with a minimum size of a few centimetres. The displacement magnitude along these zones cannot be quantified because the zones are wider than the average grain size of 0.4 mm. The fracture zones exhibit a range of internal structure from micro-breccias (where the fine-grained matrix is less than 40 per cent of the zone volume and the larger detrital grain fragments are extremely poorly sorted in terms of grain size) to well developed cemented cataclasites (where the matrix accounts for more than 85 per cent of the fracture zone volume and the detrital grain fragments are well sorted with an extremely small size range). The detrital grain fragments present in the cataclasites contain high dislocation densities greater than $3 \times 10^9 \, \text{cm}^{-2}$ (Fig, 9.8). The fine-grained matrix of the cataclastic zones is

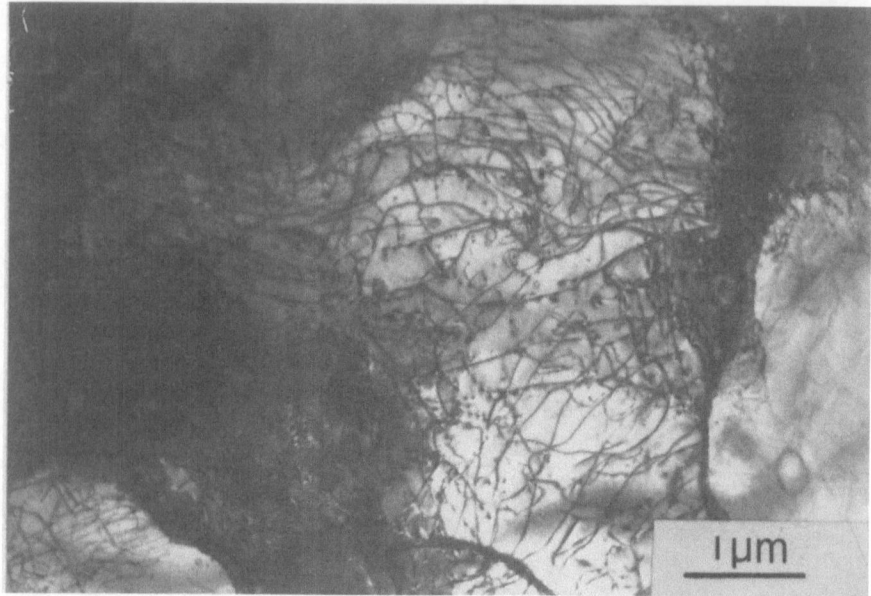

Figure 9.8 TEM micrograph of detrital grain within a cataclastic gouge zone from the Ben More Thrust sheet. Note high dislocation density within subgrains. Scale bar 1 μm.

made up of grains 2–5 μm in diameter. These grains are roughly hexagonal in shape and contain internal dislocation substructures (Fig. 9.9). However, in contrast to the detrital grain fragments the dislocation densities are usually lower than $< 3 \times 10^8$ cm^{-2} and dislocations form a more ordered array indicative of recovery.

The deformation features preserved in specimens from within the Glencoul Thrust sheet are similar to those described in the Ben More Thrust sheet but exhibit less evidence of deformation by dislocation movement. The original detrital grains exhibit undulatory extinction, deformation bands, and deformation lamellae, but rarely show evidence of recrystallization. Only very weak shape fabrics are usually present, although localized development of a strong shape fabric (axial ratios $>3:1$) by crack-seal extension of detrital grains is occasionally found near minor thrusts such as at Glas Bhein (see Fig. 9.1 for location). As with the Ben More specimens, transgranular cataclasite and micro-breccia zones are common, particularly near minor thrust planes.

The Glencoul Thrust fault which forms the base of the Assynt Thrust sheet is well exposed above the southern shore of Glencoul, where the hangingwall rocks are Lewisian gneisses, and north of Inchnadamph between Cnoc Dubh and Achmore (Map Ref. NC 256228) where the hangingwall rocks are Cambrian Quartzites. In the first of these locations the fault rocks located on the Glencoul Thrust are phyllosilicate-rich cataclasites developed by the fracturing and breakdown of the feldspar-rich Lewisian. The cataclasites contain kinematic indicators including R_1 shears and P foliations (see Rutter *et al.* 1986 for definitions) consistent with overthrusting to the WNW.

The Glencoul Thrust fault rocks derived from the Cambrian Quartzites exposed north of Inchnadamph are described in more detail here. Along this section the Glencoul fault has been rotated by the later development of duplexes in the underlying limestone and dips towards the ENE. The fault plane exposed at Cnoc Dubh is marked by a 5–10 cm thick breccia zone developed in the quartzites of the hanging wall. The breccia contains fragments up to 2 cm in diameter enclosed in anastomosing cataclasite zones (Fig. 9.10). The cataclastic zones are banded with individual bands showing different clast sizes, different degrees of sorting, and different fragment : matrix ratios. As with the Ben More fracture zones the detrital grain fragments contain high dislocation densities but also have extremely irregular margins indicative of corrosion by fluids present in the fault zone. TEM analysis of the microstructure preserved in the matrix of the cataclastic zones reveals a complex substructure where the grain size is 2–5 μm and the internal features of the larger fragmented detrital grains contrast with those of the smaller grains in the matrix. The larger grains show internal microstructures ranging from high dislocation density arrays ($> 3 \times 10^9$ cm^{-2}) to cells with lower internal dislocation densities and dislocation walls. The smaller grains contain

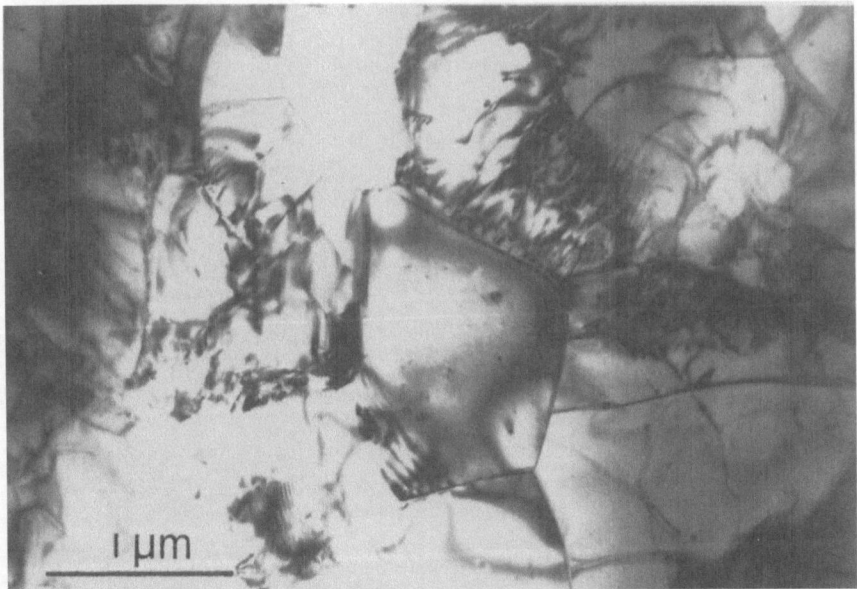

Figure 9.9 TEM micrograph of the matrix of a cataclastic zone from the Ben More Thrust sheet. Note the well organised cell structure with low dislocation densities in some areas of the specimen. Scale bar 1 μm.

Figure 9.10 Optical micrographs of the cataclastic deformation present in the Glencoul Thrust fault rocks. (A) The network of fracture zones located just above the thrust plane. (B) The microbreccia present marking the thrust plane. Note the stylolite(s) associated with the large clast in the bottom right. Scale bars 1 mm. (C) A TEM micrograph of the cataclastic matrix showing the irregular cell network present. Scale bar 1 μm.

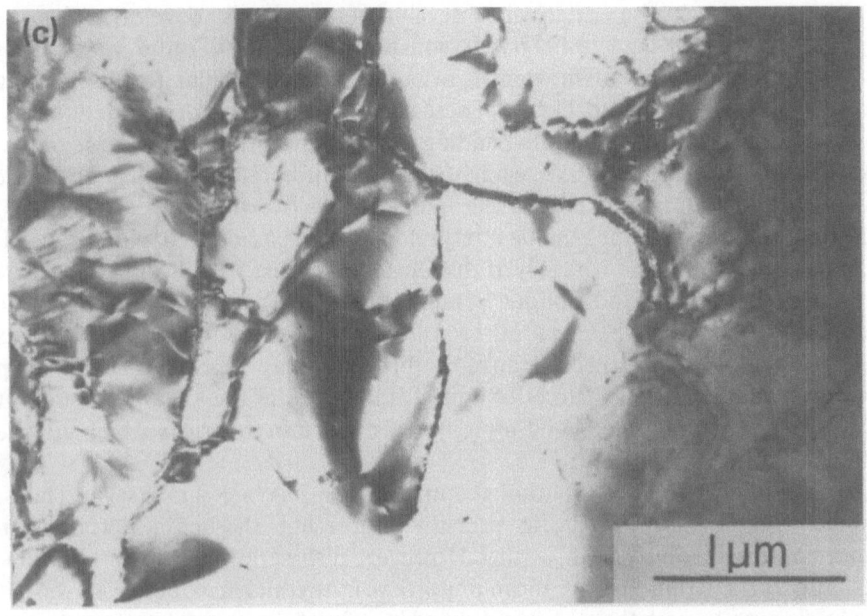

almost no defects and are separated by well defined, approximately straight grain boundaries with irregular hexagonal shapes. The small grains appear to represent the end product of a recovery process which changes the high dislocation densities into cells and then into strain-free grains. The evolution and preservation of these cataclasite microstructures is compared with those present in the other thrust sheets studied in the discussion section of this chapter.

The steepened section of the Glencoul fault, which now dips ENE, contains a number of features which truncate the cataclastic banding and postdate the main Glencoul fault displacement. The earliest of these features are thin ($\leqslant 0.1$ mm), discontinuous (< 1 mm) phyllosilicate-rich stylolites oriented obliquely ($\sim 40°$) to the cataclasite layering in thin sections cut parallel to the direction of movement and perpendicular to the fault plane. The geometry of these features indicates the same sense of shearing as the displacement inferred for the cataclastic generation. They probably represent a small amount of deformation by diffusive mass transfer processes during the final stages of WNW shearing on the Glencoul Thrust when the displacement rate and therefore the strain rate was falling (see Knipe 1989, Fig. 4). Stylolites which are approximately parallel to the cataclasite banding are also present. These structures are often concentrated near large brecciated rock fragments, but also occur at the interface between cataclastic bands. The dentate geometry of these suggests that they represent shortening perpendicular to the fault plane by diffusion mass transfer (DMT) after the shearing. A third set of features which postdates the cataclasite development and is developed in the phyllosilicate-rich cataclastic bands consists of thin (<0.1 mm wide) shear zones with directions of movement orientated perpendicular to the shearing direction of the Glencoul Thrust. These shear zones form in two orientations; one is parallel to the cataclasite banding and has an overthrusting sense of displacement towards the WSW, while the other is oriented at a high angle to the cataclastic banding and has offsets indicative of overthrusting to the ENE. Both of these shear zones can be interpreted as being associated with the development of folding and rotation of the Glencoul Thrust fault during the generation of the lateral and oblique ramps present in the underlying limestone duplexes. The minor shears parallel to the cataclasite banding with movements perpendicular to the thrust transport direction may represent flexural slip accommodation of the lateral/oblique folding, while the shears oriented at high angles to the gouge banding may represent minor thrusts which aid the fault rotation.

In summary, the deformation features of the intermediate Assynt Thrust sheet suggest that cataclastic processes represent the main deformation mechanisms involved in thrusting. However, dislocation processes also contribute to the strain and are more important in the higher structural levels. It is interesting to note that the fine-grained matrix of the gouges shows recovery features indicative of adjustments to the high dislocation contents after displacements, and that there is evidence of a change to deformation by diffusive mass transfer during the late-stage history of these fault rocks.

9.6 The Sole Thrust System

The thrust structures which occur in northern Assynt below the Glencoul Thrust are taken to comprise the Sole Thrust System. Coward (1984) has discussed the geometrical evolution of this zone in detail and suggests a minimum displacement of 3 km on the system. As noted by Coward (1984) there are thrust structures present below the Sole Thrust mapped by Peach et al. (1902) and, although these have small displacements (≪100 m) and some may be isolated faults rather than part of a simple linked system, they are included in the Sole Thrust System here.

The Cambrian Quartzites under consideration here are involved in this Sole Thrust System in the vicinity of Skiag Bridge (see Fig. 9.1). Almost all the fault zones developed in these rocks are marked by fault rocks produced by cataclastic processes. A detailed analysis of a minor backthrust which affects the Pipe Rock member of the Cambrian Quartzite is presented in a separate paper by Lloyd & Knipe (1990). This study shows the range of fracture processes involved in the initiation and linking of the fracture array associated with the fault-zone evolution and the development of cataclasites along the fault. In addition, the study highlights the different deformation-mechanism

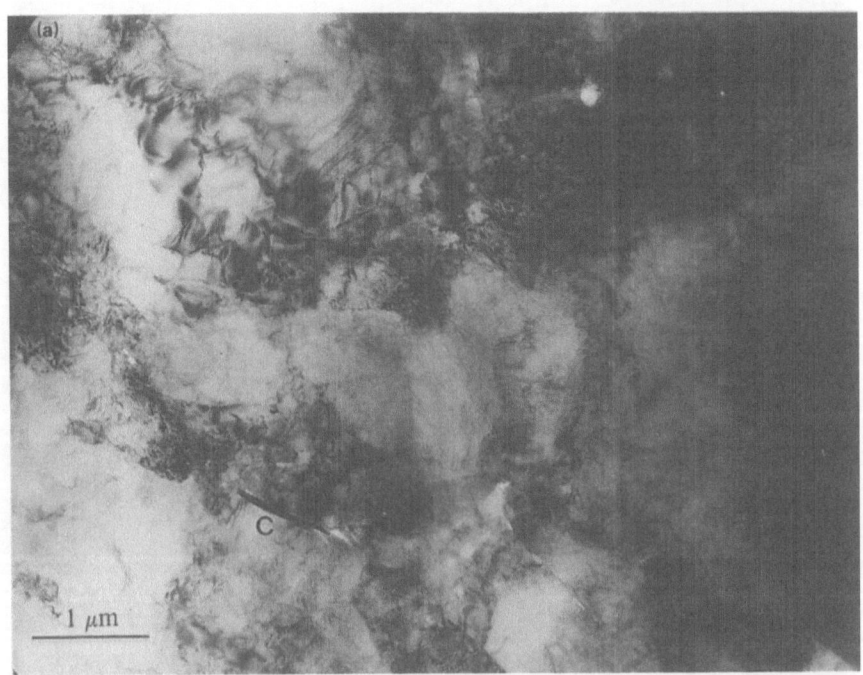

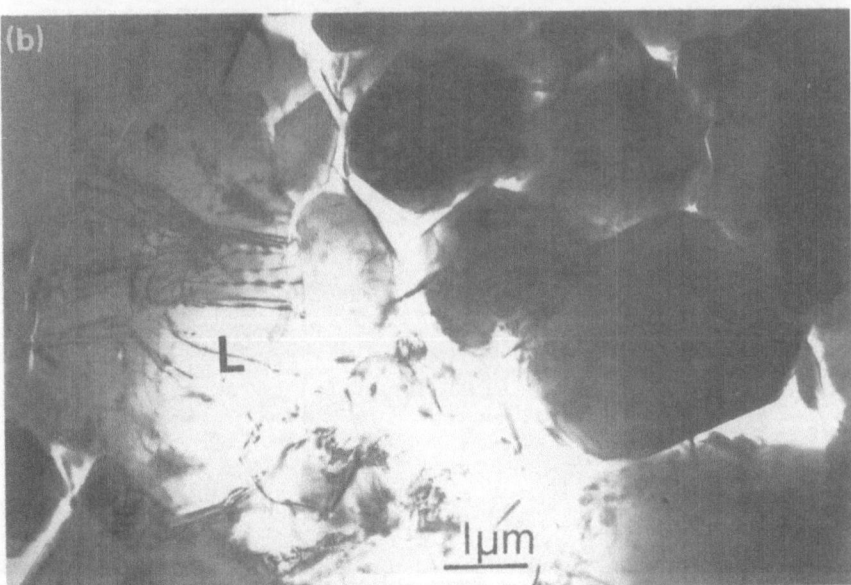

Figure 9.11 (A) TEM micrograph of a cataclastic zone matrix from the Sole Thrust sheet at Skiag Bridge. The material contains high dislocation densities and a very irregular cell substructure. Note also the chlorite inclusion C. (B). A damage region on the edge of a large fragment. Note dislocations in the border area and deformed matrix fragments.

Figure 9.12 Optical micrographs comparing the fabric outside (A) and inside (B) a tip zone developed near Skiag Bridge. Both sections cut parallel to the *x–z* surface. The shear sense in the zone is indicated in (B) and the arrow is parallel to the weak shape fabric present inside the zone. Note the increased grain contact parallel to the *x* direction.

histories experienced by rocks at different distances from the fault, and the important rôle of both dislocation and diffusive mass-transfer processes within the fault zone. The important features of the cataclasites developed along this fault recognized by Lloyd & Knipe (1990), which are important in the context of the present study are: (i) new quartz overgrowths on cataclastic fragments; (ii) larger (>100 μm^3) areas of quartz cement containing phyllosilicate inclusions (both of these growth features (i) and (ii) are undeformed and represent late-stage events in the fault evolution; (iii) deformed cement areas precipitated during an earlier deformation event which now contain extremely high dislocation densities (Fig. 9.11); (iv) detrital grain fragments also containing high dislocation densities; and (v) evidence for diffusive mass transfer creating stylolites which postdate the shearing in gouge zones.

Not all of the thrusts in the Sole Thrust System are marked by fault rocks generated by cataclastic processes. A tip zone to one thrust, preserved at map reference NC232255 is marked by a shear zone 4–15 cm wide, which deforms the worm tubes present in the Pipe Rock and shows no evidence of involving intragranular fracture. This shear zone is one of the lowest structures developed in the thrust belt at Assynt, has displacements of up to 12 cm, and induces shear strains of approximately one in the direction of 310°. Undulatory extinction within the detrital grains increases slightly within the shear zone, so that twice as many grains (~ 10 per cent) exhibit a c-axis orientation range which is greater than 6–10°. Thus, although crystal plastic deformation by dislocation processes does contribute to the deformation in the shear zone, these processes must represent a very minor contribution to the total strain. The increased contact area of grain boundaries parallel to the x–y trace inside the shear zone indicates that diffusion mass transfer is also concentrated in the shear zone (Fig. 9.12). However, the weak shape fabric present does not correspond with that predicted by the amount of shear recorded by the deformed pipes, and thus grain-boundary sliding must have operated in this zone to allow accumulation of the required strain. The absence of intragranular grain fracturing in this zone suggests that the grain disaggregation (by intergranular fracturing) and sliding accompanied by diffusive mass transfer requires slower deformation rates than those involved in the generation of the adjacent cataclastic fault zones. Thus the deformation processes preserved in this shear zone represent one of the last stages of thrusting in the Assynt area, when the displacement rate and strain rates associated with the thrusting were falling.

9.7   Discussion

Each of the major fault zones studied preserves microstructures indicative of a different range of deformation mechanisms. In addition, the deformation microstructures present reflect the different range of conditions of faulting, and both the Moine Thrust and the Assynt Thrust faults show evidence of post-displacement changes in the deformation mechanisms and strain path as well as post-displacement modifications to the earlier microstructures.

It is possible to amalgamate the deformation-mechanism histories inferred from the microstructures with the pressure–temperature–time (PTt) paths expected for the different fault zones studied. Because of the lack of detailed information on the exact conditions of deformation, especially the depth and pressure conditions, the $PT\dot{\varepsilon}t$ paths used in this discussion are speculative. They are meant only to convey the general form of the histories discussed, and to provide a framework for an assessment of the deformation mechanism histories and for the integration of the microstructural analysis into the large-scale evolution of the thrust belt. The deformation mechanism histories and the probable PTt paths for each of the fault zones studied are discussed separately below in a section on deformation-mechanism histories and thrusting, and are reviewed in Figure 9.13. The evidence for different types of cyclic deformation during the thrust evolution and the stability of the microstructures are discussed in separate sections below.

9.7.1 Deformation-mechanism histories and thrusting

The deformation mechanism history and probable PTt path for the quartzites exposed at the Stack of Glencoul are reviewed in Figure 9.13A. Johnson *et al.* (1985) have estimated that the mylonites exposed at this location developed at temperatures of 300–350°C. These conditions correspond to the first two stages of the mylonite evolution described above and represent the probable maximum temperatures experienced by these rocks located in the foot wall to the Moine Thrust (rocks in the hanging wall will have been subjected to higher temperatures). Given the thin (~1.5 km) cover of Cambro-Ordovician rocks (Swett & Smitt 1972) it is unlikely that the pre-Caledonian burial of the quartzite exceeded 3 km, and thus the pre-tectonic temperature was probably below 150°C. Displacement on the thrusts located in the Moines (e.g. the Meadie and Naver slides) above the Moine Thrust may have controlled the early tectonic burial of the quartzite at the Stack of Glencoul. Butler (1985) estimates the displacement on the thrusts located in the Moines to be above 130 km. Burial probably induced deformation by diffusive mass transfer associated with slow strain rates accompanying thrust-sheet loading. Only during the later stages of burial, when the thrusting on the Moine Thrust commences, are dislocation creep processes likely to dominate the deformation in these mylonites (Fig. 9.13A).

The probable large-scale thrust wedge geometry and the retrogression observed in the Moine rocks of the hanging wall to the Moine Thrust can be used to estimate a minimum displacement on the Moine Thrust. Assuming that the hangingwall rocks have experienced a minimum decrease in temperature of 100°C (from ~450°C to ~350°C, causing retrogression of the biotite

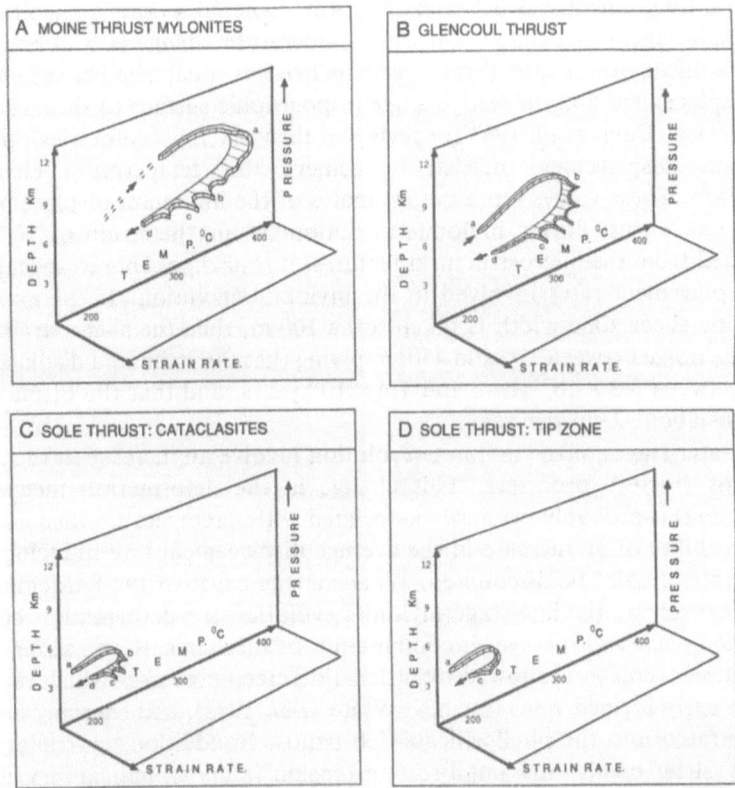

Figure 9.13 Diagrammatic representation of deformation mechanism paths and *PTt* paths for fault rocks in the Moine Thrust Zone at Assynt. The vertical plane illustrates the probable *PTt* path while the horizontal plane reviews the type of strain-rate history. Note that because of the lack of information on the pressure experienced by these rocks the depth axes have been estimated using an assumed temperature gradient of 25–30°C km⁻¹. (A) The history of quartzites located in the foot wall to the Moine Thrust at the Stack of Glencoul. The early history is associated with burial beneath higher-level thrust sheets. The high-temperature phase of the deformation (a–b) is dominated by dislocation creep and creates the mylonites. During section b–c the deformation involves alternation of fracturing and dislocation creep, while during c–d the deformation is dominated by diffusive mass transfer. The interval between b and d represents the changeover from shearing to the thinning of the thrust sheet as the displacement activity decreases on the Moine Thrust. (B) The deformation history of the Glencoul Thrust. The early history a–b is associated with burial by overriding thrust sheets and the section b–c represents the displacement period on the thrust, which is dominated by cataclastic deformation made up of transient high-strain-rate events. The period c–d involves some vertical compaction by diffusive mass transfer processes. (C) The type of history involved in the evolution of the cataclastites on the Sole Thrust. (D) The lower-strain-rate event responsible for the deformation at the tip zone at Skiag Bridge, where grain-boundary sliding and diffusive mass transfer are associated with one of the final movements on the Moine Thrust System at Assynt.

251

assemblages), during displacement on the Moine Thrust, and assuming that temperature gradient was between $25°C$ km^{-1} and $30°C$ km^{-1}, then the vertical displacement associated with this temperature change is 3–4 km. If the topographical front of the thrust wedge is fixed by a balance between erosion and displacement and the wedge angle (topographic surface to thrust surface) is 5–10° (see Davis *et al.* 1983 for review of thrust-wedge geometries), then the minimum displacement needed to achieve this temperature change is 17–45 km. These values represent estimates of the minimum displacement on the Moine Thrust during mylonite evolution. Using the strain of 10^{-10} s^{-1}, estimated from the microstructural features, it is also possible to speculate on the displacement rates involved in the mylonite evolution. If the maximum mylonitic shear zone width is taken to be 100 m, then the shear strains with the zone range between 170 and 450, implying that the estimated displacements took between 5.3×10^5 years and 1.4×10^6 years, and that the displacement rate was about 3 cm per year.

The later stages of the mylonite evolution involve an increase in the importance of fracture processes. This change in the deformation mechanisms (hardening) is probably primarily associated with a temperature decrease, but the possibility of an increase in the average displacement rate inducing faster strain rates cannot be discounted. Whatever the cause of the hardening, it is clear that during the late-stage mylonite evolution the deformation becomes more complex, and involves a combination of mechanisms and strain paths. Fracturing events are followed by dislocation creep processes which recrystallize the early formed veins (see also White *et al.* 1982), and shearing becomes concentrated into the phyllosilicate-rich bands. In addition, crystallographic fabrics, shear bands, and small-scale extension faults all indicate an increase in the importance of vertical shortening (i.e. thinning) of the thrust sheet with time.

The last stages of the deformation history in these mylonitic quartzites represent a transfer from straining dominated by shear to vertical shortening achieved by slower diffusive mass transfer processes. The thrust sheet thinning is initially interrupted by minor forethrust and backthrusting events and later by occasional vertical fracturing, which are not associated with cataclasites. The last fractures to affect these rocks may be accommodation structures related to movement on lower thrusts, or may have formed during the evolution of the post-thrusting extensional faulting.

The deformation history of the quartzites located along the Glencoul Thrust and those within the Assynt Thrust sheet are outlined in Figure 9.13B. The early history in these rocks would have been controlled by burial beneath the advancing Moine Thrust sheet. It follows from the discussion above on the evolution of the Moine Thrust fault rocks that the maximum temperatures likely to have been experienced by the rocks of the Assynt Thrust sheet were approximately those experienced by the Moine Thrust rocks just prior to the transfer of displacement to the Ben More and Glencoul Thrust faults, i.e.

300–350°C. Using the estimated displacement of 20–30 km and a minimum thrust wedge angle of 5° it is possible to estimate the vertical displacement associated with movement on the Glencoul Thrust at about 2–3 km. This geometry suggests that rocks in the hanging wall to the Assynt sheet experienced a maximum temperature drop of 60–90°C during emplacement. The displacement would have taken 20–30 Ma if the average displacement rate was 1 mm per year, or 6.6×10^5–10^6 years if the average displacement was as high as 3 cm per year.

The deformation processes associated with the emplacement of the Assynt sheet are dominated by cataclastic deformation along the main faults, generating cataclasites and breccia zones. Strain within the thrust sheets is partitioned between slip on a three-dimensional fracture network and crystal plastic deformation within the defined blocks. The crystal plastic processes are more common and pervasive in the higher structural levels, reflecting the deeper burial and higher temperatures encountered by these rocks. The prolonged period of higher temperatures experienced by these higher structural units is also important to the modification of the cataclastic microstructures. These modifications are discussed in more detail in the section on microstructural stability later (Section 9.7.3).

Deformation within thrust sheets involving small displacements on a complex three-dimensional fracture network has also been described in the southern and northern sections of the Moine Thrust Zone by Blenkinsop & Rutter (1986) and Bowler (1987). Estimation of the strain rates associated with the generation of these complex fracture networks is difficult, as it depends upon the identification of the fracture mechanism(s) involved. By assuming that the crystal plastic processes preserved in the Assynt sheet represent work hardening which controls fracture, a crude estimate of the minimum strain rates involved can be made from assessing the rates where dislocation glide dominates over dislocation creep (see Rutter 1976, Fig. 9). For the estimated temperatures of 250–300°C a minimum strain rate of $1 \times 10^{-10} \, \text{s}^{-1}$ is indicated. The strain rates involved in cataclasite formation by this mechanism are likely to have been higher than this estimate, as this rate produces a displacement rate of only $\frac{1}{3}$ mm per year in a 10 cm wide deformation zone. At present it is only possible to infer that during the emplacement of the Glencoul sheet the strain-rate history in the fault zone was probably made up of repeated, short-lived strain events where the strain rate exceeded $1 \times 10^{-10} \, \text{s}^{-1}$.

The development of pressure-solution features indicates that the slow shearing at the end of the displacement on the Glencoul Thrust induced a change in the deformation processes from cataclasis to diffusive mass transfer. The stylolites parallel to the cataclastic banding, which pre-date the rotation of the Glencoul fault, indicate that the strain path changed to include vertical shortening by diffusive mass transfer processes while the lower thrusts were active.

Faults developed as part of the Sole Thrust System of north Assynt and affecting the Cambrian Quartzites are dominated by cataclastic deformation processes (Fig. 9.13C). Crystal plastic deformation, involving dislocation movement, is far less pervasive than in the higher structural units and is confined to zones adjacent to faults (see also Lloyd & Knipe 1990). Deformation in the quartzites away from the faults is, as in the Assynt sheet above, associated with movement on a complex three-dimensional network of minor fractures. The estimated minimum displacement on the Sole Thrust System of 3 km (Coward 1984) indicates that this thrust system produces only a minor vertical displacement ($\leqslant 250$ m). The quartzites in the Sole Thrust sheet probably experienced an important temperature change during burial by the higher-level thrust sheets and, given that the maximum temperature reached was approximately 200–250°C (see Johnson *et al.* 1985), may have undergone a temperature increase of 50–100°C during this burial event. The time taken to emplace the Sole Thrust System, using average displacement rates of 1–3 mm per year was 10^5 years to 3 Ma. Not all the fault rocks in the Sole Thrust System show evidence of cataclastic deformation involving fracturing. Deformation at the fault-tip zone at Skiag Bridge involves a combination of DMT and grain-boundary sliding. As the temperature of deformation for this tip zone is estimated to have been approximately 200–250°C it is likely that the strain rate was below $\sim 10^{-12} \, \text{s}^{-1}$ for DMT to dominate over crystal plastic deformation (see Rutter 1976, fig. 9). Thus, given the shear-strain values of unity recorded in this shear zone and a shear-zone width of approximately 10 cm, the time of formation was a minimum of 3.15×10^4 years, corresponding to a displacement rate of 0.03 mm per year (Fig. 9.13D). This rate represents the last stages of movement on the Moine Thrust Zone at Assynt. The average displacement rate for the earlier deformation estimated above is much higher. A minimum figure of 3.6 mm per year is obtained from the estimate that 54 km of shortening took place in the higher levels of the Moine Thrust zone over 15 Ma, suggested by Coward (1984).

9.7.2 *Cyclic deformation*

All of the fault-rock microstructures studied show evidence of cyclic changes in strain rate during their evolution. In addition, the preservation of the low-strain-rate shear zone in the Sole Thrust System indicates the termination of the larger tectonic cycle associated with the evolution of the Moine Thrust Zone.

The duration, amplitude (in terms of stress or strain rate), and the volume of material involved in these cycles will depend upon the operating deformation mechanisms and conditions as well as the material involved (see Knipe, 1989, for a summary). For example, at high temperatures, ductile deformation by dislocation creep in a recrystallized aggregate may apparently be homogeneous on a large scale, but may involve heterogeneous flow on a finer scale. The detailed pattern of deformation in this situation may be controlled by the differences in deformability of adjacent grains which induce repeated, transient and localized strain-rate fluctuations in small domains (Knipe & Wintsch 1985; Knipe & Law 1987; Knipe 1989). The domainal microstructure in the mylonites of the Stack of Glencoul and those in the quartzite mylonites of eastern Assynt, described by Knipe & Law (1987), may be examples of this situation. In contrast, at lower temperatures and reduced confining pressures the cyclic deformation may involve seismic events and/or short-lived high-amplitude strain-rate events separated by long 'rest' periods. The repeated fracture and cementing (sealing) events indicated by the microstructures present in the cataclastic rocks of the Sole Thrust System are probably representative of such a process. The transition between these two end-member modes of cyclic deformation behaviour is associated with a change to more discontinuous deformation (the use of the term 'brittle–ductile transition' for this type of transition has been criticized by Rutter 1986). The later stages in the deformation history of the mylonites at the Stack of Glencoul, where alternations between fracturing and dislocation creep as well as cyclic fracturing and diffusive mass transfer processes represent this change in deformation behaviour.

In general, each of the different deformation cycles listed have different activity periods or frequencies which can be considered to be superimposed upon each other. The tectonic cycle is likely to be the longest-wavelength event and the rheological cycles the shortest. The deformation histories and the finite deformation features preserved in the tectonites will be dependent upon the exact form of the superimposed cycles and the conditions associated with the propagation, displacement, and post-displacement history affecting the rock.

9.7.3 *Microstructural stability*

The recognition of deformation mechanisms and the quantification of deformation conditions require a knowledge of the stability and the probable preservation of microstructures indicative of different processes and conditions (Knipe & Wintsch 1985; Knipe & Law 1986; Knipe 1989). Two of the most important factors controlling the preservation of deformation microstructures are the differential stress and temperature histories which affect the rocks after the generation of the characteristic microstructures. Features such as the dislocation density together with the subgrain and recrystallized grain-size distributions are susceptible to adjustments which depend upon the rate at which the temperature and the differential stress drops (Etheridge & Wilkie 1981; Knipe 1989). In addition, the water and impurity content of phases, together with the presence or absence of fluids in the aggregate, can all influence the preservation of microstructures (White 1979b). In thrust systems the rapid changeover of displacement accommodation from one fault to another may induce rapid decreases in the differential stress, thus promoting preservation of microstructures.

The effects both of later deformation events and adjustments of the initial deformation microstructures can be seen in the fault rocks studied here. One interesting observation is that the dislocation substructure in the quartz mylonites is only altered by later deformation events immediately adjacent (~ 2–3 mm) to the late fracture zones. This suggests that the dislocation densities and subgrain structures in the mylonite away from the later fractures are only affected by the modifications induced by the rate of differential stress and temperature drop after the main displacement on the Moine Thrust. The magnitude of these adjustments appears to be minor in the mylonites because the preserved microstructures indicate reasonable stress and strain rates.

The fault rocks which appear to have experienced the largest microstructural adjustments are the cataclasites. The cataclasites studied developed under different conditions in the different thrust sheets and have experienced a sequence of different *PTt* paths after their generation. Those located at the Stack of Glencoul have developed at approximately 350°C and been cooled during displacement on lower-level thrusts, while those in the Sole Thrust System developed at approximately 200°C. These different temperature histories, which postdate the cataclastic development, provide an ideal situation for assessing the effects of post-deformation annealing on the microstructures. Comparison of the cataclastic microstructures from the Stack of Glencoul, the Assynt Thrust sheet and the Sole Thrust (Figs 9.4, 9, 10C & 11) shows the exact progression of features expected during annealing of work-hardened material from different temperatures. This suggests that the cataclastic zones from the Stack of Glencoul initially had a microstructure similar to those preserved in the Sole Thrust System. It is the cataclasites developed at the lowest temperatures which exhibit the minimum amount of modification of the initial deformation microstructures. The larger modifications of the Stack gouges to an aggregate of strain-free, hexagonal grains reflect the prolonged annealing time from higher temperature experienced by these rocks. The intermediate stages in this adjustment, which involves the generation of crude dislocation cells by the movement of dislocations into cell configurations and their subsequent rearrangement within the wall, is well illustrated by the gouges from the Assynt sheet (compare Figure 9.11 with Figures 9.9 and 9.10C). It should also be noted that such major adjustments to the microstructure have not taken place in the mylonites, where the dislocation creep substructures are preserved. This difference in susceptibility to adjustment is considered to arise from the presence of fluids in the cataclastic zones and the increased water content of the quartz crystals grown in the fracture zones. The effect of water on promoting dislocation movement and recovery is reviewed by Kirby (1984). The presence of water along the fracture zones would also help to explain the localized grain-boundary migration features adjacent to many fracture zones.

Conclusions

(1) The microstructures preserved in the deformed quartzites of the Moine Thrust Zone in Assynt indicate the operation of a range of deformation mechanisms which reflect the range of conditions associated with displacement on different faults.

(2) There is a systematic variation in the dominant faulting processes, from displacement by dislocation creep during the emplacement of the Moine Nappe to cataclasis during the emplacement of the Assynt and Sole Thrust sheets. This sequence reflects the preservation of faulting products formed at progressively shallower depths. Microstructures generated during different parts of the Thrust Zone evolution are preserved because later faults carry the earlier, deeper-level faults.

(3) Rocks from each of the thrust sheets studied show evidence of different deformation-mechanism paths and different forms of cyclic deformation. For example, the mylonites at the base of the Moine Nappe have complex microstructures which can be related to the displacement processes becoming more heterogeneous and involving alternations of dislocation creep, fracture, and diffusive mass transfer. The mylonites also contain accommodation structures generated by movement on the underlying thrusts.

(4) Evidence for the final stages of movement on the main faults is preserved in the microstructures. The final strain-rate decay in the fault zones induces a change of deformation mechanism from either dislocation creep or cataclasis to diffusive mass transfer processes.

(5) The strain path in the higher-level thrust sheets changes from shearing-dominated deformation to vertical shortening by diffusive mass transfer. In the mylonites this change is gradational, and involves a time period during which shearing and extension in the direction of movement together with vertical shortening all alternate. The thinning in the Moine and Assynt Thrust sheets pre-dates the movement on the underlying Sole Thrust.

(6) The cataclastic fault rocks show preferential modification (annealing) of earlier microstructures, which is suggested to be aided by the high dislocation densities and higher water content of these rocks. The amount of microstructural modification varies in cataclasites developed at different structural levels, and appears to be related to the different post-displacement temperature histories experienced by the rocks.

Acknowledgements

I would like to thank Mike Coward, Rob Butler, and Rick Law for discussions and various adventures in the field, and Geoff Lloyd and Martin Casey for discussions on deformation mechanisms. Dave Rubie and David Barber are thanked for their helpful comments on a first draft. The research was sponsored by NERC grant GR3/5765.

References

Atkinson, B. K. 1987. *Fracture mechanics of rock.* London: Academic Press.

Blenkinsop, T. G. & E. H. Rutter 1986. Cataclastic deformation of quartzite in the Moine Thrust zone. *J. Struct. Geol.* **8**(6) 669–82.

Borradaile, G. J., M. B. Bayly, & C. McA. Powell, 1982. *Atlas of deformational and metamorphic rock fabrics.* Berlin: Springer.

Bowler, S. 1987. Duplex geometry: an example from the Moine Thrust Zone. *Tectonophysics* **135**, 25–35.

Boyer, S. E. & D. Elliott 1982. Thrust systems. *AAPG Bull.* **66**, 1196–230.

Brewer, J. 1981. Thermal effects of thrust faulting. *Earth Planet. Sci. Lett.* **56**, 233–44.

Brodie, K. H. & E. H. Rutter 1985. On the relationship between deformation and metamorphism, with special reference to the behaviour of basic rocks. In *Advances in physical geochemistry* **4**, A. B. Thompson & D. C. Rubie (eds), 138–79. New York: Springer.

Butler, R. W. H. 1982. The terminology of structure in thrust belt. *J. Struct. Geol.* **4**, 239–45.

Butler, R. W. H. 1984. Structural evolution of the Moine thrust belt between Loch More and Glendhu, Sutherland. *Scott. J. Geol.* **20**, 161–79.

Butler, R. W. H. 1986. Structural evolution in the Moine of N. W. Scotland – a Caledonian linked thrust system. *Geol Mag.* **123**, 1–11.

Butler, R. W. H. 1987. Thrust sequences. *J. Geol Soc. Lond.* **144**, 619–34.

Butler, R. W. H. & M. P. Coward 1984. Geological constraints, structural evolution and deep geology in N.W. Scottish Caledonides. *Tectonics* **3**, 347–65.

Christie, J. M. 1960. Mylonitic rocks of the Moine Thrust Zone in Assynt region, N.W. Scotland. *Trans Edinb. Geol Soc.* **18**, 79–93.

Christie, J. M. 1963. The Moine thrust zone in Assynt region N.W. Scotland. *Univ. Calif. Publ. Geol Sci.* **40**, 345–440.

Coward, M . P. 1982. Surge zones in the Moine thrust zone of N.W. Scotland. *J. Struct. Geol.* **4**, 247–56.

Coward, M. P. 1983. The thrust and shear zones of the Moine thrust zone and NW Scottish Caledonides. *J. Geol Soc. Lond.* **140**, 795–811.

Coward, M. P. 1984. The strain and textural history of thin skinned tectonic zones examples from the Assynt region of the Moine thrust zone, N.W. Scotland. *J. Struct. Geol.* **6**, 89–100.

Coward, M. P. 1985. The thrust structures of southern Assynt, Moine Thrust Zone. *Geol Mag.* **6**, 595–607.

Coward, M. P., J. H. Kim & J. Parke 1980. The Lewisian structures of the Moine thrust zone. *Proc. Geol Assoc. Lond.* **9**, 327–37.

Coward, M. P. & A. C. Ries 1986. *Collision tectonics.* Geol. Soc. Lond. Sp. Publ. 19.

Davis, D., J. Suppe & F. A. Dahlen 1983. Mechanics of fold and thrust belts and accretionary wedges. *J. Geophys. Res.* **88**, 1153–72.

Elliott, D. & M. R. W. Johnson 1980. The structural evolution of the north part of the Moine Thrust Zone. *Trans R. Soc. Edinb. Earth Sci.* **71**, 69–96.

Enfield, M. A. & M. P. Coward 1987. The structure of the West Orkney basin, N. Scotland. *J. Geol Soc. Lond.* **144**(6), 871–85.

Etheridge, M. A. & J. C. Wilkie 1981. An assessment of dynamically recrystallised grain size as a palaeo piezometer in quartz bearing mylonites. *Tectonophysics* **78**, 475–508.

Evans, D. J. & S. White 1984. Microstructural and fabric studies from the rocks of the Moine Nappe, Eriboll, N.W. Scotland. *J. Struct. Geol.* **6**, 369–90.

Gibson, R. G. & D. R. Grey 1985. Ductile to brittle transition in shear during thrust sheet emplacement, S. Appalachian thrust belt. *J. Struct. Geol.* **7**, 513–27.

Groshong, R. J. Jr, O. A. Pfiffner & L. R. Pringle 1984. Strain partitioning in the Helvetic thrust belt of eastern Switzerland from the leading edge of the internal zone. *J. Struct. Geol.* **6**, 5–18.

Hobbs, B. E. & H. C. Heard (eds) 1986. *Mineral and rock deformation: laboratory studies.* Geophysical Monograph **36**. Washington DC: Amer. Geophys. Union.

House, W. M. & D. R. Grey 1982. Cataclasites along the Salfville thrust, U.S.A. and their implications for thrust sheet emplacement. *J. Struct. Geol.* **4**, 257–70.

Johnson, M. R. W., S. P. Kelley, G. J. H. Oliver & D. A. Winter 1985. Thermal effects and timing of thrusting in the Moine Thrust Zone. *J. Geol Soc. Lond.* **142**, 863–74.

Kirby, S. H. 1984. Introduction and digest to the special issue on chemical effects of water on the deformation and strengths of rocks. *J. Geophys. Res.* **89**, 3991–5.

Knipe, R. J. 1985. Footwall geometry and the rheology of thrust sheets. *J. Struct. Geol.* **7**, 1–10.

Knipe, R. J. 1986. Deformation mechanism path diagrams for sediments undergoing lithification. In *Structural fabrics in DSDP cores from forearcs*, J. C. Moore (ed.), 155–60. Geol Soc. Am. Memoir 166.

Knipe, R. J. 1989. Deformation mechanisms – recognition from natural tectonites. *J. Struct. Geol.* **11**, 127–46.

Knipe, R. J. & R. P. Wintsch 1985. Heterogeneous deformation, foliation development and metamorphic processes in the polyphase mylonite. In *Advances in physical geochemistry*, A. B. Thompson & D. C. Rubie (eds), **4**, 180–210. New York: Springer.

Knipe, R. J. & R. D. Law, 1987. The influence of crystallographic orientation and grain boundary migration on microstructural and textural evolution in an S.C. mylonite. *Tectonophysics* **135**, 155–69.

Knipe, R. J. & E. H. Rutter 1990 (eds). *Deformation mechanisms, rheology and tectonics.* Special vol. London: Geol Soc., in preparation.

Kuznir, N. J. & R. G. Park 1986. Continental lithosphere strength: the critical role of lower crustal deformation. In *The nature of the lower crust*, J. B. Dawson, D. A. Carswell, J. Hall & K. H. Wedepohl (eds), 79–93. Geol Soc. Lond. Sp. Publ. 24.

Law, R. D. 1987. Heterogeneous deformation and quartz crystallographic fabric transitions: natural examples from the Moine Thrust Zone at Assynt. *J. Struct. Geol.* **9**, 819–34.

Law, R. D., R. J. Knipe & H. Dayan 1984. Strain path partitioning with thrust sheets: microstructural and petrofabric evidence from the Moine Thrust zone, Loch Eriboll. *J. Struct. Geol.* **6**, 477–97.

Law, R. D., M. Casey & R. J. Knipe 1986. Kinematic and tectonic significance of microstructures and crystallographic fabrics within the quartz mylonites from the Assynt and Eriboll regions of the Moine Thrust Zone. *Trans R. Soc. Edinb. Earth Sci.* **77**, 99–126.

Law, R. D. & G. J. Potts 1987. The Tarskavaig Nappe of Skye, N.W. Scotland. *Geol Mag.* **125**, 231–48.

Lister, G. S. & A. Snoke 1984. S–C mylonites. *J. Struct. Geol.* **6**, 617–38.

Lloyd, G. E. & R. J. Knipe 1990. Deformation mechanisms accommodating faulting of quartzite under crustal conditions. Submitted to *J. Struct. Geol.*

Mitra, G. 1984. Brittle to ductile transition due to large strains along White Rock Thrust, Wind River, Wyoming. *J. Struct. Geol.* **6**, 51–62.

Obee, H. K. & S. H. White 1985. Fault and associated rocks of the S. Aranta block, Alice Springs, Australia. *J. Struct. Geol.* **7**, 701–13.

Ord, A. & J. M. Christie 1984. Flow stresses from microstructures in mylonitic quartzites from the Assynt region, Scotland. *J. Struct. Geol.* **6**, 639–54.

Pfiffner, D. A. 1982. Deformation mechanisms and flow regimes in limestones from the Helvetics, Swiss Alps. *J. Struct. Geol.* **4**, 429–42.

Platt, J. P. 1987. The uplift and high pressure – low temperature metamorphic rocks. *Phil. Trans. Roy. Soc. Lond. A* **321**, 87–103.

Platt, J. P. & J. H. Behrmann 1986. Structures and fabrics in a crustal scale shear zone, Betic Cordilleras, S.E. Spain. *J. Struct. Geol.* **8**, 13–34.

Platt, J. P. & G. S. Lister 1985. Structural history of high pressure metamorphic rocks in the S. Vanoise Massif, French Alps and their relation to Alpine events. *J. Struct. Geol.* **7**, 19–36.

Rutter, E. H. 1976. The kinetics of rock deformation by pressure-solution. *Phil Trans R. Soc. A* **283**, 203–19.

Rutter, E. H., R. H. Maddock, S. H. Hall & S. H. White 1986. Comparative microstructures of natural and experimentally produced clay bearing fault gouges. *Pure Appl. Geophys.* **124**, 3–30.

Rutter, E. H. 1986. On the nomenclature of mode of failure transitions in rocks. *Tectonophysics* **122**, 381–7.

Schmid, S. M. 1982. Microfabric studies as indicators of deformation mechanisms and flow lavas operative in mountain building. In *Mountain building processes*, K. H. Su (ed.), 95–110. London: Academic Press.

Schmid, S. M., R. Panozzo & S. Bauer 1987. Simple shear experiments on calcite rocks: rheology and microfabric. *J. Struct. Geol.* **9**, 747–78.

Sibson, R. H. 1977. Fault rocks and faulting mechanisms. *J. Geol Soc. Lond.* **133**, 191–213.

Sibson R. H. 1980. Transient discontinuities in ductile shear zones. *J. Struct. Geol.* **2**, 165–71.

Sibson, R. H. 1982. Fault zone models, heat flow and depth distribution of earthquakes in the continental crust of the United States. *Bull. Seismol. Soc. Am.* **72**, 151–63.

Sibson, R. H. 1983. Continental fault structure and the shallow earthquake source. *J. Geol. Soc. Lond.* **140**, 741–67.

Sibson, R. H. 1985. Stopping of earthquake ruptures at dilational jogs. *Nature* **316**, 248–51.

Sibson, R. H. 1986a. Brecciation processes in fault zones: inferences from earthquake rupturing. *Pure Appl. Geophys.* **124**, 159–75.

Sibson, R. H. 1986b. Earthquakes and rock deformation in crustal fault zones. *Ann. Rev. Earth Planet. Sci.* **14**, 149–75.

Simpson, C. & S. M. Schmid 1983. An evaluation of criteria to deduce the sense of movement in sheared rocks. *Bull. Geol Soc. Am.* **94**, 1281–8.

Swett, K. & D. E. Smitt 1972. Palaeogeography and depositional environments of the Cambro-Ordovician shallow marine facies of the N. Atlantic. *Bull. Geol. Soc. Am.* **83**, 3223–48.

Thompson, A. B. & P. D. England 1984. Pressure–Temperature–time paths. *J. Petrol.* **25**, 929–55.

Thompson, A. B. & J. Ridley 1987. *P–T–t* paths for orogenic zones. In *Tectonic settings of regional metamorphism*, E. R. Oxburgh, B. W. D. Yardley & P. C. England (eds), 27–46. London: Royal Soc.

Tullis, T. E. 1986. Friction and faulting. *Pure Appl. Geophys.* **124**, 375–608.

Twiss, R. J. 1986. Variable sensitivity piezometric equations for dislocation densities and sub-grain diameters and their relevance to quartz and olivine. In *Mineral and rock deformation: laboratory studies*, B. E. Hobbs & H. C. Heard (eds), Geophys. Monogr. **36**, 247–62. Washington DC: Amer. Geophys. Union.

Wang, Ci-yuen (ed.) 1986. *Internal structure of fault zones. Pure Appl. Geophys.* Special Issue.

Weathers, M. S., J. M. Bird, R. F. Cooper & D. C. Kohlstedt 1979. Differential stress determined from deformation induced microfractures of the Moine Thrust Zone. *J. Geophys. Res.* **84**, 7495–509.

White, S. H. 1979. Grain and sub-grain size variation across a mylonite zone. *Contrib. Mineral. Petrol.* **70**, 193–202.

White, S. H. 1979b. Difficulties associated with palaeo-stress estimates. *Bull. Mineral.* 210–15.

White, S. H., D. J. Evans and D.-L. Zhong 1982. Fault rocks of the Moine Thrust zone: Micro-structures and textures of selected mylonites. *Textures Microstruct.* **5**, 33–62.

White, J. C. & S. H. White 1983. Semi-brittle deformation within the Alpine fault zone, New Zealand. *J. Struct. Geol.* **5**, 579–90.

Wojtal, S. & G. Mitra 1986. Strain hardening and strain softening in fault zones from foreland thrusts. *Bull. Geol Soc. Am.* **97**, 674–87.

CHAPTER TEN

Mechanisms of reaction-enhanced deformability in minerals and rocks

David C. Rubie

10.1 Introduction

Our current understanding of deformation mechanisms in rocks and minerals and the rheology of these materials has been obtained largely through experimental studies of single crystals and monomineralic aggregates. Such studies are essential in obtaining a basic understanding of the fundamental processes of rock deformation. However, rocks of the Earth's crust and mantle are seldom monomineralic and, therefore, to investigate deformation processes in the Earth realistically it is necessary to perform experiments on more complicated multi-phase aggregates (e.g. Tullis, this volume).

It is generally assumed that steady-state flow laws obtained in laboratory studies can be extrapolated to geological conditions in order to evaluate the rheology of the crust and mantle (albeit with large uncertainties due to the extrapolation of strain rate over several orders of magnitude). However, in some cases there could be a major problem with this assumption because of the possibility that the deformability of rocks is substantially modified when reactions occur between the constituent minerals. In most cases discussion has centred around the concept that deformability can be *enhanced* by mineral reactions (e.g. Raleigh & Paterson 1965; Heard & Rubey 1966; Gordon 1971; Sammis & Dein 1974; Murrell & Ismail 1976; White & Knipe 1978; Poirier 1982; Paterson 1983; Rubie 1983, 1984a; Brodie & Rutter 1985, 1987). However, this may not always be the case and, under some circumstances, reactions may have little effect on rheology or may even cause hardening (Rutter *et al.* 1985).

Possible effects of reaction-enhanced deformability are shown in Figure 10.1 in the form of a strain rate – time evolution for a rock undergoing a typical metamorphic cycle, assuming constant differential stress. A 'background' strain rate evolves with time as a result of progressive heating during burial followed by cooling during uplift. Superimposed on this background strain rate are peaks of enhanced strain rate that coincide with mineral reactions that

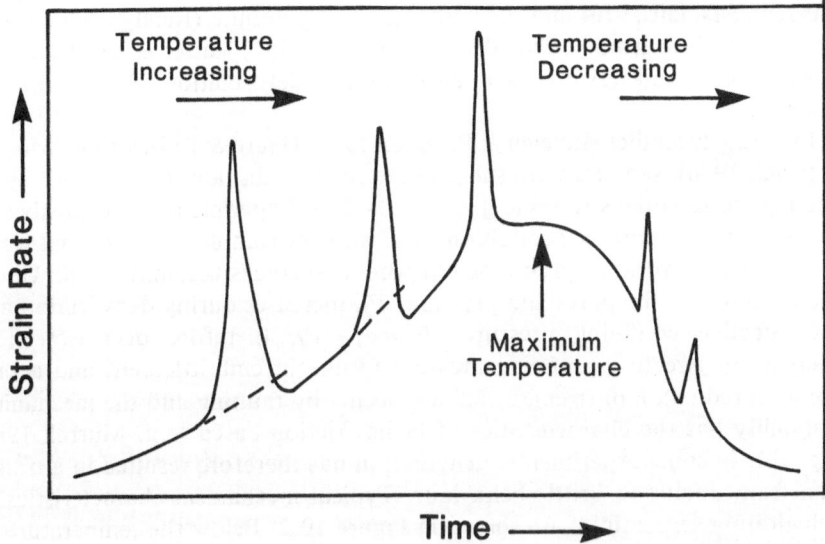

Figure 10.1 Schematic diagram showing the possible variation of strain rate with time, for constant differential stress, during a typical metamorphic cycle. Peaks of enhanced strain rate, superimposed on a 'background' strain rate, coincide with metamorphic mineral reactions. Changes in the background strain rate following a reaction are due to changes in mineralogy (see Fig. 10.3) (After Rutter *et al.* 1985).

occur in response to changing temperature (and pressure). It is obviously important to evaluate the magnitudes of the individual strain-rate peaks relative to the background. If the peaks are large, then much of the strain which can be observed in deformed rocks will have developed under conditions of enhanced deformability. In this case, the steady-state flow laws derived in laboratory studies of simple systems will not be applicable.

The purpose of this chapter is to review the evidence for enhanced deformability during mineral reactions, paying particular attention to the mechanisms involved. Data from experimental studies and from field/petrographic studies of naturally deformed rocks are discussed.

10.2 Experimental data

10.2.1 Dehydration reactions

The effects of dehydration reactions on rheology are of particular interest because of their common occurrence during prograde metamorphism of the Earth's crust (see Fyfe *et al.* 1978). Experimental studies have therefore concentrated on dehydration reactions. The main studies have been of deformation during the reaction of gypsum to anhydrite (Heard & Rubey 1966,

Murrell & Ismail 1976) and dehydration of serpentinite (Raleigh & Paterson 1965, Murrell & Ismail 1976, Brodie & Rutter 1987, Rutter & Brodie 1988). Murrell & Ismail (1976) have also studied deformation of dehydrating chloritite.

In the early studies (Raleigh & Paterson 1965, Heard & Rubey 1966, Murrell & Ismail 1976), samples were subjected to high strain rates (e.g. $3 \times 10^{-4} \, \text{s}^{-1}$) and high differential stresses (e.g. 0.1–1.0 GPa). Experiments were mostly carried out under *undrained* conditions, i.e. with the sample sealed so that water released by dehydration is confined to pores and cracks within the rock. Under such conditions the pore-fluid pressure, P_f, increases during dehydration and the effective confining pressure, $P_c = \sigma_3 - P_f$, therefore decreases. This decrease in effective confining pressure results in embrittlement and a pronounced reduction in strength. Failure occurs by faulting and the mechanical instability has the characteristics of being friction-based (e.g. Murrell 1985, Fig. 1b). In some experiments, dehydration has therefore resulted in a transition from ductile to brittle behaviour. Typical mechanical characteristics of dehydrating serpentinite are shown in Figure 10.2. Below the temperature of dehydration, the strength of serpentinite is high and increases with confining pressure (which is characteristic of brittle failure). At temperatures at which dehydration occurs, the strength drops dramatically and, being primarily dependent on P_c, shows little dependence on the confining pressure σ_3.

In experiments carried out under *drained* conditions at high strain rates, there is no reduction in strength during dehydration because the effective confining pressure remains high (Fig. 10.2b).

The mechanical behaviour of gypsum and serpentinite during heating through the respective dehydration temperatures at high strain rates is summarized in Figure 10.3, on the basis of the experimental data of Raleigh & Paterson (1965) and Heard & Rubey (1966).

The creep behaviour of dehydrating serpentinite has recently been investigated by Brodie & Rutter (1987) and Rutter & Brodie (1988), under conditions of controlled pore-water pressure, to low strain rates and low shear stresses (10–160 MPa). Below the dehydration temperature, the strength of serpentinite is relatively insensitive to changes in displacement rate, thus indicating a high stress exponent in the flow law. At 500–600°C and over a range of confining pressures between 100 and 270 MPa, dehydration occurs, and although the creep behaviour remains unchanged at high displacement rates, at low rates the strength drops very markedly and the material displays a linear viscous rheology (i.e. a stress exponent ~ 1; Fig. 10.4). Microstructural studies

---

Figure 10.2 Mechanical data for serpentinite deformed at high strain rates and various $P-T$ conditions. (a) Strength of *undrained* samples, at fracture or 2 per cent strain, at different confining pressures and temperatures. The drop of strength at 500°–600°C coincides with the onset of dehydration. (b) Stress-strain curves for undrained (U.D.) and drained (D) samples at 670°C and 0.31 GPa confining pressure. Following failure, the undrained sample showed dilatancy hardening (After Murrell & Ismail 1976).

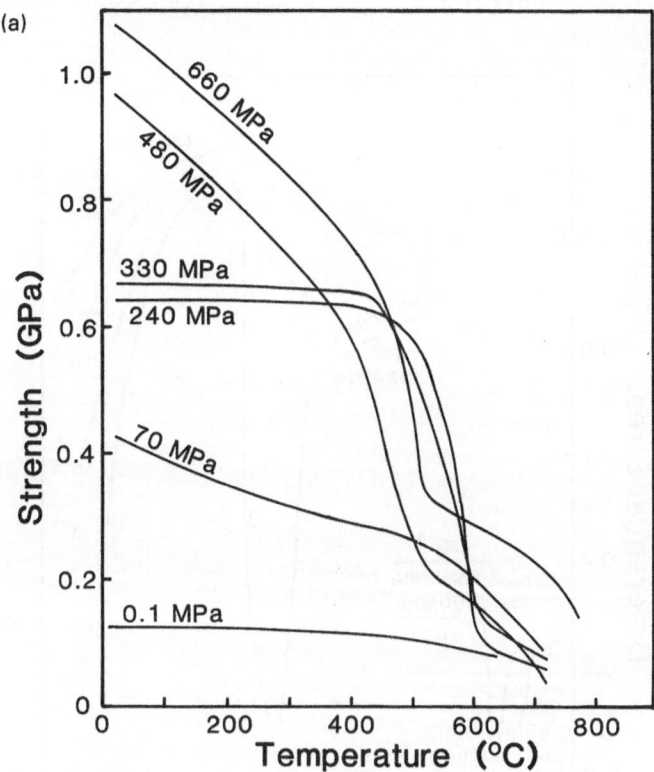

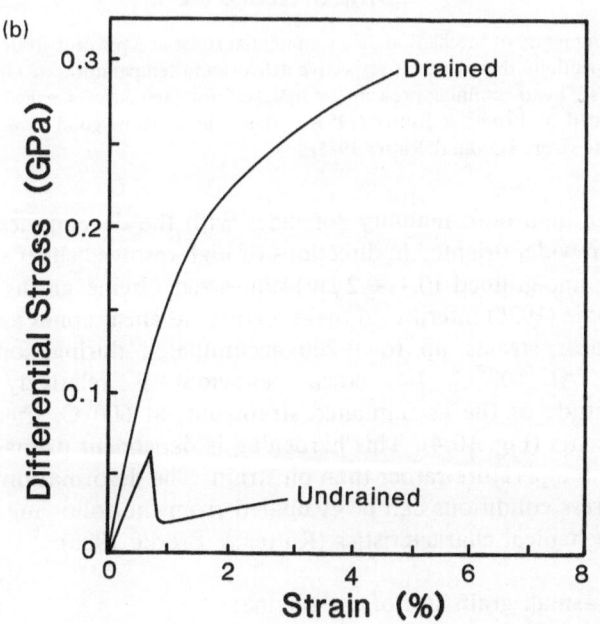

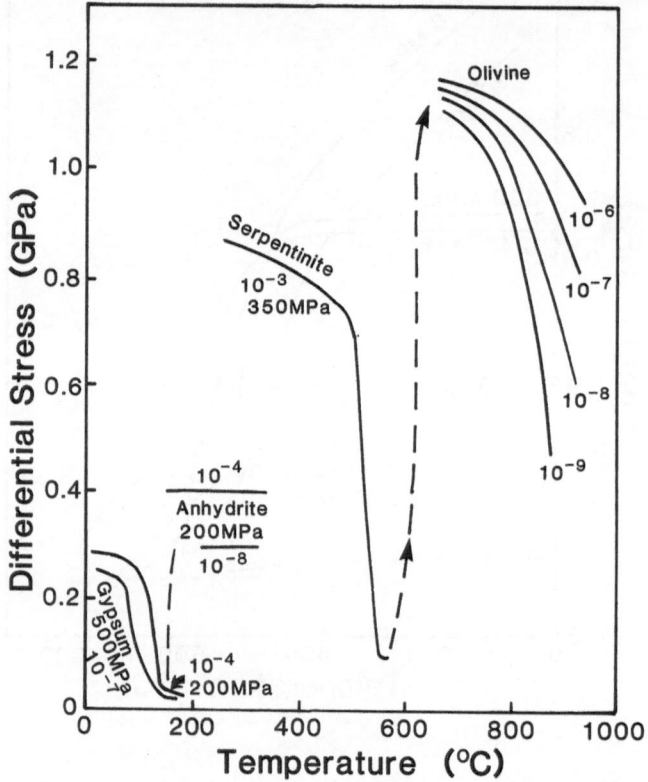

Figure 10.3 A summary of mechanical data (differential stress at 5 per cent strain or failure) for gypsum and serpentinite through their respective dehydration temperatures at high strain rates. The strain rate (s^{-1}) and confining pressure are indicated for each curve. Dashed lines show the strength, calculated by Brodie & Rutter (1985) – these curves show good agreement with the experimental data (from Brodie & Rutter 1985).

show that the high deformability coincides with the development of planar zones, ~ 5 μm wide, oriented in directions of high resolved shear stress, which contain very fine-grained (0.1–0.2 μm) equi-axed olivine grains (Fig. 10.5). Rutter & Brodie (1988) interpreted these features as shear zones and estimated that total shear strains up to ~ 200 accumulated during some of their experiments. At 500°C, the creep behaviour is relatively insensitive to the magnitude of the accumulated strain but, at 600°C, time-dependent hardening occurs (Fig. 10.4). This hardening is dependent primarily on time spent at high temperature rather than on strain. The deformation mechanism under low-stress conditions can be evaluated from the following microstructural and mechanical characteristics (Rutter & Brodie 1988):

(a) the ultra-small grain size of the olivine;

266

(b) the equi-axed shape of olivine grains and a lack of a preferred grain-shape orientation;

(c) linear-viscous creep behaviour ($n \approx 1$);

(d) insensitivity of the creep behaviour to the effective confining pressure and the pore-fluid pressure;

(e) insensitivity of the creep behaviour to strain (at $500°C$).

The time-dependent hardening which occurs at $600°C$ is thought to be related to coarsening of the fine-grained olivine in the shear zones. These characteristics are consistent with deformation by grain-size-sensitive diffusion creep. In addition, the activation energy of $240 \, kJ \, mol^{-1}$, estimated by Rutter & Brodie (1988), is close to the value estimated by Karato *et al.* (1986) for grain-size-sensitive creep of 'wet' fine-grained olivine aggregates, and is much lower than activation energies for dislocation creep in olivine. On the

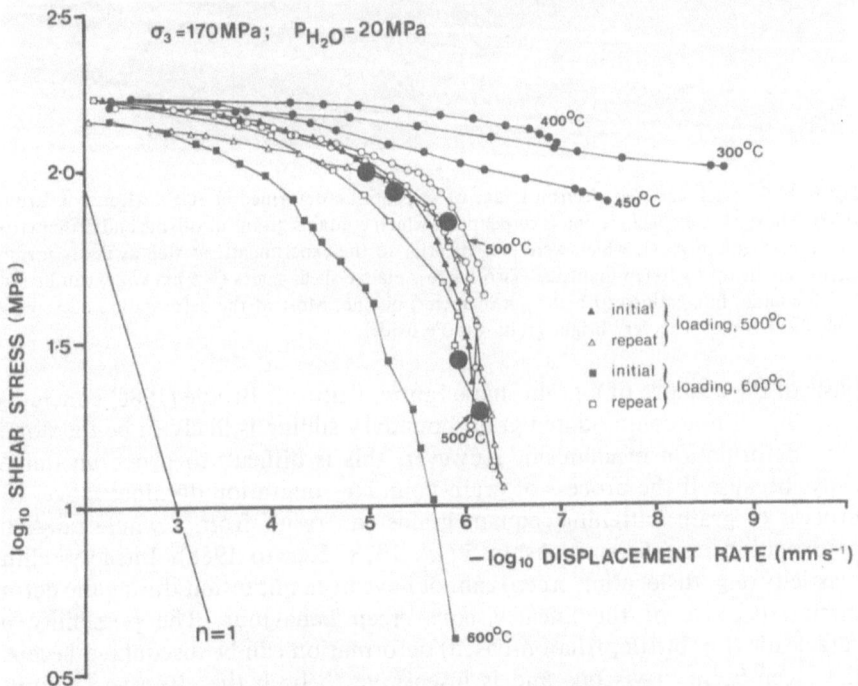

Figure 10.4 Results of stress-relaxation tests carried out on serpentinite by Brodie & Rutter (1987). Shear stress and displacement rate are both resolved along the direction of olivine-bearing shear zones which develop in the sample above $450°C$ (Fig. 10.5). This direction is oriented $\sim 35°$ to the maximum compressive stress. Dehydration (above $450°C$) results in a dramatic weakening at low displacement rates, and the creep behaviour is linear-viscous with a stress exponent $n \sim 1$. Repeat tests at $500°C$ show good reproducibility, indicating that the creep behaviour is relatively insensitive to strain and time. At $600°C$, repeat tests show evidence of time-dependent hardening, as discussed in the text (after Rutter & Brodie 1988).

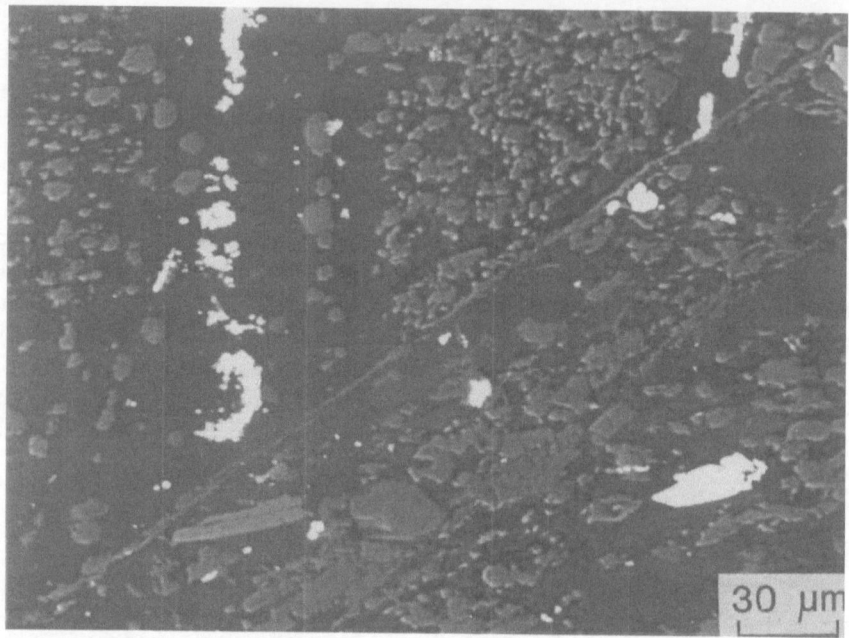

Figure 10.5 Backscattered electron image of serpentinite deformed at 500°C (Rutter & Brodie 1988). The dark grey background is serpentinite which contains grains of olivine and orthopyroxene (both medium grey), which were present prior to the experiments, as well as newly formed olivine grains up to 10 μm diameter. Two narrow planar shear zones (~ 5 μm wide) can be seen which contain fine-grained (0.1–0.2 μm diameter) olivine. Most of the deformation occurred in such shear zones. The very bright grains are Fe oxides.

basis of the absence of a grain shape fabric, Rutter & Brodie (1988) concluded that diffusion-accommodated grain-boundary sliding is likely to be the dominant deformation mechanism. However, this is difficult to prove unequivocally, because if the process of grain-boundary migration dominates over the process of grain flattening, equant grains can result from a whole range of deformation mechanisms (Ashby *et al.* 1978, Karato 1988). Intracrystalline plasticity (e.g. dislocation creep) cannot have been important during the deformation because of the linear-viscous creep behaviour. The possibility of cataclastic (i.e. brittle, friction-based) deformation can be discounted because the creep is linear-viscous and is insensitive to both the effective confining pressure and the strain.

10.2.2 Solid–solid reactions

The effects of the solid–solid reactions on creep behaviour have not received as much attention by experimentalists as the effects of dehydration reactions. However, the occurrence of solid–solid reactions is thought to be common in the Earth's mantle (Jeanloz & Thompson 1983). Such reactions may be of considerable importance in effecting rheology and therefore the dynamics of mantle convection, and may also be related to the mechanism of deep-focus earthquakes (e.g. Gordon 1971, Sammis & Dein 1974, Poirier 1982, Rubie 1984a, Christensen & Yuen 1985, Kirby 1987).

Creep experiments have been carried out on ice I_h in the stability field of ice II (Durham *et al.* 1983) and on tremolite in the stability field of diopside + talc (Burnley & Kirby 1982). In both these studies, shear instabilities developed which result in rapid stress drops and which show the following characteristics:

(a) The stress drop coincides with the development of a narrow shear zone which is oriented in the direction of the maximum resolved shear stress ($45°$ to σ_1). This orientation suggests that the deformation is not brittle because, in rocks, friction-based faults tend to develop at $\sim 30°$ to the direction of the maximum compressive stress σ_1 (e.g. Paterson 1978).

(b) Increasing the confining pressure does not suppress failure, as would be the case for brittle faulting, and in some cases actually reduces the failure stress.

(c) Faulting with the characteristics outlined in (a) and (b) only develops when the material (i.e. ice or tremolite) is deformed outside its stability field so that the potential for a phase transformation exists.

(d) The reduction in strength is a transient phenomenon. Kirby (1987) suggested that the mechanical behaviour could be due to the occurrence of phase transformations in the shear zones. The loss of strength in the shear zones could be related to (i) a transient reduction in the shear modulus, (ii) heat produced locally by exothermic reactions, (iii) formation of new phases with a reduced strength, or (iv) diffusion creep of fine-grained reaction products or transformation plasticity (Kirby 1987). Similar results have been obtained by Green & Burnley (1989) from experiments in which Mg_2GeO_4 olivine was deformed at high pressure in the spinel stability field. Faulting, which only occurred within the temperature range $800-900°C$, was found to be a direct consequence of the localized transformation of olivine to spinel.

Creep experiments of Vaughan & Coe (1981) on polycrystalline Mg_2GeO_4 olivine and Mg_2GeO_4 spinel also shed light on the possible effects of solid–solid phase transformations on rheology. Mechanical data for medium-grained (30–200 μm) Mg_2GeO_4 olivine were interpreted to indicate deformation in a dislocation creep régime ($T = 1000-1360°C$, $P = 0.6-1.0$ GPa, $\dot{\varepsilon} = 10^{-4}-10^{-8}$ s^{-1}). Polycrystalline Mg_2GeO_4 spinel, with a grain size $\sim 3\mu$m, was prepared by transforming the hot-pressed olivine aggregate at high pressure. Subsequent creep tests on the spinel showed a low stress exponent ($n \approx 2$) and a lower activation enthalpy (300 kJ mol^{-1}) than would be expected for dislocation creep. These two factors, combined with the small grain size, suggest deformation by diffusion creep. The flow laws derived by Vaughan &

Coe (1981) show that if medium-grained olivine reacts to spinel with a grain size of 3 μm, the strain rate will increase by 4–5 orders of magnitude at 900°C at a constant differential stress of 10 MPa (se also Rubie 1984a). Figure 10.6 has been constructed from these flow laws, assuming that the spinel deforms in a grain-size sensitive régime such that $\dot{e} \propto d^{-2}$, where $\dot{e}$ is strain rate and d is the grain size. It is suggested by figure 10.6 that the reaction of olivine to spinel could enhance the strain rate by a factor of 10^6–10^8, depending on temperature and the spinel grain size. It should be noted, however, that the creep data of Vaughan & Coe (1981) are subject to considerable uncertainties because there were large temperature gradients along the axis of samples during the experiments (Vaughan & Coe 1981, Fig. 2). Consequently, in some experiments, part of the sample must have been in the olivine stability field and part of it in the spinel stability field.

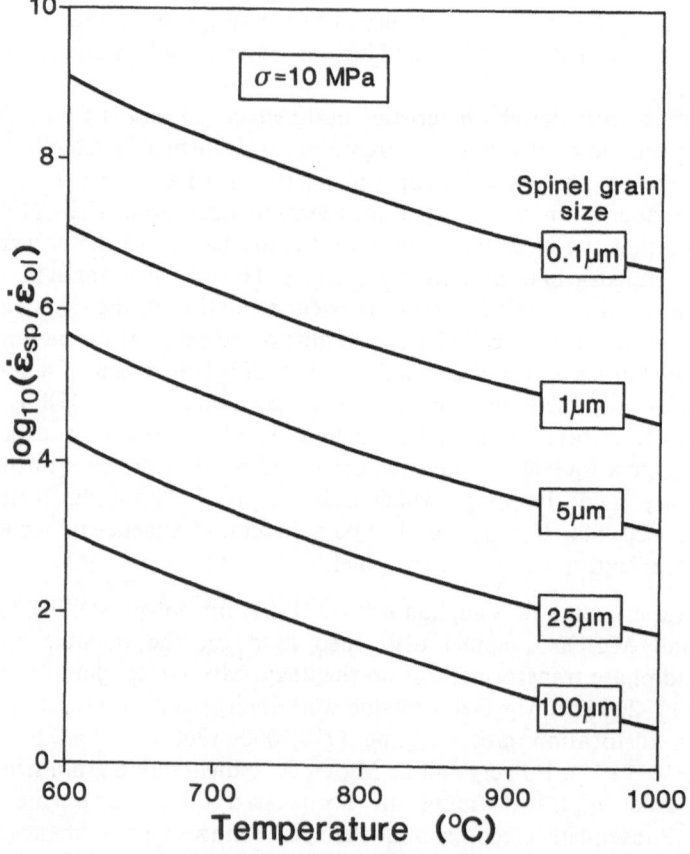

Figure 10.6 The ratio of the strain rate $\dot{\varepsilon}_{sp}$ in fine-grained Mg_2GeO_4 spinel to the strain rate $\dot{\varepsilon}_{ol}$ in medium grained Mg_2GeO_4 olivine as a function of temperature and spinel grain size with a differential stress of $\sigma = 10$ MPa (based on the data of Vaughan & Coe 1981).

Experiments have been performed recently to investigate the effect on rheology of the reaction of fine-grained (10 μm) albite to jadeite + quartz (Sotin & Tullis 1987). Preliminary results suggest that the reaction has a weakening effect. At a strain rate of $4 \times 10^{-4} s^{-1}$, for example, the strength decreases from an initial value of 600 Mpa to 250 Mpa at 70 per cent shortening.

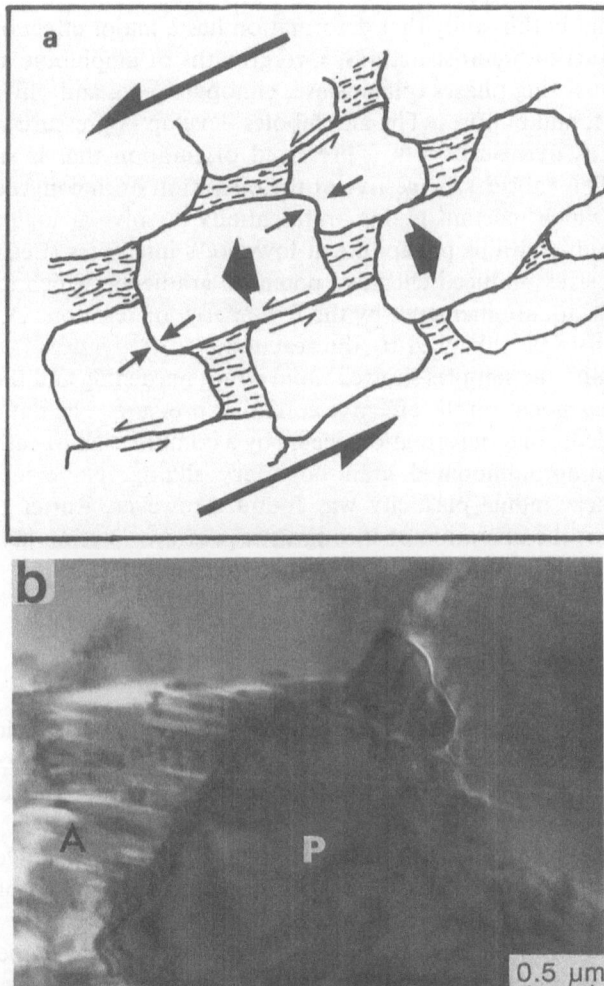

Figure 10.7 Reaction microstructures characteristic of incongruent pressure solution. (a) Schematic sketch showing dissolution of reactants at high-stress interfaces, precipitation of reaction products as fibres parallel to the dilatant direction at low-stress interfaces, and grain-boundary sliding parallel to the direction of maximum resolved shear stress (based on Rutter *et al.* 1985). (b) TEM micrograph showing an oriented overgrowth of amphibole (A) on a pyroxene grain (P) which developed during the experimental deformation of wet basalt (from Rutter *et al.* 1985).

10.2.3 Hydration reactions

The mechanical properties of basalt while undergoing hydration reactions have been studied experimentally by Rutter *et al.* (1985). Stress-relaxation tests were carried out by shearing wet crushed basalt (grain size 5–15 μm), contained in saw-cuts in intact basalt cylinders, at 600°C and various confining pressures up to 200 Mpa.

It was found in this study that deformation has a major effect on the development of reaction microstructures. Overgrowths of amphibole and feldspar form on pre-existing phases (plagioclase, clinopyroxene, and olivine) at dilatant interfaces, and in pores. The amphiboles develop in pressure shadows, on phases such as pyroxene, with a preferred orientation that is stress-related (Fig. 10.7). This fabric is suggestive of the operation of 'incongruent pressure solution' in which reactant grains preferentially dissolve at high-stress interfaces and product grains precipitate at low-stress interfaces (Beach 1982). In this process, stress-induced chemical potential gradients, which drive normal pressure solution, are increased by the free energy of reaction. Deformability should therefore be enhanced by the reaction.

Mechanically, the samples showed rapid strain hardening and the flow stress is strongly dependent on the effective confining pressure. The data were interpreted to indicate that deformation occurs by a combination of cataclastic flow and diffusion-accommodated grain-boundary sliding. No evidence of significant intracrystalline plasticity was found. However, Rutter *et al.* (1985) concluded that the operation of incongruent pressure solution has little effect on the *mechanical* properties of the basalt (see below).

10.3 Field and petrographic data

Two significant problems arise when using field and petrographic data from metamorphic terrains to evaluate the effects of mineral reactions on rheology. The first problem is actually to prove from microstructural evidence alone that mineral reactions occurring during metamorphism do produce some significant change in rheology. The second problem is to understand the mechanisms involved. In many cases, solutions to these two problems cannot be obtained because, subsequent to deformation, the necessary microstructural evidence is often destroyed. This is because prolonged periods spent under static conditions at high T and P can drastically modify the microstructure and perhaps cause further mineral reactions to occur. These comments obviously apply in particular to investigations of deformation mechanisms that operate during the early stages of a metamorphic cycle. However, several cases have been described which enable a preliminary evaluation to be made of deformability and deformation mechanisms during some stages of thermal and metamorphic evolution during orogenesis.

10.3.1 Deformation during high-pressure metamorphism

Enhanced deformability during the conversion of quartz diorite to ortho-gneiss, as a result of eclogite-facies mineral reactions, has been discussed by Rubie (1983) and Koons *et al.* (1987). This example is from the Sesia Zone (Western Alps), which consists of pre-Alpine continental basement that underwent early-Alpine high-pressure metamorphism ($T \approx 500$–$560°$C, $P > 1.4$ GPa; Compagnoni 1977, Rubie 1984b, Oberhaensli *et al.* 1985). The quartz diorite and related lithologies originated as plutons of late Hercynian age (Oberhaensli *et al.* 1985) and consisted predominantly of sodic plagio-clase + quartz + biotite ± microcline. During the early-Alpine metamorphism, these intrusive rocks were intensely deformed and the resulting orthogneiss now outcrops over an area > 100 km^2 (Koons *et al.* 1987). Rare lenses of undeformed meta-quartz diorite, generally ~ 100 m across, are preserved in the orthogneiss, and microstructural studies of gradational contacts between these two lithologies have provided the evidence for enhanced deformability. The mineralogical evolution in these rocks has been described by Koons *et al.* (1987).

Typical field relations between undeformed and deformed rocks are shown schematically in Figure 10.8. In the following discussion, it is necessary to assume that the evolution of microstructure with time is recorded by the *spatial* sequence of microstructures across an outcrop such as that shown in Figure 10.8 (see Ridley & Dixon 1984).

In the undeformed meta-quartz diorite, plagioclase was pseudomorphically replaced by jadeitic pyroxene ($\sim Jd_{0.9}$), quartz, and zoisite during high-pressure metamorphism (Fig. 10.9). Jadeite forms interlocking grains, 10–50 μm in diameter, and quartz and needles of zoisite are generally concen-trated in the intergranular regions of this mineral (Fig. 10.9b). The partial breakdown of biotite to coronas of garnet and phengite also occurred (Figure 10.9a) and, due to slow rates of diffusion, there was disequilibrium between the former biotite and plagioclase sites during metamorphism (Koons *et al.* 1987). Consequently, the jadeitic pyroxene was metastable in terms of the whole-rock composition and under the conditions of high-pressure metamor-phism, the stable pyroxene phase was omphacite (this point is important in the discussion of deformation mechanisms – see below).

A first stage of deformation is preserved in the so-called 'transition-orthogneiss' (Koons *et al.* 1987) which occurs in shear zones (~ 20 cm wide) in the meta-quartz diorite and in a wider zone situated between the unde-formed rock and the orthogneiss (Fig. 10.8). The transition orthogneiss is characterized by semi-continuous layers rich in jadeitic pyroxene ($Jd_{0.9-0.8}$) which alternate with lenses of medium-grained quartz (Fig. 10.10a). The pyroxene-rich layers were derived from the pseudomorphed plagioclase, and the quartz lenses were formed by deformation of the original igneous quartz. Comparison between Figures 10.9a and 10.10a shows that the strain produced

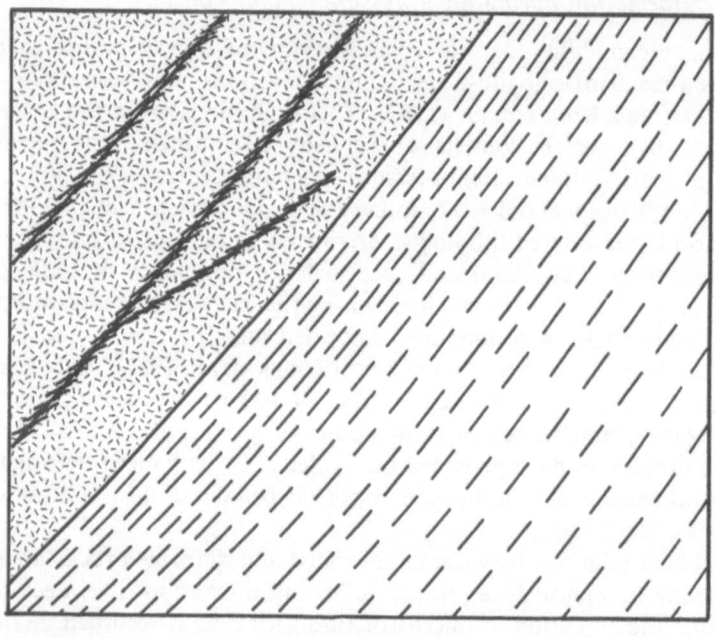

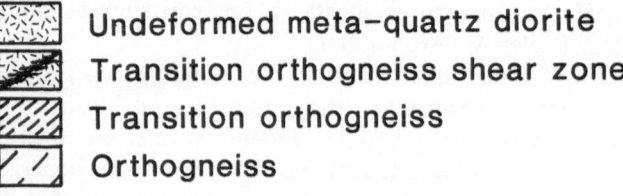

Undeformed meta-quartz diorite
Transition orthogneiss shear zone
Transition orthogneiss
Orthogneiss

Figure 10.8 Schematic sketch showing field relations between undeformed meta-quartz diorite, transition orthogneiss, and orthogneiss in the Sesia Zone at Monte Mucrone (width of field of view ~ 100 m).

Figure 10.9 Micrographs of the undeformed meta-quartz diorite from the Sesia Zone. (a) Optical micrograph showing two plagioclase crystals which have been completely pseudomorphed by fine-grained jadeite, zoisite, and quartz during high-pressure metamorphism. The remainder of the rock is mainly medium-grained quartz (Q), although biotite (b) with phengite–garnet coronas is also present on the right (plane-polarized light). (b) Backscattered electron image of a pseudomorphed plagioclase showing small (20–30 μm) jadeite grains (grey) and zoisite needles (white). Quartz (black) occurs in intergranular regions and as occasional small inclusions in jadeite.

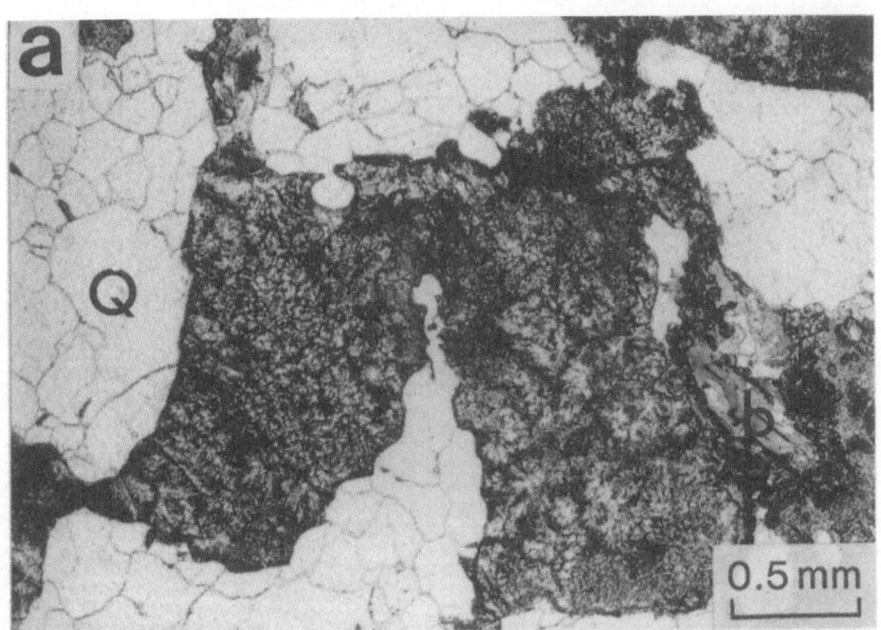

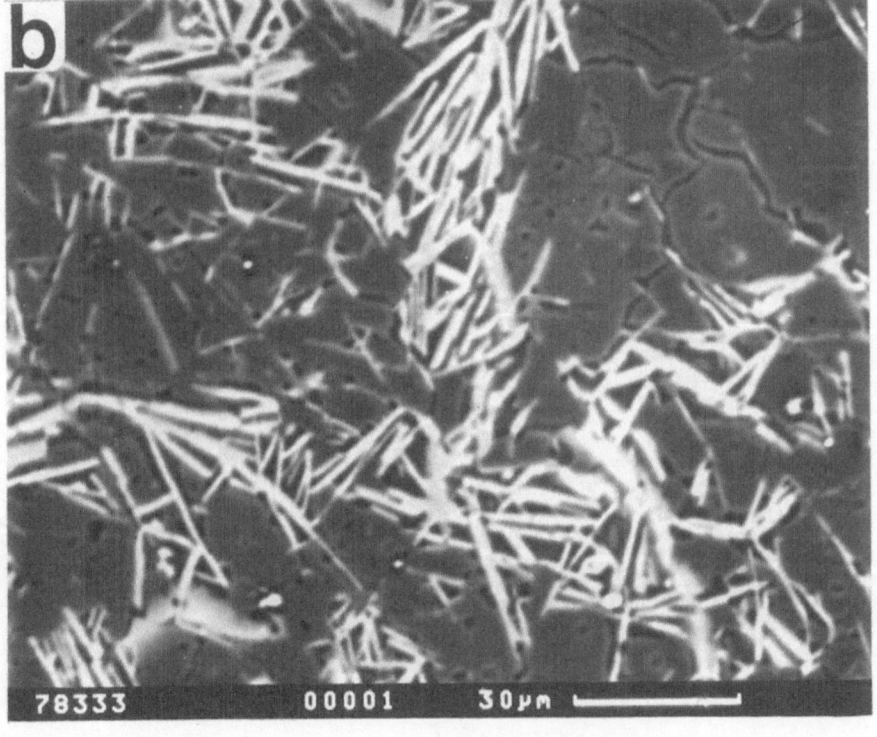

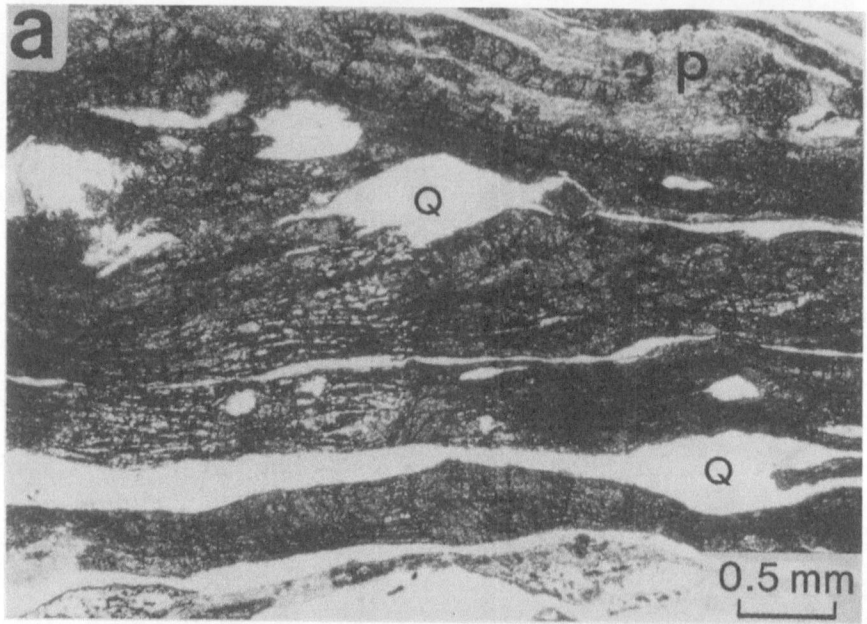

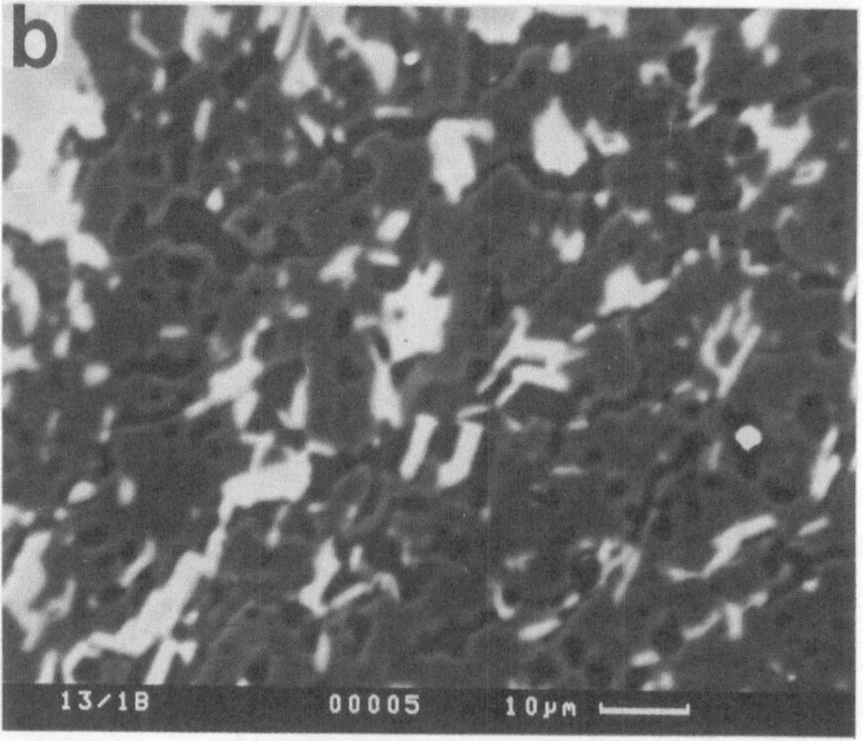

Figure 10.10 (*Continued*)

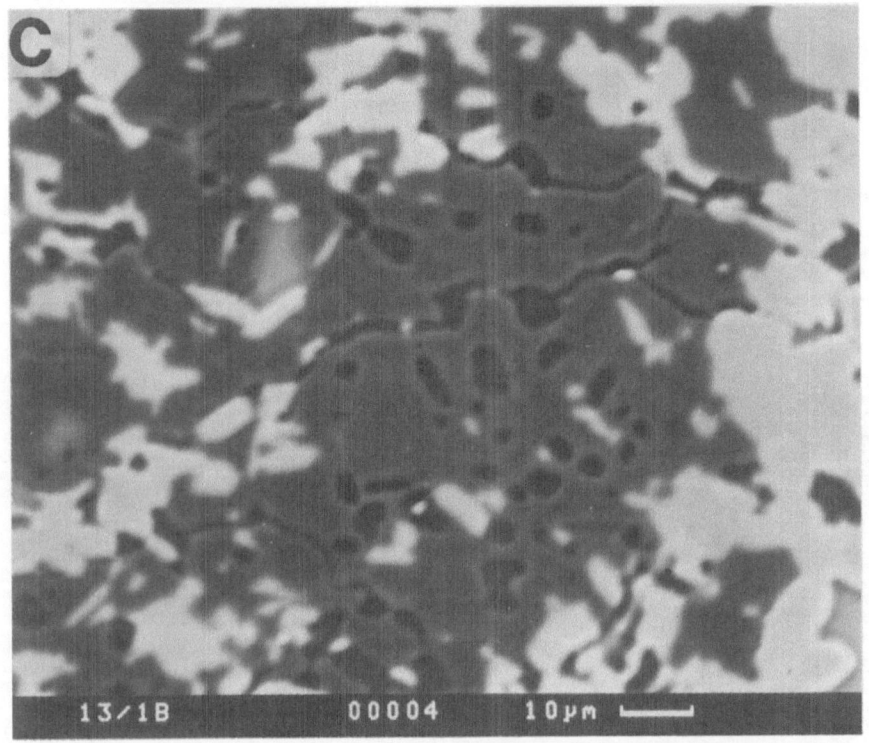

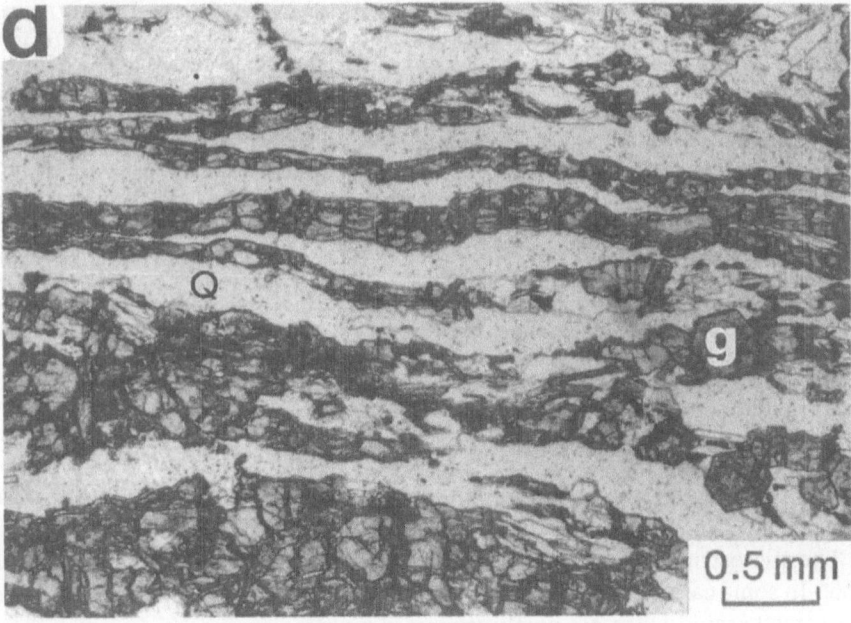

Figure 10.10 (*Continued*)

Figure 10.10 Micrographs of transition orthogneiss and orthogneiss from the Sesia Zone. (a) Optical micrograph of transition orthogneiss from a shear zone showing semi-continuous layers of jadeite + quartz + zoisite which formed by deformation of pseudomorphed plagioclase. The strain in quartz layers (Q) is generally considerably less than in the jadeite-rich layers. Partial replacement of jadeite by fine-grained phengite (P) has occurred locally (top right) (Plane-polarized light). (b) Backscattered electron image of a jadeite + quartz + zoisite layer in the transition orthogneiss. Jadeite (grey) occurs as small equant grains, which in some cases appear to have partly coalesced. Quartz (dark grey) is present in intergranular regions and effectively forms a matrix and zoisite prisms (white) have developed a crude preferred orientation. (c) Backscattered electron image of jadeite + quartz + zoisite from the transition orthogneiss. Compared with (b), coarsening of jadeite has occurred and quartz, which was probably originally intergranular to this phase, now forms inclusions within the enlarged grains. (d) Optical micrograph of transition orthogneiss in which jadeitic pyroxene has coarsened to 200–400 μm grain size. Quartz and zoisite inclusions are largely absent from the pyroxene layers in this example. Q, quartz; g, garnet (plane-polarized light). (e) Optical micrograph of orthogneiss. Pyroxene layers characteristic of the transition orthogneiss have been disrupted by deformation and may have undergone further coarsening. The strain was strongly partitioned into quartz + white mica layers during this final stage of deformation. Numerous garnets are present in the lower part of this micrograph (plane-polarized light).

by the deformation was very high, and that the greatest strain was partitioned into the pyroxene-rich layers.

The microstructure of the pyroxene-rich layers in the deformed rocks is variable. In some cases, particularly in the narrow shear zones, the pyroxene forms very small grains which are partly separated by blebs or narrow rims of quartz situated along grain boundaries (Fig. 10.10b,c). In other cases, coarsening has

278

increased the grain size to ~400 μm (i.e. the width of the pyroxene-rich layers). This coarse-grained pyroxene varies from (a) containing numerous quartz inclusions and showing strong undulatory extinction to (b) containing few inclusions of either quartz or zoisite and showing uniform extinction (Fig. 10.10d). In the shear zones, the jadeitic pyroxene is partially replaced by aggregates of fine-grained, often randomly oriented phengite (Fig. 10.10a).

The conversion of transition orthogneiss to orthogneiss involved further deformation, also under eclogite-facies conditions, during which strain was strongly partitioned into quartz + white mica layers and the pyroxene-rich layers became strongly disrupted (Fig. 10.10e). The pyroxenes developed undulatory extinction and changed in composition towards omphacite. Due to this microstructural reworking and subsequent annealing, there are no obvious indications from microstructures in the orthogneiss that the formation of this rock from the protolith quartz diorite involved the development of particularly high strains. The main indication of high strain deformation comes from the recognition that the orthogneiss evolved through the intermediate transition orthogneiss stage.

10.3.1.1 EVIDENCE FOR ENHANCED DEFORMABILITY

The microstructural evidence shows that, during the initial deformation, the highest strain was partitioned into the fine-grained pyroxene + quartz + zoisite aggregates which replaced plagioclase pseudomorphically. This material was therefore *more* deformable than medium-grained quartz. If the original quartz diorite had been deformed without metamorphic reactions occurring, it is to be expected, both from experimental data and observations on mylonites, that plagioclase would have been *less* deformable than medium-grained quartz (Tullis & Yund 1977, White *et al.* 1980). Thus, because 40–50 per cent of the original rock consisted of plagioclase, the reaction of this mineral to the high-pressure assemblage must have significantly enhanced the deformability of the quartz diorite.

10.3.1.2 DEFORMATION MECHANISMS

There are several factors to consider when evaluating deformation mechanisms.

First, medium-grained jadeite and omphacite are much less deformable than quartz. This behaviour is shown during the evolution of transition orthogneiss to orthogneiss, as described above. It is also possible, for example, to find apparently undeformed crystals of sodic pyroxene (e.g. 0.5 mm diameter) in a highly strained, dynamically recrystallized quartz matrix (Rubie 1983, Fig. 7b). It is possible, therefore, that the high ductility of the pyroxene-rich layers during the development of the transition orthogneiss was related to the very small grain size of the reaction products in the pseudomorphed plagioclases (Fig. 10.9).

Second, there is evidence that infiltration of hydrous fluid was associated with the deformation that converted the meta-quartz diorite to transition orthogneiss. Whole-rock chemical data suggest that the deformed rocks were slightly enriched in H_2O relative to the undeformed protolith (Koons *et al.* 1987). Oxygen-isotope data indicates infiltration, although fluid : rock ratios were probably spatially very variable (Frueh-Green 1985). The replacement of Na-pyroxene by phengite (see above) suggests redistribution of H_2O, at least on a localized scale. Finally, veins of zoisite + garnet ± 1 quartz, 1–2 cm wide, are frequently situated along the narrow shear zones and are likely to have developed under conditions of high fluid pressure. Unfortunately, the exact timing of infiltration is uncertain, as is often the case with shear zones (see Rutter & Brodie 1985). One possibility is that localized infiltration began prior to deformation, along cracks which subsequently became the zoisite–garnet veins. Infiltration of fluid into the surrounding rock may then have produced a weakening effect and catalysed intracrystalline plastic deformation. A complete absence of deformation microstructures in the zoisite–garnet veins, together with a variable orientation of the phengite flakes which replaced jadeite, makes this possibility unlikely. A second, more likely, possibility is that infiltration commenced either during or after the deformation as a consequence of an enhancement of permeability in the sheared rocks relative to the undeformed protolith.

On the basis of the microstructural evidence, there are several possible mechanisms which could operate to enhance the deformability of the fine-grained pyroxene-rich aggregates relative to both medium-grained plagioclase and medium-grained sodic pyroxene.

(1)   The reaction products deformed by grain-size-sensitive diffusion creep as a result of the small grain size. The mechanism may have been diffusion-accommodated grain-boundary sliding for example, as suggested by Rubie (1983).

(2)   Deformation of the reaction products by some mechanism of intra-crystalline plasticity was facilitated by the small grain size. This effect has been demonstrated for monomineralic aggregates in two experimental studies, involving a different mechanism in each case.

    (a)   Kronenberg & Tullis (1984) have shown that, during the experimental deformation of quartzite, a reduction in grain size facilitates the process of hydrolitic weakening by decreasing the diffusion distance for the uptake of H_2O from grain boundaries into the crystal structure. The same process could presumably operate as a result of fluid infiltration into rocks in which fine-grained reaction products had developed.

    (b)   Tullis & Yund (1985) found that when feldspar aggregates deform by recrystallization-accommodated dislocation creep, a reduction in grain size results in a reduction in flow stress (at constant strain rate). This is because small grains are more readily replaced by new strain-free grains (by grain-boundary migration, for example) than are large ones.

(3) The deformability was enhanced by mechanisms of transformation plasticity (e.g. Poirier 1982) as a result of the large (17 per cent) volume change.

Mechanisms which are diffusion-controlled, i.e. diffusion creep and dislocation creep, should be facilitated by the infiltration of H_2O. However, only one of the possible mechanisms listed above (2a) actually relies on fluid infiltration to catalyse deformation.

One criterion which has been used for distinguishing between dislocation creep and diffusion creep involving significant grain-boundary sliding is the presence or absence of a grain shape fabric. As emphasized elsewhere in this chapter, if rates of grain-boundary migration are fast, grains can remain equant even when deforming by dislocation creep. Grains of jadeite, which are often rather blocky and equant in shape, in a matrix of quartz, are shown in Figure 10.10b. This microstructure is suggestive of deformation of jadeite by grain-boundary sliding, accommodated by the redistribution of quartz in the intergranular regions (possibly by diffusive mass transport). Furthermore, the microstructure suggests that dislocation creep was not an important mechanism in jadeite because (1) the size and shape of jadeites are not greatly different from those in the undeformed aggregates (Fig. 10.9b) and (2) an equilibrium microstructure has not developed (Fig. 10.10b). These two observations suggest that grain-boundary migration rates were slow. It is to be expected that grain-boundary migration would be greatly inhibited in a two- or three-phase aggregate relative to a single-phase aggregate.

A further indication of the deformation mechanism is that the jadeitic pyroxenes of the meta-quartz diorite show almost no change in composition from metastable jadeite towards stable omphacite during the deformation which produced the transition orthogneiss (see Koons *et al.* 1987). The persistence of metastability must be attributed to slow rates of intracrystalline diffusion in pyroxene. During the later stage of deformation, when the transition orthogneiss evolved to orthogneiss, the pyroxenes did partially equilibrate to omphacite. In this case, as there is evidence that pyroxene deformed by dislocation creep (undulatory extinction and subgrain boundaries), diffusion rates may have been enhanced by the movement of dislocations (Koons *et al.* 1987). The fact that this did not occur during the formation of the transition orthogneiss also suggests that the earlier deformation did not occur by dislocation creep. The evidence is therefore consistent with the deformation of pyroxene by grain-boundary sliding, although in the absence of mechanical data (see below), such an evaluation of the deformation mechanism must remain somewhat speculative.

The possible role of transformation plasticity in enhancing deformability is discussed in a subsequent section.

10.3.2 Deformation during amphibolite facies metamorphism

There is evidence for enhanced deformability of some metapelites in the Adula Nappe (central Alps), during uplift from the eclogite to amphibolite facies (Heinrich 1982; Rubie & Heinrich, work in progress). Eclogite facies metapelitic schists consist of oriented flakes of phengite and paragonite in a matrix of medium-grained quartz, with occasional crystals of garnet and kyanite (Fig. 10.11a). During amphibolite facies metamorphism (of Tertiary age) this mineral assemblage reacted to biotite + plagioclase + quartz + muscovite (± garnet, staurolite, and kyanite or K-feldspar) by a reaction which involved *dehydration* (Heinrich 1982).

Initially, reaction in the eclogite-facies metapelites involved the pseudomorphic replacement of phengite and paragonite by biotite + plagioclase aggregates which now have an average grain size of 40–50 μm (Fig. 10.11b). In undeformed pseudomorphs, plagioclase forms a mosaic of equant grains and small, variably oriented biotite flakes are located on the plagioclase grain

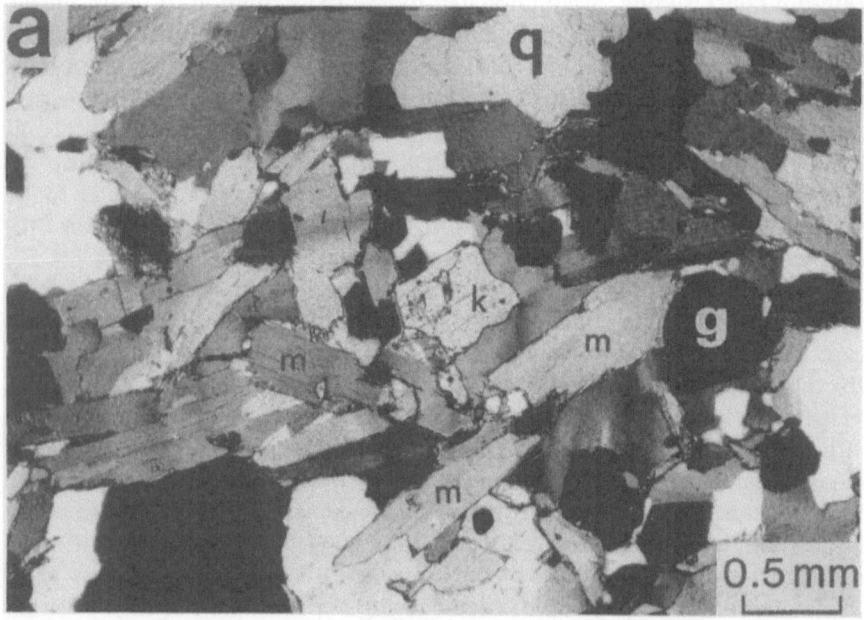

Figure 10.11 Optical micrographs of metapelites from the Adula Nappe, Central Alps (crossed polars). (a) Eclogite facies metapelite consisting of flakes of white mica (m) (phengite and paragonite) in a matrix of medium-grained equant quartz (q). Kyanite (k) and garnet (g) are also present. (b) Eclogite facies metapelite which has partially reacted to an amphibolite facies assemblage. Flakes of white mica (m) have been variably replaced pseudomorphically by fine-grained aggregates of equant plagioclase grains and variably oriented biotite flakes. (c) Deformed partially reacted metapelite. The strain is strongly partitioned into zones containing primarily the fine-grained plagioclase + biotite reaction products. (q) is quartz.

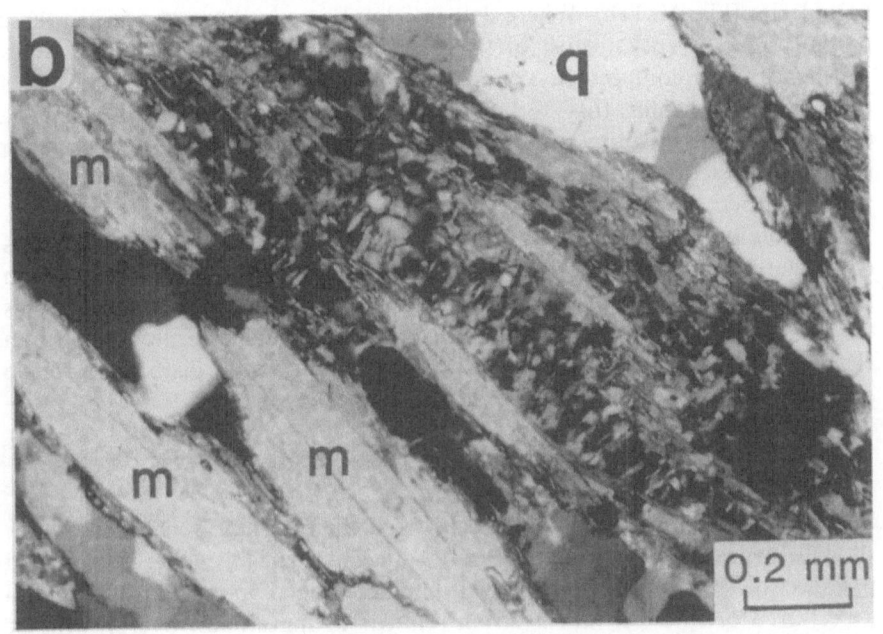

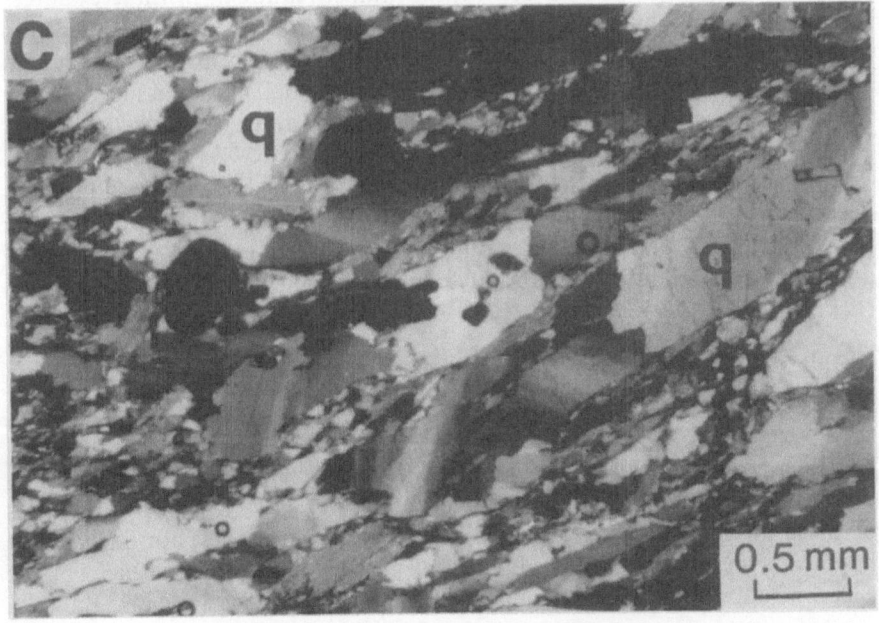

boundaries (Fig. 10.11b). Where deformation has subsequently affected these partially reacted rocks, most of the strain is partitioned into narrow zones which primarily contain fine-grained biotite + plagioclase (Fig. 10.11c). This observation indicates that the fine-grained reaction products were more deformable than medium-grained quartz + white mica in the same rock, even though the reaction reduced the amount of sheet silicates. The reaction therefore enhanced the deformability of these rocks.

The dominant deformation mechanism is difficult to evaluate on the basis of currently available microstructural data. The small grain size of the reaction products was probably an important factor. Possible deformation mechanisms for the reaction products include diffusion creep, cataclastic flow due to an elevated pore-fluid pressure during dehydration, and dislocation creep enhanced by the small grain size (as discussed above).

10.3.2   Other examples

Beach (1982) has described cover rocks from the external French Alps which show evidence of incongruent pressure solution. Typically, overgrowths and pressure shadows of reaction products have developed on reactant grains with a preferred orientation that is strain related. In limestones, overgrowths of ferroan-calcite have developed on calcite grains, and in greywackes, overgrowths of quartz + muscovite have developed on quartz and feldspar. Brodie & Rutter (1985) have described comparable microstructures in a hornblendite from Loch Alsh, Scotland, in which oriented overgrowths of chlorite and actinolite have developed on hornblende grains. Although in these examples the deformation has had a very significant effect on controlling reaction microstructures, the effect of the reactions on rheology is very difficult to assess (see below and Rutter *et al.* 1985). Also, the bulk strain associated with the overgrowths and pressure shadows is generally low (e.g. Brodie and Rutter 1985, Plate 3c).

An example of enhanced deformability of fine-grained reaction products has been described by Brodie & Rutter (1987). Clinopyroxene in a metagabbro from north-west Spain has partially reacted to fine-grained hornblende + quartz and, apparently, as a result, high-strain zones have developed in the rock.

10.4   Deformation mechanisms

Possible mechanisms of reaction-enhanced deformability have been reviewed by White & Knipe (1978) and Brodie & Rutter (1985). The effectiveness of these mechanisms and the conditions under which they are likely to be important are discussed and summarized in this section on the basis of the experimental and field/petrographic data described above. Because the data are currently very limited, the range of mechanisms discussed here should not be regarded as comprehensive.

10.4.1 Grain-size-sensitive diffusion creep

Mechanical and microstructural data from some of the examples described above are consistent with deformation of fine-grained reaction products by grain-size-sensitive diffusion creep. Here the term 'diffusion creep' is used in a broad sense to include Coble creep, Nabarro–Herring creep, and diffusion-accommodated grain-boundary sliding, i.e. mechanisms with varying contributions from grain-boundary sliding. In geological literature, the phenomenological term 'superplasticity' has also been used to describe deformation by such mechanisms (e.g. Boullier & Gueguen 1975, Schmid *et al.* 1977, Rubie 1983), although such usage has been criticized by Poirier (1985, p. 205).

The characteristics of diffusion creep involving grain-boundary sliding have been extensively studied in fine-grained metals which show superplastic behaviour (e.g. Edington *et al.* 1976; Poirier 1985; see also the volume edited by Baudelet & Suery 1985). Microstructural evidence for the operation of this type of mechanism in rocks has been discussed by Boullier & Gueguen (1975). Flow laws have the form

$$\dot{\varepsilon} \propto \sigma^n / d^m \qquad (10.1)$$

where $\dot{\varepsilon}$ is the strain rate, σ is the differential stress, d is the grain size, n is the stress exponent which lies in the range 1–2, and m lies in the range 2–3 (e.g. Gifkins 1976). Unfortunately, the mechanical behaviour (i.e. the low stress exponent in the flow law and the grain-size sensitivity) is the only conclusive indication of diffusion creep (Schmid *et al.* 1977, Karato *et al.* 1986, Rutter & Brodie 1988). An absence of a crystallographic preferred orientation would also distinguish this mechanism from dislocation creep. Conclusive evidence of grain-boundary sliding is very difficult to obtain, even in experimental studies (e.g. Edington *et al.* 1976). One technique for estimating the fraction of the total strain contributed by grain-boundary sliding involves the examination of topographical relief on pre-polished split cylinders after deformation (Schmid *et al.* 1977). Although a predominance of equant grains in a deformed aggregate is sometimes regarded as evidence for diffusion-accommodated grain-boundary sliding, this criterion is not reliable. This is because, if rates of grain-boundary migration are fast, equant grains and textural equilibration can be maintained during deformation even when other mechanisms, such as Coble creep, Nabarro–Herring creep, or dislocation creep, operate (Ashby *et al.* 1978; Karato 1988; see also Tullis & Yund 1985). In the absence of mechanical data, the operation of diffusion-accommodated grain-boundary sliding in natural rocks (e.g. Boullier & Gueguen 1975, Rubie

1983) must also be difficult to prove unequivocally. This is particularly so when grain coarsening and other microstructural readjustments have occurred subsequent to deformation.

Potentially, deformation of fine-grained reaction products by diffusion creep may be the most effective mechanism of reaction-enhanced deformability. The possibility of a strain-rate enhancement of 10^6–10^8 (at constant stress) is demonstrated by Figure 10.7 (see also Rubie 1984a). Very high strains can develop by this mechanism provided that grain coarsening does not occur too rapidly (Rutter & Brodie 1988). The potential for the development of high strains is well demonstrated by the superplastic behaviour of fine-grained metals in which elongations of 1000 per cent can develop, without necking, during tensile tests (Edington et al. 1976, Poirier 1985). Factors which are of major importance in controlling rheology and the strain rate – time evolution are (a) the initial grain size of the reaction products and (b) the rate of grain coarsening.

10.4.1.1 INITIAL GRAIN SIZE OF REACTION PRODUCTS

It is demonstrated in Figure 10.6 that the initial enhancement of strain rate, at constant stress, depends greatly on the grain size of the reaction products during and at the termination of reaction. Unfortunately, the magnitude of this grain size cannot be determined reliably in natural rocks because of the possibility that grain coarsening has occurred. Thus, even in rocks described above, from the Sesia Zone and the Adula Nappe, which show fine-grained (10–60 μm) reaction products pseudomorphing precursor minerals (Figs 10.9 & 11b), considerable coarsening *may* have occurred subsequent to reaction. Rubie (1983) has argued that reaction resulting from rapid rates of change of P and/or T should favour a small grain size in product phases. There is some evidence that this is the case, because reaction products formed during contact metamorphism, under conditions of rapid heating followed by rapid cooling, often have a submicron grain size (e.g. Brearley 1986, Worden et al. 1987). Similarly, in a recent experimental study of the breakdown of muscovite + quartz, in which the sample was heated rapidly to a temperature ~ 200°C above equilibrium, K-feldspar + biotite + sillimanite developed with a grain size 0.3–3 μm (Rubie & Brearley 1987). However, to understand the evolution of grain size resulting from reactions occurring during very slow (geological)

Figure 10.12 Electron micrographs of spinel which has been produced by transformation from hot-pressed Mg_2GeO_4 olivine (grain size ~ 30 μm) at 900°C and 2 GPa. (a) The microstructure of spinel after complete transformation from olivine. The spinel forms equant grains, separated mostly by high-angle grain boundaries, and has a very small grain size compared with the original olivine. {110} stacking faults are common. (b) An early stage in the transformation showing two spinel grains (containing stacking faults) which have nucleated in olivine (ol). The largest grain (1) has grown to the maximum observed size (~ 1 μm diameter). The second, smaller grain (2) is thought to have nucleated at the phase boundary between the first spinel and olivine in a different crystallographic orientation. The olivine contains a high density of dislocations.

286

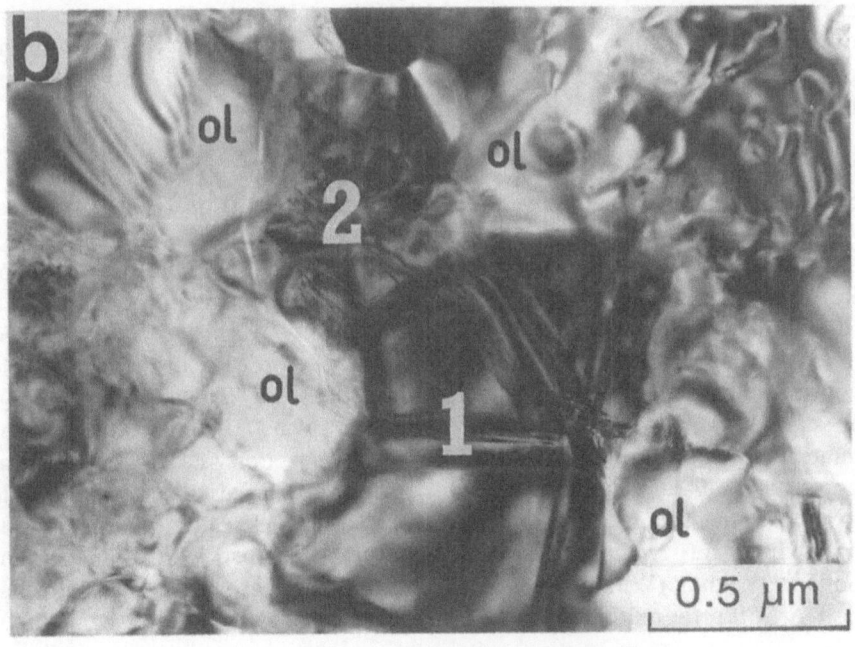

rates of heating or cooling (e.g. $1°C$ per 10^5 years) is obviously problematic because it cannot be investigated experimentally!

Factors which control the grain size of reaction products during a polymorphic phase transformation involving a significant volume change have been investigated by Rubie & Champness (1987). A hot-pressed Mg_2GeO_4 olivine aggregate was reacted to spinel at $900°C$ and 2 GPa under near-hydrostatic conditions. These conditions overstep equilibrium by ~ 1.7 GPa. The parent olivine had an average grain size $\sim 30 \mu m$, but the grain size of the product spinel was $\leqslant 1 \mu m$ (Fig. 10.12a). Rubie & Champness (1987) discussed several models to explain the grain-size refinement. In their preferred model, the growth of an individual spinel grain becomes progressively inhibited by the stress field which develops in the surrounding olivine as a result of the 8 per cent volume decrease. When the spinel reaches a certain size ($\sim 1 \mu m$), growth effectively ceases. New spinels nucleate at the interphase boundary with a different orientation from the original grain (Fig. 10.12b). The new orientation takes advantage of the anisotropic stress field in the olivine. Growth of these new spinels is also inhibited by the developing stress field. Thus the flow stress of the olivine and its ability to deform to accommodate the volume change are important factors which influence the spinel grain size at conditions of large overstepping.

The grain size of reaction products may be reduced if they nucleate and grow in an actively deforming shear zone. In the serpentinite dehydration experiments of Rutter & Brodie (1988), equant grains of olivine up to $10 \mu m$ diameter developed in the undeformed regions of the sample (Fig. 10.5). However, in the narrow shear zones, in which high strains developed, the grain size of the olivine was only $0.1-0.2 \mu m$. Evidently, at an early stage of growth, the olivine grains rolled or slid away from the nucleation sites, thus inhibiting further growth (Rutter & Brodie 1988).

10.4.1.2 GRAIN COARSENING

Because of the grain-size sensitivity of diffusion creep (Eqn 10.1), a progressive increase in grain size during deformation, as a result of grain coarsening, must cause time-dependent hardening (Rubie 1983, 1984a; Karato *et al.* 1986; Rutter & Brodie 1988). The theoretical basis for the coarsening of metal and ceramic aggregates, under hydrostatic conditions, is fairly well established (e.g. Martin & Doherty 1976). The kinetics of the process, in monomineralic quartz and calcite aggregates, have been investigated by Tullis & Yund (1982), Joesten (1983), Rutter (1984), and Olgaard & Evans (1986), but the level of understanding of grain growth in rocks is still very poorly developed. In addition, it is doubtful whether the results of these studies are directly applicable to rocks deforming by diffusion creep, for two reasons. (1) In metals, rates of coarsening are *enhanced* by deformation mechanisms involving grain-boundary sliding (Wilkinson 1985). Abnormal grain growth during deformation of Carrara marble has been demonstrated by Schmid *et al.* (1980).

(2) Rates of coarsening are slower in multiphase aggregates than in single-phase aggregates. Olgaard & Evans (1986) have investigated the effect of second-phase Al_2O_3 particles on grain growth in calcite. Because calcite grain boundaries are pinned by the second-phase particles, boundary migration is inhibited and a stable grain size develops which depends on the volume fraction of the second phase. Coarsening kinetics of two phase aggregates (e.g. plagioclase + biotite) in which the second phase (e.g. biotite) can also coarsen have not been investigated but would be more relevant geologically (see Rubie & Thompson 1985, p. 67). Evidence for coarsening of a multiphase aggregate is seen in Fig. 10.10.

10.4.2 Cataclastic flow

It has been suggested, for example by Murrell (1985, 1986), that distributed cataclastic flow should result during metamorphism of the Earth's crust as a result of an increase in the pore-fluid pressure during dehydration reactions. In the experimental studies described above, brittle failure and cataclastic flow only resulted when dehydrating samples were deformed at high strain rates (e.g. $> 10^{-4} s^{-1}$) under undrained conditions. Such results have led to models in which cataclastic flow occurs in dehydrating rock volumes which are capped by low-permeability layers so that a high fluid pressure is maintained during deformation (Murrell 1985, 1986).

At present there is a lack of field evidence supporting the occurrence of wide-scale cataclastic flow in crustal rocks. In addition, in experimental studies of dehydrating serpentinite carried out at *low strain rates*, both undrained and with a controlled pore-fluid pressure, there is no mechanical evidence for any contribution from cataclastic flow (Rutter & Brodie 1988). Such results cast uncertainty on the importance of this process in the Earth's crust where strain rates are likely to be lower than in any of the experimental studies. However, the possibility of cataclastic flow making at least some contribution during periods of high fluid pressure cannot be excluded. It has been emphasized by Borradaille (1981) and Rutter *et al.* (1985) that there can be a complete spectrum of variation between pressure-insensitive diffusion-accommodated grain-boundary sliding and pressure-sensitive cataclastic flow. Unfortunately, in the absence of mechanical data, it is probably impossible to evaluate the relative contributions of such mechanisms in nature from microstructures alone. Presumably the relative contributions will be controlled by the relative rates of fluid production (i.e. reaction kinetics), and fluid loss (permeability-dependent) as well as factors such as strain rate and temperature.

10.4.3 Incongruent pressure solution

This mechanism, which produces oriented pressure shadows and overgrowths of product phases on reactant grains (Fig. 10.7), operates by the enhancement of stress-induced chemical potential gradients during metamorphic reactions.

Although this mechanism undoubtedly has a major effect on reaction microstructures (Fig. 10.7), there is currently no experimental evidence to demonstrate its effect on rheology. Rutter et al. (1985) concluded that its effect is small, with a strain rate enhancement of less than a factor of 2–3. This conclusion was based on both mechanical data and theoretical considerations. For normal pressure solution, the chemical potential difference which drives inter-granular diffusion from dissolution sites at high-stress interfaces to precipitation sites at low-stress interfaces can be approximated by

$$\Delta\mu \approx \bar{V}\Delta\sigma_n \qquad (10.2)$$

where $\bar{V}$ is the molar volume of the solid and $\Delta\sigma_n$ is the normal stress difference (Paterson 1973). Because the chemical potential difference depends on the molar volume $\bar{V}$ of the solid, Rutter et al. (1985) argued that diffusion must be faster along the interface where hydration reactions have occurred because these must increase $\bar{V}$ and therefore $\Delta\mu$. According to their model, the enhancement of deformability by incongruent pressure solution is small because it is only dependent upon the molar volume difference between reactant and product phases. An alternative hypothesis is that the stress-induced chemical potential difference is increased by the total free energy of reaction, which, for small departures from equilibrium, can be approximated by

$$\Delta G_r \approx \Delta S(T_e - T) \qquad (10.3)$$

where ΔS is the entropy change, T_e is the equilibrium temperature, and T is temperature. For a differential stress of 10 MPa and a molar volume of 50 cm^{-3} mol^{-1}, from (10.2), $\Delta\mu \approx 500$ J mol^{-1}. With $(T_e - T) = 10°C$, ΔG_r might typically vary from ~ 200 to 2000 J mol^{-1}, depending on the nature of the reaction (Ridley 1985, p. 83), and would thus enhance the stress-induced chemical potential difference by a maximum of ~ 5 (see also Fyfe 1976). With a large value of $(T_e - T)$, e.g. $> 100°C$, ΔG_r becomes very large relative to the stress-induced chemical potential difference. Thus, incongruent pressure solution may only significantly affect deformability at conditions distant from equilibrium. Such conditions are most likely to occur during retrograde metamorphism when reactions occur in response to the infiltration of fluid, in ductile shear zones for example (see Rubie 1986).

10.4.4   The rôle of fluids

It is well known that H_2O enhances rates of both intracrystalline and grain-boundary diffusion, and therefore facilitates both dislocation-accommodated and diffusion-accommodated creep. This effect may be of considerable importance when water is released during dehydration reactions. It must also be important if fluid-infiltration occurs during deformation that was originally catalysed by solid–solid reactions. Available evidence suggests that strain rates during diffusion-accommodated creep, for example, may be enhanced by orders of magnitude by the addition of H_2O (Karato *et al.* 1986, see also Rubie 1986).

10.4.5 Transformation plasticity

Transformation plasticity is a weakening which has frequently been observed in metals as they pass through a phase transformation (Edington *et al.* 1976; Poirier 1982, 1985). In the model of Greenwood & Johnson (1965) the volume change of the transformation creates internal stresses which overcome the yield strength of the material so that only a small externally applied stress is required for creep. In the models of Poirier (1982) and Paterson (1983), dislocations form as a result of stress caused by the volume change, and consequently the material deforms more easily and the strain rate increases (at constant applied stress). Volume changes in many reactions of geological importance are large. For example, the albite $\rightarrow$ jadeite + quartz reaction involves a 17 per cent volume decrease. The development of high differential stresses (400–800 MPa) during the transformation of Mg_2GeO_4 olivine to spinel has been described by Rubie & Champness (1987) and is probably related to the 8 per cent volume decrease. Potentially, transformation plasticity should therefore be of some importance in the Earth (see also Gordon 1971, Sammis & Dein 1974). However, in metals, high strains generally only develop during *repeated* cycling across a phase boundary, and it is therefore possible that the phenomenon only makes a small contribution to the total strain in rocks. Currently, there appear to be no known microstructural features which would indicate the operation of transformation plasticity in naturally deformed rocks.

10.5 Conclusions

The examples of both naturally and experimentally deformed rocks described in this chapter demonstrate that mineral reactions can greatly enhance deformability. Under some circumstances, the strain rate can increase by a factor of 10^6–10^8, at constant stress, when reactions occur (Figs 10.4 & 6).

In experimental studies, the combination of mechanical and microstructural data generally enables the deformation mechanisms which produce the enhanced deformability to be evaluated (e.g. Rutter & Brodie 1988). However, because studies of naturally deformed rocks rely on microstructural data alone, it is often difficult or impossible to determine unequivocally the important mechanisms which operate in nature. In some cases, two or more mechanisms probably operate in conjunction to enhance deformability. During dehydration reactions, for example, it is possible that important mechanisms which operate simultaneously include (a) diffusion-accommodated grain-boundary sliding of fine-grained reaction products, (b) cataclastic flow due to an elevated pore-fluid pressure, (c) enhancement of grain-boundary diffusion rates due to the presence of water produced by dehydration, and (d) transformation plasticity. To understand the rôle of one particular mechanism, further carefully designed experiments will be required. This is particularly so for incongruent pressure solution and transformation plasticity because of current uncertainties concerning their effectiveness.

In natural examples in which high strains developed under conditions of enhanced deformability, the small grain size of reaction products appears to be an important factor. In such cases, the dominant mechanism may be diffusion creep, possibly with a large contribution from grain-boundary sliding. Provided that the grain size is stabilized, this mechanism can continue to operate long after reactions have reached completion. The importance of contributions to reaction-enhanced deformability in the Earth from other mechanisms such as cataclastic flow during dehydration, incongruent pressure solution and transformation plasticity is, at present, uncertain.

Acknowledgements

I would like to thank S. Karato, S. A. F. Murrell, and E. H. Rutter for constructive reviews, and K. H. Brodie and E. H. Rutter for kindly providing micrographs and a preprint. Backscattered electron micrographs (Figs 10.9 & 10) were taken by G. E. Lloyd in the Electron Optics Unit of the Metallurgy Department at the University of Leeds. Support from the National Environment Research Council (Grant GR3/6409) is gratefully acknowledged.

References

Ashby, M. F., G. H. Edward, J. Davenport & R. A. Verrall 1978. Application of bound theorems for creeping solids and their application to large strain diffusional flow. *Acta Metall.* **26**, 1379–88.

Bauchette, B. & M. Suery (eds) 1985. *Superplasticity*. Paris: Centre National de la Recherche Scientifique.

Beach, A. 1982. Deformation mechanisms in some cover thrust sheets from the external French Alps. *J. Struct. Geol.* **4**, 137–49.

Borradaille, G. J. 1981. Particulate flow of rock and the formation of cleavage. *Tectonophysics* **72**, 305–21.

Boullier, A. M. & Y. Gueguen 1975. SP-mylonites: origin of some mylonites by superplastic flow. *Contrib. Mineral. Petrol.* **50**, 93–104.

Brearley, A. J. 1986. An electron optical study of muscovite breakdown in pelitic xenoliths during pyrometamorphism. *Mineral. Mag.* **50**, 385–397.

Brodie, K. H. & E. H. Rutter 1985. On the relationship between deformation and metamorphism with special reference to the behaviour of basic rocks. In *Advances in physical geochemistry* **4**, A. B. Thompson & D. C. Rubie (eds), 138–79. Berlin: Springer.

Brodie, K. H. & E. H. Rutter 1987. The role of transiently fine grained reaction products in syntectonic metamorphism: natural and experimental examples. *Can. J. Earth Sci.* **24**, 556–64.

Burnley, P. C. & S. H. Kirby 1982. Pressure-induced embrittlement of polycrystalline tremolite $Ca_2Mg_5Si_8O_{22}(OH_2F)_2$. *Trans Am. Geophys. Union* **63**, 1095.

Christensen, U. R. & D. A. Yuen 1985. Layered convection induced by phase transformations. *J. Geophys. Res.* **90**, 10 291–300.

Compagnoni, R. 1977. The Sesia–Lanzo Zone: high pressure – low temperature metamorphism in the Austroalpine continental margin. *Rend. Soc. Ital. Mineral. Petrol.* **33**, 281–334.

Durham, W. B., H. C. Heard & S. H. Kirby 1983. Experimental deformation of polycrystalline ice: preliminary results. *J. Geophys. Res.* **88**, B377–92.

Edington, J. W., K. N. Melton & C. P. Cutler 1976. Superplasticity. *Progr. Mat. Sci.* **21**, 61–158.

Frueh-Green, G. 1985. Stable isotope indications of fluid involvement in ductile shear zones during eclogite facies metamorphism. *Trans Am. Geophys. Union* **66**, 1126–7.

Fyfe, W. S. 1976. Chemical aspects of rock deformation. *Phil Trans R. Soc. Lond.* A **283**, 221–8.

Fyfe, W. S., N. J. Price & A. B. Thompson 1978. *Fluids in the Earth's crust*. Amsterdam: Elsevier.

Gifkins, R. C. 1976. Grain boundary sliding and its accommodation during creep and superplasticity. *Metall. Trans* 7A., 1225–32.

Gordon, R. B. 1971. Observation of crystal plasticity under high pressure with application to the Earth's mantle. *J. Geophys. Res.* **76**, 1248–54.

Green, H. W. & P. C. Burnley, 1989. A new self-organizing mechanism for deep-focus earthquakes. *Nature* **341**, 733–7.

Greenwood, G. W. & R. H. Johnson 1965. The deformation of metals under small stresses during phase transformations. *Proc. R. Soc. Lond.* A **283**, 403–22.

Heard, H. C. & W. W. Rubey 1966. Tectonic implications of gypsum dehydration. *Geol. Soc. Am. Bull.* **77**, 741–60.

Heinrich, C. A. 1982. Kyanite–eclogite to amphibolite facies evolution of hydrous mafic and pelitic rocks, Adula Nappe, Central Alps. *Contrib. Mineral. Petrol.* **81**, 30–8.

Jeanloz, R. & A. B. Thompson 1983. Phase transitions and mantle discontinuities. *Rev. Geophys. Space Phys.* **21** 51–74.

Joesten, R. 1983. Grain growth and grain boundary diffusion in quartz from the Christmas Mountains (Texas) contact aureole *Am. J. Sci.* **283A**, 233–54.

Karato, S. 1988. Defects and plastic deformation in olivine. In *Rheology of solids and of the Earth*, S. Karato & M. Toriumi (eds). Oxford: Oxford University Press.

Karato, S., M. S. Paterson & J. D. Fitzgerald 1986. Rheology of synthetic olivine aggregates: influence of grain size and water. *J. Geophys. Res.* **91**, 8151–76.

Kirby, S. H. 1987. Localized polymorphic phase transformations in high-pressure faults and applications to the physical mechanism of deep earthquakes. *J. Geophys. Res.* **92**, 13 789–800.

Koons, P. O., D. C. Rubie & G. Frueh-Green 1987. The effects of disequilibrium and deformation on the mineralogical evolution of quartz diorite during metamorphism in the eclogite facies. *J. Petrol.* **28**, 679–700.

Kronenberg, A. K. & J. Tullis 1984. Flow strengths of quartz aggregates: grain size and pressure effects due to hydrolytic weakening. *J. Geophys. Res.* **89**, 4281–97.

Martin, J. W. & R. D. Doherty 1976. *Stability of microstructure in metallic systems.* Cambridge: Cambridge University Press.

Murrell, S. A. F. 1985. Aspects of relationships between deformation and prograde metamorphism that causes evolution of water. In *Advances in physical geochemistry* **4**, A. B. Thompson & D. C. Rubie (eds), 211–41. Berlin: Springer.

Murrell, S. A. F. 1986. The role of deformation, heat, and thermal processes in the formation of the lower continental crust. In *The nature of the lower continental crust*, J. B. Dawson, D. A. Carlswell, D. A. Hall & K. W. Wedopohl (eds), 107–17. Geol. Soc. Sp. Publ. No. 24.

Murrell, S. A. F. & I. A. H. Ismail 1976. The effect of decomposition of hydrous minerals on the mechanical properties of rocks at high pressures and temperatures. *Tectonophysics* **31**, 207–58.

Oberhaensli, R., J. C. Hunziker, G. Martinotti & W. B. Stern 1985. Geochemistry, geochronology and petrology of Monte Mucrone: an example of Eo-Alpine eclogitization of Permian Granitoids in the Sesia–Lanzo Zone, Western Alps, Italy. *Chem. Geol. (Isotope Geosci. Section)* **52**, 165–84.

Olgaard, D. L. & B. Evans 1986. Effect of second-phase particles on grain growth in calcite. *J. Am. Ceram. Soc.* **69**, C272–7.

Paterson, M. S. 1973. Nonhydrostatic thermodynamics and its geologic applications. *Rev. Geophys. Space Phys.* **11**, 355–89.

Paterson, M. S. 1978. *Experimental rock deformation – the brittle field.* Berlin, Springer.

Paterson, M. S. 1983. Creep in transforming polycrystalline materials. *Mech. Mat.* **2**, 103–9.

Poirier, J.-P. 1982. On transformation plasticity. *J. Geophys. Res.* **87**, 6791–7.

Poirier, J.-P. 1985. *Creep of crystals.* Cambridge: Cambridge University Press.

Raleigh, C. B. & M. S. Paterson 1965. Experimental deformation of serpentinite and its tectonic implications. *J. Geophys. Res.* **70**, 3965–85.

Ridley, J. 1985. The effect of reaction enthalpy on the progress of a metamorphic reaction. In *Advances in physical geochemistry* **4**, A. B. Thompson & D. C. Rubie (eds), 80–97. Berlin: Springer.

Ridley, J. & J. E. Dixon 1984. Reaction pathways during the progressive deformation of a blueschist metabasite: the role of chemical disequilibrium and restricted range equilibrium. *J. Metamorph. Geol.* **2**, 115–28.

Rubie, D. C. 1983. Reaction-enhanced ductility: the role of solid–solid univariant reactions in deformation of the crust and mantle. *Tectonophysics* **96**, 331–52.

Rubie, D. C. 1984a. The olivine → spinel transformation and the rheology of subducting lithosphere. *Nature* **308**, 505–8.

Rubie, D. C. 1984b. A thermal-tectonic model for high-pressure metamorphism and deformation in the Sesia Zone, Western Alps. *J. Geol.* **92**, 21–36.

Rubie, D. C. 1986. The catalysis of mineral reactions by water and restrictions on the presence of aqueous fluid during metamorphism. *Mineral. Mag.* **50**, 399–415.

Rubie, D. C. & A. J. Brearley 1987. Metastable melting during the breakdown of muscovite + quartz at 1 kbar. *Bull. Mineral.* **110**, 533–49.

Rubie, D. C. & P. E. Champness 1987. The evolution of microstructure during the transformation of Mg_2GeO_4 olivine to spinel. *Bull. Mineral.* **110**, 471–80.

Rubie, D. C. & A. B. Thompson 1985. Kinetics of metamorphic reactions at elevated temperatures and pressures: an assessment of available experimental data. In *Advances in physical geochemistry* **4**, A. B. Thompson & D. C. Rubie (eds), 27–79. Berlin: Springer.

Rutter, E. H. 1984. The kinetics of grain coarsening in calcite rocks. In *Progress in experimental petrology* 6, C. M. B. Henderson (ed.), 245–9. The Natural Environment Research Council Publications Series D, No. 25.

Rutter, E. H. & K. H. Brodie 1985. The permeation of water into hydrating shear zones. In *Advances in physical geochemistry* 4, A. B. Thompson & D. C. Rubie (eds), 242–50. Berlin: Springer.

Rutter, E. H. & K. H. Brodie 1988. Experimental 'syntectonic' dehydration of serpentinite under conditions of controlled pore water pressure. *J. Geophys. Res.* 93, 4907–32.

Rutter, E. H., C. J. Peach, S. H. White & D. Johnson 1985. Experimental 'syntectonic' hydration of basalt. *J. Struct. Geol.* 7, 251–66.

Sammis, C. G. & J. L. Dein 1974. On the possibility of transformational superplasticity in the Earth's mantle. *J. Geophys. Res.* 79, 2961–5.

Schmid, S., J. N. Boland & M. S. Paterson 1977. Superplastic flow in fine grained limestone. *Tectonophysics* 43, 257–91.

Schmid, S. M., M. S. Paterson & J. N. Boland 1980. High temperature flow and dynamic recrystallization in Carrara marble. *Tectonophysics* 65, 245–80.

Sotin, C. & J. Tullis 1987. Interaction between plastic deformation and phase transformation: The case of albite → jadeite + quartz. *Trans Am. Geophys. Union* 68, 1454.

Tullis, J. 1990. Experimental studies of deformation mechanisms and microstructures in quartzofeldspathic rocks. This volume 190–227.

Tullis, J. & R. A. Yund 1977. Experimental deformation of dry Westerly granite. *J. Geophys. Res.* 82, 5705–18.

Tullis, J. & R. A. Yund 1982. Grain growth kinetics of quartz and calcite aggregates. *J. Geol.* 90, 301–18.

Tullis, J. & R. A. Yund 1985. Dynamic recrystallization of feldspar: a mechanism for ductile shear zone formation. *Geology* 13, 238–41.

Vaughan, P. J. & R. S. Coe 1981. Creep mechanism in Mg_2GeO_4: effects of a phase transition. *J. Geophys. Res.* 86, 389–404.

White, S. H. & R. J. Knipe 1978. Transformation- and reaction-enhanced ductility in rocks. *J. Geol. Soc. Lond.* 135, 513–16.

White, S. H., S. E. Burrows, J. Carreras, N. D. Shaw & F. J. Humphreys 1980. On mylonites in ductile shear zones. *J. Struct. Geol.* 2, 175–88.

Wilkinson, D. S. 1985. Grain size effects in superplasticity. In *Superplasticity*, B. Baudelet & M. Suery (eds), 6.1–13. Paris: Centre National de la Recherche Scientifique.

Worden, R. H., P. E. Champness & G. T. R. Droop 1987. Transmission electron microscopy of the pyrometamorphic breakdown of phengite and chlorite. *Mineral. Mag.* 51, 107–22.

CHAPTER ELEVEN

Thermodynamics of rock deformation by pressure solution

Florian K. Lehner

11.1 Introduction

Geologists are indebted to Sorby (1863) for the recognition of the fact that rock deformation in the presence of water is often accomplished by processes in which 'mechanical force is resolved into chemical action'. Sorby later coined the term 'pressure solution' in ascribing phenomena such as the pitting of pebbles (Mosher 1981) to stress-enhanced solubility, with which he was familiar from contemporary work in physical chemistry (Durney 1978). In recent work on deformation mechanisms, the terms 'solution precipitation creep' and 'solution transfer' or 'transport creep' have become customary, the former being more familiar from the metallurgical literature. In this chapter the term 'pressure solution' will be used to examine a principal mechanism of ductile rock deformation in the upper crust, which operates under diagenetic and low metamorphic grade conditions up to 200–400°C, depending upon grain size and mineralogy (Weyl 1959, Durney 1972, Elliott 1973, Rutter 1983).

Pressure solution is well documented by field observations as giving rise to a highly localized mode of rock deformation in the form of stylolitic or smooth solution seams (Stockdale 1922; Dunnington 1967; Wanless 1984, and others in the same publication) and also to a macroscopically pervasive mode in which the grains in a rock exhibit either irregular, 'microstylolitic', or smooth sutured contacts (Waldschmidt 1941; Wilson & Sibley 1978; House-knecht 1984, 1988). This pervasive intergranular mode of pressure solution gives rise to a 'fitted fabric' (Buxton & Sibley 1981) rock texture, in which individual grains typically appear flattened, presumably perpendicular to the largest compressive stress, while re-precipitation in 'pressure shadows' may be variable in importance. Other microstructural expressions of pressure solution include cleavage formation (Alvarez *et al*. 1976), fibre-coated slickensided sliding surfaces (Elliott 1976), and other indicators as discussed by Rutter (1983).

N.B. The scheme of symbols in this chapter differs from the others in this book.

This chapter deals exclusively with the deformation mechanism aspect of pervasive, intergranular pressure solution. As such, pressure solution remains poorly understood for two principal reasons. First, it is difficult to obtain experimental creep data free from ambiguities in interpretation and supported by the production of convincing microstructures. This difficulty was discussed by Rutter (1983) who, in reviewing earlier experimental work, emphasized the particular problem that arises with the low activation enthalpies characteristic of the kinetics of this process. The experimental investigations on quartz sand by de Boer *et al.* (1977) and, recently, by Gratier & Guiget (1986) have nevertheless produced clear microstructural evidence for intergranular pressure solution. However, the most striking experimental evidence has been obtained recently by Spiers & Schutjens (this volume), who studied single-crystal halite aggregates under isotropic densification creep conditions.

An interesting experiment has been carried out by Tada & Siever (1986), who subjected single crystals of halite to a knife-edge load in a saturated brine environment in order to observe the detailed evolution of pressure solution contacts and the rates of pressure solution. The response to such a loading of halite was found, not surprisingly perhaps, to be governed by a combination of crystal plastic deformation and 'free face' pressure solution of extruded material. This modified version of Bathurst's (1958) undercutting hypothesis is contrasted by these authors with Weyl's (1959) grain-boundary diffusion hypothesis. Tada & Siever recognize, however, the need to incorporate this view into a model of the sutured grain-to-grain contacts that are observed in granular aggregates. They therefore postulate a grain-boundary structure that is qualitatively indistinguishable from the 'island' or 'asperity' structure considered by Raj (1982), Lehner & Bataille (1984), Spiers & Schutjens (this volume), and that permits solid–solid contacts to transmit intergranular stresses while a fluid-solution phase is allowed to permeate the grain boundary.

This brings into focus the second reason why pressure solution has remained poorly understood, namely the lack of a thermodynamically sound, comprehensive description of grain-boundary processes based upon an accepted physical model. In almost all existing theoretical work it is taken for granted that pressure-solution creep operates by liquid-phase-enhanced grain-boundary diffusion in a manner akin to Coble creep (Coble 1963, Raj & Ashby 1971, Elliott 1973), allowing for both grain-boundary sliding and indentation (or densification) modes of creep. A recent exception is the work of Pharr & Ashby (1983), in which grain boundaries are treated as being impermeable to fluids and instead plastically deforming grain-contact neck regions are emphasized. During densification creep, these necks are continuously undercut by pressure solution and recreated by plastic deformation, in a manner that again resembles the earlier ideas of Bathurst (1958). While it has been argued by some investigators that plastic deformation along highly stressed grain-to-grain contacts should effectively disrupt any continuous grain-boundary solution film, thus creating isolated fluid inclusions, others (e.g. Rutter 1976,

1983) have invoked special strength properties for adsorbed fluid films, or have advocated the fluid-permeated island model mentioned above or an asperity structure that would support local solid–solid contact stresses. Accordingly, there exist at present rather different treatments of grain boundaries, and these reflect certain fundamental differences in the application of thermodynamics to non-hydrostatically loaded grain-to-grain contacts and in the identification of an appropriate thermodynamic driving force for the process of pressure solution.

Existing work on pressure-solution creep is concerned mostly with two-component solid/diluent systems and the simplest situation of an homogeneous solid in contact with its own aqueous solution phase, when bulk diffusion within the solid may be ignored. Consistent with the idea of diffusion-controlled creep, the transport of solute in a grain-boundary fluid phase is assumed to be driven by gradients in its chemical potential, which in turn is derived from a condition of local thermodynamic equilibrium between the grain-boundary fluid phase and the embedding non-hydrostatically stressed grain-boundary solid phase. The required equilibrium condition is usually viewed as being given by Paterson's (1973) condition

$$f^s - \sigma_n/\rho^s = \tilde{\mu}_1$$

in which f^s and ρ^s denote the mass-specific Helmholtz free energy and the density of the solid phase, respectively, $-\sigma_n$ is the (compressive) normal stress across the grain-to-grain contact, and $\tilde{\mu}_1$ denotes the mass specific chemical potential of the dissolved solid (component 1) in the grain-boundary fluid, the solvent (component 2) being water. If f^s is considered to be negligible in comparison with $-\sigma_n/\rho^s$ and is ignored, Paterson's condition assumes the form of a familiar boundary condition for Nabarro–Herring diffusion or grain-boundary diffusion. A scalar 'creep law' for diffusion-rate-limited pressure-solution creep is then readily derived for simple grain geometries by linking the simplified equilibrium condition to mass transfer and balance equations. For a plane, circular grain-to-grain contact, Rutter (1976, 1983) has obtained in this manner the unidirectional indentation creep rate

$$\dot{e} = 32\,\overline{V}^s c_1 \tilde{\rho}^f D_{gb} \bar{\sigma}_n / \rho^s R T d^3$$

where $\overline{V}^s$ is the molar volume of the solid, c_1 is the solute concentration in the pore fluid, $\tilde{\rho}^f$ is the density of the grain-boundary fluid, D_{gb} is the effective grain-boundary diffusivity (in ms^{-1}), $\bar{\sigma}_n$ is the surface-averaged grain-to-grain normal stress, R is the gas constant, T is the absolute temperature, and d is the grain diameter.

The line of reasoning leading to this creep law on the basis of Paterson's equilibrium condition has become widely accepted in the geological literature, but recently it has also been criticized, for example by Lehner & Bataille

298

(1984), for the following reasons. In the first place, Paterson's condition is based upon an *a priori* assumption of thermodynamic equilibrium for a grain-to-grain contact zone that is treated by Paterson as an 'inert loading frame' capable of transmitting an arbitrary intergranular normal stress σ_n while permitting the loaded solid phase to dissolve in a grain-boundary fluid phase at pressure p. The main interest in this conceptual device derives from its physical· interpretation as a fluid-permeated grain-to-grain contact zone that enables two adjacent grains to be in solid–solid contact across islands or asperities. If this interpretation is made, however, the stresses and also the strain energy densities in the solid material within this zone must be highly inhomogeneous, so that interfacial equilibrium between the solid and the grain-boundary fluid will be impossible, as was shown by Gibbs in 1876. Instead, there will be local dissipative processes of dissolution, diffusive transport and re-precipitation which, if a grain contact zone were isolated and allowed to reach equilibrium, would tend to smooth out irregularities and eventually to trap any mobile fluid phase in isolated inclusions, equilibrated individually with the surrounding solid. Using transmission electron microscopy, White & White (1981) have indeed been able to demonstrate the existence of such bubble arrays at grain boundaries, and similar arrays of isolated fluid inclusions are known to form from fluid-filled microcracks. The latter observation, as Urai *et al.* (1986) also point out, refers to an evolution towards equilibrium, in the course of which a continuous fluid film breaks up into isolated inclusions. The fact that here one is dealing with a transient state has been emphasized by Urai (1983), who cautioned that the bubble arrays observed by White & White (1981) may indeed be typical for the final state in the evolution of a grain boundary, while a continuous liquid phase may be present during deformation (see also Spiers & Schutjens, this volume).

In this chapter a theoretical treatment of pressure solution in porous and permeable rocks is presented, in the spirit of the classical thermodynamic theory of irreversible processes. The first section is intended as a reasonably self-contained review of the fundamental theoretical relationships that are assumed to characterize the thermodynamics of a pure solid substance, of its own aqueous solution phase, and of the phase boundary. This is followed by a derivation of a macroscopic Gibbs equation for a granular aggregate that exhibits inelastic creep due to intergranular pressure solution. A technique of spatial averaging will be used for this purpose, which serves to clarify the manner in which internal displacements of grain boundaries and pore walls and their conjugate thermodynamic forces enter into a macroscale continuum theory of the 'internal variable' type for a representative rock sample. The development of macroscopic theory is not pursued in this chapter beyond establishing a Gibbs equation. While there exists a general framework in the work of Rice (1971, 1975, 1977) and of Heidug (1985), which could guide the further development of a macroscopic constitutive theory of pressure-solution-controlled deformation, such a development would remain somewhat formal without the prior resolution of the existing difficulties in the description of grain-boundary processes. The third section of this chapter therefore concentrates on this kinetic aspect of the theory. In keeping with the phenomenological approach, this section offers a thermodynamic analysis of pressure solution in a fluid-infiltrated representative volume element of a grain-to-grain contact zone, leading to the formulation of phenomenological rate equations for grain-boundary indentation displacements. This analysis also aims to clarify the nature of the thermodynamic driving forces in the limiting cases of interface-reaction-controlled and diffusion-controlled pressure solution. The chapter closes with the construction of a macroscopic 'creep law' for dense aggregates that remain permeable to fluids while compacting along grain boundaries or larger-scale pressure-solution seams.

11.2   Microscale continuum theory of constituent behaviour

The fundamental entity selected for study consists of a representative elementary volume (REV) of rock, e.g. a granular sandstone. At some arbitrarily

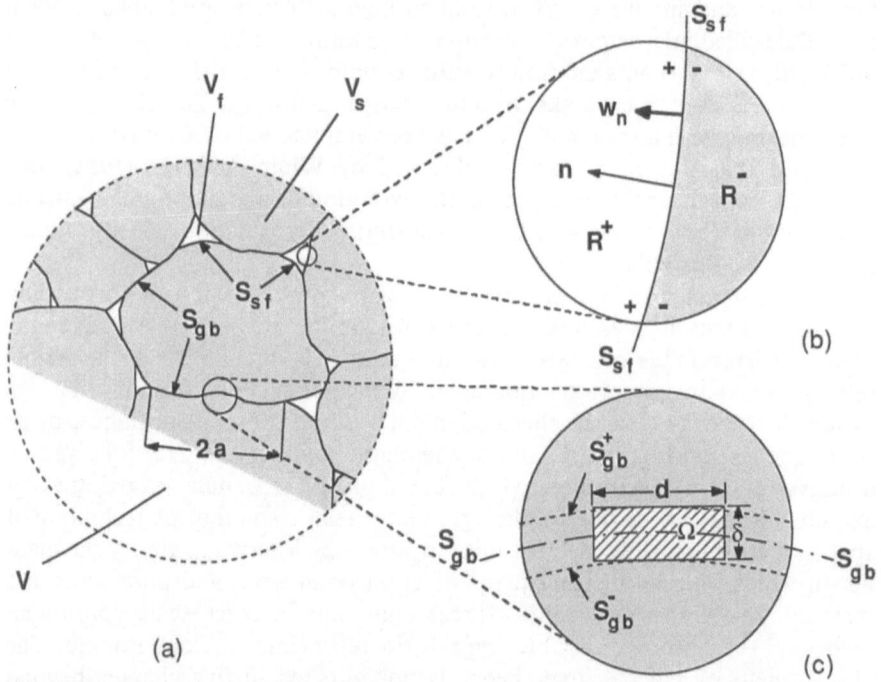

Figure 11.1   (a) Grain-scale view of representative elementary volume, V, of grain aggregate, consisting of solid matrix V_s and fluid-saturated interconnected pore space V_f, which may extend (not visible on this scale) into grain-to-grain contacts S_{gb}. Enlargements serve to define (b) pore walls (solid/fluid-phase boundary S_{sf}) and (c) control volume Ω for contact zone (shown in detail in Fig. 11.2).

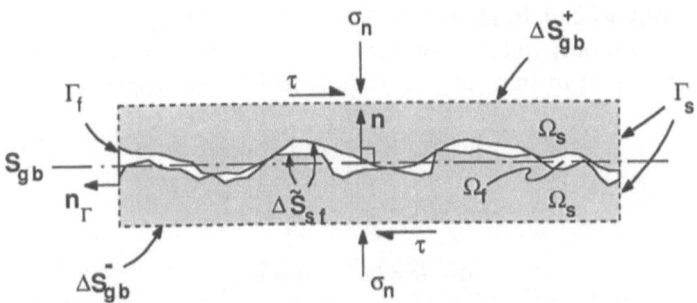

Figure 11.2 Fine-scale view of section through pillbox-shaped representative elementary volume Ω of a contact zone (see Fig 11.1); solid material occupies Ω_s, and interstitial aqueous solution occupies interconnected region Ω_f. Note the assumed 'island' or 'asperity' structure.

fixed instant such an REV may be identified, as illustrated in Figure 11.1, by cutting out a sufficiently large region of the rock and by using solid material 'marker points' lying in the smooth intersections of this cut with individual grains to define the geometry of the boundary of the REV. The circumference of an REV is thus to be viewed as a material boundary for the solid phase contained within the REV, but since part of it cuts across an effective pore space saturated by mobile fluids, the boundary of an REV is essentially semi-permeable, i.e. impermeable to the solid phase and permeable to the fluid phase.

It will be appropriate to distinguish three scales of description in this discussion. On the coarsest scale, one is concerned with the macroscopic, or bulk, behaviour of an element of rock material. The local values of field variables defined on this 'macroscale' will be constructed as spatial averages, taken over the REV, of fields on the next finer scale, which is the 'grain scale' shown in Figure 11.1a. For consideration of pore walls and grain-to-grain contacts this picture must be further enlarged, so as to permit a theoretical treatment of processes on an appropriate 'fine scale' or 'microscale', which is the scale of Figures 11.1b and 11.1c and of Figure 11.2. The fine-scale features of Figure 11.2 include a pure solid substance that coexists with an aqueous solution of the solid within a certain grain-to-grain contact zone. This fluid phase is mobile, although the presence of an adsorbed film of water on the grain periphery is not excluded as long as no peculiar effects on the processes to be studied are ascribed to this film.

11.2.1 Balance equations and thermodynamic preliminaries

All theoretical developments in this chapter are undertaken within a framework of continuum thermodynamics, and start from a set of local balance equations for mass, partial mass of a component, momentum, energy, and entropy. On the microscale these must hold for any mass point in the interior of the regions V_f and V_s shown in Figure 11.1a. If body forces (gravity) are disregarded and only quasi-static processes are admitted, they take the form (the notation used in this chapter is explained in the Appendix)

$$\partial_t \rho + \text{div } \rho \mathbf{v} = 0, \tag{11.1}$$

$$\partial_t \rho_i + \text{div } \rho_i \mathbf{v} + \text{div } \mathbf{J}_i = 0, \tag{11.2}$$

$$\text{div } \boldsymbol{\sigma} = \mathbf{0}, \qquad \boldsymbol{\sigma} = \boldsymbol{\sigma}^T, \tag{11.3}$$

$$\partial_t(\rho u) + \text{div}(\rho u \mathbf{v} - \boldsymbol{\sigma} \cdot \mathbf{v} + \mathbf{q}) = 0, \tag{11.4}$$

$$\partial_t(\rho s) + \text{div}(\rho s \mathbf{v} + \boldsymbol{\Phi}) = \gamma \tag{11.5}$$

Equations (11.4) and (11.5) also yield a balance equation for the Helmholtz free energy, $f = u - Ts$. Under isothermal conditions, which are considered below, this equation reads

$$\partial_t(\rho f) + \text{div}(\rho f \mathbf{v} - \boldsymbol{\sigma} \cdot \mathbf{v} + \mathbf{q} - T\boldsymbol{\Phi}) = -T\gamma \leqslant 0 \tag{11.6}$$

Here $\rho(\mathbf{x}, t)$, $\rho_i(\mathbf{x}, t)$, $u(\mathbf{x}, t)$, $s(\mathbf{x}, t)$, $\boldsymbol{\sigma}(\mathbf{x}, t)$, $\mathbf{q}(\mathbf{x}, t)$, $\boldsymbol{\Phi}(\mathbf{x}, t)$, $\gamma(\mathbf{x}, t)$, and T denote the fields of the mass density, partial mass density of component i, specific internal energy, specific entropy, Cauchy stress, energy flux vector, entropy flux vector, rate of entropy production per unit volume, and the temperature. The second law of thermodynamics is expressed by the inequality in (11.6). The diffusive mass flux vector $\mathbf{J}_i$ of the ith component, appearing in (11.2), is defined by

$$\mathbf{J}_i = \rho_i(\mathbf{v}_i - \mathbf{v}) = \rho c_i(\mathbf{v}_i - \mathbf{v}) \tag{11.7}$$

in terms of the component velocity $\mathbf{v}_i$, the barycentric velocity $\mathbf{v} = \Sigma_i c_i \mathbf{v}_i$, and the mass fraction $c_i = \rho_i/\rho$. For n components these must satisfy the relations

$$\sum_{i=1}^{n} \mathbf{J}_i = 0 \qquad \text{and} \qquad \sum_{i=1}^{n} c_i = 1 \tag{11.8}$$

In this chapter the pore fluid is assumed to be composed of the neutral 'solvent' H_2O and a binary electrolyte, i.e. the 'solute' consisting of the dissolved solid material. In the absence of an electric current, the solute and the solvent may thus be treated as component 1 and component 2, in the sense of the phase rule, of a binary mixture. The solid phase is assumed to be pure and homogeneous and consists solely of component 1 in the solid state.

An essential assumption implicit in (11.5) and (11.6) in the classical theory of non-equilibrium thermodynamics postulates the existence of a local thermodynamic state (see De Groot & Mazur 1962). According to this assumption, a set of values u, ρ, c_i at any point within the *fluid phase* determines well defined local values of the temperature T, the pressure p, and the mass-specific chemical potential μ_i, and thereby a well defined value of the specific entropy s as the associated thermodynamic potential. In other words, the fluid is assumed to be locally in thermodynamic equilibrium. Changes in s are thus related to those in u, ρ, and c_i by the Gibbs equation

$$T \, \mathrm{d}s = \mathrm{d}u + p \, \mathrm{d} \frac{1}{\rho} - \sum_{i=1}^{2} \mu_i \, \mathrm{d}c_i \tag{11.9}$$

Equivalently, the thermodynamic relationship

$$\sum_{i=1}^{2} c_i \mu_i = f + p \frac{1}{\rho} \tag{11.10}$$

leads to the Gibbs equation in the Helmholtz free energy density

$$\mathrm{d}(\rho f) = \sum_{i=1}^{2} \mu_i \, \mathrm{d}\rho_i \tag{11.11}$$

under the assumption of isothermal conditions. The entropy flux vector $\boldsymbol{\Phi}$ is given by (Meixner & Reik 1959, De Groot & Mazur 1962)

$$\boldsymbol{\Phi} = \begin{cases} \mathbf{q}/T & \text{in the solid phase} \\[2ex] \dfrac{\mathbf{q}}{T} - \displaystyle\sum_{i=1}^{2} \frac{\mu_i}{T} \mathbf{J}_i & \text{in the fluid phase} \end{cases} \tag{11.12}$$

The rate of dissipation per unit volume of the fluid phase is then found from (11.6) with the aid of the last expression for $\boldsymbol{\Phi}$ and the Gibbs equation (11.11)

$$T\gamma = (\boldsymbol{\sigma} + p\mathbf{1}) : \operatorname{grad} \mathbf{v} - \sum_{i=1}^{2} \mathbf{J}_i \cdot \operatorname{grad} \mu_i \tag{11.13}$$

This result, together with the inequality in (11.6), furnishes an example of the general expression

$$T\gamma = \sum_{k=1}^{m} X_k Y_k \geqslant 0$$

for the non-negative rate of dissipation, which occupies a central position in the classical theory of irreversible processes (De Groot & Mazur 1962). The theory is concerned mostly with establishing functional relationships between the thermodynamic forces X_k and fluxes Y_k when these vanish simultaneously in equilibrium. The only result needed here is the linear phenomenological law governing binary diffusion

$$\mathbf{J}_1 = -\mathbf{J}_2 = -l \; \mathrm{grad}(\mu_1 - \mu_2)$$

where $l > 0$ is a phenomenological coefficient. The Gibbs–Duhem relationship

$$\sum_{i=1}^{2} c_i (\mathrm{d}\mu_i)_T = \frac{1}{\rho} \, \mathrm{d}p \qquad (11.14)$$

which may be deduced from (11.10) and (11.11), implies that $\mathrm{grad}(\mu_1 - \mu_2) = (1 - c_1)^{-1} \, \mathrm{grad} \, \mu_1 - [\rho(1 - c_1)]^{-1} \, \mathrm{grad} \, p$. In the present context the microscale fluid pressure gradients will be negligible in comparison with chemical potential gradients, thus giving

$$\mathbf{J}_1 = -\frac{l}{1 - c_1} \; \mathrm{grad} \; \mu_1 \qquad (11.15)$$

The theory of solutions yields the following expression for the mass-specific chemical potential of the solute (see, for example, Denbigh 1971):

$$\mu_1 = \mu_1^0(p, T) + RT/M_1 \ln c_1 \gamma_1 \qquad (11.16)$$

where $\mu_1^0 \, (p, T)$ is a function independent of concentration, R is the gas constant, M_1 is the molar mass of the solute, and γ_1 is an activity coefficient. This allows (11.15) to be written in the familiar form of Fick's law as

$$\mathbf{J}_1 = -\rho D \; \mathrm{grad} \; c_1 \qquad (11.17)$$

where the diffusion coefficient D (in $\mathrm{m}^2\mathrm{s}^{-1}$) is related to the coefficient l by

$$D = \frac{l(\partial \mu_1 / \partial c_1)_{p, T}}{\rho(1 - c_1)} = \frac{lRT}{\rho M_1 c_1 (1 - c_1)} \left[1 + \frac{c_1 \, \mathrm{d}(\ln \gamma_1)}{\mathrm{d}c_1} \right] \qquad (11.18)$$

For small concentrations D may be approximated by $lRT/(\rho M_1 c_1)$.

The thermomechanical behaviour of the *solid phase* will be characterized as inelastic. This behaviour can be modelled in a general way by allowing the specific Helmholtz free energy f of the solid to depend upon an unspecified number of scalar internal state variables, ξ_α, $\alpha = 1, ..., N$. Here N denotes the number of such variables needed to define the thermodynamic state of a given material sample, i.e. to determine a unique value of f at given values of the temperature T and strain $\mathbf{E}$, as measured with respect to some fixed reference state. Thus, f is assumed to satisfy a fundamental equation of the form

$$f = f(T, \mathbf{E}, \xi_1, ..., \xi_N) \qquad (11.19)$$

Each variable ξ_α measures the extent of some local structural rearrangement at one of N distinct sites. The use of the vector $\boldsymbol{\xi} = \{\xi_1, ..., \xi_N\}$ as a state variable leads to a simple thermodynamic framework that, despite certain limitations, has led to important advances in the theory of viscoelasticity and metal plasticity (Meixner 1960; Kestin & Rice 1970; Rice 1971, 1975). A brief summary of this formalism will now be given to provide the necessary background for the discussion of thermodynamic equilibrium along phase boundaries. The same formalism will reappear later and receive a specific interpretation in the description of the macroscopic, aggregate behaviour of a porous medium deforming by pressure-solution creep.

The introduction of internal state variables leads to the concept of constrained equilibrium states. According to this concept, each internal variable ξ_α is assumed to evolve in time at a finite rate as governed by some internal dissipative relaxation process. An appropriate kinetic rate equation describing this process must therefore be added to (11.19) to complete the thermodynamic description of the system, i.e. a representative sample of material. Instantaneous values of $\boldsymbol{\xi}$ may then be viewed as being maintained by an appropriate set of constraints, which allow the solid to attain a state of constrained equilibrium such that $\boldsymbol{\xi}$ together with T and $\mathbf{E}$ determine a well defined value of the specific free energy f. Here the rate of change of $\boldsymbol{\xi}$ is thought to be governed by specific kinetic processes that operate at the sites characterized by $\boldsymbol{\xi}$. The imagined constraints thereby receive some physical justification in terms of barriers to be overcome in thermally activated processes.

Associated with (11.19) is a Gibbs equation for the free energy of a mass $\rho_0 V_0$, where ρ_0 and V_0 denote the bulk density and volume of the sample in an arbitrarily fixed reference state. For isothermal conditions this equation becomes

$$d(\rho_0 V_0 f) = V_0 \mathbf{T} : d\mathbf{E} - \sum_{\alpha=1}^{N} A_\alpha \, d\xi_\alpha \qquad (11.20)$$

where N denotes the number of internal variables needed to characterize the (homogeneous) state of the mass $\rho_0 V_0$. In addition,

$$\mathbf{T} = \rho_0 \left. \frac{\partial f}{\partial \mathbf{E}} \right|_{T, \, \boldsymbol{\xi}} \qquad (11.21)$$

is the stress, defined at fixed $\boldsymbol{\xi}$ as in thermoelasticity, and

$$A_\alpha = - \rho_0 V_0 \left. \frac{\partial f}{\partial \xi_\alpha} \right|_{T, \mathbf{E}, \xi_\beta} \qquad (11.22)$$

is the thermodynamic force or 'affinity' (at fixed values of all ξ_β, $\beta \neq \alpha$) associated with ξ_α.

The tensor $\mathbf{E}$ appearing in (11.19)–(11.21) represents the Lagrange strain, as defined by

$$\mathbf{E} = \tfrac{1}{2}(\mathbf{F}^\mathrm{T} \cdot \mathbf{F} - 1) \qquad (11.23)$$

in terms of the deformation gradient $\mathbf{F} = \partial x(\mathbf{X}, t)/\partial \mathbf{X}$, where the function $\mathbf{x} = \mathbf{x}(\mathbf{X}, t)$ describes the motion of the solid by tracing the spatial position $\mathbf{x}$ of material 'particles' labelled by their initial position $\mathbf{X}$. The stress $\mathbf{T}$ is the work conjugate of $\mathbf{E}$, i.e. the symmetric Piola–Kirchhoff stress, as defined in terms of the Cauchy stress σ by

$$\mathbf{T} = J\mathbf{F}^{-1} \cdot \sigma \cdot (\mathbf{F}^{-1})^\mathrm{T} \qquad (11.24)$$

where $J = \det \mathbf{F} > 0$.

The fact that the increment $\mathrm{d}\xi_\alpha$ in (11.20) cannot be controlled by any external manipulation of a sample of material clearly indicates the need for additional, kinetic relations. These can be explored to some extent within a purely thermodynamic framework, as has been shown by the authors cited above and by others. The key is again held by the expression for the dissipation rate that is obtained by substituting the Gibbs equation (11.20) into the balance equation (11.6). To accomplish this, (11.6) will first be cast in terms of a referential or 'material' description, corresponding to a change from the spatial $\mathbf{x}$ co-ordinate system to the referential (material) $\mathbf{X}$ co-ordinate system. Let V accordingly denote the current volume of a material element that occupies the volume V_0 in a chosen reference configuration. Their ratio satisfies

$$V/V_0 = \rho_0/\rho = J \qquad (11.25)$$

Let $\dot{J}$, $\dot{\mathbf{F}}$, etc. denote the 'material' time derivatives of $J(\mathbf{X}, t)$, $\mathbf{F}(\mathbf{X}, t)$, etc., performed while following the motion of a fixed material element or 'mass point', i.e. holding the position $\mathbf{X}$ fixed. It can be shown (see Chadwick 1976) that

$$\dot{J} = J\dot{\mathbf{F}} : \mathbf{F}^{-1} = J \operatorname{div} \mathbf{v} \qquad (11.26)$$

This also implies that

$$(J\psi)^{\boldsymbol{\cdot}} = \partial_t(J\psi) + \mathbf{v} \cdot \operatorname{grad}(J\psi) = J\{\partial_t\psi + \operatorname{div}\,\psi\mathbf{v}\} \qquad (11.27)$$

where ψ represents the scalar density of any extensive quantity dealt with in the present context. With the aid of these kinematic relationships, and after substitution of the expression $\mathbf{q}/T$ for the entropy flux, (11.6) can now be written in the form

$$(\rho_0 f)^{\boldsymbol{\cdot}} - \mathbf{T} : \dot{\mathbf{E}} = -JT\gamma \qquad (11.28)$$

Here the identity $\operatorname{div}(\boldsymbol{\sigma} \cdot \mathbf{v}) = \mathbf{v} \cdot \operatorname{div} \boldsymbol{\sigma} + \boldsymbol{\sigma} : \operatorname{grad} \mathbf{v}$ has been used, together with the fact that $\operatorname{div} \boldsymbol{\sigma} = \mathbf{0}$ and $J\boldsymbol{\sigma} : \operatorname{grad} \mathbf{v} = \mathbf{T} : \dot{\mathbf{E}}$.

A comparison of (11.20) and (11.28) shows that the rate of dissipation per unit volume in the reference state is, in the present case,

$$JT\gamma = \frac{1}{V_0} \sum_{\alpha=1}^{N} A_\alpha \dot{\xi}_\alpha \geqslant 0 \tag{11.29}$$

In the absence of permanent constraints, both A_α and $\dot{\xi}_\alpha$ must vanish in a state of unconstrained equilibrium. The inequality therefore demands a functional dependence of the $\dot{\xi}_\alpha$ upon the set of thermodynamic forces $\mathbf{A}$, corresponding to the above kinetic rate equations. Given these, together with the equations of state (11.21) and (11.22), the thermomechanical behaviour of a sample of solid material is completely specified.

11.2.2 Jump conditions at the solid/fluid phase boundary

Equations (11.1)–(11.6) are assumed to hold at points in the interior of the phases, but they generally fail to hold at points located on a phase boundary such as S_{sf} in Figure 11.1. At such boundaries certain thermodynamic variables will exhibit steep gradients in the normal direction across a narrow transition layer. In the following these are treated on the level of a continuum description as finite jump discontinuities in the relevant variable across the 'singular surface' S_{sf}. Thus, if $\psi(\mathbf{x}, t)$ denotes some function, continuous in some regions R^+ and R^- that extend from the positive and negative side of S_{sf}, as oriented by its surface normal $\mathbf{n}$, the jump in $\psi(\mathbf{x}, t)$ across S_{sf} is defined – using the customary bracket notation – by

$$[\psi](\mathbf{x}, t) = \psi^+(\mathbf{x}, t) - \psi^-(\mathbf{x}, t), \qquad \mathbf{x} \text{ on } S_{sf}$$

where the superscripts denote the values attained by the variable $\psi(\mathbf{x}, t)$ on the positive and negative sides respectively of the oriented interface S_{sf}. With this definition, the balance equations (11.1)–(11.5) are now supplemented by equivalent balance statements expressed in the form of jump conditions that must hold at points on the phase boundary (for a derivation, see, for example, Chadwick 1976). Under isothermal and quasistatic conditions these have the form

$$[\rho(w_n - \mathbf{v} \cdot \mathbf{n})] = 0, \tag{11.30}$$

$$[\rho_i(w_n - \mathbf{v} \cdot \mathbf{n}) - \mathbf{J}_i \cdot \mathbf{n}] = 0, \tag{11.31}$$

$$[\mathbf{n} \cdot \boldsymbol{\sigma}] = 0, \tag{11.32}$$

$$[\rho(w_n - \mathbf{v} \cdot \mathbf{n})u + \mathbf{n} \cdot \boldsymbol{\sigma} \cdot \mathbf{v} - \mathbf{q} \cdot \mathbf{n}] = 0, \tag{11.33}$$

$$[\rho(w_n - \mathbf{v} \cdot \mathbf{n})s - \boldsymbol{\Phi} \cdot \mathbf{n}] = -\sigma \leqslant 0 \tag{11.34}$$

Here w_n denotes the speed of displacement of the phase boundary in the direction of its surface normal. Condition (11.34) is the only condition that contains a surface excess term, the superficial rate of entropy production σ that characterizes the kinetics of solution and precipitation and must be nonnegative. No other surface-specific terms appear on the right-hand sides of these jump conditions, because all other intrinsic properties of the interface S_{sf}, including interfacial tension, and all interface transport processes (such as surface diffusion) are disregarded in the present analysis. As will be seen subsequently, however, a more general form of these jump conditions, allowing for interfacial transport, will apply to fluid-infiltrated grain boundaries when these are viewed as interfaces on a continuum scale.

The jump condition corresponding to (11.6) is obtained directly upon multiplying (11.34) by T and subtracting the result from (11.33), giving

$$[\rho(w_n - \mathbf{v} \cdot \mathbf{n})f + \mathbf{n} \cdot \boldsymbol{\sigma} \cdot \mathbf{v} - \mathbf{n} \cdot (\mathbf{q} - T\boldsymbol{\Phi})] = T\sigma \geqslant 0 \qquad (11.35)$$

Using the continuity of w_n and $\mathbf{n} \cdot \boldsymbol{\sigma}$ and writing $\{\rho(w_n - v_n)\} = \rho^+(w_n - v_n^+) = \rho^-(w_n - v_n^-)$, one has

$$[\mathbf{n} \cdot \boldsymbol{\sigma} \cdot \mathbf{v}] = \sigma_n[v_n] + \tau[v_t] = -\{\rho(w_n - v_n)\}\sigma_n\left[\frac{1}{\rho}\right] + \tau[v_t]$$

where σ_n and τ are the normal and tangential components of the traction $\mathbf{n} \cdot \boldsymbol{\sigma}$ acting on S_{sf}, and v_n and v_t are the components of $\mathbf{v}$ in the same directions. Moreover, in this discussion the entropy flux $\boldsymbol{\Phi}$ is given by (11.12), in accordance with the constraint that there be no diffusion of fluid-phase components into the solid. By defining the positive side of S_{sf} as facing the fluid phase, it follows from (11.10), (11.12), and (11.31) that

$$[\mathbf{n} \cdot (\mathbf{q} - T\boldsymbol{\Phi})] = \sum_{i=1}^{2} \mu_i \mathbf{J}_i^+ \cdot \mathbf{n} = \sum_{i=1}^{2} \mu_i[\rho_i(w_n - v_n)]$$

$$= \{\rho(w_n - v_n)\} \sum_{i=1}^{2} \mu_i[c_i] = \{\rho(w_n - v_n)\}\left(\sum_{i=1}^{2} \mu_i c_i - \mu_1\right)$$

$$= \{\rho(w_n - v_n)\}(f^+ + p/\rho^+ - \mu_1)$$

where use has been made of the fact that $c_1^- = 1 - c_2^- = 1$. Substitution of these expressions in (11.35) now yields

$$\{\mu_1 - f^- + \sigma_n/\rho^- - (\sigma_n + p)/\rho^+\}\{\rho(w_n - v_n)\} + \tau[v_t] = T\sigma \geqslant 0$$

This condition equates the work terms in the left-hand side to the total rate of dissipation by corresponding interfacial kinetic processes. Here the term $\tau[v_t]$ may safely be ignored, since, in the present context of exceedingly slow fluid motion, fluid-phase dissipative stresses at the phase boundary contribute negligibly to the total interfacial entropy production during solution and precipitation. When this assumption is made, σ_n is equated to $-p$ and the result is

$$(\mu_1 - f^s - p/\rho^s) \{\rho(w_n - v_n)\} = T\sigma \geqslant 0 \qquad (11.36)$$

where f^s and ρ^s denote the quantities f^- and ρ^- associated with the solid phase. This condition can be interpreted (see Lehner & Bataille 1984) within the framework of non-equilibrium thermodynamics as determining the thermodynamic force $X := \mu_1 - f^s - p/\rho^s$ and conjugate flux $Y := \{\rho(w_n - v_n)\}$ which govern the solution/precipitation process. As was discussed by these authors, the inequality (11.36) implies that X and Y must be functionally related and vanish simultaneously in equilibrium, when $T\sigma = 0$. Close enough to equilibrium, the smooth function $Y(X)$ may be linearized to yield

$$\{\rho(w_n - v_n)\} = L(\mu_1 - f^s - p/\rho^s) \qquad (11.37)$$

where $L > 0$ is a phenomenological coefficient that may depend on temperature and further variables determining the state of the phase boundary. It follows that phase equilibrium at the boundary between a pure solid and a binary solution of the solid in a neutral solvent is characterized by the condition

$$\mu_1 = f^s + p/\rho^s \qquad (11.38)$$

which was first obtained by Gibbs in 1876 (Gibbs 1906, Equation 388) by his classical variational method, based on a minimal energy principle. Condition (11.38) pertains to the special case considered by Gibbs among more general situations in which the phase boundary remains impermeable to any solvent components and where solid state diffusion is not taken into account. In the modern view, these constraints express the observation that certain processes are slowed down sufficiently (for example, at low enough temperatures) to appear as inhibited during a time-span of interest. Failure to recognize the existence of constrained equilibria has led to some confusion in the past concerning the possible existence of equilibrium states in non-hydrostatically stressed solids. A detailed treatment of constraints may be found in recent work by Mullins & Sekerka (1985) on the thermodynamics of crystalline solids, which deals with systems composed of both mobile and immobile species such as interstitial solutes or point defects.

A further example of the operation of constraints within a phase has already been encountered in the above characterization of the solid phase. The presence of these constraints implies that the value of f^s in condition (11.38) may correspond to a constrained equilibrium state within an inelastic solid. This provides the justification for including energies associated with crystal plastic deformation in f^s, as has occasionally been discussed (Bosworth 1981, Urai 1983, Green 1984). A strained solid of the type described may be imagined to undergo a slow stress relaxation process while being kept in local contact equilibrium with a solvent by μ_1 and p being adjusted at a given interfacial location so that (11.38) is satisfied continuously while f^s and ρ^s are allowed to evolve through a sequence of constrained local equilibria. An important point concerning condition (11.38), therefore, is that the vanishing of the thermodynamic force $\mu_1 - f^s - p/\rho^s$ prevents the interfacial process of solution or precipitation from occurring locally, but remains perfectly compatible with entropy, producing non-equilibrium processes at interior points of the phases.

To ensure a global state of unconstrained equilibrium in some solid/fluid two-phase system, condition (11.38) must necessarily be satisfied along the entire phase boundary. Global unconstrained equilibrium is of little importance, however, since only one stable case has been shown to exist, i.e. that of hydrostatically stressed homogeneous and isotropic phases (Gibbs 1906, Kamb 1961). However, global constrained equilibria in solid/fluid two-phase systems are of particular interest in the present context. Indeed, on a global level interfacial rate processes and also rates of diffusive transport through the fluid phase may be treated on an equal footing with relaxation processes in the interior of the solid phase, and it is this enlarged scope of internal variable theories that will become more apparent below.

11.3 Thermodynamic framework for a macroscale theory of aggregate behaviour

The preceding microscale phenomenological description must now be translated into a useful macroscale theory of the aggregate behaviour of a representative sample of a porous sedimentary rock that exhibits pressure-solution creep behaviour. A method well suited to this task is that of spatial averaging. Several variants of this technique have been used in the literature on transport through porous media (for example, Slattery 1972), the rheology of suspensions (Batchelor 1970, Brenner 1970), and the mechanics of heterogeneous solids (Hill 1972). In the simplest version, spatial averages are defined as analytically convenient volume averages over a representative elementary volume (REV) of the material. This volume is typically large in comparison with the microscale dimensions of grains or pores, but small enough to enable volume-averaged variables to function as meaningful macroscale field variables. The values of volume-averaged variables are assumed to be independent of the size of the REV on a local macroscopic scale, reflecting a postulated property of local statistical homogeneity of the material or two-phase medium. Here it will suffice to refer to the literature (see Beran 1968) for a precise discussion of these assumptions, and also to note that their validity must ultimately be judged by the practical consequences of the theory founded on them. It should also be recognized that volume averages may alternatively, and more appropriately, be replaced by equivalent surface averages when the variables to be averaged represent surface densities of fluxes (Brenner 1970).

The method of volume averaging involves a few essential definitions and analytical steps. These will be dealt with briefly for a single *generic* balance equation. The intended outcome is a macroscale description for a solid aggregate, in which grain boundary displacements appear in the rôle of internal variables with identifiable conjugate thermodynamic forces. The results extend those obtained in earlier work by Lehner & Bataille (1984) and Heidug (1985).

11.3.1 Spatial averaging of microscale balance equations

Consider the *generic* balance equation

$$\partial_t \psi + \mathrm{div}(\psi \mathbf{v} - \mathbf{P} \cdot \mathbf{v} + \mathbf{f}) + r = 0 \qquad (11.39)$$

which is assumed to hold at interior points of the phases, and the associated *generic* jump condition

$$[\psi(w_n - \mathbf{v} \cdot \mathbf{n}) + \mathbf{n} \cdot (\mathbf{P} \cdot \mathbf{v} - \mathbf{f})] = \Delta \qquad (11.40)$$

which holds along the solid/fluid phase boundary S_{sf}. The variables occurring in these balances may be so interpreted as to recover any one of the specific balances (11.1)–(11.6), (11.30)–(11.34), and (11.35) or (11.36). These interpretations will be needed subsequently and are summarized in Table 11.1.

The volume averages that will subsequently serve as macroscale variables all involve integration over subregions of an REV that are occupied by individual phases. Here V_s and V_f are used to denote both the open regions occupied by the phases and also their volumes. They exclude the narrow contact zones between individual grains (see Figure 11.2), which are assumed to contribute

Table 11.1 Interpretation of generic variables in equations (11.39) and (11.40).

Balanced quantity	ψ	$\mathbf{P}$	$\mathbf{f}$	r	Δ
total mass	ρ	$\mathbf{0}$	$\mathbf{0}$	0	0
mass of component i	ρ_i	$\mathbf{0}$	$\mathbf{J}_i$	0	0
momentum*	$\mathbf{0}$	$\mathbf{0}$	σ	$\mathbf{0}$	$\mathbf{0}$
free energy	ρf	σ	$\begin{cases} 0 \text{ in solid} \\ \sum \mu_i \mathbf{J}_i \text{ in} \\ \text{fluid} \end{cases}$	$T\gamma$	$T\sigma$

*Only quasi-static processes are considered.

negligibly to the total volume V of the REV; thus $V = V_s + V_f$. Accordingly,

$$\phi_\nu = V_\nu/V, \qquad \nu = s, f \qquad (11.41)$$

denotes the phase volume fractions, which satisfy $\phi_s + \phi_f = 1$.

The quantity defined by

$$\psi_\nu := \frac{1}{V} \int_{V_\nu} \psi \, dv, \qquad \nu = s, f \qquad (11.42)$$

is called a 'phase average', while

$$\psi^\nu := \frac{1}{V_\nu} \int_{V_\nu} \psi \, dv, \qquad \nu = s, f \qquad (11.43)$$

is often referred to as the 'intrinsic phase average'. These conditional averages are related simply by $\psi_\nu = \phi_\nu \psi^\nu$. The unconditional (or 'bulk') volume average defined by

$$\psi := \frac{1}{V} \int_V \psi \, dv = \sum_\nu \psi_\nu = \sum_\nu \phi_\nu \psi^\nu, \qquad \nu = s, f \qquad (11.44)$$

is simply written without a suffix, but no confusion should arise from the omission of a special notation for averaged variables, since the distinction between microscale and macroscale variables will always be clear from the context. All averages are assumed to behave as smooth point functions on the macroscale.

When operation (11.42) is applied to equation (11.39) for the solid phase, for example, a characteristic problem arises with the conversion of phase averages of spatial or time derivatives into derivatives of phase averages. These operations do not generally commute, but are related by the following two theorems: the transport theorem (see Chadwick 1976), which for the region V_s with fixed outer boundary states that

$$\int_{V_s} \partial_t \psi \, dv = \partial_t \int_{V_s} \psi \, dv - \int_{S_{sf}} \psi^- w_n \, da + \int_{S_{gb}} [\psi] w_n \, da \qquad (11.45)$$

and the related 'averaging theorem' (see Slattery 1972), as applied to the divergence of some tensor field $\mathbf{Q}$ of rank 1 or 2,

$$\int_{V_s} \text{div} \, \mathbf{Q} \, dv = \text{div} \int_{V_s} \mathbf{Q} \, dv + \int_{S_{sf}} \mathbf{n} \cdot \mathbf{Q}^- da - \int_{S_{gb}} [\mathbf{n} \cdot \mathbf{Q}] \, da \qquad (11.46)$$

The first of the two surface integrals in each of these theorems arises because the phase boundary S_{sf} changes both in time and with the position of the REV;

these integrals are evaluated at the solid (here negative) side of S_{sf}, where w_n again represents the speed of displacement of S_{sf} in the direction of its normal. The second type of surface integral, taken over the grain-to-grain contact surface S_{gb}, allows for the expected discontinuities in the field variables across that non-material boundary that is propagating at the speed w_n. The jump quantities $[\psi] w_n$ and $[\mathbf{n} \cdot \mathbf{Q}]$ may be interpreted in terms of the fine-scale picture of Figure 11.2 as differences between surface averages taken over ΔS_{gb}^{+} and ΔS_{gb}^{-} (see the discussion of grain contact zones given below, which is based on the model of Lehner & Bataille 1984). On a coarser scale, at which grain boundaries are treated as sharp interfaces, these jump quantities are conceived as point functions that vary continuously along S_{gb}.

Equation (11.39) may now be integrated over V_s, using (11.45) and (11.46), to give

$$\partial_t \int_{V_s} \psi \, dv + \mathrm{div} \int_{V_s} (\psi \mathbf{v} - \mathbf{P} \cdot \mathbf{v} + \mathbf{f}) \, dv$$

$$- \int_{S_{sf}} \{\psi(w_n - v_n) + \mathbf{n} \cdot (\mathbf{P} \cdot \mathbf{v} - \mathbf{f})\}^{-} \, da$$

$$+ \int_{S_{gb}} [\psi(w_n - v_n) + \mathbf{n} \cdot (\mathbf{P} \cdot \mathbf{v} - \mathbf{f})] \, da + \int_{V_s} r \, dv = 0 \qquad (11.47)$$

The first two terms involve averages of the types (11.42) and (11.43). With the assumption of local statistical homogeneity and in the absence of non-linear (inertia) terms, the spatial derivative of the average of a product $\psi \mathbf{v}$ or $\mathbf{P} \cdot \mathbf{v}$ may be equated to the derivative of a product of averages. (For a justification of this important step, the reader is referred to discussions of averaging procedures by Batchelor (1970) and Hill (1972).) This enables one to write

$$\mathrm{div} \, \frac{1}{V} \int_{V_\nu} \psi \mathbf{v} \, dv = \mathrm{div}(\psi \mathbf{v})_v = \mathrm{div} \, \psi_\nu \mathbf{v}^\nu = \mathrm{div} \, \psi^\nu \mathbf{v}_\nu \qquad (11.48)$$

and identical relationships for the product $\mathbf{P} \cdot \mathbf{v}$. The averaged balance equation thus becomes

$$\partial_t \psi_s + \mathrm{div}(\psi_s \mathbf{v}^s - \mathbf{P}_s \cdot \mathbf{v}^s + \mathbf{f}_s) - \frac{1}{V} \int_{S_{sf}} \{\psi(w_n - v_n) + \mathbf{n} \cdot (\mathbf{P} \cdot \mathbf{v} - \mathbf{f})\}^{-} \, da$$

$$+ \frac{1}{V} \int_{S_{gb}} [\psi(w_n - v_n) + \mathbf{n} \cdot (\mathbf{P} \cdot \mathbf{v} - \mathbf{f})] \, da + \frac{1}{V} \int_{V_s} r \, dv = 0 \quad (11.49)$$

Using Table 11.1, one can now make appropriate substitutions in this equation so as to obtain specific macroscale balance equations for the densities of the mass, linear momentum (equation of equilibrium), and free energy of the solid

313

phase. With regard to the latter, it is henceforth assumed that the grain interiors deform elastically and that inelastic deformation remains restricted to the narrow grain-to-grain contact zones, i.e. the region Ω_s shown in Figure 11.2, which lies outside the region of integration V_s of (11.49). The three balance equations then become

$$\partial_t(\rho^s\phi_s) + \text{div } \rho^s\phi_s\mathbf{v}^s - \frac{1}{V}\int_{S_{sf}} \{\rho(w_n - v_n)\}\, da$$

$$+ \frac{1}{V}\int_{S_{gb}} [\rho(w_n - v_n)]\, da = 0, \quad (11.50)$$

$$\text{div } \sigma_s + \frac{1}{V}\int_{S_{sf}} \mathbf{n}\cdot\sigma\, da - \frac{1}{V}\int_{S_{gb}} [\mathbf{n}\cdot\sigma]\, da = 0, \quad (11.51)$$

$$\partial_t(\rho^s f_s) + \text{div}(\rho^s f_s\mathbf{v}^s) - p^f(\partial_t\phi_f + \text{div } \phi_f\mathbf{v}^s) - \sigma:\text{grad } \mathbf{v}^s$$

$$- \frac{1}{V}\int_{S_{sf}} \{\rho(w_n - v_n)\}(f^- + p/\rho^-)\, da$$

$$+ \frac{1}{V}\int_{S_{gb}} [\rho(w_n - v_n)f + \mathbf{n}\cdot\sigma\cdot(\mathbf{v} - \mathbf{v}^s)]\, da = 0 \quad (11.52)$$

In the last equation the traction $\mathbf{n}\cdot\sigma$ acting on the interface S_{sf} is approximated by $-p\mathbf{n}$, as in (11.36), while the equilibrium condition for the bulk stress is approximated correspondingly by

$$\text{div } \sigma = \text{div } \sigma_s + \text{div } \sigma_f \approx \text{div } \sigma_s - \text{grad } \phi_f p^f = 0 \quad (11.53)$$

which is appropriate for slow viscous flow through a porous medium (see Lehner 1979); here div σ_s is expressed subsequently through (11.51). Also used in the derivation of (11.52) were the relationships

$$\partial_t\phi_f = -\partial_t \frac{1}{V}\int_{V_s} dv = -\frac{1}{V}\int_{S_{sf}} w_n\, da \quad (11.54)$$

and

$$\text{grad } \phi_f = \frac{1}{V}\int_{S_{sf}} \mathbf{n}\, da \quad (11.55)$$

which are readily obtained by applying the theorems (11.45) and (11.46) to

314

$\psi = 1$ and $\mathbf{Q} = 1$, respectively. Furthermore, a term

$$\frac{1}{V} \int_{S_{sf}} (p - p^f)(w_n - \mathbf{v}^s \cdot \mathbf{n}) \, \mathrm{d}a$$

has been deleted from (11.52). The term represents a negligibly small work rate associated with porosity changes effected by pressure fluctuations about the macroscopic mean pressure p^f.

11.3.2 A macroscale Gibbs equation for the solid skeleton

Equation (11.52) may now be cast in terms of a referential description, following the steps discussed in deriving the local balance equation (11.28) for the solid phase. Accordingly, let V_0 now denote the region occupied by a representative sample and its volume in a fixed reference configuration, and let V denote the same entities in the current configuration, when the sample occupies the REV. This region is conceived as 'material' by defining its boundary as a closed surface passing through a set of solid particles that were already part of the solid phase and of the boundary of V_0 in the reference configuration. The existence of such sets of particles that remain within the solid phase can always be assured for arbitrarily large finite deformations by an appropriate division into finite deformation steps. The displacement of a macroscale mass point and its velocity can then be defined in terms of the mean displacements and mean velocities of the particles forming this boundary set. This is consistent with the definition of the mean velocity $\mathbf{v}^s$ of the solid phase as a volume average (Brenner 1970), and by considering neighbouring REVs a macroscale deformation gradient $\mathbf{F}$ may be defined from their motion. With these reinterpretations in mind and with $\mathbf{v}$ replaced by $\mathbf{v}^s$, (11.23)–(11.27) may be used to transform (11.52) into

$$\dot{W}_s = p^f \dot{v}_f + \mathbf{T} : \dot{\mathbf{E}} + \frac{1}{V_0} \int_{S_{sf}} \{\rho(w_n - v_n)\}(f^- + p/\rho^-) \, \mathrm{d}a$$

$$- \frac{1}{V_0} \int_{S_{gb}} [\rho(w_n - v_n) + \mathbf{n} \cdot \boldsymbol{\sigma} \cdot (\mathbf{v} - \mathbf{v}^s)] \, \mathrm{d}a \quad (11.56)$$

where the abbreviations $V\rho^s f_s / V_0 \to W_s$ and $V\phi_f / V_0 \to v_f$ have been introduced for the free energy and void volume per unit reference volume.

From the manner in which (11.56) has been deduced directly from (11.6), (11.12), and the restriction to elastic behaviour, it follows that this equation represents a macroscale Gibbs equation for a solid skeleton from which grain contact zones are imagined to have been removed. The first two terms on the right-hand side are familiar from the non-linear theory of elastic porous solids (Biot 1972, 1973). The third term appears in the work of Heidug (1985) dealing

315

with the stress-enhanced solution of a contiguous solid matrix in a pore fluid, while the fourth term makes allowance for both pressure solution and frictional sliding processes along grain boundaries.

Equation (11.56) can be brought into a form in which dissipative work terms associated with solution/precipitation processes appear explicitly, in accord with the general structure of the internal variable theory outlined in the above. Thus, if the mass balance (11.50) is multiplied by the factor $J\mu_1^f$ and the result added to (11.56), while use is made of the abbreviation $m_s = J\rho^s\phi_s$, one obtains

$$\dot{W}_s = p^f\dot{v}_f + \mathbf{T}:\dot{\mathbf{E}} + \mu_1^f\dot{m}_s + \frac{1}{V_0}\int_{S_{sf}}\{\rho(w_n - v_n)\}(f^- + p/\rho^- - \mu_1^f)\,\mathrm{d}a$$

$$-\frac{1}{V_0}\int_{S_{gb}}[\rho(w_n - v_n)(f - \mu_1^f) + \mathbf{n}\cdot\boldsymbol{\sigma}\cdot(\mathbf{v} - \mathbf{v}^s)]\,\mathrm{d}a \quad (11.57)$$

The first of the two surface integral terms in this equation is readily interpreted, with reference to (11.36), as the rate at which energy is dissipated in solution/precipitation processes along the pore walls S_{sf}. The second surface integral term represents the rate at which energy is stored or released and dissipated by interfacial and diffusive mass transfer processes in grain-to-grain contact zones. Since the chemical potential μ_1^f represents an intrinsic average taken over the pore space, it enters as a constant into the last two terms. Over the length scale of an REV, the chemical potential of the solute in the interconnected pore space is thus treated as uniform in exactly the same manner as the pore fluid pressure. This corresponds to an assumption of local equilibrium on the macroscale for the pore fluid, which clearly does *not* apply to the pore walls and grain-to-grain contacts. The interface reaction and grain-boundary diffusion processes that occur there furnish the relaxation mechanism by which the internal state of the open system comprised by the REV tends to evolve towards an unconstrained state of equilibrium. The set of internal variables that controls this evolution turns out to be infinite-dimensional in the present case, in that it is fully determined only by a complete specification of the position of the pore walls and grain boundaries within an REV. The reason for this complication is seen to lie in the spatial variation of the integrands in (11.57), i.e. the dependence of the density, free energy, and traction on position along the interfaces S_{sf} and S_{gb}. Consider the surface integral over S_{sf}. Upon multiplication by an increment in time, $\mathrm{d}t$, its integrand may be written as

$$(\rho^- f^- + p - \rho^- \mu_1^f)\{(w_n - v_{\bar{n}})\,\mathrm{d}t\} = -A^{sf}\,\mathrm{d}\xi$$

i.e. as a product of an affinity A^{sf} and its conjugate normal displacement $\mathrm{d}\xi$ (of the phase boundary with respect to the solid material). While p may be viewed as constant and equal to p^f, ρ^- and f^- and hence A^{sf} will in general vary along S_{sf}.

Similarly, one may write

$$[(\rho f - \sigma_n - \rho\mu_1^f)(w_n - v_n)] + \tau[v_t] = [A^{gb}\dot{\xi}] + \tau[v_t]$$

for the integrand of the grain-boundary term in (11.57). Here $[v_t]$ represents the slip displacement rate and τ its conjugate shear stress, which is assumed to be continuous across S_{gb}, although a small jump in τ could arise there from fluid pressure gradients (see (11.68)). In terms of the affinities so defined, the Gibbs equation (11.57) now reads

$$\dot{W}_s = p^f\dot{v}_f + \mathbf{T}:\dot{\mathbf{E}} + \mu_1^f\dot{m}_s - \frac{1}{V_0}\int_{S_{sf}} A^{sf}\dot{\xi}\,da - \frac{1}{V_0}\int_{S_{gb}}\{[A^{gb}\dot{\xi}] + \tau[v_t]\}da$$

It is quite clear that the last two terms in this Gibbs equation are of the same kind as the finite sum $V_0^{-1}\sum_\alpha A_\alpha\dot{\xi}_\alpha$ in (11.20), while the third term expresses the fact that the system under consideration, i.e. the solid skeleton phase within an REV, is an open system.

Rice (1975) has presented a general thermodynamic formalism for obtaining the equations of state that are associated with a Gibbs equation of the type considered here. Central to this theory are reciprocity relations that yield expressions for strain increments due to incremental changes in the internal variables, corresponding, in the present case, to the increments $d\xi$ along S_{sf}, and $d\xi^\pm$ and $v_t^\pm\,dt$ along S_{gb}. In a first application of this theory to pressure solution, Heidug (1985) did not take into account grain-boundary sliding or indentation displacements, but demonstrated the construction of a complete constitutive theory from a Gibbs equation without a grain-boundary term. An essential part in such a theory is played by the kinetic relationships that govern solution/precipitation processes at phase boundaries, i.e. at the pore walls and, in general, within fluid-infiltrated grain-to-grain contact zones. An example of such a relationship is furnished by the phenomenological rate equation (11.37), which may be viewed as linking a local value of the affinity A^{sf} to its conjugate displacement rate $\dot{\xi}$. As was pointed out in the introduction, it is this part of the theory that leads to difficulties with the grain-boundary term, where the correct interpretation of the thermodynamic driving force, i.e. of the affinity A^{gb}, has remained a much-debated question in the geological literature. Any attempt to generalize the constitutive framework established by Heidug (1985), by including pressure solution along grain-to-grain contacts, must first resolve the issue of the thermodynamic driving force. This problem was therefore chosen as the main topic for the remainder of this chapter, which will be concerned in particular with the formulation of approximate relationships for the rate of grain indentation due to intergranular pressure solution.

11.4 Intergranular pressure solution as a stationary non-equilibrium process

A proper reformulation and interpretation of the last term in the Gibbs equation (11.57) can be obtained only from a more detailed analysis of grain-boundary processes, based on the assumed solid- and fluid-phase behaviour, to which inelastic solid-state deformation mechanisms will now be admitted. The jump quantity appearing in the last term of (11.57) has already been characterized as a grain-scale average associated with the surface element ΔS_{gb}, as shown in Figure 11.2. It is now necessary to express this quantity as part of a free energy balance for an elemental volume Ω of the grain boundary, and thereby to relate it to the rate of energy dissipation within the volume element, for it is this connection that will yield the desired phenomenological rate equations for the grain-boundary displacements.

Consider therefore the pillbox-shaped volume element Ω, as shown in Figure 11.2, which is assumed to propagate with the median surface S_{gb} at the speed w_n. The median surface S_{gb} and its unit normal $\mathbf{n}$ are defined at the grain scale, where $\mathbf{n}$ is allowed to vary in a piecewise continuous manner over S_{gb}. On the fine scale, the element Ω is assumed to be bounded on either side of S_{gb} by planar surface elements, ΔS_{gb}^{+} and ΔS_{gb}^{-} which, on the grain scale, must be viewed as collapsed into a single infinitesimal surface element ΔS_{gb} of orientation $\mathbf{n}$. No limitation will arise from viewing Ω as flat on the fine scale since the equations obtained by averaging the fine-scale description over Ω will make full allowance for surface curvature of S_{gb} on the grain scale.

A straightforward application of the transport theorem (11.45) to the *generic* balance equation (11.39) leads to the following expression for the rate of change of the content $\int_{\Omega}\psi \, dv$ of some quantity ψ within the propagating volume element Ω:

$$\partial_t \int_{\Omega} \psi \, dv = \int_{\Delta S_{gb}^{+}} \{\psi(w_n - v_n) + \mathbf{n} \cdot (\mathbf{P} \cdot \mathbf{v} - \mathbf{f})\} \, da$$

$$- \int_{\Delta S_{gb}^{-}} \{\psi(w_n - v_n) + \mathbf{n} \cdot (\mathbf{P} \cdot \mathbf{v} - \mathbf{f})\} \, da$$

$$- \int_{\Gamma} \{\psi\mathbf{v} \cdot \mathbf{n}_\Gamma - \mathbf{n}_\Gamma \cdot (\mathbf{P} \cdot \mathbf{v} - \mathbf{f})\} \, da - \int_{\Omega_f} r \, dv - \int_{\Omega_s} r \, dv - \int_{\Delta \tilde{S}_{sf}} \Delta \, da$$

$$(11.58)$$

Here the *generic* jump condition (11.40) has been applied along the solid/fluid phase boundary $\tilde{S}_{sf}$ within the grain contact zone, $\Delta \tilde{S}_{sf}$ being the increment of $\tilde{S}_{sf}$ covered by the volume element Ω (see Fig. 11.2). Equation (11.58) may now e cast in a form appropriate to the local description of grain-boundary processes on the grain scale of Figure 11.1 and equation (11.57). This can be ccomplished by a systematic rescaling procedure, based on the use of a small arameter $\varepsilon = \delta/a$ to define a set of stretched co-ordinates $x' = x/\varepsilon$ such that he x' system is used when viewing the grain contact at the fine scale of Figure 1.2, while the x system is associated with Figure 11.1. Here δ denotes the hickness of the pillbox-shaped volume element Ω, while a denotes the linear limension of a typical grain-to-grain contact. The scale change contemplated orresponds to taking the limit $\varepsilon \to 0$ while holding fixed the ratio δ/d, where l is the diameter of the cylindrical region Ω. This scale change converts the urface integrals over ΔS_{gb}^+ and ΔS_{gb}^- into a local jump quantity defined on S_{gb}, while the integrals over $\Gamma = \Gamma_s + \Gamma_f$ yields surface divergences of flux uantities associated with the two phases. Equation (11.58), after division by $\Delta S = \pi d^2/4$ and some rearrangement, is transformed by this fine-scale averging into

$$\psi(w_n - v_n) + \mathbf{n} \cdot (\mathbf{P} \cdot \mathbf{v} - \mathbf{f})] = \delta \partial_t \tilde{\psi} + \delta \sum_v \mathrm{div}(\tilde{\psi}_v \tilde{\mathbf{v}}^v - \tilde{\mathbf{P}}_v \cdot \tilde{\mathbf{v}}^v + \tilde{\mathbf{f}}_v)$$

$$+ \frac{1}{\Delta S} \int_{\Omega_f} r \, dv + \frac{1}{\Delta S} \int_{\Omega_s} r \, dv + \frac{1}{\Delta S} \int_{\Delta \tilde{S}_{sf}} \Delta \, da \quad (11.59)$$

he grain-scale averages and surface divergence appearing in this equation are lefined by (with $v = s, f$)

$$\tilde{\psi}_v = \frac{1}{\delta \Delta S} \int_{\Omega_v} \psi \, dv, \qquad \tilde{\psi} = \sum_v \tilde{\psi}_v = \sum_v \delta_v \tilde{\psi}^v / \delta \quad (11.60)$$

$$\mathrm{div} \, \tilde{\mathbf{q}}_v = \frac{1}{\delta \Delta S} \int_{\Gamma_v} \mathbf{q} \cdot \mathbf{n}_\Gamma \, da, \qquad \tilde{\mathbf{q}} = \sum_v \tilde{\mathbf{q}}_v = \sum_v \delta_v \tilde{\mathbf{q}}^v / \delta \quad (11.61)$$

$$\tilde{\mathbf{q}}_v \cdot \mathbf{n} = 0 \quad (11.62)$$

vhere $\mathbf{q}$ stands for any of the flux vectors $\mathbf{v}$, $\mathbf{f}$, $\psi\mathbf{v}$, and $\mathbf{P} \cdot \mathbf{v}$, and a decomposiion of the type (11.48) is assumed for the average of a product, i.e.

$$\mathrm{div}(\widetilde{\psi\mathbf{v}})_v = \mathrm{div} \, \tilde{\psi}_v \tilde{\mathbf{v}}^v = \mathrm{div} \, \tilde{\psi}^v \tilde{\mathbf{v}}_v,$$

$$\mathrm{div}(\widetilde{\mathbf{P} \cdot \mathbf{v}})_v = \mathrm{div}(\tilde{\mathbf{P}}_v \cdot \tilde{\mathbf{v}}^v) = \mathrm{div}(\tilde{\mathbf{P}}^v \cdot \tilde{\mathbf{v}}_v) \quad (11.63)$$

imilarly, the jump quantities forming the left-hand member of (11.59) have he significance of surface averages taken over ΔS_{gb}^+ and ΔS_{gb}^-, and must here e viewed as grain-scale point functions defined on S_{gb}. The width δ of the conact zone is assumed to be uniform along S_{gb}, comprising the sum of the mean fluid-film thickness $\delta_f = \text{vol}(\Omega_f)/\Delta S$ and mean solid boundary layer thickness $\delta_s = \text{vol}(\Omega_s)/\Delta S$, the latter extending over the disturbed zones of any pair of contacting grains, as shown in Figure 11.2.

The further analysis will now be simplified by restricting this discussion to grain-boundary states that evolve slowly enough to permit a process description in terms of a succession of steady states, when viewed by an observer moving in the tangent plane of S_{gb} at a speed equal to the projection of the *uniform* (on the grain scale) macroscopic solid phase velocity $\mathbf{v}^s$. Moreover, it will be assumed that in this frame the surface divergences of all convective flux quantities vanish in the solid grain-boundary phase. These assumptions are expressed by the conditions

$$\partial_t \tilde{\psi}^\nu + \mathbf{v}^s \cdot \text{grad } \tilde{\psi}^\nu = 0, \quad \nu = s, f \quad \text{div}\{(\tilde{\psi}^s \mathbf{1} - \tilde{\mathbf{P}}^s) \cdot (\tilde{\mathbf{v}}^s - \mathbf{v}^s)\} = 0$$

which when introduced into (11.59), yield

$$[\psi(w_n - v_n) + \mathbf{n} \cdot (\mathbf{P} \cdot \mathbf{v} - \mathbf{f})] =$$

$$- \delta \mathbf{v}^s \cdot \text{div } \tilde{\mathbf{P}} + \delta_f \text{ div}\{(\tilde{\psi}^f \mathbf{1} - \tilde{\mathbf{P}}^f) \cdot (\tilde{\mathbf{v}}^f - \mathbf{v}^s)\} + \delta \text{ div } \tilde{\mathbf{f}}$$

$$+ \frac{1}{\Delta S} \int_{\Omega_f} r \, dv + \frac{1}{\Delta S} \int_{\Omega_s} r \, dv + \frac{1}{\Delta S} \int_{\Delta \tilde{S}_{sf}} \Delta \, da \quad (11.64)$$

This equation is now used together with Table 11.1 to derive the following specific grain-scale jump conditions, valid at any point along the grain boundary S_{gb}:

$$[\rho(w_n - v_n)] = \delta_f \text{ div } \tilde{\rho}^f(\tilde{\mathbf{v}}^f - \mathbf{v}^s), \qquad \text{total mass} \quad (11.65)$$

$$[\rho(w_n - v_n)] = \delta_f \text{ div}\{\tilde{\rho}_1(\tilde{\mathbf{v}}^f - \mathbf{v}^s) + \tilde{\mathbf{J}}_1\}, \qquad \text{solute mass} \quad (11.66)$$

$$0 = \delta_f \text{ div}\{\tilde{\rho}_2(\tilde{\mathbf{v}}^f - \mathbf{v}^s) + \tilde{\mathbf{J}}_2\}, \qquad \text{solvent mass} \quad (11.67)$$

$$- [\mathbf{n} \cdot \sigma] = \delta \text{ div } \tilde{\sigma}, \qquad \text{momentum} \quad (11.68)$$

$$[\rho(w_n - v_n)f + \mathbf{n} \cdot \sigma \cdot \mathbf{v}] =$$

$$- \delta \mathbf{v}^s \cdot \text{div } \tilde{\sigma} + \delta_f \text{ div}\{(\tilde{\rho}^f \tilde{f}^f \mathbf{1} - \tilde{\sigma}^f) \cdot (\tilde{\mathbf{v}}^f - \mathbf{v}^s) + \sum_i \tilde{\mu}_i \tilde{\mathbf{J}}_i\}$$

$$+ \frac{1}{\Delta S} \int_{\Omega_f} T\gamma \, dv + \frac{1}{\Delta S} \int_{\Omega_s} T\gamma \, dv + \frac{1}{\Delta S} \int_{\Delta \tilde{S}_{sf}} T\sigma \, da, \qquad \text{free energy} \quad (11.69)$$

A comparison of these conditions with their equivalent among conditions (11.30)–(11.35) again reveals the fundamental difference between the two types of singular surfaces — pore walls (S_{sf}) and grain boundaries (S_{gb}) — that are distinguished in this analysis: grain boundaries are the site of an autonomous fluid phase and of associated interfacial transport phenomena. The diffusive mass fluxes $\tilde{\mathbf{J}}_i$ are defined here, as in (11.7), in terms of the partial densities $\tilde{\rho}_i$ and component velocities $\tilde{\mathbf{v}}_i$ by $\tilde{\mathbf{J}}_i = \tilde{\rho}_i \, (\tilde{\mathbf{v}}_i - \tilde{\mathbf{v}}^f)$, with $\tilde{\mathbf{v}}^f = \Sigma_i \, \tilde{\rho}_i \tilde{\mathbf{v}}_i / \tilde{\rho}^f$.

For the purpose of obtaining phenomenological rate equations for intergranular pressure solution, condition (11.69) is of prime interest. As a balance equation, it asserts that the net influx of energy and momentum into a grain-boundary element (of volume $\delta \Delta S$ on the microscale) equals the energy dissipated by irreversible deformation, transport, and interface reaction processes. However, a more useful statement can be extracted from condition (11.69) upon making use of the explicitly known expressions for rates of energy dissipation in the last three terms. In preparation for this step, the term $-\delta \mathbf{v}^s \cdot \text{div } \tilde{\sigma}$ is absorbed in the left-hand member of (11.69), through use of condition (11.68), as is the first term of the expansion

$$\delta_f \text{ div}\{(\tilde{\rho}^f \tilde{f}^f \mathbf{1} - \tilde{\sigma}^f) \cdot (\tilde{\mathbf{v}}^f - \mathbf{v}^s)\} = \tilde{\mu}_1 \, [\rho(w_n - v_n)]$$

$$+ \delta_f(\tilde{\mathbf{v}}^f - \mathbf{v}^s) \cdot \text{grad } \tilde{p}^f + \delta_f \sum_i \tilde{\mathbf{J}}_i \cdot \text{grad } \tilde{\mu}_i$$

which is obtained upon putting $\tilde{\rho}^f \tilde{f}^f \mathbf{1} - \tilde{\sigma}^f \approx (\tilde{\rho}^f \tilde{f}^f + \tilde{p}^f)\mathbf{1} = \Sigma_i \, \tilde{\rho}_i \tilde{\mu}_i \mathbf{1}$ and subsequent use of conditions (11.66) and (11.67). Condition (11.69) now reads

$$[\rho(w_n - v_n)(f - \tilde{\mu}_1) + \mathbf{n} \cdot \sigma \cdot (\mathbf{v} - \mathbf{v}^s)] = \delta_f(\tilde{\mathbf{v}}^f - \mathbf{v}^s) \cdot \text{grad } \tilde{p}^f$$

$$+ \delta_f \sum_i \tilde{\mathbf{J}}_i \cdot \text{grad } \tilde{\mu}_i + \frac{1}{\Delta S} \int_{\Omega_f} T\gamma \, dv + \frac{1}{\Delta S} \int_{\Omega_s} T\gamma \, dv + \frac{1}{\Delta S} \int_{\Delta \tilde{S}_{sf}} T\sigma \, da \quad (11.70)$$

Now, by averaging expression (11.13), the surface-specific, grain-scale rate of dissipation in the fluid phase is found to be

$$\frac{1}{\Delta S} \int_{\Omega_f} T\gamma \, dv = -\delta_f(\tilde{\mathbf{v}}^f - \mathbf{v}^s) \cdot \text{grad } \tilde{p}^f - \delta_f \sum_i \tilde{\mathbf{J}}_i \cdot \text{grad } \tilde{\mu}_i$$

$$- \frac{1}{\Delta S} \int_{\Delta \tilde{S}_{sf}} \{\rho(w_n - v_n)\}(\mu_1 - \tilde{\mu}_1) da \quad (11.71)$$

Here the first two terms represent the rate of dissipation due to viscous fluid flow and molecular diffusion along the grain boundary, while the last term represents dissipation associated with fine-scale diffusion fluxes. These fluxes may exist, even if grain-scale gradients in $\tilde{p}^f$ and $\tilde{\mu}_1$ vanish; for example, in a pure sliding mode of grain-boundary displacement with no net mass transport along

S_{gb} through a volume element Ω but with local transfer within Ω from solution to nearby precipitation sites (see the discussion by Lehner & Bataille 1984/85). The spatial fine-scale fluctuations $\mu_1 - \tilde{\mu}_1$ in the chemical potential are, of course, due to the irregular grain-boundary structure.

Substitution of the expressions (11.29), (11.36), and (11.71) in (11.70) now gives

$$[\rho(w_n - v_n)(f - \tilde{\mu}_1) + \mathbf{n} \cdot \boldsymbol{\sigma} \cdot (\mathbf{v} - \mathbf{v}^s)] = T\tilde{\sigma} \qquad (11.72)$$

with

$$T\tilde{\sigma}: = -\frac{1}{\Delta S} \int_{\Delta \tilde{S}_{sf}} \{\rho(w_n - v_n)\}(\mu_1 - \tilde{\mu}_1)\mathrm{d}a + \frac{1}{\Delta S} \sum_{a=1}^{N} A_\alpha \dot{\xi}_\alpha$$

$$-\frac{1}{\Delta S} \int_{\Delta \tilde{S}_{sf}} \{\rho(w_n - v_n)\}(f^- + p/\rho^- - \mu_1) \, \mathrm{d}a \geqslant 0 \quad (11.73)$$

The inequality holds separately for each of the three terms. This follows from (11.29) and (11.36) for the second and for the last term, and since it is readily justified for the sum of the first and last, it also applies separately to the first term.

Condition (11.72) has the significance of a Gibbs equation for a volume element of a grain boundary that constitutes an open system and is the site of a stationary non-equilibrium process. It is fundamental to obtaining the phenomenological rate equations that govern grain-boundary sliding and indentation displacements, and will furnish the desired expression for the last term in the Gibbs equation (11.57).

In reality, grain-boundary sliding in an aggregate is likely to be constrained by grain indentation, which then becomes the mode of pressure solution of primary interest in that it will control both the compaction and deviatoric creep rates. By assuming this to be the case and by restricting attention to flat grain-to-grain contacts that are symmetric in the sense that – on the grain scale – f and ρ may be considered to be continuous across S_{gb}, condition (11.72) is now reduced to

$$[\rho(w_n - v_n)](f - \sigma_n/\rho - \tilde{\mu}_1) = T\tilde{\sigma} \geqslant 0 \qquad (11.74)$$

where $f - \sigma_n/\rho$ may be defined on either side of S_{gb}. The limitation to symmetric grain-to-grain contacts is, of course, severe, even for monocrystalline aggregates, and will exclude from consideration important crystal orientation effects that are visibly present in the experiments of Spiers & Schutjens (this volume). While it is dictated at this point – as is the earlier restriction to binary systems – by the need to establish a basic theoretical framework, lifting this restriction in subsequent studies should result in important quantitative insights into grain anisotropy and inhomogeneity effects on pressure-solution textures.

When (11.74) applies, the last term in (11.57) may be written as the sum $\Sigma = \Sigma_I + \Sigma_D$ of two contributions to the total rate of dissipation along grain boundaries, where

$$\Sigma_I: = \frac{1}{V_0} \int_{S_{gb}} [\rho(w_n - v_n)](f - \sigma_n/\rho - \tilde{\mu}_1)\mathrm{d}a \qquad (11.75)$$

and

$$\Sigma_D: = \frac{1}{V_0} \int_{S_{gb}} [\rho(w_n - v_n)](\tilde{\mu}_1 - \mu_1^f)\mathrm{d}a \qquad (11.76)$$

are associated with interface reactions and grain-boundary diffusion, respectively. The important limiting cases of diffusion-rate-limited and interface-reaction-rate-limited pressure solution can now be defined succinctly by the requirements $\Sigma_I << \Sigma_D$ and $\Sigma_I >> \Sigma_D$, respectively. In the first case, the dissipation $T\tilde{\sigma}$ associated with the local grain-boundary processes, as given by (11.73), is considered to be negligible in comparison with dissipation associated with diffusive solute transport along grain boundaries, driven by the potential difference $\tilde{\mu}_1 - \mu_1^f$. This case is therefore treated as if $T\tilde{\sigma} \simeq 0$ in (11.74), while $[\rho(w_n - v_n)] \neq 0$, which amounts to putting

$$f - \sigma_n/\rho = \tilde{\mu}_1 \qquad \text{on } S_{gb} \qquad (11.77)$$

This is Paterson's (1973) 'equilibrium condition'. It emerges here as an approximation by which the solution/precipitation reaction is treated as non-dissipative, precisely as is done for Stefan problems of propagating melting/solidification boundaries across which a Gibbs free energy is assumed to remain continuous. This is the implication of Paterson's a priori assumption of the existence of a local equilibrium state at points along a fluid-infiltrated, non-hydrostatically stressed grain-to-grain contact. However, the boundary condition (11.77) obtained in this manner – unlike Gibbs' condition (11.38) – does not represent a sufficient condition of thermodynamic equilibrium in the sense that satisfying it would bring the solution/precipitation process to a halt. This is readily understood for the grain-boundary structure shown in Figure 11.2, which cannot be in equilibrium unless $-\sigma_n = \tilde{p}^f$, so that condition (11.77) reduces to Gibbs' condition (11.38). Indeed, if $-\sigma_n$ exceeds $\tilde{p}^f$, which is the situation of interest, compression of the highly irregular contact zone will lead to non-uniformities in the free energy and specific volume of the solid phase along the phase boundary $\tilde{S}_{sf}$. According to (11.36) and (11.37) these non-uniformities must generate thermodynamic forces which will drive fine-scale dissipative processes of interface reaction and diffusion. Since (11.73) demands that $\tilde{\sigma} > 0$ in the presence of such fine-scale fluctuations, it follows that (11.77) cannot qualify as an equilibrium condition for the assumed grain-boundary structure.

When the rate of pressure solution is controlled by the interface reaction, that is when $\Sigma_I \gg \Sigma_D$, the chemical potential along the grain boundaries will be approximately uniform and equal to the potential in the free pore space, i.e.

$$\tilde{\mu}_1 \simeq \mu_1^f \tag{11.78}$$

The last term in (11.57), which under the above assumptions may be written as

$$\Sigma := \frac{1}{V_0} \int_{S_{gb}} [\rho(w_n - v_n)] (f - \sigma_n/\rho - \mu_1^f) \mathrm{d}a \tag{11.79}$$

is now seen to contain the product of the mass transfer rate and the appropriate thermodynamic driving force for each of the limiting cases characterized by (11.77) and (11.78), respectively. The difference lies in the kinetic relations that link the force and the flux; these will now be considered.

11.4.1   Interface-reaction-controlled pressure solution

As long as the deformation behaviour of the grain interior remains in the elastic range, it may be assumed that the rates of inelastic deformation of the solid contact-zone material are determined by the interface mass transfer rate. For reaction-rate-limited pressure solution the term of importance in expression (11.73) is therefore the interfacial dissipation term, so that (11.73), (11.74), and (11.78) yield the approximate relationship

$$[\rho(w_n - v_n)] (f - \sigma_n/\rho - \mu_1^f) =$$

$$- \frac{1}{\Delta S} \int_{\Delta S_{sf}} \{\rho(w_n - v_n)\} (f^- + p/\rho^- - \mu_1^f) \mathrm{d}a \geqslant 0 \tag{11.80}$$

Viewed as a balance statement, the equality part of this relationship shows how the higher work rate of the normal stress $-\sigma_n$, as compared with that of the fluid pressure p, is compensated for by a transfer of excess free energy $f^- - f$ across the phase boundary, as highly stressed or plastically deforming asperities are removed. The use of (11.80) comes from the observation that if the force and the flux under the surface integral are related by (11.37), then a relationship

$$[\rho(w_n - v_n)] = L^*(f - \sigma_n/\rho - \mu_1^f) \tag{11.81}$$

324

will exist (in which L^* differs from L by a geometric factor) as long as the 'open' grain-boundary structure of Figure 11.2 prevails. The latter quali-fication is essential and relates to the observation made above that the island structure of Figure 11.2, with its interconnected fluid-invaded pore space, can-not be an equilibrium structure. The implication of this is that relation (11.81) cannot be expected to hold continuously down to vanishing dissolution rates. Indeed, the physical state of the grain-to-grain contact zone is likely to change in approaching equilibrium, with mobile grain-boundary fluid being squeezed into the open pore space or else forming disconnected fluid inclusions (Urai *et al.* 1986; Spiers & Schutjens, this volume; and other investigations men-tioned in the introduction) which can become equilibrated individually with the solid phase. A full discussion of this transition from the above-postulated statistically stationary non-equilibrium grain-boundary structure to an equilib-rium structure will not be attempted here, partly because interfacial tension effects (which here have been disregarded) are essential to its understanding. It will be clear, however, that pressure solution at grain-to-grain contacts may, in general, cease to operate at a finite value of the force $f - \sigma_n/\rho - \mu\{$. Evidently, it is difficult to associate a definite value for this quantity with the re-initiation of pressure solution at a grain boundary, for example by some process of marginal dissolution and/or fluid bubble coalescence that will depend on the initial state of the grain boundary and may be driven, in part, by shear stresses. This unresolved question will nevertheless be crucial to stu-dies concerned with porosity–depth relationships in sedimentary rocks. In this situation it is reasonable to introduce a tentative assumption that may subse-quently be verified. The simplest one to make is clearly the assumption that the range validity of (11.81) may be extended towards simultaneously vanish-ing values of the force and the flux. More precisely, it is assumed tentatively that the condition

$$\bar{f} - \bar{\sigma}_n/\bar{\rho} = \mu\{ \tag{11.82}$$

is a necessary and sufficient condition of equilibrium for interior points along a flat grain-to-grain contact, where the bar denotes surface averages taken over a particular grain-to-grain contact.

Condition (11.82) must be distinguished carefully from Paterson's condition (11.77) which implicitly demands that the inequality $f - \sigma_n/\rho > \mu\{$ be maintained at all contact points during a diffusion-rate-controlled process. Condition (11.82) is in fact readily understood when a granular aggregate is considered for which $\bar{f}$ and $\bar{\rho}$ may be taken as approximately uniform and equal to f^s and ρ^s, so that $\mu_1^{\text{eq}} := f^s + p^f/\rho^s$ defines a uniform equilibrium chemical potential for the solute in the free pores in accordance with Gibbs' condition (11.38). Condition (11.82) may then be written

$$- (\bar{\sigma}_n + p^f)/\rho^s = \mu\{ - \mu_1^{\text{eq}} \tag{11.83}$$

where the right-hand side may be re-expressed in terms of a supersaturation, using (11.16). It is thus seen that condition (11.82) specifies a supersaturation – proportional to the effective grain-boundary normal stress – needed to halt pressure solution at a particular grain boundary. Once this supersaturation has been attained, grain boundaries will become closed by precipitation. Along the free pore walls, however, precipitation could continue thereafter, thereby reducing the supersaturation without re-initiation of pressure solution at grain contacts. For this reason condition (11.82) is unlikely to provide a reliable limit condition for predicting the initiation of pressure solution at closed grain boundaries.

With these restrictions in mind, the only application of (11.81) and (11.82) contemplated here is to wet grain boundaries removed from, but possibly approaching, a state of equilibrium.

When f, ρ, σ_n are approximated by uniform values f^s, ρ^s, $\bar{\sigma}_n$, relation (11.81) can be replaced by

$$[\rho(w_n - v_n)] = -L^*\{(\bar{\sigma}_n + p^f)/\rho^s + \mu\{-\mu_1^{eq}\}\} \tag{11.84}$$

11.4.2 Diffusion-controlled pressure solution

Here the rate of dissolution $[\rho(w_n - v_n)]$ in (11.79) must be obtained by solving an appropriate grain-boundary diffusion equation, which is derived from the mass balance (11.66) and from Fick's law (11.15). If, consistent with the assumption of diffusion rate control, the convective flux $\tilde{\rho}_1(\tilde{\mathbf{v}}^f - \mathbf{v}^s)$ is negligible in comparison with $\tilde{\mathbf{J}}_1$, this equation becomes

$$[\rho(w_n - v_n)] = -\operatorname{div} l\delta_f/(1 - \tilde{c}_1)\operatorname{grad} \tilde{\mu}_1 \tag{11.85}$$

where $\tilde{\mu}_1$ must be substituted from Paterson's Stefan-type boundary condition (11.77) to obtain a mass transfer rate compatible with the solution of the mechanical contact problem that determines the left-hand side of (11.77). This last point appears to have been overlooked in earlier work by Rutter (1976) and Robin (1978). Thus, Rutter solves a diffusion equation for $\tilde{\mu}_1$, assuming a uniform dissolution rate $[\rho(w_n - v_n)]$ and hence indentation creep rate, and subsequently determines the contact stress σ_n from an approximate form $\sigma_n = -\rho_s\tilde{\mu}_1$ of Paterson's condition (11.77). There is clearly no reason for this stress distribution to be always consistent with the solution of the appropriate mechanical contact problem.

A possible resolution of this difficulty, which appears attractive but must be considered as hypothetical at this stage, could lie in a transient, periodically inward-progressing grain-boundary dissolution process, such that the solution rate $[\rho(w_n - v_n)]$ would be uniform in the mean, and (11.77) and (11.85) would hold in a time-averaged sense. The interesting implication of such a periodic process for the normal stress σ_n is that its value could at any instant atisfy the solution of the relevant mechanical contact problem, while the time verage of this periodically varying stress would be distributed over the grain-o-grain contact as demanded by (11.77) and the solution of (11.85). The esulting stress distribution would then again be of the type obtained by Rutter 1976), whose approach to diffusion-rate-limited pressure solution would thus ppear justified on a more consistent, although still hypothetical, basis.

The nature of this contact stress distribution and the general form of the henomenological rate law for grain-boundary diffusion-controlled pressure olution are made clear by Rutter's example of a plane, circular contact of adius a (see also Spiers & Schutjens, this volume). Thus, by introducing an ffective grain boundary diffusivity $D_{gb} = \delta_f lRT/\{\tilde{\rho}^f M_1 \tilde{c}_1 (1 - \tilde{c}_1)\}$ in (11.85) nd approximating a factor $\tilde{c}_1$ by the free pore space concentration c_1, a solu-ion for the chemical potential in the radial co-ordinate $0 \leqslant r \leqslant a$, subject to $\tilde{\mu}_1|_a = \mu_1^f$ and $(d\tilde{\mu}_1/dr)|_0 = 0$, yields

$$\tilde{\mu}_1(r) - \mu_1^f = \frac{[\rho(w_n - v_n)]RT}{4\tilde{\rho}^f M_1 c_1 D_{gb}} (a^2 - r^2) \qquad (11.86)$$

vhere $[\rho(w_n - v_n)]$ is treated as a constant. Taking the surface average over he contact of this expression and substituting for $\tilde{\mu}_1$ from condition (11.77), one finds

$$[\rho(w_n - v_n)] = \frac{8\tilde{\rho}^f M_1 c_1 D_{gb}}{RTa^2} (\bar{f} - \bar{\sigma}_n/\bar{\rho} - \mu_1^f) \qquad (11.87)$$

urther, if $\bar{f}$ and $\bar{\rho}$ are reasonably approximated by uniform values f^s and ρ^s, hen this relationship can be written

$$[\rho(w_n - v_n)] = - \frac{8\tilde{\rho}^f M_1 c_1 D_{gb}}{RTa^2} \{(\bar{\sigma}_n + p^f)/\rho^s + \mu_1^f - \mu_1^{eq}\} \qquad (11.88)$$

If, as is appropriate for diffusion rate control, the supersaturation $\mu_1^f - \mu_1^{eq}$ n the pore space is assumed to be negligible in this expression, then the rate of dissolution simply becomes proportional to the effective normal stress

$$[\rho(w_n - v_n)] = - \frac{8\tilde{\rho}^f M_1 c_1 D_{gb}}{\rho^s RTa^2} (\bar{\sigma}_n + p^f) \qquad (11.89)$$

To ascertain the applicability of this result one should, however, consult a criterion for rate control obtained from (11.79), (11.84), (11.88), i.e. the ratio

$$\frac{\Sigma_I}{\Sigma_D} = \frac{[\rho(w_n - v_n)]_I}{[\rho(w_n - v_n)]_D} = \frac{L^* RTa^2}{8\tilde{\rho}^f M_1 c_1 D_{gb}} = \frac{k^* a^2}{8c_1 \tilde{\rho}^f D_{gb}} \qquad (11.90)$$

327

Here $k^* = L^* RT/\rho^s \bar{V}^s$ is an effective transfer coefficient and $\bar{V}^s$ is the molar volume of the solid. It should be observed, however, that this criterion is linked to the assumption of linear rate laws. In general, therefore, the criterion for rate control may well involve a dependence on the magnitude of the thermodynamic driving force and, inasmuch as the rate laws (11.84) and (11.87) are to be viewed only as the simplest possible choices, no statement is implied here about their range of validity.

11.4.3 Interface-reaction-controlled volumetric creep of dense aggregates

An inspection of relation (11.88) suggests that even the simplest type of aggregate geometry will, in general, exhibit non-linear creep behaviour, primarily because particle contact areas and contact stresses will change during deformation. While a treatment of these geometrical non-linearities is beyond the scope of this chapter, there is at least one fairly simple but instructive case that lends itself to analysis without the need to employ a complete constitutive description. This is the geologically interesting case of volumetric creep in aggregates that have retained only an intergranular porosity in the form of 'sheet pores'. The process of pervasive mass removal by pressure solution can continue here, provided that there is some grain-boundary permeability. Creep velocities are then typically determined by the interface reaction rate, with the chemical potential of the solute in the grain-boundary fluid being controlled by long-range advective transfer of material.

The loss in bulk volume per unit time, which is the quantity of interest, is now simply equal to $\dot{V}_s$, the rate of change of the solid volume, assuming that changes in the small grain-boundary porosity may be disregarded. Under conditions of constant stress and pore-fluid pressure, the density of the solid phase may be taken as constant. One therefore has the simple kinematic relationship

$$ j = \frac{1}{V_0} \int_{S_{gb}} [v_n] \, \mathrm{d}a \tag{11.91} $$

The phenomenological rate equation (11.84) is now substituted for $[v_n]$, and it will be assumed that the grain-boundary stresses may be set equal to resolved macrostresses. This gives

$$ j = \frac{k^* \bar{V}^s}{\rho^s RT} \{ (\sigma + p^f \mathbf{1}) + \rho^s (\mu_1^f - \mu_1^{eq}) \mathbf{1} \} : \frac{1}{V_0} \int_{S_{gb}} \mathbf{nn} \, \mathrm{d}a $$

or

$$ \frac{1}{V} \frac{\mathrm{d}V}{\mathrm{d}t} = \tfrac{1}{3} \lambda : \{ (\sigma + p^f \mathbf{1}) + \rho^s (\mu_1^f - \mu_1^{eq}) \mathbf{1} \} \tag{11.92} $$

where

$$\lambda = \frac{k^* \overline{V}^s}{\rho^s RTd} \frac{3}{S_{gb}} \int_{S_{gb}} \mathbf{nn} \, da, \qquad d = V/S_{gb} \qquad (11.93)$$

Thus, the rate of volume loss will, in general, be governed by a tensorial viscosity coefficient, which reflects the current grain-boundary fabric. For large strains this tensor would have to be recalculated continuously by integrating the local rates of grain-boundary displacement. In the special case of isotropic stressing of an isotropic aggregate, λ will be isotropic and equal to $\lambda = \{k^* \overline{V}^s / \rho^s RTd\}\mathbf{1} = \lambda\mathbf{1}$, so that

$$\frac{1}{V} \frac{dV}{dt} = \lambda\{ (\tfrac{1}{3}\sigma_{ii} + p^f) + \rho^s(\mu_1^f - \mu_1^{eq}) \} \qquad (11.94)$$

where the coefficient λ remains constant if the grain size is reduced in a self-similar fashion while the aggregate bulk volume is shrunk.

Since the volumetric creep rate is controlled by the chemical potential μ_1^f of the solute, which in turn is determined by large-scale transport in a given hydrólogical régime, it is clear that volume creep and transport are intimately coupled, as has long been recognized.

11.5 Concluding summary

A thermodynamic theory of the mechanism of rock deformation by pressure-solution processes has been presented. A case was made for an approach that uses averaging techniques to derive a macroscale aggregate description from a postulated microscale continuum formulation of the underlying physical processes. It has been shown in this manner how 'internal displacements' that result from intergranular or pore-wall dissolution enter into a rigorously formulated macroscopic Gibbs equation of the 'internal variable' type. The same averaging procedure, when applied to a representation grain-boundary element, has enabled the identification of the thermodynamic driving forces conjugate to grain-boundary displacements and has shed light on the nature of a widely used 'equilibrium condition' proposed by Paterson. This condition was found to have the significance of a Stefan-type boundary condition in which the interface reaction is treated as non-dissipative, as will be appropriate for diffusion-controlled pressure solution. Since grain-boundary processes were treated as stationary in this chapter and since surface-tension effects were disregarded, the important problem of the nature of transient interface states – particularly in the approach towards equilibrium – could

329

not be examined. It was observed, however, how these more general transient states of a grain boundary may modify the phenomenological rate equations obtained from steady-state theory, and indeed invalidate these relations near equilibrium, when interfacial energies are likely to become important. These limitations of the present analysis are significant, and point in the direction of further theoretical and experimental work that is needed.

Acknowledgements

I would like to thank Professor D. Barber and Dr C. Spiers for their careful reading and constructive criticism of the manuscript. This chapter is published by the kind permission of Shell Research B.V.

Appendix

Throughout this chapter direct (bold face) notation is used. In terms of Cartesian components with respect to an orthonormal basis $\mathbf{e}_k$ ($k = 1, 2, 3$), the following definitions apply:

$\mathbf{v} = v_k \mathbf{e}_k$ $\qquad\qquad$ vector

$\boldsymbol{\sigma} = \sigma_{kl} \mathbf{e}_k \mathbf{e}_l$ $\quad$ ($\boldsymbol{\sigma}^{\mathrm{T}} = $ transpose of $\boldsymbol{\sigma}$) $\qquad$ tensor of rank two

$\mathbf{1} = \delta_{kl} \mathbf{e}_k \mathbf{e}_l$ $\qquad\qquad$ unit tensor

$\mathbf{u} \cdot \mathbf{v} = u_k v_k$ $\qquad\qquad$ inner product of pair of vectors

$\mathbf{T} : \mathbf{E} = T_{kl} E_{kl},$ $\qquad \mathbf{T} : \mathbf{1} = T_{kl} \delta_{kl} = T_{kk}$ $\qquad$ inner product of pair of tensors

$\text{grad } \mathbf{v} = \dfrac{\partial v_l}{\partial x_k} \mathbf{e}_k \mathbf{e}_l$ $\qquad\qquad$ gradient of $\mathbf{v}$

$\text{div } \boldsymbol{\sigma} = \dfrac{\partial \sigma_{kl}}{\partial x_k} \mathbf{e}_l$ $\qquad\qquad$ divergence of $\boldsymbol{\sigma}$

$\partial_t \psi = \dfrac{\partial \psi}{\partial t}$ $\qquad\qquad$ partial derivative of ψ with respect to time

References

Alvarez, W., T. Engelder & W. Lowrie 1976. Formation of spaced cleavage and folds in brittle limestone by dissolution. *Geology* 4, 698–701.

Batchelor, G. K. 1970. The stress system in a suspension of force-free particles. *J. Fluid Mech.* **41**, 543–70.

Bathurst, R. C. G. 1958. Diagenetic fabrics in some British Dinantian limestones. *Liverpool Manchester Geol. J.* **2**, 11–36.

Beran, M. J. 1968. *Statistical continuum theories*. New York: Wiley.

Biot, M. A. 1972. Theory of finite deformations of porous solids. *Ind. Univ. Math. J.* **21**, 597–620.

Biot, M. A. 1973. Nonlinear and semilinear rheology of porous solids. *J. Geophys. Res.* **78**, 4924–37.

Bosworth, W. 1981. Strain-induced partial dissolution of halite. *Tectonophysics* **78**, 509–25.

Brenner, H. 1970. Rheology of two-phase systems. *Ann. Rev. Fluid Mech.* **2**, 137–76.

Buxton, T. M. & D. F. Sibley 1981. Pressure solution features in a shallow buried limestone. *J. Sediment. Petrol.* **51**, 19–26.

Chadwick, P. 1976. *Continuum mechanics*. New York: Wiley.

Coble, R. L. 1963. A model for boundary diffusion controlled creep in polycrystalline materials. *J. Appl. Phys.* **34**, 1679–82.

De Boer, R. B., P. J. C. Nagtegaal & E. M. Duyvis 1977. Pressure solution experiments on quartz sand. *Geochim. Cosmochim. Acta* **41**, 257–64.

De Groot, S. R. & P. Mazur 1962. *Non-equilibrium thermodynamics*. Amsterdam: North Holland.

Denbigh, K. 1971. *The principles of chemical equilibrium*, 3rd edition. Cambridge: Cambridge University Press.

Dunnington, H. V. 1967. Aspects of diagenesis and shape change in stylolitic limestone reservoirs. In *Proceedings Seventh World Petroleum Congress Mexico*, Vol. 2, 339–458.

Durney, D. W. 1972. Solution-transfer, an important geological deformation mechanism. *Nature* **237**, 315–17.

Durney, D. W. 1978. Early theories and hypotheses on pressure-solution-redeposition. *Geology* **6**, 369–72.

Elliott, D. 1973. Diffusion flow laws in metamorphic rocks. *Geol Soc. Am. Bull.* **84**, 2645–64.

Gibbs, J. W. 1906. On the equilibrium of heterogeneous substances. In *The scientific papers of J. Willard Gibbs*, Vol. 1, 55–353, Toronto: Longman (New York; Dover, 1961).

Gratier, J. P. & R. Guiget 1986. Experimental pressure solution on quartz grains; the crucial effect of the nature of the fluid. *J. Struct. Geol.* **8**, 845–56.

Green, H. W. 1984. 'Pressure solution' creep: some causes and mechanisms. *J. Geophys. Res.* **89**, 4313–18.

Heidug, W. 1985. *A thermodynamic theory of fluid infiltrated porous media undergoing large deformations and changes of phase*. PhD thesis, Division of Engineering, Brown University.

Houseknecht, D. W. 1984. Influence of grain size and temperature on intergranular pressure solution, quartz cementation, and porosity in a quartzose sandstone. *J. Sediment. Petrol.* **54**, 348–61.

Houseknecht, D. W. 1988. Intergranular pressure solution in four Quartzose sandstones. *J. Sediment. Petrol.* **58**, 228–46.

Hill, R. 1972. On constitutive macro-variables for heterogeneous solids at finite strain. *Proc. R. Soc. Lond.* A **326**, 131–47.

Kamb, W. B. 1961. The thermodynamic theory of non-hydrostatically stressed solids. *J. Geophys. Res.* **66**, 259–71.

Kestin, J. & J. R. Rice 1970. Paradoxes in the application of thermodynamics to strained solids. In *A critical review of thermodynamics*, E. B. Stuart, B. Gal-or & A. J. Brainard (eds), 275–98. Baltimore, Maryland: Mono Book Corp.

Lehner, F. K. 1979. A derivation of the field equations for slow viscous flow through a porous medium. *Indust. Engng Chem. Fundament.* **18**, 41–5.

Lehner, F. K. & J. Bataille 1984. Non-equilibrium thermodynamics of pressure solution. *Pure Appl. Geophys.* **122**, 53–85.

Meixner, J. 1960. Relaxationserscheinungen und ihre thermodynamische Behandlung. *Ned. T. Natuurk.* **26**, 259–73.

Mosher, S. 1981. Pressure solution deformation of the purgatory conglomerate from Rhode Island. *J. Geol.* **89**, 37–55.

Mullins, W. W. & R. F. Sekerka 1985. On the thermodynamics of crystalline solids. *J. Chem. Phys.* **82**, 5192–202.

Paterson, M. S. 1973. Nonhydrostative thermodynamics and its geologic applications. *Rev. Geophys. Space Phys.* **11**, 355–89.

Pharr, G. M. & M. F. Ashby 1983. On creep enhanced by a liquid phase. *Acta Metall.* **31**, 129–38.

Raj, R. & M. F. Ashby 1971. On grain boundary sliding and diffusional creep, *Metall. Trans.* **2**, 1113–27.

Raj, R. 1982. Creep in polycrystalline aggregates by matter transport through a liquid phase. *J. Geophys. Res.* **87**(B6), 4731–9.

Rice, J. R. 1971. Inelastic constitutive relations for solids: an internal variable theory and its application to metal plasticity. *J. Mech. Phys. Solids* **19**, 433–55.

Rice, J. R. 1975. Continuum mechanics and thermodynamics of plasticity in relation to micro-scale deformation mechanisms. In *Constitutive equations in plasticity*, A. S. Argon (ed.), 23–75. Cambridge, Mass.: MIT Press.

Rice, J. R. 1977. Pore pressure effects in inelastic constitutive formulations for fissured rock masses. In *Advances in civil engineering through engineering mechanics*, 360–3. New York: American Society of Civil Engineers.

Robin, P. Y. F. 1978. Pressure solution at grain-to-grain contacts. *Geochim. Cosmochim. Acta* **42**, 1383–9.

Rutter, E. H. 1976. The kinetics of rock deformation by pressure solution. *Phil Trans R. Soc. Lond.* A **283**, 203–19.

Rutter, E. H. 1983. Pressure solution in nature, theory and experiment. *J. Geol. Soc. Lond.* **140**, 725–40.

Slattery, J. C. 1972. *Momentum energy and mass transfer in continua*. New York: McGraw-Hill.

Sorby, H. C. 1863. On the direct correlation of mechanical and chemical forces. Proc. Roy. Soc. Lond. **12**, 538–50.

Spiers, C. J. & P. M. T. M. Schutjens 1990. Densification of crystalline aggregates by fluid-phase diffusional creep. This volume, 334–531.

Stockdale, P. B. 1922. Stylolites: their nature and origin. *Indiana Univ. Stud.* **9**, 1–97.

Tada, R. & R. Siever 1986. Experimental knife-edge pressure solution of halite. *Geochim. Cosmochim. Acta* **50**, 29–36.

rai, J. L. 1983. *Deformation of wet salt rocks.* PhD thesis, Inst. Earth. Sci., University of Utrecht.

rai, J. L., W. D. Means & G. S. Lister 1986. Dynamic recrystallization of minerals. In *Mineral and rock deformation: laboratory studies,* B. E. Hobbs & H. C. Heard (eds), 161–99. *Geophys. Monogr.* **36,** Washington, DC.: American Geophysical Union.

ɫaldschmidt, W. A. 1941. Cementing materials in sandstones and their probable influence on migration and accumulation of oil and gas. *AAPG Bull.* **25,** 1839–79.

ɫanless, H. R. 1984. Styles of pressure dissolution. In *Stylolites and associated phenomena – relevance to hydrocarbon reservoirs,* 81–105. Special Publication of Abu Dhabi National Reservoir Research Foundation (ADREF), Abu Dhabi.

ɫeyl, P. K. 1959. Pressure solution and the force of crystallisation – a phenomenological theory. *J. Geophys. Res.* **64,** 2001–25.

ɫhite, J. C. & S. H. White 1981. On the structure of grain boundaries in tectonites. *Tectonophysics* **78,** 613–28.

ɫilson, T. V. & D. F. Sibley 1978. Pressure solution and porosity reduction in shallow buried quartz arenite. *AAPG Bull.* **62,** 2329–34.

CHAPTER TWELVE

Densification of crystalline aggregates by fluid-phase diffusional creep

Christopher J. Spiers & Peter M. T. M. Schutjens

12.1 Introduction

It is widely accepted in the literature that polycrystalline minerals, ceramics, and rocks containing subcontinuous intergranular films of solvent or melt can deform by diffusive mass transfer through the grain-boundary liquid phase (Durney 1972, Robin 1978, Raj 1982, Rutter 1983, Cooper & Kohlstedt 1984). This type of mechanism is generally referred to as fluid-phase diffusional creep (Stocker & Ashby 1973), solution-precipitation creep (Raj 1982), or pressure solution (e.g. Rutter 1983), and is of well known interest both in materials science (see Lange *et al.* 1980, Raj & Chyung 1981, Raj 1982) and in the Earth sciences (see Robin 1978, Rutter & Mainprice 1978, Rutter 1983, Green 1984, Urai *et al.* 1986). In the present chapter we shall use the term 'fluid-phase diffusional creep' (FPDC) for the process. For mechanisms involving coupled solid-state flow plus dissolution at grain-contact margins (Pharr & Ashby 1983, Green 1984) we shall use the term 'dissolution-coupled creep'.

In recent years, there have been numerous attempts to develop theoretically based constitutive models for fluid-phase diffusional creep (Stocker & Ashby 1973; Rutter 1976, 1983; Raj & Chyung 1981; Raj 1982). In these models, grain boundaries are considered to contain fluid in some interconnected form which cannot be squeezed out, i.e. in a strongly adsorbed thin film (Rutter 1976, 1983; Robin 1978) *or* in a fine-scale island-channel network containing fluid at uniform pressure p (Elliot, 1973, Raj & Chyung 1981, Raj 1982). Creep is envisaged to occur via a fluid-enhanced grain-boundary diffusional mechanism involving dissolution of material at interfaces under high mean normal stress (σ_n), diffusion through the grain-boundary 'fluid', and precipitation at interfaces under low mean normal stress. In this manner, both deviatoric and densification (compaction) creep are thought to be possible (see Fig. 12.1).

In almost all FPDC models, mass transfer is considered to be driven by

334

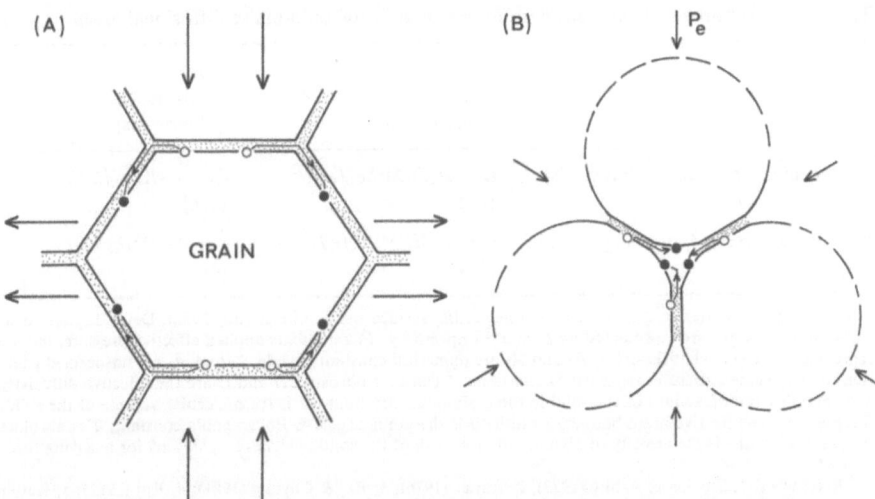

Figure 12.1 Mass transfer steps thought to be involved in fluid-phase diffusional creep. (A) Creep in dense polycrystalline material. (B) Densification or compaction creep in a fluid-saturated porous aggregate subjected to effective pressure P_e.

stress-induced gradients in chemical potential, defined by

$$\Delta \bar{\mu}_s = \Delta \sigma_n \Omega^s \tag{12.1}$$

where $\Delta \bar{\mu}_s$ is the chemical potential drop (per molecule) between individual source and sink sites, $\Delta \sigma_n$ is the corresponding difference in mean inter-granular or interfacial normal stress, and Ω^s is the molecular volume of the solid phase (Paterson 1973, Robin 1978, Raj & Chyung 1981). An expression for creep rate is then obtained from a consideration of the kinetics of the three serial processes of dissolution at source sites, diffusion through the grain-boundary fluid, and precipitation at sink sites, the slowest of these steps being rate controlling (Raj & Chyung 1981, Raj 1982). Constitutive equations obtained in this manner are summarized in Table 12.1.

This theoretical approach is widely accepted and to some extent supported by experimental evidence (Raj & Chyung 1981, Spiers *et al.* 1986). However, it has recently been shown by Lehner & Bataille (1984) that the concept of driv-ing force represented by equation (12.1) lacks a sound theoretical basis and that physically realistic models require formulation within a more general (non-equilibrium) thermodynamic framework than the classical equilibrium approach (Paterson 1973, Robin 1978) upon which (12.1) is based.

In the present chapter, we develop a model for densification by FPDC using a thermodynamic dissipation approach. This avoids (12.1), and an expression for creep rate is obtained by assuming that all work done during compaction is dissipated driving the FPDC mechanism. We go on to report a series of com-paction creep experiments performed on wet NaCl powder. The results agree

Table 12.1 Summary of previous constitutive models for fluid-phase diffusional creep.

	Diffusion-controlled case [References][*]	Dissolution/precipitation-controlled case [References][*]
deviatoric creep (dense polycrystals)	$\dot{\varepsilon}_{ij} = A_1 DC\Omega^s S\sigma_{ij}/kTd^3$ [1-5]	$\dot{\varepsilon}_{ij} = A_2 I\Omega^s\sigma_{ij}/kTd$ [3, 4]
densification creep (porous aggregates)	$\dot{\beta} = A_3 DC\Omega^s SP_e/kTd^3$ [2, 4, 5]	$\dot{\beta} = A_4 I\Omega^s P_e/kTd$ [4]

The above are written in generalized form neglecting surface energy effects (Raj 1982). Deviatoric and densification creep rates are represented by $\dot{\varepsilon}_{ij}$ and $\dot{\beta}$ respectively. P_e represents applied effective pressure, and σ_{ij} represents applied deviatoric stress. A_1 and A_2 are numerical constants, while A_3 and A_4 are unspecified functions of aggregate structure, regarded as constants at constant porosity. D and C are the effective diffusivity and solubility (mol fraction) of the solid in the grain-boundary fluid, Ω^s is the molecular volume of the solid, S represents the effective grain-boundary width, d is the grain size, k is Boltzmann's constant, T is absolute temperature, and I is the velocity of dissolution or growth of the solid (whichever is slower) for a driving force of 1 kT.

[*]References: 1, Stocker & Ashby (1973); 2, Rutter (1976); 3, Raj & Chyung (1981); 4, Raj (1982); 5, Rutter (1983).

favourably with our model and, in contrast to previous work (Raj 1982), provide strong evidence that densification of wet NaCl occurs by a diffusion-controlled FPDC mechanism.

12.2 The model

12.2.1 Grain-boundary structure

For fluid-phase diffusional creep to be a physically feasible mechanism of densification, grain boundaries must possess a stable structure allowing wetting by the solvent while simultaneously transmitting intergranular tractions. In the present model we assume an island-channel grain-boundary structure with the generalized configuration illustrated in Figure 12.2. Following Raj (1982) and Lehner & Bataille (1984), this structure is assumed to be 'dynamically stable' in the sense that during continuous dissolution or precipitation within an individual element, the average structure remains constant.

This assumption is based on the following argument (see also Lehner & Bataille 1984). Consider Figure 12.2: within such an element, the condition of chemical equilibrium between solid and fluid at any point on the phase boundary is given (neglecting surface energy terms) by

$$\mu_s = f^s + p/\rho^s \tag{12.2}$$

where μ_s is the mass specific chemical potential of the solid in solution, f^s is the mass specific Helmholtz free energy of the solid at its surface (including

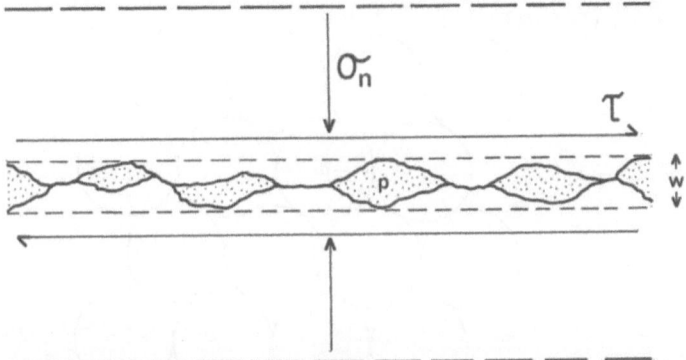

Figure 12.2 Section through representative grain-boundary element with island-channel structure. The structure is characterized by the parameters w and α, where w is the thickness of the grain-boundary zone, and α is the volume fraction occupied by islands (i.e. by solid). The channels contain fluid at pressure $p < \sigma_n$, where σ_n is the *mean* normal stress transmitted across the element. In the present model, the structure is assumed to be 'dynamically stable' (see text).

contributions related to elastic and inelastic deformation), p is the pressure in the fluid, and ρ^s is the density of the solid (Lehner & Bataille 1984). When the grain-boundary element is loaded (Fig. 12.2), stress concentrations at islands will cause preferential deformation at these sites plus corresponding increases in f^s due to elastic and defect-stored energies (in the general case). The resulting gradients in f^s will also be accompanied by gradients in ρ^s, although these will usually be negligibly small. Applying (12.2) to the phase boundary, it now becomes clear that *equilibrium cannot exist* within the zone considered. Instead, the island structure will tend to be smoothed out under the influence of gradients in f^s, and the grain-boundary fluid will be expelled or trapped in isolated inclusions. In general, however, the presence of defects in the solid (e.g. dislocations, impurity atoms), heterogeneities in solid deformation, and crystallographically controlled interface kinetics will lead to perturbations in the rate of dissolution and precipitation along the phase boundary. Provided that the fluid wets the solid and that the grain-boundary element constitutes an open system, this process will maintain a rough surface with some average structure, i.e. it will maintain a 'dynamically stable' island structure characterized by the parameters w and α defined in Figure 12.2. This assumed, non-equilibrium structure forms the basis of the present model for fluid-phase diffusional creep.

12.2.2 The creep model

Consider now a representative volume of idealized crystal aggregate consisting of a close-packed (fcc) array of spherical grains, each of diameter d (Fig. 12.3). This representative volume is subjected to an external hydrostatic pressure P and flooded with a pore-fluid solution phase at pressure $p = P$, the

(A)

(B)

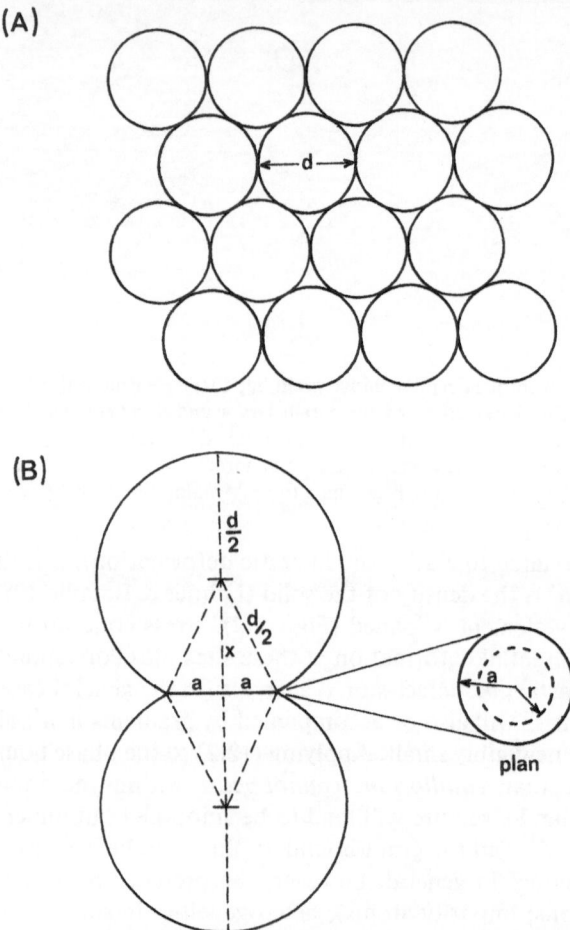

plan

Figure 12.3 Geometric aspects of the idealized crystal aggregate considered in the present model for densification by FPDC. (A) Aggregate in initial uncompacted state, i.e. volumetric strain $e_v = 0$. (B) Grain-contact geometry after a small volumetric compaction $e_v \simeq 3e$, where e is the conventional strain in any principal direction. Note that the radius $(d/2)$ of free grain surfaces is assumed constant. This is a reasonable approximation provided that $e_v < 20$ per cent. Note also that $x = d(1 - e)/2$, and $a^2 = d^2(2e - e^2)/4 \simeq d^2 e_v/6$ when e is small.

solution phase being in chemical equilibrium (i.e. saturated) with respect to the solid. The value of P is now increased, while p is held constant, so that the aggregate experiences an applied effective pressure $P_e = (P - p)$ tending to drive compaction. In developing our model for the response of the aggregate, we proceed by making the following principal assumptions:

(1) The representative volume of aggregate plus enclosed fluid is assumed to form a thermodynamically homogeneous system.

(2) After an initial pseudo-Hertzian deformation, grain contacts are assumed to develop the fine-scale dynamically stable island-channel structure discussed above, with the channels containing fluid at the pore pressure p. Within the system thus defined, radial gradients in average interfacial tractions will in general exist between the centre of grain contacts and free pore walls, these being accompanied by concomitant gradients in surface deformation (elastic and/or plastic), in f^s at the phase boundary, and in the grain-boundary structural parameters α and w within grain contacts (see also Lehner & Bataille 1984; Lehner, this volume). In the present treatment, we assume that gradients in α and w are negligible (i.e. that α and w are constants). Under circumstances for which this assumption is broadly valid, the *above-mentioned gradients in f^s* will provide a driving force for mass transfer from contacts to pores and thus for densification by an FPDC mechanism. Note that this driving force is *not* equivalent to that assumed in most previous treatments (e.g. Rutter 1976, Raj & Chyung 1981, Raj 1982), i.e. it is not due to gradients in the '*PV*' term $\sigma_n \Omega^s$ represented in (12.1). Indeed, (12.1) cannot strictly apply to the island-structured grain boundary, since its derivation (Paterson 1973, Robin 1978) is based on an *a priori* assumption of equilibrium within individual grain-boundary elements. This point is discussed in detail by Lehner & Bataille (1984) and Lehner (this volume).

(3) Densification of the aggregate is assumed to occur purely by the above FPDC mechanism, with *all* work done by the applied effective pressure $P_e = (P - p)$ being dissipated in driving dissolution within grain contacts, diffusion into the pores, and precipitation on pore walls.

(4) During compaction, grain contacts are assumed to remain macroscopically flat, with free grain surfaces remaining spherical (see Fig. 12.3). Inwardly progressive dissolution effects at grain-contact margins (i.e. marginal dissolution) are assumed to be negligible in comparison with the rate of removal of material from grain contacts.

(5) Strains are assumed to be small. This simplifies the analysis greatly, introducing significant errors only when the volumetric strain of the aggregate (defined as $e_v = -\Delta V / V_0$) exceeds 20 per cent.

On the basis of assumption (12.3) we can now write

$$\dot{W}_v = \Delta_v \qquad (12.3)$$

where W_v represents the rate at which work is done on the compacting aggregate, and Δ_v is the total rate of dissipation associated with the FPDC mechanism (both quantities volume specific). In the present case $W_v = -P_e \dot{v}/V = P_e \dot{\beta}$, where V is the instantaneous volume of the element of aggregate considered and $\dot{\beta}$ is the volumetric strain (compaction) rate (Raj 1982). Hence,

$$\Delta_v = P_e \dot{\beta} \qquad (12.4)$$

per unit volume. From a consideration of the geometry of the compacting aggregate (Fig. 12.3) and assuming equal dissolution rates at all contacts (hence hydrostatic strain), this can be rewritten in terms of the dissipation (Δ_c) *per grain contact*. The result obtained is

$$\Delta_c = \frac{1}{6\sqrt{2}} (1 - e_v)d^3 P_e \dot{\beta} \qquad (12.5)$$

Attention is now restricted to the case of diffusion-controlled rates of FPDC, i.e. to the limiting case in which the grain-boundary diffusional step is the only significant dissipative process (all other processes such as dissolution and precipitation being negligibly easy). In this case, the rate of dissipation (Δ_{gb}) associated with diffusion through each grain contact becomes equal to Δ_c. We now seek an expression for Δ_{gb} from a consideration of diffusional dissipation within the grain boundary fluid. First, the rate of dissipation (per unit volume) at any point in the fluid is given (Meixner & Reik 1959) by

$$T\gamma^f = - \mathbf{J}_s \cdot \mathrm{grad}(\mu_s - \mu_d) \qquad (12.6)$$

where T is temperature, γ^f is the rate of entropy production per unit volume of solution, $\mathbf{J}_s$ is the diffusive mass flux vector of the solute, and μ_s and μ_d are the mass-specific chemical potentials of the solute and solvent. Coupling this with Fick's law written in the form (Lehner & Bataille 1984)

$$\mathbf{J}_s = - L \; \mathrm{grad}(\mu_s - \mu_d) \qquad (12.7)$$

and the Gibbs–Duhem equation

$$d\mu_s/d\mu_d = (C_s - 1)/C_s \qquad (12.8)$$

leads to the relation

$$T\gamma^f = L \,|\, \mathrm{grad} \; \mu_s \,|^2 / (1 - C_s)^2 \qquad (12.9)$$

where C_s is the concentration (mass fraction) of solute in solution, and L is a phenomenological coefficient related to solute diffusivity (D) and the density (ρ^f) of the solution via the relation

$$L(\partial \mu_s / \partial C_s) =)(1 - C_s)D\rho^f \qquad (12.10)$$

Consider now the geometry of a *single* grain-contact region (Fig. 12.3). The concentration profile within such a contact region, when the rate of dissolution (A) of the contact walls is uniform (i.e. when the contact remains *perfectly flat*), is given (Rutter 1976) by

$$C_s = C_s^0 + A(a^2 - r^2)/4D\rho^f \qquad (12.11)$$

where C_s now represents the solute concentration at radius r within the contact, a is the outer radius of the contact zone, C_s^0 is the solute concentration at the contact margin, and A is expressed in units of mass rate per unit volume of grain-boundary fluid. In the present model, grain contacts are assumed to possess a dynamically stable island structure while remaining *macroscopically flat*. This requires (12.11) to be satisfied in some time-statistical sense, implying a time-statistically radial distribution of C_s within grain contacts. Coupling (12.11) with the relation

$$\text{grad } \mu_s = (\partial \mu_s / \partial C_s) \text{grad } C_s \qquad (12.12)$$

and with (12.10), we obtain the expression

$$\text{grad } \mu_s = [(1 - C_s)D\rho^f / L](- Ar/2D\rho^f) \qquad (12.13)$$

which on substitution into (12.9) yields

$$T\gamma^f = A^2 r^2 / 4L \qquad (12.14)$$

for the dissipation rate per unit volume at any point within the grain-boundary fluid. The dissipation occurring in an annular element (width δr) of the contact (Fig. 12.3) then takes the form $T\gamma^f[2\pi rw(1 - \alpha)]\delta r$, and the total dissipation occurring due to diffusion through an entire single contact is given by

$$\Delta_{gb} = 2\pi w(1 - \alpha) \int_0^a (A^2 r^3 / 4L) dr \qquad (12.15)$$

Here, A is independent of r, and L is approximately constant (for concentrated solutions); thus

$$\Delta_{gb} = \pi A^2 w(1 - \alpha) a^4 / 8L \qquad (12.16)$$

Substituting into this for $A = d\dot{\beta}\rho^s / 3w(1 - \alpha)$ (Rutter 1976) and for $a^2 \simeq d^2 e_v / 6$ (see Fig. 12.3), and equating Δ_{gb} with Δ_c (the total work dissipated per contact – see (12.5) – now leads to the creep equation

$$\dot{\beta} = ZSLP_e(1 - e_v)/(\rho^s)^2 \, d^3 e_v^n \qquad (12.17)$$

where $Z = 216\sqrt{2}/\pi$, $S = w(1 - \alpha)$, and $n = 2$. To obtain an approximation for L, (12.10) is now coupled with the relation

$$\mu_s = \mu_s^0 + (kT/\rho^s\Omega^s) \ln \gamma_s C_s \qquad (12.18)$$

341

(where μ_s^0 is a reference potential, γ_s is a normalized activity coefficient assumed to be independent of C_s, and k is Boltzmann's constant), yielding

$$L \simeq (\rho^s \Omega^s C_s / kT)(1 - C_s) D \rho^f \qquad (12.19)$$

where C_s now takes the significance of the average solubility of the solute within the grain-boundary fluid. Substituting this into (12.17) gives

$$\dot{\beta} \simeq \frac{ZSC_s(1 - C_s)D\Omega^s \rho^f}{kT\rho^s} \frac{P_e(1 - e_v)}{d^3 e_v^n} \qquad (12.20)$$

These results (i.e. (12.17) and (12.20)) are closely similar to previous expressions for densification by diffusion-controlled FPDC (see Table 12.1). However, they possess the advantage that the influence of e_v (hence porosity) is explicitly accounted for. The manner in which (12.17) and (12.20) have been derived also possesses the advantage that since (12.1) is not used, the constitutive response of the solid at the grain contact scale is not explicitly tied to the distribution of μ_s (see Lehner, this volume).

Finally, we consider the influence of grain packing and shape on creep laws (12.17) and (12.20). Essentially identical results are obtained for non-fcc packing geometries, the final expressions for $\tilde{\beta}$ varying by a numerical factor less than 2. The influence of grain shape is harder to evaluate. However, a crude indication of the effect of considering cubic rather than spherical grains can be obtained by assuming that cube corners are truncated by dissolution. It is easily shown that this creates contact zones with average radius $a = \eta d e_v$, where η is a mean orientation factor of value approximately 2/3. When this relation is used in (12.16), n in (12.17) and (12.20) takes the value 4.

12.3 Experiments on sodium chloride

We now report experiments designed to test the applicability of the above model to the case of densification creep in brine-saturated NaCl aggregates. The approach adopted involved truly hydrostatic compaction creep experiments in which the applied effective pressure (P_e) and sample grain size (d) were independently varied, from test to test, in order to determine the dependence of volumetric compaction rate ($\dot{\beta}$) on P_e, d, and on volumetric strain (e_v).

12.3.1 Experimental method

The starting material was prepared by sieving oven-dried, analytical grade NaCl powder into controlled grain-size fractions of 180–212 μm, 212–250 μm, 250–300 μm, and 300–355 μm. Individual 100 g samples of these grain-size fractions were jacketed in rubber 'balloons' and tested at room temperature using the servo-controlled hydrostatic compaction apparatus illustrated in Fig. 12.4. In all experiments, the controlled grain size starting powder was first compacted dry at an effective pressure of 2.15 Mpa for ~30 min, venting the pores to atmosphere via valve A (Fig. 12.4). In each case, this led to an essentially time-independent volumetric compaction of 2–3 per cent, producing a well controlled 'starting aggregate' with a porosity of $\phi_0 = 40.5 \pm 1.5$ per cent.

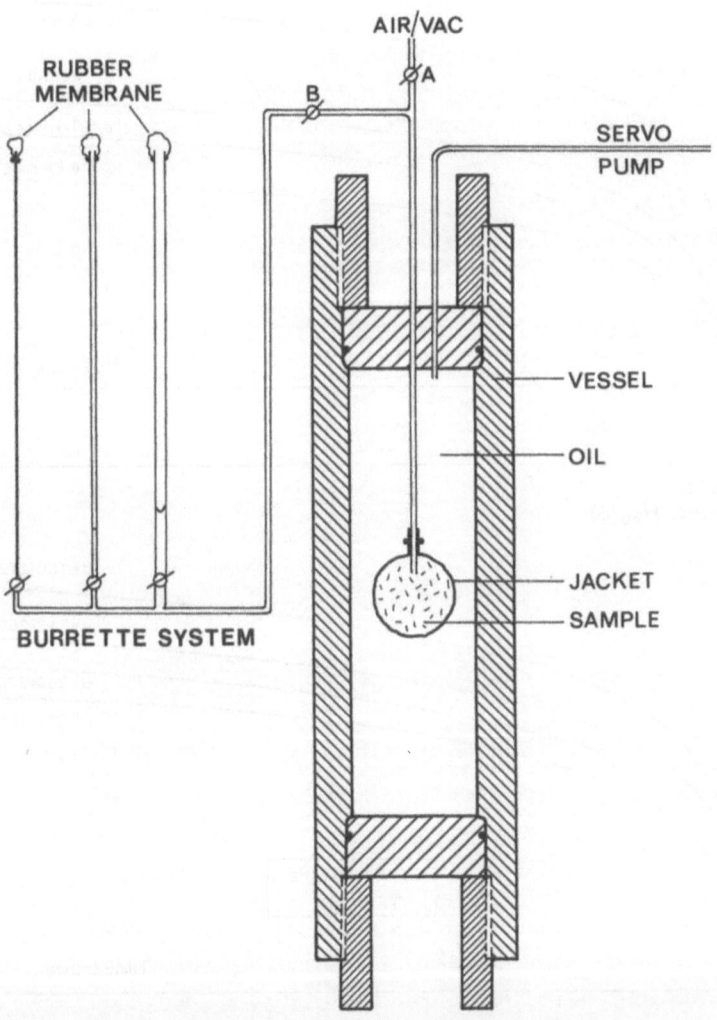

Figure 12.4 Semi-schematic diagram illustrating the hydrostatic compaction apparatus used in the present experiments on brine-saturated NaCl aggregates. The compaction-creep behaviour of each sample was monitored by measuring the volume of brine displaced from the sample into the burette system.

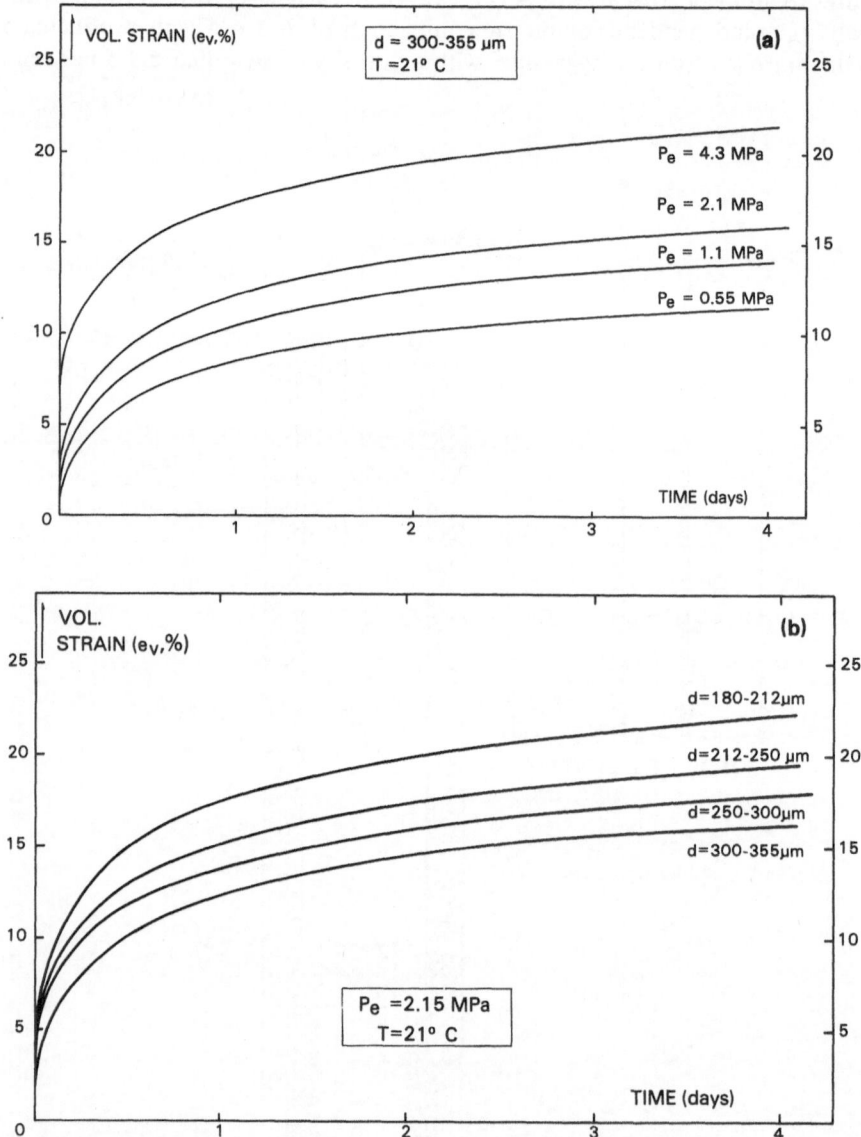

Figure 12.5 Typical densification creep curves obtained for brine-saturated samples. (A) Influence of effective pressure (P_e) at constant grain size. (B) Influence of grain size at constant effective pressure. (Note that data are shown to 4 days only, whereas a typical test duration is 10 days.)

The dry-compacted samples were then unloaded (P_e reduced to 0.05 MPa), evacuated via valve A (Fig. 12.4), and flooded with saturated brine from the burrette measuring system via valve B (Fig. 12.4). Samples were then reloaded by increasing P_e to the desired test value (the brine pressure being always equal to ~1 atm; see Fig. 12.4), and the compaction creep behaviour was monitored as a function of time by measuring the volume of brine displaced from each sample into the burrette system. The experiments were performed at constant effective pressures in the range $0.5 < P_e < 4.5$ MPa, using samples of each available grain-size fraction. Volumetric strains of up to 25 per cent were achieved. Tests were terminated by reducing P_e to zero and removing the sample from the compaction vessel as quickly as possible. The compacted sample was then clamped lightly into a rubber tube, and the remaining pore brine was carefully flushed out using compressed air and evaporating oil. This was done to minimize corruption of the microstructure by post-test evaporation and precipitation effects. Finally, the samples were impregnated to allow sectioning and subsequent microstructural study.

Note that up to four samples could be simultaneously tested using the apparatus illustrated in Figure 12.4, each sample being connected to its own system of burrettes. Although seemingly crude, these systems provided an extremely accurate method of measuring volume change, with a proven resolution of around 10^{-5} of the sample volume. The servo-pump used to control test

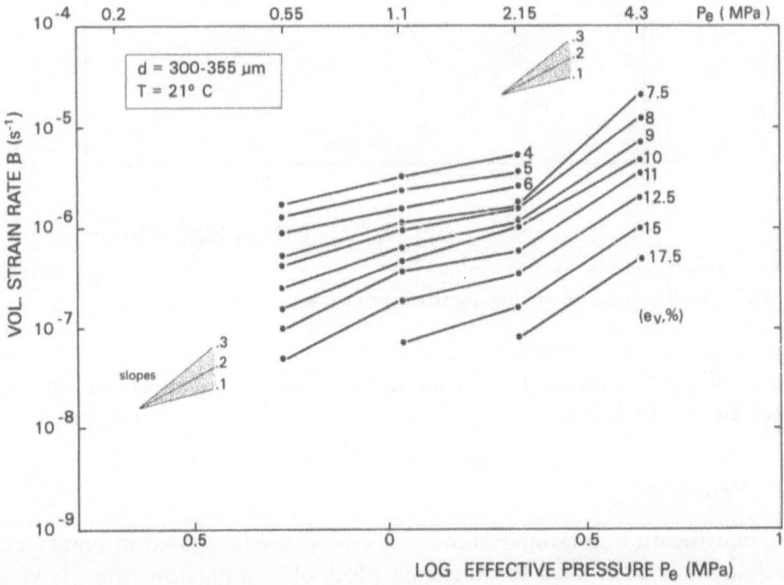

Figure 12.6 Log–log plot of volumetric compaction rate ($\dot{\beta}$) vs. effective pressure (P_e) constructed from the compaction-creep data of Figure 12.5A, for the values of volumetric strain (e_v) shown.

345

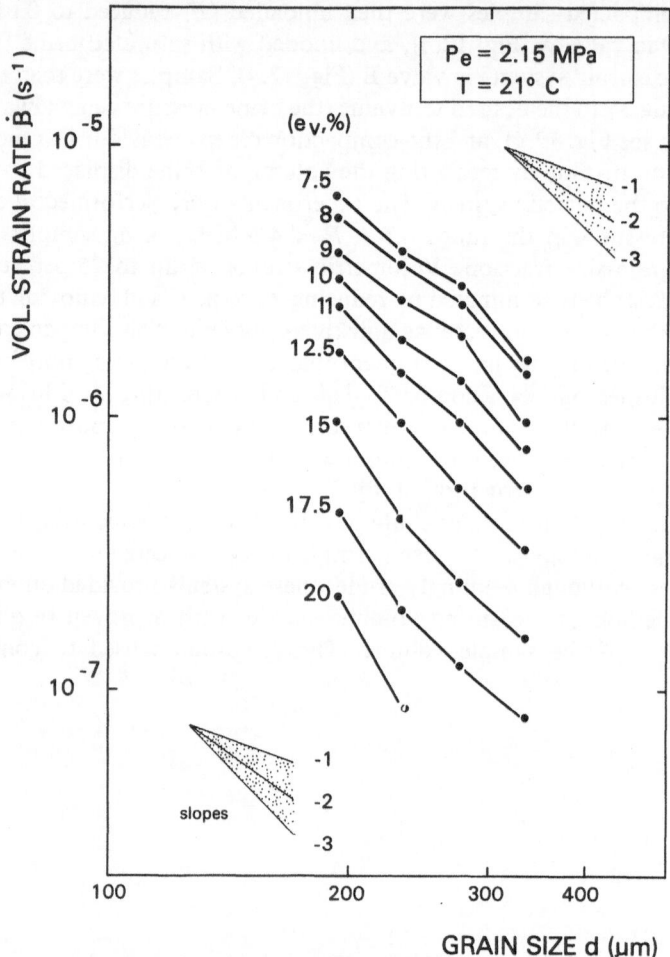

Figure 12.7 Log–log plot of compaction rate ($\dot{\beta}$) vs. grain size (d) constructed from the data of Figure 12.5B, for the values of volumetric strain (e_v) shown.

pressures allowed control to within about 0.005 MPa, with an absolute accuracy of ~0.01 MPa.

12.3.2 Results

Typical densification or compaction creep curves are presented in Figure 12.5. These data have been used to construct plots of compaction rate ($\dot{\beta}$) versus effective pressure (P_e), grain size (d), and volumetric strain (e_v); see Figures 12.6, 7 & 8 respectively. From Figure 12.6, it is clear that at constant volumetric strain (e_v) and grain size (d), $\dot{\beta}$ is more or less linearly related to

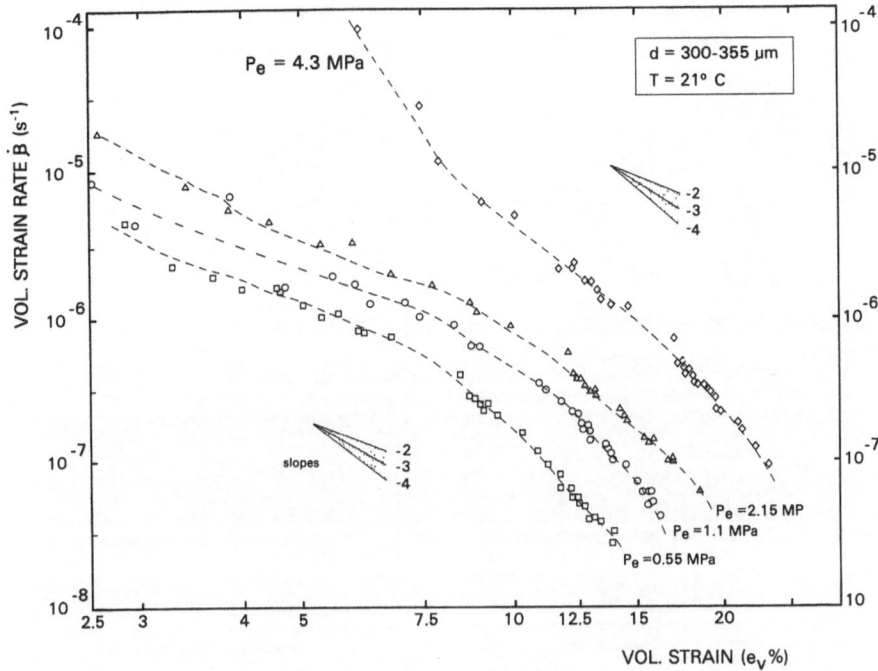

Figure 12.8 Log–log plot of compaction rate ($\dot{\beta}$) vs. volumetric strain (e_v) constructed from the creep data of Figure 12.5A, for the various effective pressures (P_e) investigated.

P_e (i.e. $\dot{\beta} \propto P_e^1$) in the region $P_e \leqslant 2.15$ MPa. At higher values of P_e, the exponent to P_e increases sharply. From Figure 12.7, it is evident that at constant effective pressure and volumetric strain, $\dot{\beta}$ is approximately proportional to d^{-3}. It is shown in Figure 12.8 that at constant effective pressure (P_e) and grain size (d), $\dot{\beta}$ is proportional to e_v^{-2} to e_v^{-5}, at least when $P_e \leqslant 2.15$ MPa.

12.3.3 Microstructural observations

The general microstructural nature of the *sieved salt fractions* prior to compaction is illustrated in Fig. 12.9A. Optical examination of thin sections of *dry-compacted samples* (i.e. material produced from the dry loading stage of the tests) revealed a highly porous aggregate structure consisting of a more or less randomly packed array of cubes, with little optically visible evidence for plastic deformation or indentation at grain contacts (Fig. 12.9B).

In the *wet-compacted samples* a strikingly different microstructure was observed. These samples showed a marked decrease in porosity plus abundant evidence for grain-to-grain indentation (Fig. 12.9C), contact truncation (Figs. 12.9D,E), and euhedral overgrowths within the pores (Fig. 12.9F). In the samples deformed at $P_e > 2.15$ MPa, occasional evidence was also found

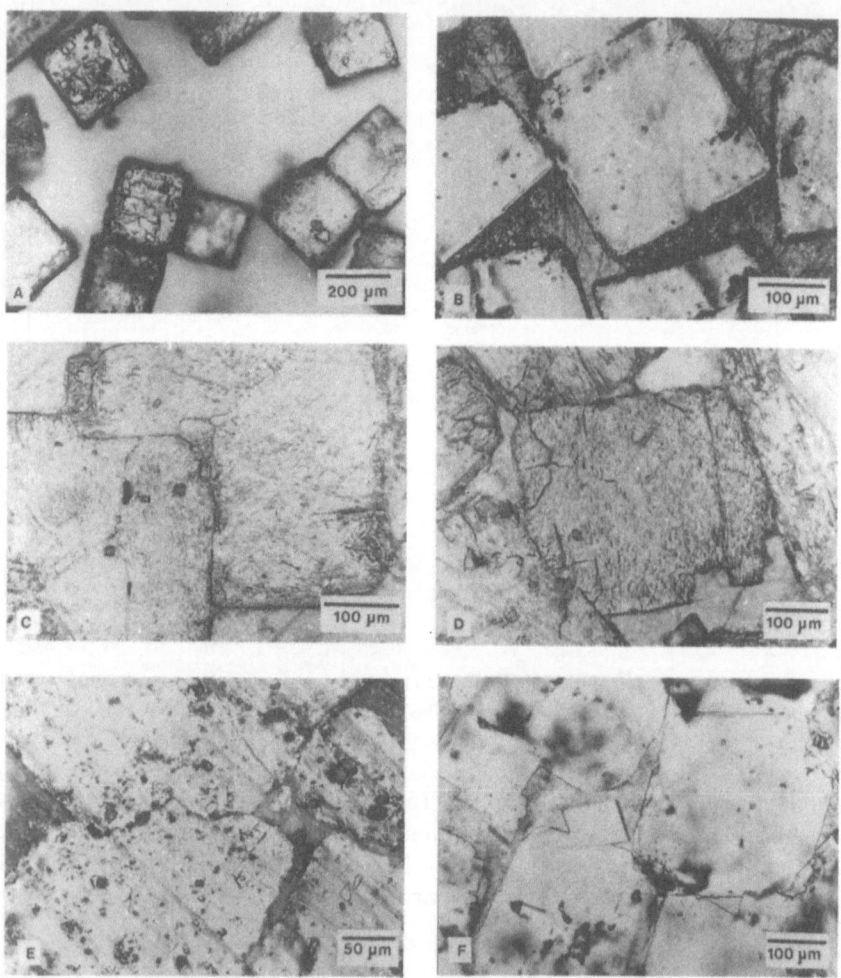

Figure 12.9 Optical microstructures. (A) Micrograph showing general nature of the sieved salt starting material used in the present experiments. (B) Transmission micrograph illustrating the microstructure of samples compacted dry at $P_e = 2.15$ MPa. (C–E) Typical indentation and truncation microstructures seen in wet-compacted samples. The particulate matter visible in micrograph (E) is Al_2O_3 powder added to this sample in an (unsuccessful) attempt to mark grain boundaries. (F) Transmission micrograph showing euhedral overgrowth features developed on free pore walls.

for fluid-assisted grain-boundary migration (Spiers *et al.* 1986, Urai *et al.* 1986), indicating that at least some plastic deformation had occurred. No clear evidence was found for the development of 'necks' at grain contacts, of the type expected if dissolution-coupled mechanisms or marginal dissolution (Pharr & Ashby 1983) were operative. However, optical examination of thick sections did reveal the presence of connected channel-like arrays of fluid (gas) inclusions within most grain contacts (Figs 12.10A,B). These channels showed a clear tendency to be crystallographically controlled and to 'neck-down' into negative crystal form. SEM studies confirmed the presence of these structures on mechanically parted grain contacts (Figs 12.10C,D), and showed the average thickness of grain-contact zones to be generally less than 100–200 nm.

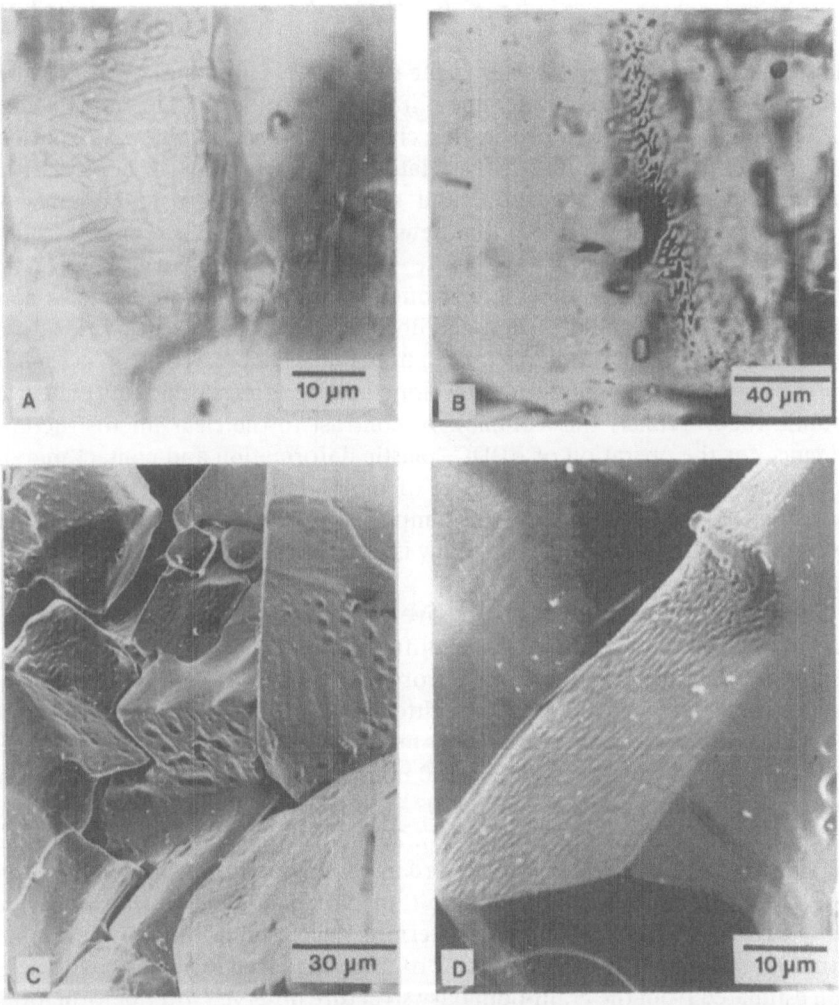

Figure 12.10 Structure of grain contacts in wet-compacted samples. (A, B) Transmission optical micrographs of grain contacts, in plan and oblique views respectively, showing channel-like arrays of fluid inclusions. (C) SEM micrograph of disaggregated sample showing grain-boundary channel structures and inclusions with nega-crystal form. (D) SEM micrograph of mechanically separated grain contact showing island-channel-like structure.

12.4 Discussion

12.4.1 Behaviour at $P_e \leqslant 2.15$ MPa

The experimental data reported above indicate that the densification creep behaviour of the brine-saturated samples, at effective pressures in the range $P_e \leqslant 2.15$ MPa, can be described by a constitutive relation of the form

$$\dot{\beta} = K^* P_e^{-1}/d^{-3} e_v^{\ n} \qquad (12.21)$$

where K^* is a constant and n falls in the range 2–5. For values of $e_v < 20$ per cent, this agrees favourably with the diffusion-controlled FPDC model (Eqns (12.17) and (12.20)) developed in this chapter. (Note that when $e_v < 20$ per cent, the term $(1 - e_v)$ in (12.17) is relatively unimportant.) In addition, the observed indentation, truncation, and overgrowth microstructures provide classical evidence (Rutter 1983) that densification occurred predominantly by 'pressure solution', i.e. by fluid-phase diffusional creep. Furthermore, the observed grain-boundary structure strongly suggests that grain contacts contained brine in some kind of non-equilibrium island-channel network during compaction, although it is important to note that the observed structure might represent a relaxed configuration developed on or after unloading (Lehner & Bataille 1984; Lehner, this volume). In contrast to the clear microstructural evidence for the operation of FPDC, plastic deformation and contact margin dissolution effects seem to have been of minor importance in the tests performed at $P_e \leqslant 2.15$ MPa. Since all samples were compacted dry at 2.15 MPa *before* wet testing, pure particulate flow is also thought to have been unimportant.

On this basis, we infer that at effective pressures $P_e \leqslant 2.15$ MPa, densification of our samples occurred by a diffusion-controlled FPDC mechanism, similar to that described by our theoretical model. Rough fitting of our experimental data to the model as written in (12.17), for the region $e_v < 10$ per cent, yields a value of $\sim 10^{-20}$ kg s m^{-2} for the phenomenological kinetic coefficient SL. Assuming that $S \simeq 100$ nm (as suggested by microstructural observations), this implies that $L \simeq 10^{-13}$ kg s m^{-3}. The fluid diffusivity term $C_s(1 - C_s)D$ (see Eqns (12.19) and (12.20)) must then take a value around 3×10^{-12} m^2 s^{-1}, which is about two orders of magnitude lower than the value of $C_s(1 - C_s)D$ for bulk NaCl solution (Spiers *et al.* 1986). This is thought to imply either that the fluid diffusivity term is decreased in the thin fluid films within grain contacts, *or* that the value of S *during* creep is substantially lower than estimated from the grain-boundary structure observed *after* deformation.

Finally, we compare our findings for the régime $P_e \leqslant 2.15$ MPa with densification data obtained by Raj (1982) for wet NaCl aggregates. Raj reports a d^{-1} dependence of creep rate, which he interprets as an indicator of interface kinetic control by comparison with the models given in Table 12.1. However, the dependence of $\dot{\beta}$ on e_v (i.e. on aggregate structure) is not taken into account in Raj's analysis, and his experimentally obtained relation between $\dot{\beta}$ and d appears to have been determined from pseudo-steady-state values of $\dot{\beta}$ rather than from values determined at constant e_v. This would lead to a systematic underestimation of grain-size sensitivity which, we believe, explains the discrepancy between Raj's findings ($\dot{\beta}$ proportional to d^{-1}) and those reported here ($\dot{\beta}$ proportional to d^{-3}).

12.4.2   Behaviour at $P_e > 2.15$ MPa

The behaviour observed in this régime clearly differs from that seen at $P \leqslant 2.15$ MPa and from the theoretical model. From microstructural evidence, this seems to be partly due to increased plasticity and minor fluid-assisted recrystallization (Spiers *et al.* 1986, Urai *et al.* 1986) occurring alongside FPDC. Further, since all samples were initially dry-compacted at $P_e = 2.15$ MPa, the wet runs at $P_e > 2.15$ MPa almost certainly involved a component of particulate sliding not seen in the low-pressure tests.

12.5   Summary

(1)   A theoretical model has been developed for densification of crystalline aggregates by fluid-phase diffusional creep. The approach adopted avoids the usual assumption that solution transfer is driven by chemical potential gradients defined by $\Delta \mu_s = \Delta \sigma_n \Omega^s$, and an expression for creep rate is obtained by assuming that all work done on the aggregate is dissipated by FPDC. For the grain-boundary diffusion-controlled case, this approach leads to the (isothermal) constitutive model $\dot{\beta} = KP_e(1 - e_v)/d^3 e_v{}^n$, where $n = 2$ for spherical grains and ~ 4 for cubes. The model is derived assuming a dynamically stable island-channel grain-boundary structure.

(2)   Densification creep experiments have been performed on brine-saturated NaCl aggregates at room temperature. Before testing in the brine-saturated state, all samples were compacted dry at an effective pressure (P_e) of 2.15 MPa. At effective pressures at and below 2.15 MPa, the mechanical data for the wet material match the theoretical model well. Microstructural evidence indicates that densification occurred predominantly by FPDC, and that grain boundaries contained a fluid-filled channel network during deformation. The implication is that at $P_e \leqslant 2.15$ MPa (the dry-compaction stress), densification of the wet NaCl powder occurred by a diffusion-controlled FPDC mechanism similar to our model. At effective pressures in excess of 2.15 MPa, deformation by FPDC seems to have been accompanied by plasticity, fluid-assisted recrystallization and particulate sliding.

Acknowledgements

We thank F. K. Lehner for discussions and advice regarding the theoretical treatment of solution-transfer processes. C. J. Peach is thanked for discussions and help in the laboratory. SEM work was performed under the supervision of M. R. Drury, using facilities supported by the Netherlands Organisation for the Advancement of Pure Science (NWO). P. M. T. M. Schutjens is supported by a generous research studentship awarded by Royal Dutch Shell (Koninklijke/Shell Exploratie en Produktie Laboratorium, Rijswijk, Netherlands). The authors benefitted significantly from discussions with B. K. Smith, made possible by NATO Grant RG.86/0148.

References

Cooper, R. F. & D. L. Kohlstedt 1984. Solution-precipitation enhanced diffusional creep of partially molten olivine basalt aggregates during hot-pressing. *Tectonophysics* **107**, 207–33.

Durney, D. W. 1972. Solution transfer, an important geologic deformation mechanism. *Nature* **237**, 315–17.

Elliot, D. 1973. Diffusion flow laws in metamorphic rocks. *Bull. Geol Soc. Am.* **84**, 2645–64.

Green, H. W. 1980. On the thermodynamics of non-hydrostatically stressed solids. *Phil Mag. A.* **41**, 637–47.

Green H. W. 1984. 'Pressure-solution' creep: some causes and mechanisms. *J. Geophys. Res.* **89**, 4313–18.

Lange, F. F., B. I. Davis & D. R. Clarke 1980. Compressive creep of Si_3N_4/MgO alloys. *J. Mat. Sci.* **15**, 601–10.

Lehner, F. K. 1990 Thermodynamics of rock deformation by pressure solution. This volume, 296–333.

Lehner, F. K. & J. Bataille 1984. Non-equilibrium thermodynamics of pressure solution. *Pure Appl. Geophys.* **122**, 53–85.

Meixner, J. & H. G. Reik 1959. Thermodynamik der irreversibelen Prozesse. In *Handbuch der Physik*, Vol. III/2: *Prinzipien der Thermodynamik und Statistik*. S. Flügge (ed), 413–523. Berlin: Springer.

Paterson, M. S. 1973. Nonhydrostatic thermodynamics and its geologic applications. *Rev. Geophys. Space Phys.* **11**, 355–89.

Pharr, G. M. & M. F. Ashby 1983. On creep enhanced by a liquid phase. *Acta Metall.* **31**, 129–38.

Raj, R. 1982. Creep in polycrystalline aggregates by matter transport through a liquid phase. *J. Geophys. Res.* **87**(B6), 4731–9.

Raj, R. & C. K. Chyung 1981. Solution precipitation creep in glass ceramics. *Acta Metall.* **29**, 159–66.

Robin, P.-Y. F. 1978. Pressure-solution at grain-to-grain contacts. *Geochim. Cosmochim. Acta* **42**(9), 1383–98.

Rutter, E. H. 1976. The kinetics of rock-deformation by pressure-solution. *Phil Trans R. Soc. A* **283**, 203–19.

Rutter, E. H. & D. H. Mainprice 1978. The effect of water on the stress relaxation of faulted and unfaulted sandstone. *Pure Appl. Geophys.* **116**, 634–54.

Rutter, E. H. 1983. Pressure solution in nature, theory and experiment. *J. Geol Soc. Lond.* **140**, 725–40.

Spiers, C. J., J. L. Urai, G. S. Lister, J. N. Boland & H. J. Zwart 1986. *The influence of fluid–rock interaction on the rheology of salt rock*. Commission of the European Communities, Nuclear Science and Technology, EUR 10399 EN.

Stocker, R. L. & M. F. Ashby 1973. On the rheology of the upper mantle. *Rev. Geophys. Space Phys.* **11**, 391–426.

Urai, J. L., C. J. Spiers, H. J. Zwart & G. S. Lister 1986. Weakening of rock salt by water during long-term creep. *Nature* **324**, 554–7.

CHAPTER THIRTEEN

Dynamic recrystallization and grain size

Brian Derby

13.1 Introduction

The idea that information about the stress history of a geological sample can be extracted from measurements made of the microstructure in the laboratory is attractive. While not attempting to be a review of palaeopiezometers, this chapter will present evidence for a 'universal relation' between deformation stress and the steady-state recrystallized grain size resulting from stress-induced boundary migration. This law is, of course, derived from measurements from laboratory experiments. Any application to naturally deformed rocks and minerals should only be carried out if all possible sources of confusion and error are fully understood. In particular, problems could arise from the effects of stress changing with time, and from subsequent deformation in other stress and temperature régimes leading to static recrystallization. More lengthy discussions of these problems are presented elsewhere (Mercier *et al.* 1977, Poirier 1985).

Studies of dynamic recrystallization are complicated by two distinct mechanisms, bearing similar names, which result in very different deformation microstructures: (1) rotation recrystallization (sometimes called *in situ* recrystallization) in which a gradual increase in sub-boundary misorientation after initial polygonization of dislocation structures, finally results in new high-angle grain boundaries; (2) migration recrystallization (or recrystallization by stress-induced boundary migration) in which a migration of existing grain boundaries sweeps through the deformed microstructure, replacing the polygonized material with new strain-free regions. This chapter will concentrate on the second mechanism.

It is generally accepted that, during elevated temperature deformation, the dislocation substructure forms a cellular or polygonized configuration with relatively strain-free cells or subgrains separated by dislocation walls or sub-boundaries, such that stress (σ) is related to the mean subgrain size, d:

$$\sigma d^m = \text{constant} \qquad (13.1)$$

with $m \approx 1$. This relation, which is also discussed by Barber (this volume), has been confirmed in metals and geological materials by many authors, and further refined by Takeuchi & Argon (1976) to:

$$\sigma d / G b = K_m \qquad (13.2)$$

where b is Burgers vector, G is the shear modulus, and K_m is a materials constant (specifically, $K_m \approx 10$ for metals and $K_m \approx 25-80$ for ionic crystals).

There has been very little consideration of the stress dependence of grain size during rotation dynamic recrystallization other than in individual material studies. The early work of White (1973) and Schmid *et al.* (1980) indicated the grain size to be indistinguishable from that of the prior subgrains. Tungatt & Humphreys (1984) show a dependence similar to that of subgrains (Eqn (13.1)), but with an exponent not equal to 1; however, the ranges of recrystallized grain sizes and prior subgrain sizes were very similar. Guillopé & Poirier (1979) found the relation given in (13.1), but with the new grains a little larger than the subgrains. We can only conclude that there is a close relation between rotation-recrystallized grain size and the prior subgrain structure.

The relation between deformation stress and migration-recrystallized grain size (D) has been studied by many authors in a number of metallurgical and geological systems. It is generally agreed that the relation

$$\sigma D^m = \text{constant} \qquad (13.3)$$

applies to each material, but the exponent m has been observed in a range of approximately $0.4 < m < 1.0$ The significance of this equation has been a topic of discussion in recent years. Twiss (1977) plotted recrystallized grain size data, normalized by the Burgers vector, against applied stress normalized by the shear modulus (i.e. D/b against σ/G) for a number of materials, and inferred a common exponent of $m \approx 0.8$ and a constant $(\sigma/G)(D/b)^m \approx 15$. Poirier (1985), however, produced a similar plot and showed that although the gradients on a log–log scale were in a small range, there was little evidence for a universal constant. The work of Drury *et al.* (1985) is confused by the plotting of a mixture of rotation- and migration-recrystallized material. Finally, a recent compilation of data (Derby & Ashby 1987) seemed to agree more with the hypothesis of Twiss. As a further criticism of the case for a 'law of recrystallized grain size', Poirier (1985) questioned the validity of such a law as an empirical fit of data in the absence of a mechanism. In particular, the rather arbitrary choice of a universal exponent m was criticized.

Here we present the fullest possible analysis of migration-recrystallization data and assess the evidence for a universal law. A review of the current understanding of this recrystallization mechanism will be presented. Finally, an attempt will be made to justify the validity of such a relation in the light of knowledge of physical constants within classes of materials.

13.2 Dynamic recrystallization by migration: mechanisms and models

The mechanism of dynamic recrystallization by stress-induced boundary migration is generally agreed to be identical to that accepted for static recrystallization (Bailey & Hirsch 1962). Recrystallization is nucleated when the energy stored by dislocations accumulated during deformation is sufficient to nucleate a small section of grain boundary to bow out and migrate, leaving a recrystallized strain-free region in its wake. The energy is stored primarily in the dislocations which make up the sub-boundaries. This mechanism gives no insight as to why a relation between stress and steady-state grain size should occur. A more detailed modelling of the process is required.

Sandström & Lagneborg (1975) produced a complex numerical model of dynamic recrystallization. This developed the Bailey & Hirsch idea (1962) along with a model of dislocation multiplication and predicted that grain size and strain rate ($\dot{\varepsilon}$) were related, with $D^2 \propto 1/\dot{\varepsilon}$. This model is considered to be unsatisfactory because of the need to allocate arbitrary values to a number of constants. Twiss (1977) presented a thermodynamic argument that balances stored energy of deformation against the boundary energy of the steady-state microstructure to predict a grain size – stress relationship. However, as pointed out by Poirier (1985), the argument is based on an unstable thermodynamic equilibrium, and is not a dynamic mechanism in accordance with observations of recrystallization in transparent materials.

A recent model of dynamic recrystallization by boundary migration (Derby & Ashby 1987) is derived from consideration of the steady-state microstructure. During steady-state recrystallization, it is known from observations of analogue transparent organic materials (Urai et al. 1975, Means 1983), that the constant mean grain size is achieved by cycles of nucleation and growth events occurring out of phase throughout the material. Nucleation events occur only on grain boundaries; hence we can assume a mean constant nucleation rate on all grain boundaries of $\dot{N}$ (per unit area) and a mean boundary-migration rate of g. For a steady state to exist, there must be a balance between nucleation rate and growth rate such that one nucleation event will occur per volume equivalent to one grain, in the time taken for a similar volume to be swept out by a recrystallization front (grain boundary), i.e.

$$\pi D^3 \dot{N}/2g = 1 \qquad (13.4)$$

Thus an estimate of mean grain size can be made if laws governing grain-boundary migration and nucleation rates can be derived. A more complete derivation of this model is presented elsewhere (Derby & Ashby 1987). Here only a brief description will be given.

The Bailey & Hirsch criterion for recrystallization states that nucleation will occur if the dislocation density, ρ, in the region ahead of a bowing grain boundary disc of diameter l is greater than a critical value ρ_c, defined as

$$\rho_c = 2E/(T_d l) \tag{13.5}$$

where T_d is the dislocation line tension, and E is the grain-boundary energy per unit area. Richardson *et al.* (1967) suggest that the distance l is defined by the spacing of dislocation sub-boundary walls acting as pinning points. These walls also act as the reservoir of stored energy during recrystallization. If the subgrains are of mean size d_s and of wall energy E_s, then the driving force (N m^{-2}) for recrystallization will be:

$$F = E_s/d_s \tag{13.6}$$

If we consider a two-dimensional analogue of the microstructure, a nucleus of diameter l can form by the bulging of a grain boundary if

$$E_s/d_s > E/l \quad \text{or} \quad l/d_s > E/E_s \tag{13.7}$$

Thus a nucleation criterion is defined as a length l of grain boundary pinned between two adjacent sub-boundaries on one side, which has at least n sub-boundaries (where $n = l/d_s$) intersecting on the other.

If we assume the sub-boundaries to randomly impinge the grain boundary with a mean spacing of $\bar{l}$, then the probability of there being a segment of length l free of sub-boundaries on a given side will be given by the exponential distribution

$$P(l) = (1/\bar{l}) \exp(-l/\bar{l}) \tag{13.8}$$

Similarly, in any given length of grain boundary l, the probability that n sub-boundaries intersect from the other side will be given by the Poisson distribution

$$P(n) = (1/n!) \, (l/\bar{l})^n \, \exp(-l/\bar{l}) \tag{13.9}$$

The fraction of grain-boundary segments between l and $l + \delta l$ is $P(l)\delta l$. Thus the number of pinned segments per unit length of boundary with n sub-boundaries impinging from the other side is approximately

$$N(n) = \frac{1}{n!\bar{l}} \int_0^\infty \left(\frac{l}{\bar{l}}\right)^n \exp\left(\frac{-2l}{\bar{l}}\right) \frac{dl}{\bar{l}} \tag{13.10}$$

which integrates to

$$N(n) = 1/(2^{n+1}\bar{l}) \tag{13.11}$$

357

The number of nuclei per unit length is the sum of $N(n)$ for all values of n above the critical value n_c, if we set $\bar{l}$ to $\bar{d}$, the mean sub-grain diameter, and then

$$N = \sum_{n_c}^{\infty} N(n) = 1/(2^{n_c}\bar{d}) \tag{13.12}$$

Hence, nucleation rate per unit length is

$$\dot{N}_L = -\dot{\varepsilon}(2^{n_c}\bar{d}^2)d(\bar{d})/\,\mathrm{d}\varepsilon \tag{13.13}$$

or, in these dimensions, the equivalent nucleation rate per unit area is

$$\dot{N} = -2\dot{\varepsilon}/(2^{n_c}\bar{d}^3)d(\bar{d})/\,\mathrm{d}\varepsilon \tag{13.14}$$

The nucleation rate is clearly dependent on the rate of subgrain formation. If recrystallization *does not* occur, (13.2) defines a limiting subgrain size of $d^* = K_m Gb/\sigma$. We postulate that subgrains approach this size from a limit of the grain size $\bar{D}$, such that

$$\frac{\mathrm{d}(\bar{d})}{\mathrm{d}t} = S(d^* - \bar{d})\frac{\mathrm{d}\varepsilon}{\mathrm{d}t} \tag{13.15}$$

where $1/S$ is a characteristic strain (typically 5 per cent) required to achieve steady state. Solving this for d and substituting into (13.14) gives

$$\dot{N} = \frac{S\dot{\varepsilon}(\bar{D}/d^* - 1)\exp(-S\varepsilon)}{2^{n_c-1}\,d^{*2}\,\{1 + (\bar{D}/d^* - 1)\exp(-S\varepsilon)\}^3} \tag{13.16}$$

During steady state, the dynamic balance criterion requires that in the time taken for a migrating grain boundary to sweep the mean grain volume, one new nucleation event must occur per equivalent volume of the material. This can be rewritten in terms of a recrystallization strain, ε_r, characteristic of the deformation cycle, defined by the integral of (13.16):

$$N = \int_0^{\varepsilon_r} \dot{N}G'\bar{D}^2\,\mathrm{d}\varepsilon =$$

$$\frac{(\bar{D}/d^* - 1)\bar{D}^2 J'}{2^{n_c-1}\,d^{*2}}\left[\frac{1}{\{1 + (\bar{D}/d^* - 1)\exp(-S\varepsilon_r)\}^2} - \frac{1}{(\bar{D}/d^*)^2}\right] = 1 \tag{13.17}$$

here J' is a geometric term related to grain shape. If the mean grain size is considerably larger than the steady-state subgrain size (i.e. $\bar{D} \gg d^*$), and if the recrystallization strain ε_r is less than the steady-state recovery strain $1/S$, (13.17) can be expanded thus

$$J' \bar{D} S \varepsilon_r / 2^{n_c - 1} d^* = 1 \qquad (13.18)$$

ε_r, the recrystallization strain, is the strain during one cycle of recrystallization as defined by the dynamic balance (13.4), so

$$\varepsilon_r = \dot{\varepsilon}_{ss} \bar{D} / g \qquad (13.19)$$

here $\dot{\varepsilon}_{ss}$ is the mean strain rate during dynamic recrystallization, which from (13.18) gives

$$D^2 = J d^* g / \dot{\varepsilon}_{ss} \qquad (13.20)$$

where $j = 2^{n_c - 1} / J' S$.

The growth rate is related to the stored energy, F, by a mobility term such that $g = MF$. This mobility is defined as a conventional grain-boundary mobility, i.e. $M = \delta D_B b / kT$; here δD_B is the product of boundary thickness and diffusion coefficient, and kT has its usual meaning. Now we assume that the stored energy is held by the subgrain structure, taking a mean boundary misorientation of θ and $T_d \approx G b^2$ for dislocation line tension then, from (13.6), $F \approx 6 G b \theta / d^*$. if we assume $\dot{\varepsilon}_{ss}$ to be similar to that determined by recovery creep, then, to a reasonable approximation

$$\dot{\varepsilon} = \frac{A D_v G b}{kT} \left(\frac{\sigma}{G} \right)^n \qquad (13.21)$$

in which A and n are constants and D_v is the volume diffusion coefficient. Substituting all the above into (13.20) gives:

$$\frac{D^2}{b^2} = \left(\frac{\delta D_B}{b D_v} \right) \left[\frac{6 J \theta}{A (\sigma/G)^n} \right] \qquad (13.22)$$

which can be rewritten in the form of (13.3), i.e.

$$\frac{\sigma}{G} \left(\frac{D}{b} \right)^m = C \qquad (13.23)$$

where $m = 2/n$ (n is the creep power-law exponent) and C is a constant containing creep, diffusion, and grain-shape terms.

13.3 Microstructural measurements as a function of stress

One must exercise due caution in reading too much into a single set of experimental measurements. There is a great deal of scatter and bias intrinsic in the methods of measuring mean grain and subgrain diameters. Hence it is best to consider data from tests on many materials when reduced to a dimensionless form by suitable normalizing materials' constants. For this work, the microstructural scale is normalized by a Burgers vector, while in the case of non-cubic materials the easiest slip system has been chosen (e.g. basal plane slip for Mg). The deformation stress has been normalized by a shear modulus

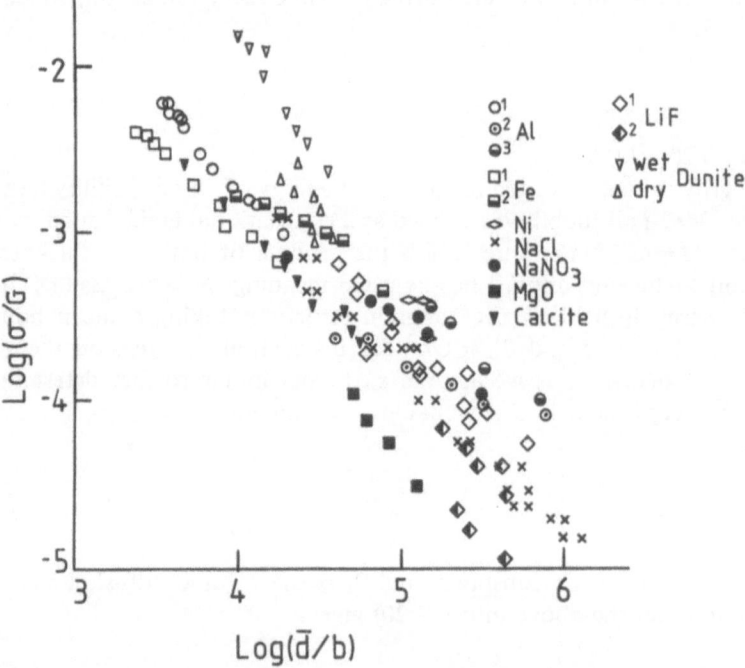

Figure 13.1 Mean deformation stress normalized by shear modulus as a function of measured mean subgrain diameter normalized by Burgers vector. Al[1], McQueen & Hockett (1970); Al[2], Miller & Sherby (1976); Al[3], Servi & Grant (1951); Fe[1], Goldberg (1966); Fe[2], Barrett *et al.* (1967); Ni, Richardson *et al.* (1967); NaCl/halite, Guillopé & Poirier (1979); NaNO₃/soda nitre, Tungatt & Humphreys (1984) – see (1) below; MgO, Hüther & Reppich (1973); LiF[1] Streb & Reppich (1973); LiF[2], Cropper & Pask (1973); calcite, Schmid *et al.* (1980) – see (2) below; dunite/olivine, Ross *et al.* (1980).

(1) Burgers vector from crystallographic data (Robie *et al.* 1966), with slip behaviour assumed to be identical to that of calcite. Elevated temperature elastic modulus from Birch (1966), with extrapolation law assumed to be identical to that for calcite.

(2) Burgers vector (*r*-slip) from crystallographic data (Robie *et al.* 1966); elevated temperature elastic modulus from Birch (1966).

All other Burgers vectors and shear moduli from Frost & Ashby (1982).

(G) taken as the c_{44} term of the compliance tensor. This shear modulus has a weak dependence on temperature, and the normalizing modulus used was either that measured at test temperature or else extrapolated. This temperature variation of shear modulus has not always been accounted for in previous compilations of recrystallization data, and thus only material where the test temperature was defined has been used in this study.

In Figures 13.1 & 2 the fullest compilation of reliable measurements of subgrain and migration-recrystallized grain size is presented as a function of deformation stress. Although there is a wide scatter of gradients for individual materials within both plots there is a distinct trend of $\sigma d \approx$ constant for the subgrain data (Fig. 13.1) and of $\sigma D^{0.67} \approx$ constant for the dynamic recrystallization by boundary-migration data (Fig. 13.2).

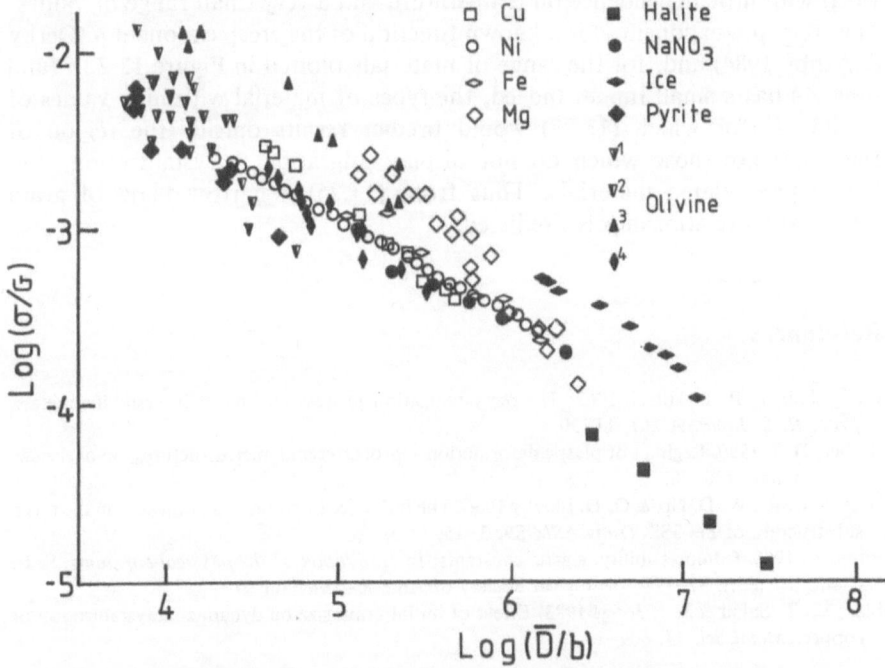

Figure 13.2 Mean deformation stress normalized by shear modulus as a function of measured migration-recrystallized mean grain diameter normalized by Burgers vector. Cu, Blaz *et al.* (1983); Ni, Sah *et al.* (1974); Fe, Glover & Sellars (1973); Mg, Drury *et al.* (1985); NaCl/halite, Guillopé & Poirier (1979); NaNO$_3$/soda nitre, Tungatt & Humphreys (1984) – see (1) below; ice, Steinneman, 1958); pyrite, Cox *et al.* (1981) – see (2) below; olivine/dunite[1] deformed wet at 1100°C, olivine/dunite[2] deformed wet at 1225°C, olivine/dunite[3] deformed dry at 1100°C, Ross *et al.* (1980); olivine/dunite[4] deformed at 1650°C, Karato *et al.* (1980).

(1) See Figure 13.1 for normalizing data.

(2) Burgers vector from crystallographic data (Robie *et al.* 1966) assuming the same slip system as NaCl structure; elevated temperature shear modulus from Birch (1966).

All other Burgers vectors and shear moduli from Frost & Ashby (1982).

Considering Figure 13.2 in more detail, it can be seen that all the grain size – stress data is within a region bounded by two lines of the form

$$\log(\sigma/G) = K - 0.667 \log(D/b) \qquad (13.24)$$

with an upper bound of $K \approx 20$ and a lower bound of $K \approx 1$. This region contains the relation inferred by Twiss (1977). How can this narrow region of results be reconciled with (13.23) and the materials constant C? From (13.22),

$$C = \left(\frac{\delta D_B}{bD_v} \frac{6J\theta}{A}\right)^{1/n} \qquad (13.25)$$

$\delta D_B/bD_v$ is a dimensionless grouping of materials constants (Frost & Ashby 1982) with little dependence on temperature and a very small range of values. The creep pre-exponent A is a known function of the creep exponent n (Derby & Ashby 1984) and, for the range of materials plotted in Figure 13.2, n (and thus A) has a small range. Indeed, the types of material with high values of n and A (for which (13.23) would predict results outside the region of Fig. 13.2) are those which do not display migration recrystallization, e.g. highly precipitated materials. Thus from (13.25) a narrow band of grain size – stress relationship is predicted.

References

Bailey J. E. & P. B. Hirsch 1962. The recrystallization process in some polycrystalline metals. *Proc. R. S. Lond.* A **267**, 11–30.

Barber, D. J. 1990. Regimes of plastic deformation – processes and microstructures; an overview. This volume, 138–78.

Barrett, C. R., W. D. Nix & O. D. Sherby 1966. The influence of strain and grain size on the creep substructure of Fe–3Si. *Trans ASM* **59**, 3–15.

Birch, F. 1966. Compressibility: elastic constants. In *Handbook of the physical constants*, S. P. Clark, Jr. (ed.), 97–174. Geological Society of America, Memoir **97**.

Blaz, L., T. Sakai & J. J. Jonas 1983. Effect of initial grain size on dynamic recrystallization of copper. *Metal Sci.* **17**, 609–16.

Cox, S. F., M. A. Etheridge & B. E. Hobbs, 1981. The experimental ductile deformation of polycrystalline and single crystal pyrite. *Econ. Geol.* **76**, 2105–17.

Cropper, D. R. & J. A. Pask 1973. Creep of lithium fluoride single crystals at elevated temperatures. *Phil Mag.* **27**, 1105–24.

Derby, B. & M. F. Ashby 1984. Power-laws and the $A-n$ correlation in creep. *Scripta Metall.* **18**, 1079–84.

Derby, B. & M. F. Ashby 1987. On dynamic recrystallization. *Scripta Metall.* **21**, 832–7.

Drury, M. R., F. J. Humphreys & S. H. White 1985. Large strain deformation studies using polycrystalline magnesium as a rock analogue, part II: dynamic recrystallization mechanisms at high temperatures. *Phys. Earth Planet. Inter.* **40**, 208–22.

Frost, H. J. & M. F. Ashby 1982. *Deformation-mechanism maps.* Oxford: Pergamon.

Glover, C. & C. M. Sellars 1973. Recovery and recrystallization during high temperature deformation of α-iron. *Metall. Trans* **4**, 765–75.

Goldberg, A. 1956. Influence of prior cold work on the creep resistance and microstructure of a 0.05% carbon steel. *J. Iron Steel Inst.* **204**, 268–77.

Guillopé, M. & J.-P. Poirier 1979. Dynamic recrystallization during creep of single crystal halite: an experimental study. *J. Geophys. Res.* **84**, 5557–67.

Hüther, W. & B. Reppich 1973. Dislocation structure during creep of MgO single crystals. *Phil Mag.* **28**, 363–71.

Karato, S., M. Toriumi & T. Fugii 1980. Dynamic recrystallization of olivine single crystals during high temperature creep. *Geophys. Res. Lett.* **7**, 649–52.

McQueen, H. J. & J. E. Hockett 1970. Microstructures of aluminium compressed at various rates and temperatures. *Metall. Trans.* **1**, 2997–3004.

Means, W. D. 1983. Microstructure and micromotion in recrystallization flow of octachloropropane: a first look. *Geol. Rund.* **72**, 511–28.

Mercier, J. C. C., D. A. Anderson & N. L. Carter 1977. Stress in the lithosphere: inference from the steady state flow of rocks. *Pure Appl. Geophys.* **115**, 199–226.

Miller, A. K. & O. D. Sherby 1976. On sub-grain strengthening at high temperatures. *Scripta Metall.* **10**, 311–17.

Poirier, J.-P. 1985. *Creep of crystals.* Cambridge: Cambridge University Press.

Richardson, G. J., C. M. Sellars & W. J. McTegart 1966. Recrystallization during creep of nickel. *Acta Metall.* **14**, 1225–36.

Robie, R. A., P. M. Bethke, M. S. Toulmin and J. L. Edwards 1966. X-ray crystallographic data, densities and molar volumes of minerals. In *Handbook of the physical constants*, S. P. Clark, Jr (ed.) 27–74. Geological Society of America, Memoir **97**.

Ross, J. V., H. G. Avé Lallemant & N. L. Carter 1980. Stress dependence of recrystallized grain and subgrain size in olivine. *Tectonophysics* **70**, 39–61.

Sah, J. P., G. J. Richardson & C. M. Sellars 1974. Grain size effects during dynamic recrystallization of nickel. *Metal Sci.* **8**, 325–31.

Sandström, R. & R. Lagneborg 1975. A model for hot working occurring by recrystallization. *Acta Metall.* **23**, 387–98.

Schmid, S. M., M. S. Paterson & J. N. Boland 1980. High temperature flow and dynamic recrystallization in Carrara marble. *Tectonophysics* **65**, 245–80.

Servi, I. S. & N. J. Grant 1951. Structure of aluminium deformed in creep at elevated temperatures. *Trans Am. Inst. Mech. Engrs* **191**, 917–22.

Steinneman, Von S. 1958. Experimentelle Untersuchungen zur Plastizitat von Eis. *Beitrage zur Geologie der Schweiz, Hydrologie* no. 10, 1–71.

Streb, G. & B. Reppich 1973. Steady state deformation and dislocation structure of pure and Mg-doped LiF single crystals. *Physica Status Solidi* (a)16, 493–505.

Takeuchi, S. & A. S. Argon 1976. Steady-state creep of single phase crystalline matter at high temperatures. *J. Mat. Sci.* **11**, 1547–55.

Tungatt, P. D. & F. J. Humphreys 1984. The plastic deformation and dynamic recrystallization of polycrystalline sodium nitrate. *Acta Metall.* **32**, 1625–35.

Twiss, R. J. 1977. Theory and applicability of a recrystallized grain size palaeopiezometer. *Pure Appl. Geophys.* **115**, 227–44.

Urai, J. L., F. J. Humphreys & S. E. Burrows 1980. *In-situ* studies of the deformation and dynamic recrystallization of rhombohedral camphor. *J. Mat. Sci.* **15**, 1231–40.

White, S. H. 1973. Syntectonic recrystallization and texture development in quartz. *Nature* **244**, 267–8.

CHAPTER FOURTEEN

Simulation of dislocation-assisted plastic deformation in olivine polycrystals

Toru Takeshita, Hans-Rudolf Wenk, Gilles R. Canova & Alain Molinari

14.1 Introduction

Studies on mantle convection provide insight into the geodynamics of the Earth. Convection models have relied on flow laws for olivine (e.g. Hager 1984, Christensen 1987), which is believed to be the major constituent of the upper mantle. The mantle is assumed to deform in a régime of steady-state creep in which diffusion and dislocation glide occur. When dislocation glide is present, polycrystalline materials develop anisotropy due to rotations of crystals during straining, which produce a preferred orientation distribution. Strong seismic anisotropy has been observed in the upper mantle underneath oceanic crust (e.g. Hess 1964) and continental crust (e.g. Fuchs 1983), and it is generally accepted that at high pressure where microfractures are closed, seismic anisotropy can be due to crystallographic preferred orientation or texture (e.g. Christensen 1984). Whereas Arrhenius-type flow laws are adequate to describe the diffusion-controlled plastic deformation (e.g. Karato *et al.* 1986), they do not take account of plastic anisotropy due to slip. We attempt to introduce a generalized description of anisotropic plastic flow of olivine based on polycrystal plasticity theory which takes into account the deformation history. It should be emphasized that all arguments brought forward relate to processes in which dislocation movements are involved and do not apply to recrystallization, which is also important in deformation of olivine (e.g. Avé Lallemant & Carter 1970). Deformation by grain-boundary sliding, such as during superplastic flow, does not produce texture.

Deformation of polycrystals by intracrystalline slip was first explained quantitatively by Sachs (1928). He assumed that in each crystal the slip system which has the highest resolved shear stress is activated. Such a model allows for a continuous stress state, but since each crystal deforms differently, it creates openings or overlaps at grain boundaries. This is generally not the case in coherent real materials, and Taylor (1938) suggested that deformation is homogeneous, i.e. each component crystal undergoes the same shape change as the whole polycrystal. To accommodate an arbitrary shape change requires activation of up to five independent slip systems. The Taylor theory has been highly successful in modelling texture development in cubic metals and many minerals (e.g. Lister *et al.* 1978, Van Houtte 1978, Takeshita *et al.* 1987); but some minerals such as quartz have a very anisotropic single-crystal yield surface and crystals in unfavourable orientations are much stronger than others (Takeshita & Wenk 1988). Such strong grains are likely to deform less than soft ones. In olivine, with less than five independent slip systems, the situation is worse. The yield surface is not closed, and in order to obtain a solution, the assumption of homogeneous strain needs to be partly abandoned. In the 'relaxed constraints' models (e.g. Honneff & Mecking 1978, Kocks & Canova 1981, Van Houtte 1982) some of the five equations which describe the strain caused by the slips are dropped.

We first apply a relaxed Taylor theory to olivine. Next we use a self-consistent theory in which grains have more freedom to deform, achieving better stress equilibrium at the sacrifice of local strain continuity (Molinari *et al.* 1987, Wenk *et al.* 1989a). We deal only with high-temperature conditions which prevail in the mantle and lower crust, neglecting low-temperature olivine fabrics ('(100)-type'; e.g. Möckel 1969, den Tex 1970).

14.2 Slip systems

Numerous single-crystal deformation experiments have been carried out to determine slip systems of olivine at different temperatures (e.g. Raleigh 1968, Carter & Avé Lallemant 1970, Phakey *et al.* 1972, Durham & Goetze, 1977, Mackwell *et al.* 1985). In summary, at moderate temperatures, slip on systems with a [001] direction is most important; at higher temperatures {0kl} [100] slip (so-called pencil glide) dominates, and at highest temperatures (010) [100] slip becomes predominant (Nicolas & Poirier 1976). We have estimated critical resolved shear stresses (CRSS) for two conditions representative of lower temperature (A) and higher temperature (B) (Table 14.1). Model B is based on the exact values of experimentally determined CRSS's (Durham & Goetze 1979, Mackwell *et al.* 1985). Model A is preliminary, and equal ease of {110} [001] and {0kl} [100] (represented by {011} [100]) slip is assumed. It turns out that texture development is not very sensitive to these assumptions. Since all slip directions are parallel to crystallographic axes [100] and [001], the normal-strain components ε_{11}, ε_{22}, and ε_{33}, cannot be accommodated by dislocation glide on existing slip systems, whereas the shear-strain components ε_{12}, ε_{13}, and ε_{23}, can (strain components are referred to the principal crystallographic axes). During flow in the asthenosphere both slip and climb are active (e.g. Nicolas *et al.* 1971, Green & Radcliffe 1972), and slip and climb on

Table 14.1 Slip systems in olivine and assumed normalized CRSS used for models A and B.

Slip system {hkl}⟨uvw⟩	CRSS used in Taylor calculations A	B
{110}[001]	1.0	∞
{011}[100]	1.0	∞
(010)[100]	∞	1.0
(001)[100]	∞	1.3
(010)[001]	∞	2.7

existing slip systems are sufficient to create an arbitrary strain. We propose that shear-strain components are accommodated mainly by slip and normal-strain components by climb (or diffusion in general). This conforms with experiments on olivine single crystals which indicate that a significant amount of strain is created by dislocation climb in the [101] orientation (Durham & Goetze 1977). Other mechanisms to accommodate minor strain incompatibility are grain-boundary diffusion and grain-boundary sliding. In mathematical terms of the Taylor model, this assumption means that equation 5.17 in Gil Sevillano *et al.* (1980) is only satisfied for the shear components:

$$d\varepsilon_{ij} = \frac{1}{2} \sum_{s=1}^{3} (b_i^s n_j^s + b_j^s n_i^s)\, d\gamma^s, \qquad ij = 12, 13 \text{ or } 23 \tag{14.1}$$

where $d\varepsilon_{ij}$ is the imposed strain-increment tensor on each crystal, b_i^s and n_i^s are the components of slip direction and slip plane normal unit vectors, and $d\gamma^s$ is the macroscopic shear-strain increment for each slip system s. If normal-strain components are 'relaxed' or omitted, three independent slip systems suffice. Whereas slip produces a rigid body rotation of crystals, climb does not and therefore does not add to texture development. If the normal-strain components accommodated by climb are smaller than those required to satisfy homogeneous deformation, local heterogeneous deformation has to occur to prevent openings and overlaps across grain boundaries. We assume that the additional plastic work and lattice rotation accompanying such heterogeneous deformation is negligible.

14.3 Texture predictions from Taylor theory

Rotations due to activation of slip systems lead to preferred orientation which we have analysed for pure and simple shear deformation for both models described in Table 14.1. In each case ten steps of a von Mises strain increment

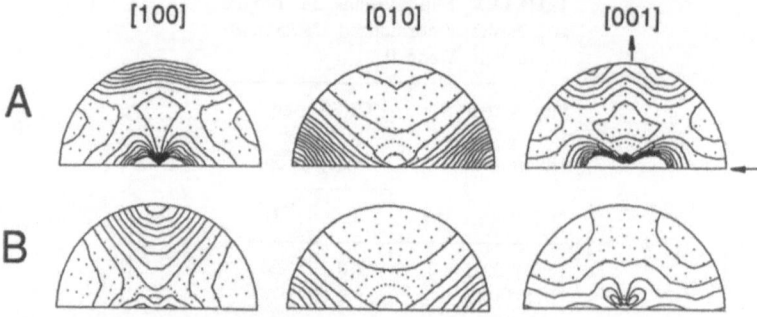

Figure 14.1 Axis diagrams displaying Taylor simulated pure shear textures for model A and model B conditions (Table 14.1). Shortening and elongation directions are indicated (equal-area projection). Contour interval 0.25 m.r.d., dotted below 1 m.r.d. (multiples of a random distribution).

of 0.05 were applied, leading to a 40 per cent thickness reduction in pure shear. The orientation patterns are shown in [100]-, [010]- and [001]-axis diagrams (Figs 14.1 & 2). Note that despite different slip systems in Models A and B, similar textures result, which reassures us that uncertainties in CRSS values will not cause significant changes.

In *pure shear* (Fig. 14.1), [010] shows the strongest preferred orientation, with a maximum in the compression direction for both A and B conditions. This is consistent with textures in many naturally deformed peridotites (e.g. Christensen 1984, Mercier 1985) and with textures observed in compression experiments (e.g. Nicolas *et al.* 1973).

In *simple shear* (Fig. 14.2), the [010] maximum is rotated about 30° from the shear plane normal against the sense of the macroscopic shear. A similar behaviour has been observed in Taylor simulations for calcite (Wenk *et al.* 1987), quartz (Lister & Williams 1979), and halite (Wenk *et al.* 1989a), and also in experiments [e.g. Kern & Wenk 1983, Dell'Angelo 1985, Franssen

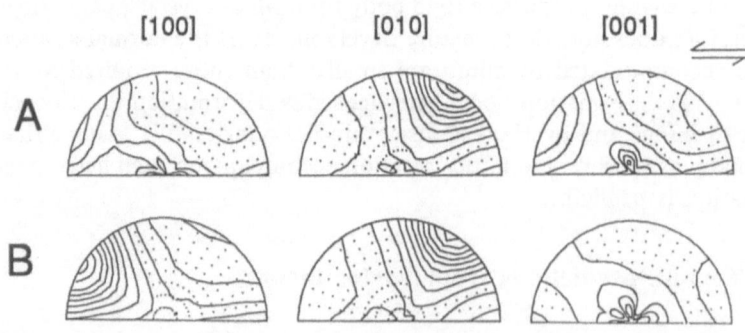

Figure 14.2 Axis diagrams displaying Taylor simulated simple shear textures. Total shear is 1.0 The shear plane and shear sense are indicated (equal-area projection). Contour interval 0.25 m.r.d., dotted below 1 m.r.d.

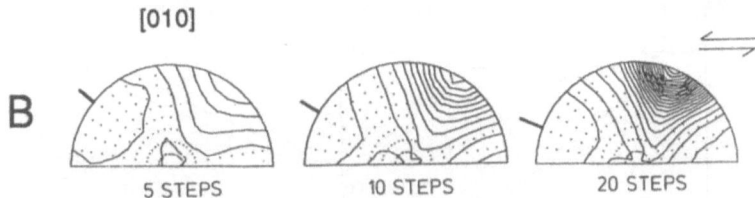

Figure 14.3 [010] axis diagrams for simple shear, model B, illustrating the effect of increasing shear. Each step corresponds to an incremental shear of 0.1. The direction of the longest finite strain axis is indicated. Contour interval 0.25 m.r.d., dotted below 1 m.r.d.

1987). The [010] diagrams in Figure 14.3 illustrate that the maximum does rotate slightly with respect to the shear plane in the sense of the shear during increasing deformation, and becomes stronger. The rotation of the [010] maximum corresponds closely to the rotation of the finite strain ellipsoid. (The long axis is marked in Figure 14.3).

With the Taylor theory we can determine not only changes in orientation distribution but also the work on slip systems that is necessary to accomplish a deformation step. Depending on the evolution of the orientation distribution a polycrystal may plastically harden or soften (e.g. Wenk *et al.* 1986).

In evaluating this plastic anisotropy for olivine, it is necessary to assess both the work necessary to accommodate the shear-strain components by slip and the work necessary to accommodate the normal-strain components by climb. Since the latter is not very well known, we have considered two alternatives.

(a) *Slip is rate controlling.* If slip is more difficult than climb, the geometrical hardening due to orientation changes is assessed by changes of work done by slip systems (SS) with increasing strain. This work corresponds to the average Taylor factor, except that for our relaxed model three instead of five slip systems are used. The work for different deformation modes decreases significantly (Fig. 14.4a), particularly for axisymmetric elongation. One may therefore expect that, analogous to quartzites, extension is the preferred strain mode (Takeshita & Wenk 1988). However, simple shear, the hardest strain mode, is often observed in peridotites (Nicolas & Poirier 1976) and therefore the assumption that climb is much easier than slip is probably incorrect.

(b) *Climb is rate controlling.* If climb is more difficult than slip and our assumption that climb accommodates most of the normal strains applies, then geometrical hardening can be assessed by changes of the averaged sum of normal-strain components over all grains (*NS*) with increasing strain:

$$NS = \sum_{g=1}^{n} \left\{ |\, d\varepsilon_{11}\,| + |\, d\varepsilon_{22}\,| + |\, d\varepsilon_{33}\,| \right\}_g /n\, d\varepsilon_{VM} \qquad (14.2)$$

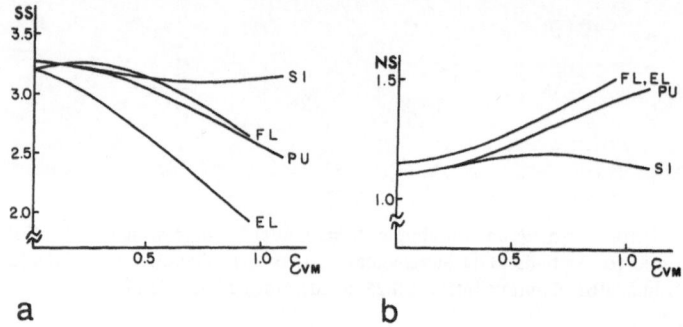

Figure 14.4 Olivine model B (high-temperature). Plots of the non-dimensional work necessary to accommodate the shear strains (*SS*) (a) and the averaged sum of the normal strains (*NS*) (b) versus von Mises strain ε_{VM} for four different deformation modes: FL, axisymmetric flattening; EL, axisymmetric elongation; PU, pure shear; SI, simple shear. See text for definitions of *SS* and *NS*.

where n is the number of grains. *NS* is normalized to the von Mises strain ($d\varepsilon_{VM}$) to allow comparisons of different deformation modes. In this model, olivine polycrystals harden except in simple shear (Fig. 14.4b), which appears to be the preferred strain mode in natural peridotites. It is important that in simple shear the material neither hardens nor softens significantly as preferred orientation develops. Therefore for olivine deforming in simple shear, 'steady-state' Arrhenius-type flow models may be a good approximation.

Our model of polycrystal deformation of olivine which is based on the relaxed Taylor theory predicts textures which are in fair agreement with those observed in experimentally and naturally deformed peridotites, although experiments are admittedly incomplete and so far confined to axial compression. If these predictions are correct, we can predict the sense of shear from the asymmetric disposition of the [010] (or [100]) maximum relative to the shear plane (Fig. 14.3).

14.4 Texture predictions from self-consistent theory

There have always been objections that Taylor's condition of homogeneous strain does not apply to strongly anisotropic minerals. Therefore we have also used a new, large-strain, viscoplastic self-consistent theory (Molinari *et al.* 1987) to model deformation of polycrystalline olivine. In this theory, each crystal is embedded in a homogeneous anisotropic medium. Deformation which is assumed to be homogeneous in each grain is affected both by compatibility and equilibrium conditions with the neighbourhood. This maintains the macroscopic compatibility but locally – from grain to grain – strain is heterogeneous. Stress is closer to equilibrium than in the Taylor theory. Under self-consistent conditions, favourably oriented crystals are allowed to deform fast and unfavourably oriented ones may not deform at all (Wenk *et al.* 1989a). There is no restriction as to the minimum number of slip systems. Also the theory assumes a viscoplastic behaviour which accounts for the strain-rate dependence, whereas the classical Taylor theory models rigid plastic deformation. This is significant for olivine with a small stress exponent and therefore a strong rate sensitivity (we used a power law with $n = 5$).

Results for simple shear deformation to $\gamma = 1$ of 200 grains are shown in Figure 14.5. The predictions of the self-consistent theory for simple shear are very similar to those of the relaxed Taylor theory (Fig. 14.2). [100] and [010] pole figures show the most pronounced textures; the [001] pole figures are less distinct. In the self-consistent model individual grains can deform at different rates. We have indicated the relative change in grain shape by the size of symbols. Some crystals have not deformed at all (small symbols), whereas others have an average von Mises strain of 3 (largest symbols). The distribution of

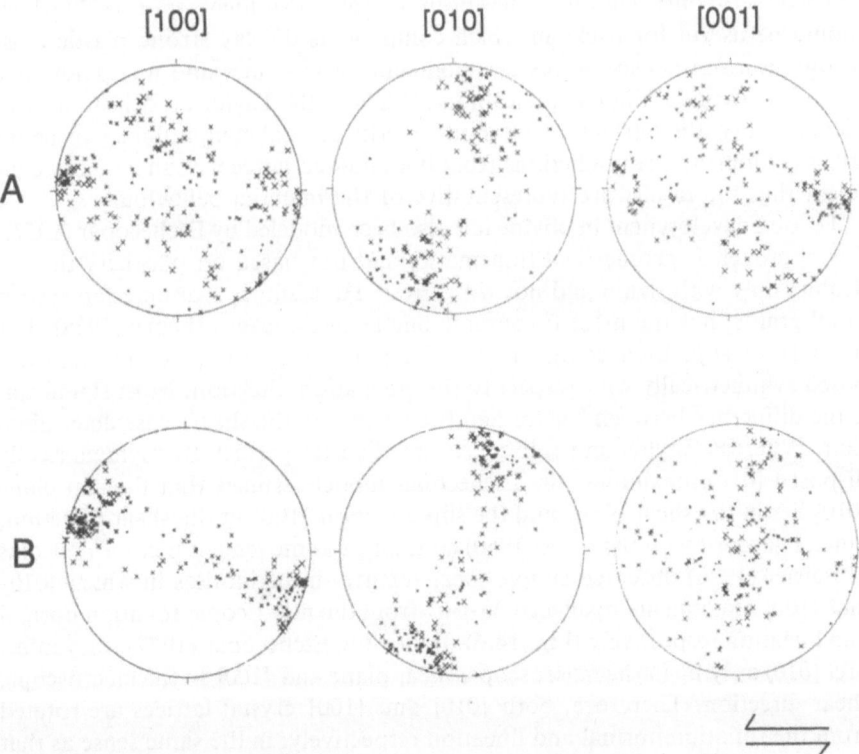

Figure 14.5 Axis diagrams predicted from self-consistent deformation for simple shear of model A and model B conditions. Total shear is 1.0. The size of the symbols indicates the relative deformation of individual grains (200 grains, equal-area projection).

deformed and undeformed grains is similar. The average number of significantly activated slip systems is between 2 and 2.5 compared to 3.0 for Taylor. As for Taylor, the simple-shear flow stress barely changes during deformation.

14.5 Discussion

Different theories have been developed to describe the plastic deformation of polycrystals. Some assume that the material is isotropic, which is a good approximation for diffusion processes and for grain-boundary sliding. Other theories take account of the anisotropic nature of constituent crystals, which is essential if dislocation glide plays an important rôle. These theories differ in the way they satisfy strain compatibility between grains and/or stress equilibrium. Some theories, such as that of Sachs (1928), have been largely dismissed because they fail to predict experimentally observed textures. A modified Taylor (1938) theory is most widely accepted today. Recently a viscoplastic self-consistent theory has been proposed (Molinari *et al.* 1987) which should be useful for rocks in which components display strong plastic anisotropy. But self-consistent texture modelling is fairly new and we do not have much experience about its applicability. Whereas the Taylor model emphasizes compatibility, the self-consistent model sacrifices local compatibility for better stress equilibrium. If predictions from both models agree we can be fairly confident that the results are representative of the material behaviour.

Texture development in olivine has also been modeled by Etchecopar (1977). His approach is geometrical (kinematic) and not based on plasticity theory, dealing only with strain and not with stress. He assumes a unique slip system in all grains, not the most favourable one as in the Sachs theory (1928). For (010) [100] slip, Etchecopar predicts for pure shear two [100] maxima disposed symmetrically with respect to the elongation direction. Most significant is the difference between Taylor and Etchecopar in the shear sense determination. Whereas Taylor and self-consistent theories predict an asymmetrically disposed (010) maximum, the Etchecopar model assumes that the slip plane (010) lies in the shear plane and the slip direction [100] in the shear direction, much in accord with Sander's (1950) concept of a 'movement picture'. Nicolas & Poirier (1976) observed simple shear textures in peridotites in which [010] and [100] maxima are displaced 20–30° from the mesoscopic foliation normal and lineation respectively (Fig. 14.6). Following Etchecopar (1977), they interpret (010) as lying in the macroscopic shear plane and [100] in the macroscopic shear direction. Therefore, both [010] and [100] crystal lattices are rotated from the foliation normal and lineation respectively, in the same sense as that of the macroscopic shear (Fig. 14.7a). In our simulations the [010] is asymmetric with respect to the shear plane and displaced in a direction opposite to the sense of shear (Fig. 14.7b). The sense of shear predicted from the Etchecopar

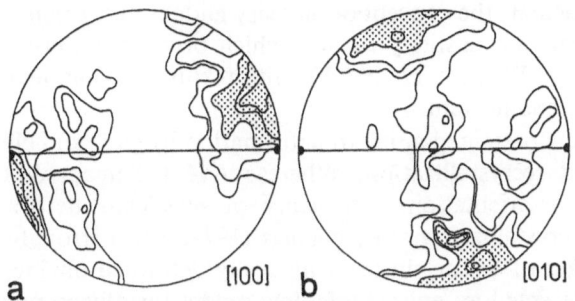

Figure 14.6 Fabric diagrams for olivine aggregate from the Bay of Islands ophiolite complex, Newfoundland, from Mercier (1985): (a) [100]; (b) [010]. Foliation is horizontal (straight line), and lineation trends E–W (dots) (equal-area projection).

model is just the opposite from that predicted with the Taylor model. This indicates some controversy with far-reaching geological implications. Does olivine deform in single slip or in polyslip? The answer is not conclusive, but slip occurs on different crystallographic planes both in natural rocks (Nicolas *et al.* 1971) and in experimentally deformed single crystals (Durham & Goetze 1977, Mackwell *et al.* 1985). A polyslip model therefore seems applicable. Unfortunately there are so far no simple-shear experiments on olivine with which texture predictions could be compared, and natural fabrics rely on interpretation. In these the identification of the shear plane and the foliation plane can be extremely difficult and uncertain (e.g. Lister & Snoke 1984). The question of which of the deformation models is more applicable remains undecided. In our view, it seems unlikely that olivine behaves differently from any other material of geological or engineering interest which has been studied

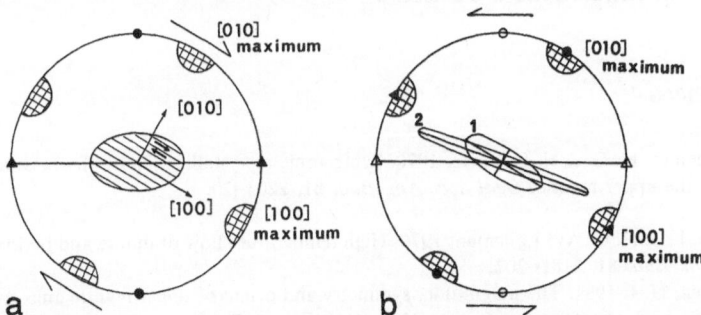

Figure 14.7 Determination of the sense of shear from fabric diagrams of olivine aggregates deformed in simple shear based on (a) interpretation by Nicolas & Poirer (1976) and (b) our calculations. Note that the interpretation by Nicolas & Poirier leads to a dextral sense of shear, whereas our calculation leads to a sinistral sense of shear. Solid circles and triangles indicate foliation normal and lineation, respectively. Open circles and triangles in (b) indicate shear plane normal and direction, respectively, assumed to be parallel to the shortest and longest axes of the finite strain ellipsoid. Arrows indicate the macroscopic shear plane and sense of shear.

373

so far. In particular, the concept of an 'easy glide' stable orientation in simple shear is contrary to plasticity theory, which predicts a constant rollover of orientations (e.g. Wenk *et al.* 1989b), with texture maxima merely representing a dynamic equilibrium.

All arguments presented apply to deformation by glide and climb, and have not considered recrystallization. Whereas Avé Lallemant & Carter (1970) observed a near-orientation coincidence of plastically deformed and syntectonically recrystallized grains, Nicolas (1971, 1973) strongly argues for a kinematic origin of texture in most naturally deformed olivines.

We have presented an anisotropic flow model for olivine polycrystals that allows us to predict deformation in this important system which has fewer than five independent slip systems, thus precluding application of the classical Taylor theory. A relaxed Taylor theory and a self-consistent model predict similar textures for pure shear and simple shear. Simple-shear textures can be used to infer the sense of shear. It is assumed that deformation which can not be achieved by means of slip systems occurs by climb. For simple shear the predicted flow stress does not change greatly during deformation, suggesting that in this case steady-state flow laws may apply reasonably well.

Acknowledgements

We appreciate support from grants NSF EAR 87-09378 and IGPP-LANL. H.R.W. acknowledges the generous hospitality of the Technical University of Hamburg–Harburg through the A. von Humboldt Foundation. Stimulating discussions with S. I. Karato, U. F. Kocks, and L. E. Weiss were most valuable. P. Van Houtte kindly provided a Taylor computer code on which part of our simulations are based.

References

Avé Lallemant, H. G. & N. L. Carter 1970. Syntectonic recrystallization of olivine and modes of flow in the upper mantle. *Geol Soc. Am. Bull.* **81**, 2203–20.

Carter, N. L. & H. G. Avé Lallemant 1970. High temperature flow of dunite and peridotite. *Geol Soc. Am. Bull.* **81**, 2181–202.

Christensen, N. I. 1984. The magnitude, symmetry and origin of upper mantle anisotropy based on fabric analyses of ultramafic tectonites. *Geophys. J. R. Astron. Soc.* **16**, 89–111.

Christensen, U. R. 1987. Some geodynamical effects of anisotropic velocity. *Geophys. J. R. Astron. Soc.* **91**, 711–36.

Dell'Angelo, L. 1985. Quartz *c*-axis preferred orientation in experimentally deformed shear zone. *Geol Soc. Am. Abstr. with Program* **17**, 562.

Den Tex, E. 1970. Principal olivine fabrics: their tectonic and metamorphic significance. In *Experimental and natural rock deformation*, P. Paulitsch (ed.), 486–95. Berlin: Springer.

Durham, W. B. & C. Goetze 1977. Plastic flow of oriented single crystals of olivine, 1. Mechanical data. *J. Geophys. Res.* **82**, 5131–753.

Etchecopar, A. 1977. A plane kinematic model of progressive deformation in a polycrystalline aggregate. *Tectonophysics* **39**, 121–39.

Franssen, R. C. M. W. 1987. The influence of the deformation path on mechanical behavior. *EOS, Trans Am. Geophys. Union* **68**, 1454.

Fuchs, K. 1983. Recently formed elastic anisotropy and petrological models for the continental subcrustal lithosphere in southern Germany. *Phys. Earth Planet. Inter.* **31**, 93–118.

Gil Sevillano, J., P. Van Houtte & E. Aernoudt 1980. Large strain work hardening and textures. *Progr. Mat. Sci.* **25**, 69–412.

Green, H. W. & S. V. Radcliffe 1972. Deformation processes in the upper mantle. In *Flow and fracture of rocks*, H. C. Heard, I. Y. Borg, N. L. Carter & C. B. Raleigh (eds), 139–56. Geophys. Monogr. **16**. Washington, DC: American Geophysical Union.

Hager, B H. 1984. Subducted slabs and the geoid constraints on mantle rheology and flow. *J. Geophys. Res.* **89**, 6003–15.

Hess, H. H. 1964. Seismic anisotropy of the uppermost mantle under oceans. *Nature* **203**, 629–31.

Honneff, H. & H. Mecking 1978. A method for the determination of the active slip systems and orientation changes during single crystal deformation. In *Proceedings of the 5th International Conference on Texture of Materials*, G. Gottstein & K. Lücke (eds), 265–75. Berlin: Springer.

Karato, S.-I., M. S. Paterson & J. D. Fitzgerald 1986. Rheology of synthetic olivine aggregates: influence of grain size and water. *J. Geophys. Res.* **91**, 8151–76.

Kern, H. & H.-R. Wenk 1983. Calcite texture development in experimentally induced ductile shear zones. *Contrib. Mineral. Petrol.* **83**, 231–6.

Kocks, U. F. & G. R. Canova 1981. How many slip systems, and which? In *Deformation of polycrystals: mechanisms and microstructures*, Proc. 2nd Risø Symp., Risø Nat. Lab., Roskilde, Denmark.

Lister, G. S., M. S. Paterson & B. E. Hobbs 1978. The simulation of fabric development during plastic deformation and its application to quartzite: the model. *Tectonophysics* **45**, 107–158.

Lister, G. S. & A. W. Snoke 1984. S–C mylonites. *J. Struct. Geol.* **6**, 617–38.

Lister, G. S. & P. F. Williams 1979. Fabric development in shear zones: theoretical controls and observed phenomena. *J. Struct. Geol.* **1**, 283–97.

Mackwell, S. J., D. L. Kohlstedt & M. S. Paterson 1985. The role of water in the deformation of olivine single crystals. *J. Geophys. Res.* **90**, 11 319–33.

Mercier, J.-C. C. 1985. Olivine and pyroxene. In *Preferred orientation in deformed metals and rocks: an introduction to modern texture analysis*, H.-R. Wenk (ed.), 407–30. Orlando, Florida: Academic Press.

Möckel, J. R. 1969. Structural petrology of the garnet peridotite of Alpe Arami (Ticino, Switzerland). *Leidse Geol. Med.* **42**, 61–130.

Molinari, A., G. R. Canova & S. Ahzi 1987. A self-consistent approach of the large deformation polycrystal viscoplasticity. *Acta Metall.* **35**, 2983–94.

Nicolas, A., J. L. Bouchez, F. Boudier & J.-C. C. Mercier 1971. Textures, structures and fabrics due to solid state flow in some European lherzolites. *Tectonophysics* **12**, 55–85.

Nicolas, A., F. Boudier & A. M. Boullier 1973. Mechanisms of flow in naturally and experimentally deformed peridotites. *Am. J. Sci.* **273**, 853–76.

Nicolas, A. & J.-P. Poirier 1976. *Crystalline plasticity and solid state flow in metamorphic rocks.* New York: Wiley.

Phakey, P., G. Dollinger & J. Christie 1972. Transmission electron microscopy of experimentally deformed olivine crystals. In *Flow and fracture of rocks*, H. C. Heard, I. Y. Borg, N. L. Carter & C. B. Raleigh (eds), 117–38. Geophys. Monogr. **16**. Washington, DC: American Geophysical Union.

Raleigh, C. B. 1968. Mechanisms of plastic deformation of olivine. *J. Geophys. Res.* **73**, 5391–406.

Sachs, G. 1928. Zur Ableitung einer Fließbedingung. *Z. Verein. Dtsch. Ing.* **12**, 134–6.

Sander, B. 1950. *Einführung in die Gefügekunder der geologischen Körper.* Vienna: Springer.

Takeshita, T., C. Tomé, H.-R. Wenk & U. F. Kocks 1987. Single crystal yield surface for trigonal lattices: application to texture transitions in calcite polycrystals. *J. Geophys. Res.* **92B**, 12 917–30.

Takeshita, T. & H.-R. Wenk 1988. Plastic anisotropy and geometrical hardening in quartzites. *Tectonophysics* **149**, 345–61.

Taylor, G. I. 1938. Plastic strain in metals. *J. Inst. Metals* **62**, 301–24.

Van Houtte, P. 1978. Simulation of the rolling and shear texture of brass by the Taylor theory adapted for mechanical twinning. *Acta Metall.* **26**, 591–604.

Van Houtte, P. 1982. On the equivalence of the relaxed Taylor theory and the Bishop–Hill theory for partially constrained plastic deformation of crystals. *Mat. Sci. Engng* **55**, 69–77.

Wenk, H.-R., T. Takeshita, P. Van Houtte & F. Wagner 1986. Plastic anisotropy and texture development in calcite polycrystals. *J. Geophys. Res.* **91**, 3861–3869.

Wenk, H.-R., G. R. Canova, A. Molinari & H. Mecking 1989a. Texture development in halite: comparison of Taylor model and self-consistent theory. *Acta Metall.* **37**, 2017–29.

Wenk, H.-R., G. R. Canova, A. Molinari & U. F. Kocks 1989b. Viscoplastic modelling of texture development in quartzite. *J. Geophys. Res.* **94**, 17 895–906.

Wenk, H.-R., T. Takeshita, E. Bechler, B. G. Erskine & S. Matthies 1987. Pure shear and simple shear calcite textures: comparison of experimental, theoretical and natural data. *J. Struct. Geol.* **9**, 731–45.

CHAPTER FIFTEEN

On the slip systems in uranium dioxide

Arthur H. Heuer, Robert J. Keller & Terry E. Mitchell

15.1 Introduction

Uranium dioxide (UO_2) exists over a wide range of O : U ratios ($UO_{2 \pm x}$) at elevated temperatures as shown in Fig. 15.1 (Nadeau 1969). The variation of various physical properties with the O : U ratio has considerable technological significance, as use of UO_2-based nuclear fuels necessarily encompasses a large region of the $T–x$ space of Fig. 15.1. Plastic deformation in UO_2 has been of interest since the earliest uses of UO_2 as a nuclear fuel, since fuel pins are subjected to large thermomechanical stresses. In this chapter, we consider the *slip systems* in UO_2 as a function of T and x. The way in which x is accommodated, particularly in hyperstoichiometric UO_2, is also of importance to the mechanical behaviour. Isolated point-defect models are clearly not appropriate when 5–10 per cent excess oxygen can be added to the basic fluorite structure (Fig. 15.2) as interstitial species. Defect clusters (Willis 1978), an example of which is shown in Fig. 15.3, are stable over a range of T and x, and by analogy with interstitial solutes in metals and alloys, could be expected to be potent hardeners. However, such is not the case!

There is general agreement (Seltzer *et al*. 1972) that some softening accompanies increasing x in polycrystalline UO_2 undergoing diffusion-controlled creep, because of the increase in both oxygen and uranium self-diffusion kinetics with increasing non-stoichiometry. For single-crystal UO_2, the literature (Nadeau 1969, Ronchi & Blank 1970) on plastic deformation indicates softening on some but not all slip systems with increasing non-stoichiometry, and a sensitivity to O : U ratio concerning the slip system which has the lowest critical resolved shear stress (CRSS).

At the time the present work was begun, there was general agreement that the primary slip system of UO_2 was $\{100\}$, a common slip system for both oxides and fluorides with the fluorite structure, but reports existed for both $\{110\} \langle 0\bar{1}1 \rangle$ and $\{111\} \langle 0\bar{1}1 \rangle$ slip and, indeed, of non-crystallographic slip with an $\langle 0\bar{1}1 \rangle$ slip direction (Nadeau, 1969, Ronchi & Blank 1970; Yust 1971, 1973). The choice of $1/2\langle 0\bar{1}1 \rangle$ for the Burgers vectors of dislocations in

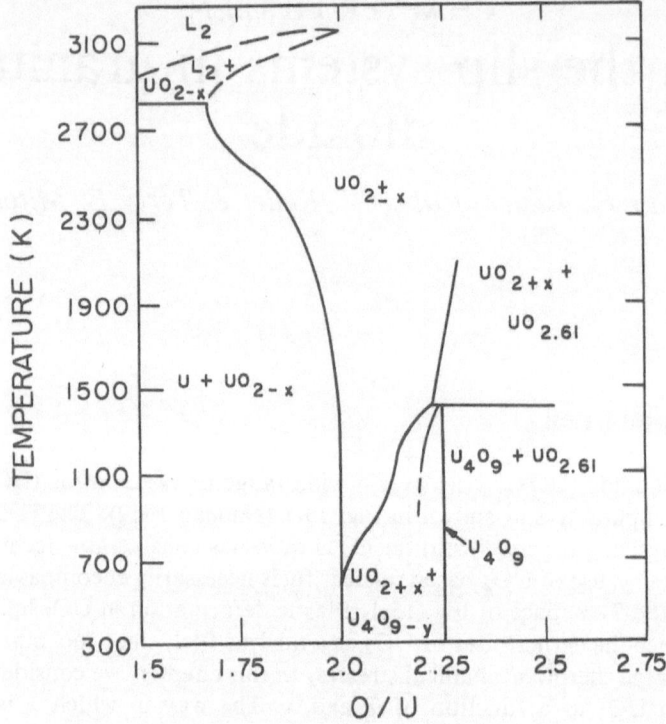

Figure 15.1 Phase diagram of the U–O system (after Nadeau 1969).

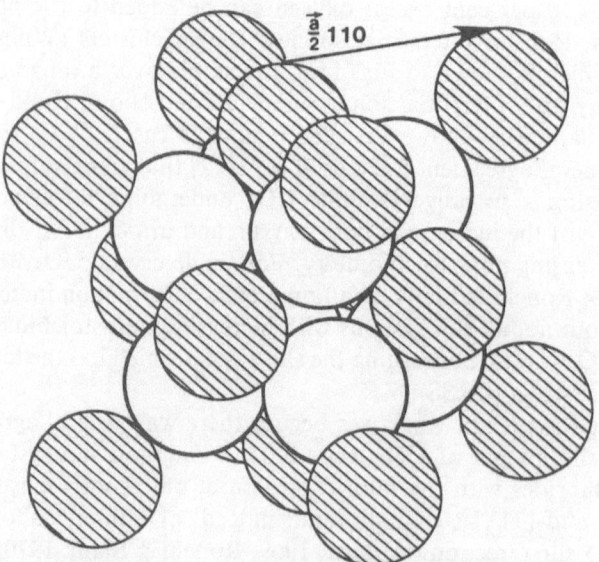

Figure 15.2 Crystal structure of stoichiometric UO_2.

□ ANION VACANCY

◉ <011> INTERSTITIAL

◉ <111> INTERSTITIAL

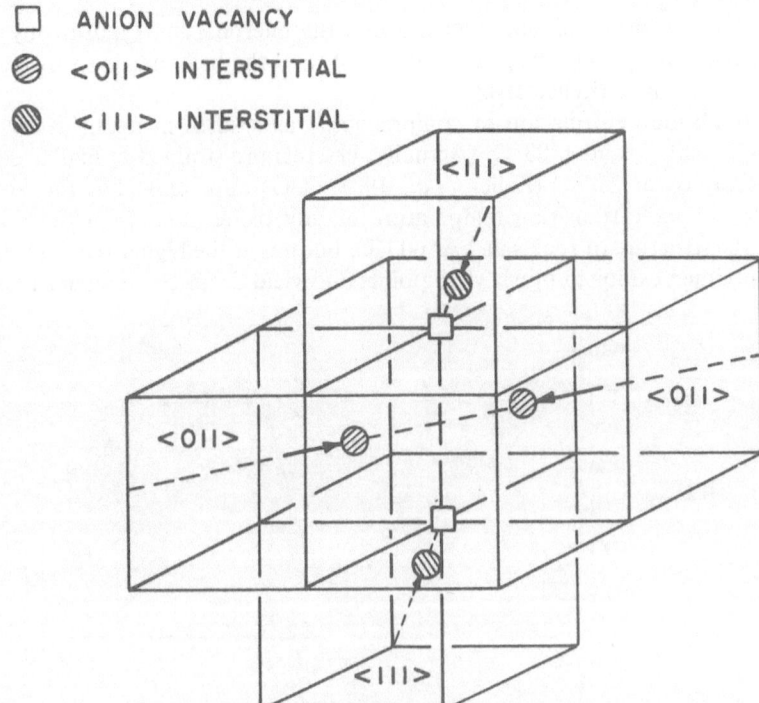

Figure 15.3 2:2:2 Willis defect cluster in hyper-stoichiometric UO_{2+x} (Willis 1978). The two oxygen interstitials do not occupy the centre of the alternate empty subcells but relax along $\langle 011 \rangle$ towards two lattice oxygens, which in turn relax along $\langle 111 \rangle$ away from the two interstitials, resulting in two oxygen vacancies (hence the 2:2:2 notation).

crystals with the fluorite structure is obvious from inspection of the crystal structure (Fig. 15.2), as it is the shortest lattice vector. On the other hand, the choice of the preferred slip plane is not at all obvious.

In this chapter we report slip systems determined from slip trace analysis for UO_{2+x} crystals deformed over ranges of temperature, O : U ratio, and orientation such that a variety of slip systems have been activated; experimental details have been reported elsewhere (Keller 1982, Keller *et al.* 1988). The principal aim of the discussion section is to attempt a simple-minded but atomistic interpretation of these data.

15.2 Effects of non-stoichiometry on microhardness, flow stress, and preferred slip system

Single crystals were grown by the directional solidification of internally zone-melted powder compacts (Chapman & Clark 1965), cut into one of several orientations, and annealed in various CO/CO_2 mixtures to establish the

379

desired O : U ratio. At room temperature, the microhardness anisotropy was consistent with {111} ⟨0$\bar{1}$1⟩ slip (Fig. 15.4), and the peak hardness increased with increasing x (Keller 1982).

Bulk-plastic deformation in compression was studied in detail at 600°C, 800°C, 1000°C, and 1400°C. (Actually, crystals pre-strained at 600°C could be deformed at 250°C (Keller *et al.* 1988); UO$_2$ apparently has the lowest brittle-to-ductile transition temperature of any oxide other than those that have the structure of rock salt.) At 600°C, but not at the higher temperatures, the specimen exhibited upper yield points and yield drops. At temperatures of

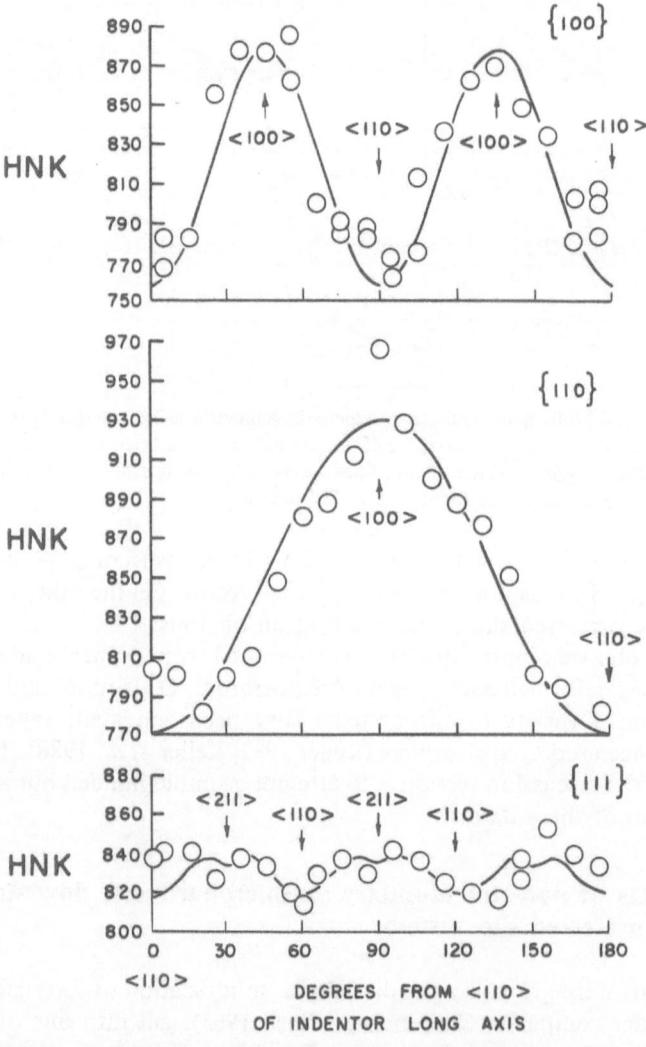

Figure 15.4 Knoop hardness anisotropy on {110} as a function of O : U ratio for UO$_{2.001}$.

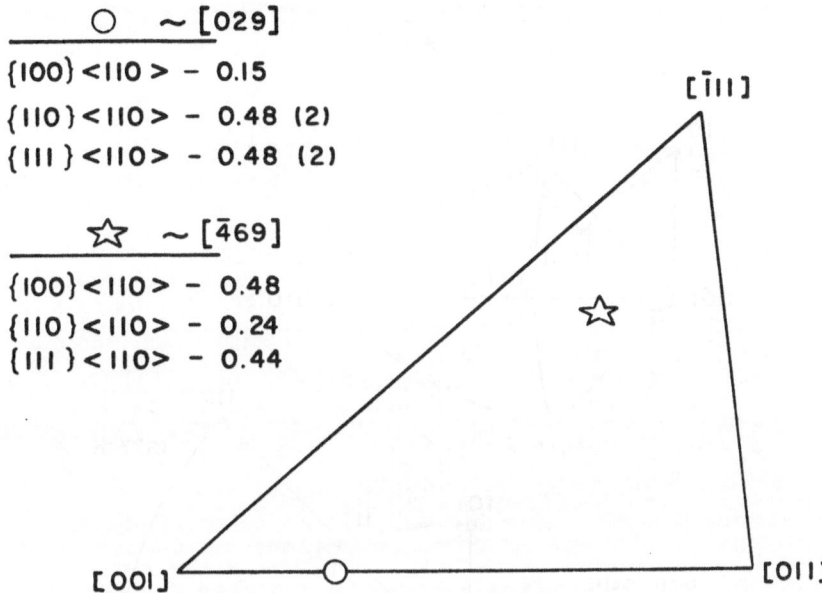

O ~ [029]

{100} <110> − 0.15

{110} <110> − 0.48 (2)

{111} <110> − 0.48 (2)

☆ ~ [4̄69]

{100} <110> − 0.48

{110} <110> − 0.24

{111} <110> − 0.44

Figure 15.5 Compression orientations used and the corresponding Schmid factors.

1000°C and below, the yield stress was surprisingly insensitive to the O : U ratio but was widely scattered (the range at 600°C was 45–95 MPa, and lower at the higher temperatures), which was attributed to extrinsic impurities present in variable amounts in the crystals studied (Keller et al. 1988). At 1400°C, the same was true for crystals undergoing {100} slip, but not for those undergoing {111} slip – they softened with increasing x (Keller et al. 1988). Apparently, defect clusters of the type shown in Fig. 15.3 are 'transparent' to moving dislocations; reorientation of the vacancies and interstitials of the defect must occur with sufficient rapidity not to impede dislocation motion and may even facilitate {111} slip at 1400°C (Keller 1982). The lack of any effect of non-stoichiometry on {100} slip at 1400°C rules out explanations attributing non-stoichiometric softening to enhanced diffusion kinetics.

On the other hand, there was a marked dependence of the preferred slip system on T and x. This was checked using two compression orientations: $\langle 4̄69 \rangle$, which provides nearly equal Schmid factors for both {100} $\langle 0\bar{1}1 \rangle$ and {111} $\langle 0\bar{1}1 \rangle$ slip (0.48 and 0.46, respectively); and $\langle 029 \rangle$, which provides an equal Schmid factor of 0.49 for a pair of {110} $\langle 1\bar{1}0 \rangle$ systems and a pair of {111} $\langle 0\bar{1}1 \rangle$ systems (Fig. 15.5). At 600°C, slip trace analysis showed that {111} slip predominated for both orientations and for all O : U ratios.

At 1000 and 1400°C, the data were more complex (Keller et al. 1988). For the $\langle 4̄69 \rangle$ orientation, the more nearly stoichiometric crystals ($x \leqslant 10^{-3}$) deformed by {100} $\langle 0\bar{1}1 \rangle$ slip, while the more oxygen-rich crystals exhibited non-crystallographic slip on a maximum resolved shear stress (MRSS) plane

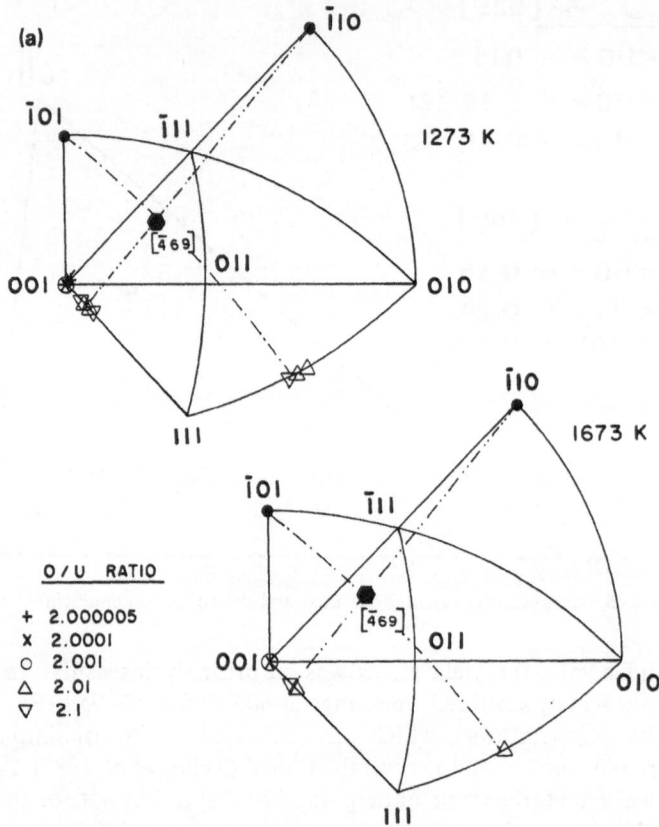

Figure 15.6 Primary slip planes for specimens deformed along (a) [$\overline{4}$69] and (b) [029] as a function of O:U ratio and deformation temperature, showing slip directions (solid circles) and the presence of primary slip planes on or near the corresponding maximum resolved shear stress planes.

that combined {100} and {111} slip (Fig. 15.6a). For the ⟨029⟩ orientation, a similar transition was seen: {111} slip for the most oxygen-rich crystals ($x \geqslant 10^{-2}$), and non-crystallographic MRSS slip on a combination of {110} and {111} planes for more nearly stoichiometric crystals (Fig. 15.6b). Clearly, {111} slip is preferred at low temperatures, and {100} slip is preferred at high temperatures for stoichiometric crystals, while plastic anistropy nearly disappears at high temperatures for non-stoichiometric crystals.

15.3 The structure of half-slipped dislocations

We attempt to explain the evidence just summarized by focusing on the structure of ⟨0$\overline{1}$1⟩ dislocations on {100}, {110}, and {111}.

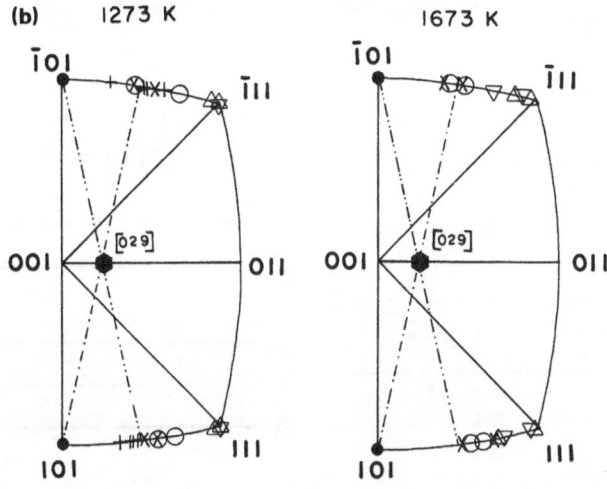

(b) 1273 K   1673 K

O / U   RATIO

+ 2.000005
× 2.0001
○ 2.001
△ 2.01
▽ 2.1

Figure 15.6   (*Continued*)

The first issue is one of charge. It was recognized quite early (Ashbee & Frank 1970, Brantley & Bauer 1970) that if straight- and jogged-edged dislocations on {100} carry no intrinsic charge, anions must move normal to the Burgers vector along the dislocation line for the dislocation to move. Alternatively, anion shuffling can be avoided (Evans & Pratt 1969) at the expense of a charge of $\pm 2e$ per atom plane intersected by the dislocation line (for UO_2 or $\pm e$ for CaF_2); this charge can be avoided by alternating positively and negatively charged segments, resulting in an atomically jogged dislocation. Straight-screw dislocations are neutral.

Straight-edge and screw dislocations on {110} are uncharged (Sawbridge & Sykes 1970), since the two extra half-planes of the edge dislocation carry equal and opposite charges at their edges. Dislocations on {111} can move between two oxygen planes, or between an oxygen and a uranium plane (Keller *et al*. 1988, Keller 1982). In the first case, each half-plane ends in the stoichiometric ratio of O to U so each half-plane is neutral, as is the dislocation, which is therefore equivalent to an edge dislocation on {110} in a rock-salt crystal. If the dislocation moves between an oxygen and uranium plane, each half-plane has equal and opposite charge so the dislocation is again neutral. In summary, no intrinsic charge problem exists for dislocations on either {110} or {111}

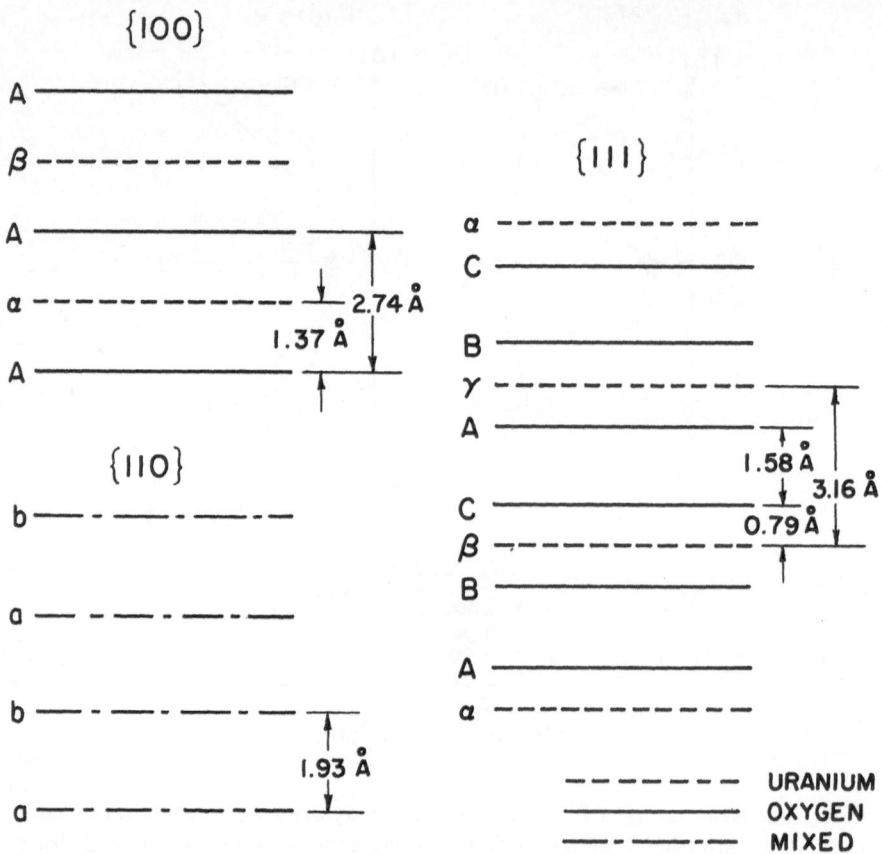

Figure 15.7 Planar spacing along $\langle 100 \rangle$, $\langle 110 \rangle$, and $\langle 111 \rangle$.

planes, but dislocations on {100} may have to assume an unusual core structure.

Next consider the spacing of planes. Elementary dislocation theory suggests that the most widely spaced planes should offer the least resistance to the glide of dislocations (in simple metallic structures, this means the most close packed, but a more general definition is needed for compounds). The planar spacing along $\langle 100 \rangle$, $\langle 110 \rangle$, and $\langle 111 \rangle$ is shown in Fig. 15.7; {110} is clearly the most widely spaced, and this, combined with the charge neutrality of edge dislocations gliding on {110}, makes the absence of {110} slip surprising.

One probable answer lies in considering the structure of the dislocation in a half-slipped position, as was done many years ago by Gilman (1960) to explain the difficulty of activating {100} $\langle 0\bar{1}1 \rangle$ slip in ionic rock-salt structure crystals. (This is actually tantamount to studying the core structure of a dislocation in an ionic crystal at the zeroth level of atomic simulation.)

The perfect lattice and the corresponding structure of the dislocation in its half-slipped position is shown in Fig. 15.8 for dislocations on {100}, {110}, and {111} planes. Consider the {110} case first (Fig. 15.8a). As pointed out by Sawbridge & Sykes (1971), it is immediately obvious that planes of highly charged anions and cations are present in the half-slipped position. This is equivalent to bringing unscreened like charges near one another at the dislocation core, which must contribute to a sizeable Peierls barrier and prevent the occurrence of {110} slip.

Similar planes of anions and cations are not present in the half-slipped positions for glide on {100} and {111} planes. It can be seen in Figures 15.8b and 15.8c, however, that unscreened anion–anion separations increase in the half-slipped position on {100} planes, while they decrease (to 0.273 nm) across the slip plane in the half-slipped position on {111} planes. This favours slip on {100} and apparently is the dominant effect in the deformation of CaF_2, where

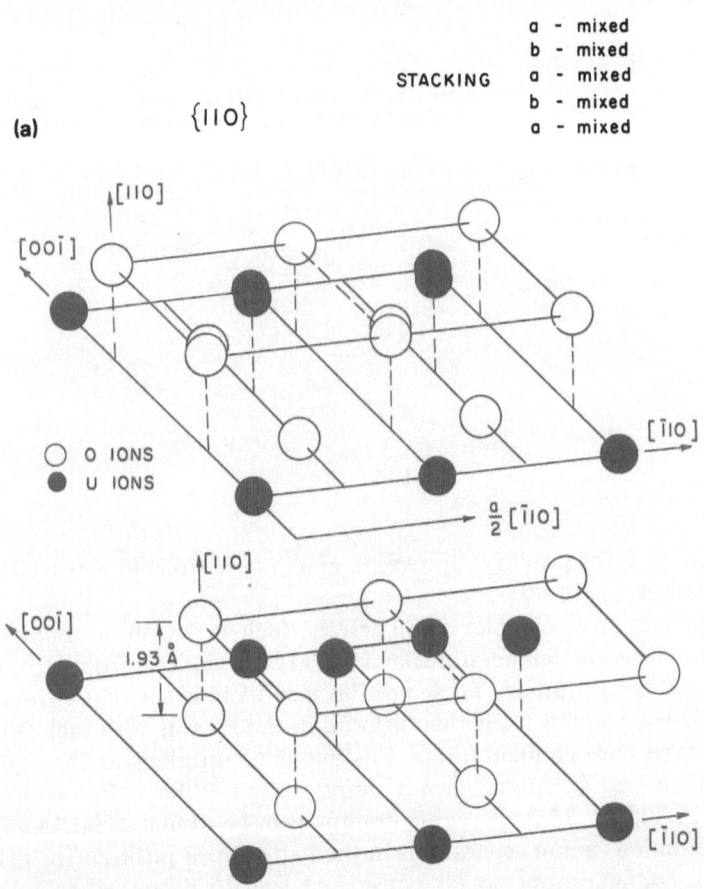

Figure 15.8 Orientation of perfect unit cell, and after one-half unit of shear, along (a) {110}, (b) {100}, and (c) {111}.

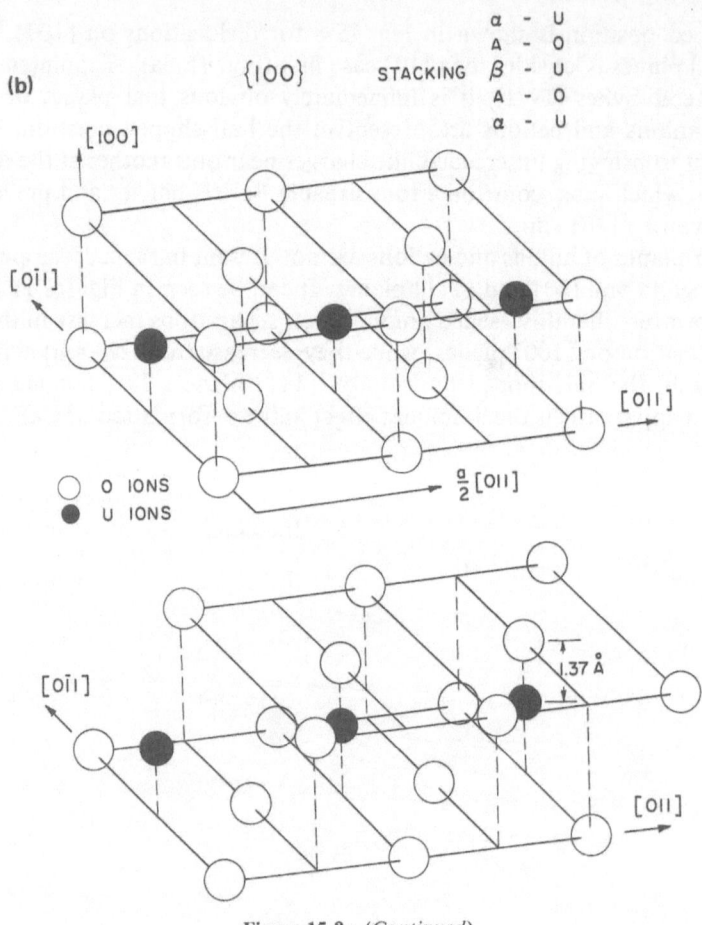

Figure 15.8 *(Continued)*

{100} ⟨0̄11⟩ is the primary slip system, and in near-stoichiometric UO₂ at high temperatures.

The formation of complex anion defects, such as that shown in Figure 15.3, is favoured by low temperatures and by large deviations from stoichiometry at elevated temperatures. These are the same conditions that favour slip on {111} planes, and it is likely that dislocations interacting with such defects will have altered core configurations. Possible configurations at the equilibrium and half-slipped positions when a simple 1 : 1 : 1 Willis defect is present are shown in Figure 15.9. A configuration can be found (Fig. 15.9b) which increases anion–anion separations in the half-slipped position for {111} slip, while all configurations so far considered lead to decreased separation for {100} slip. Although the 1 : 1 : 1 defect is simpler than the defects likely to be present at low temperatures and in highly non-stoichiometric crystals at

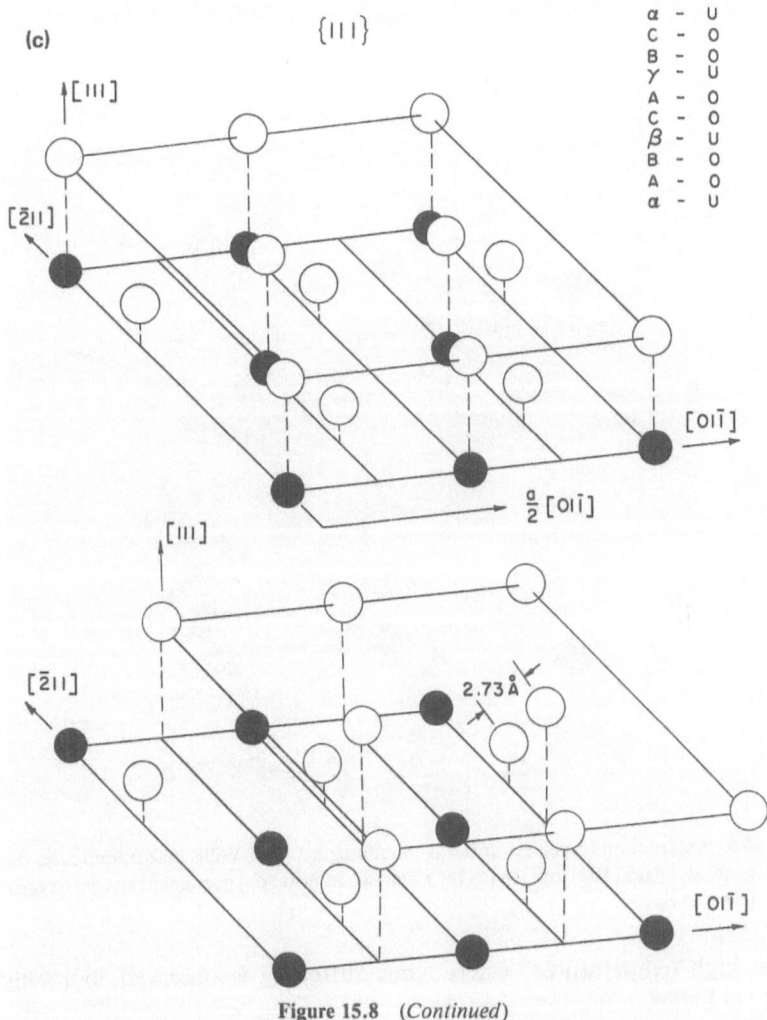

Figure 15.8 *(Continued)*

elevated temperatures, we suggest that interactions of this type cause the dominance of {111} slip.

15.4 Questions for future work

While this study has clarified some of the issues in the deformation of UO_2, a number of important questions remain which we now summarize:

(1) Why is the flow stress so relatively insensitive to the $O:U$ ratio, except

387

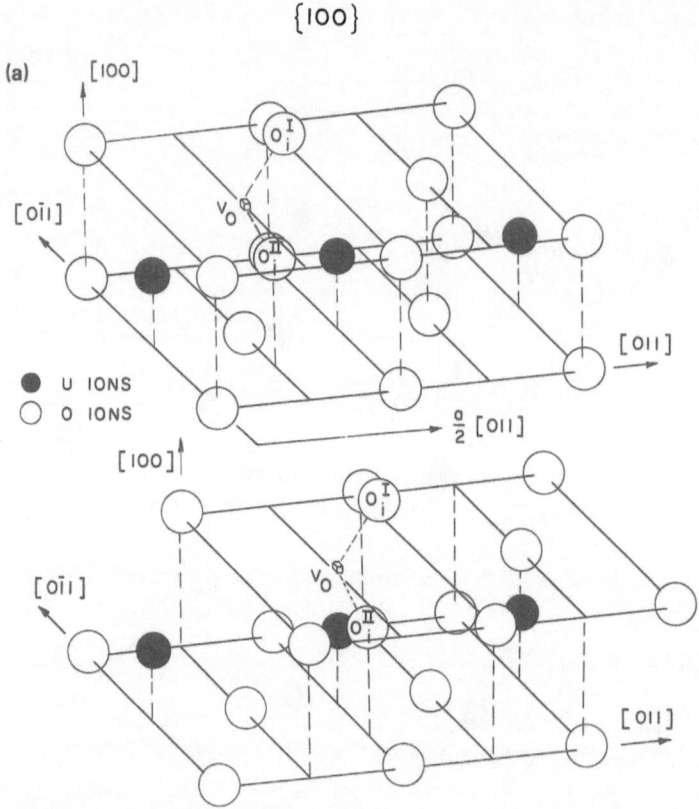

Figure 15.9 Orientation of perfect unit cell containing a $1:1:1$ Willis defect, and after one-half unit of shear, along (a) $\{100\}$ and (b) $\{111\}$. V_0 and O_i indicate oxygen vacancies and oxygen interstitials, respectively.

at high temperatures, where some softening is observed, but only for $\{111\}$ slip?

(2) What is the nature of the interaction between dislocations and complex Willis defects known to exist in highly non-stoichiometric material?

(3) Why does UO_2 have such a low brittle-to-ductile transition temperature $(T_{B-D} \sim 0.15 T_m)$, whereas other isostructural oxides, e.g. stabilized ZrO_2, have a T_{B-D} that is more nearly equal to $0.5 T_m$?

It is likely that theoretical work is required on all these issues, probably involving computer simulation.

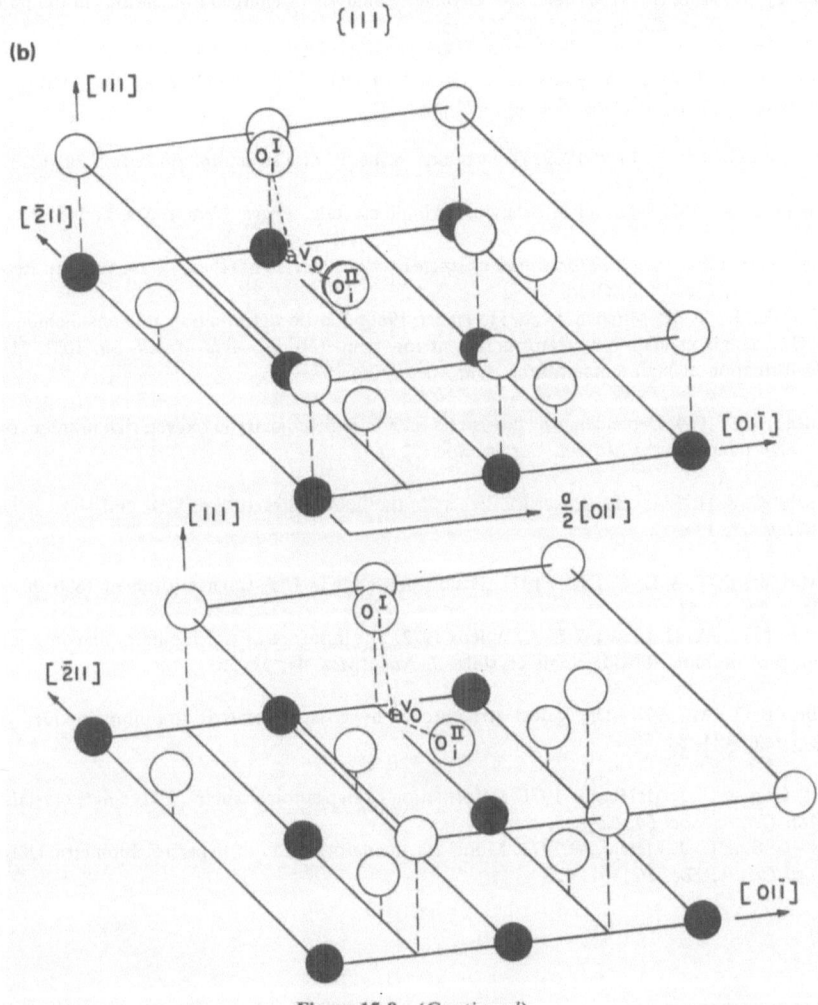

Figure 15.9 (*Continued*)

Acknowledgement

This work was supported by the U.S. Department of Energy, Basic Energy Science – Material Science, under contract no. DEAC0277ER042217.

References

Ashbee, K. H. G. & F. C. Frank 1970. Dislocations in the fluorite structure. *Phil Mag.* **21**, 211–13.

Brantley, W. A. & Ch. L. Bauer 1970. Geometric analysis of charged dislocations in the fluorite structure. *Phys. Status Solidi* **40**, 707–15.

Chapman, A. T. & G. W. Clark 1965. Growth of UO_2 single crystals using the floating zone technique. *J. Am. Ceram. Soc.* **48**, 494–5.

Evans, A. G. & P. L. Pratt 1969. Dislocations in the fluorite structure. *Phil Mag.* **20**, 1213–37.

Gilman, J. J. 1961. Mechanical behavior of ionic crystals. *Progr. Ceram. Sci.* **1**, 146–99.

Keller, R. J. 1982. *Plastic deformation of single crystal UO_{2+x}*. PhD thesis, Case Western Reserve University, Cleveland, Ohio.
Keller, R. J., T. E. Mitchell & A. H. Heuer 1988. Plastic deformation in nonstoichiometric UO_{2+x} single crystals – I. Deformation at low temperatures. *Acta Metall.* **36**, 1061–71; II. Deformation at high temperature. *Acta Metall.* **36**, 1073–83.

Nadeau, J. S. 1969. Dependence of flow stress on non-stoichiometry in oxygen rich uranium oxide at high temperatures. *J. Am. Ceram. Soc.* **52**, 1–7.

Ronchi, C. & H. Blank 1970. Lattice defects in the fluorite structure of UO_2 and UO_{2+x}. *Nucl. Metall.* **17**, 174–82.

Sawbridge, P. T. & E. C. Sykes 1971. Dislocation glide in UO_2 single crystals at $1600°$ K. *Phil Mag.* **24**, 33–53.
Seltzer, M. S., A. H. Clauer & B. A. Wilcox 1972. The influence of stoichiometry on compression creep of uranium dioxide single crystals. *J. Nucl. Mat.* **44**, 43–56.

Willis, B. T. M. 1978. The defect structure of hyper-stoichiometric uranium dioxide. *Acta Crystall.* **A34**, 88–90.

Yust, C. S. & C. J. McHargue 1971. Deformation of hyperstoichiometric UO_2 single crystals. *J. Am. Ceram. Soc.* **54**, 628–35.
Yust, C. S. & C. J. McHargue 1973. Model for the deformation of hyperstoichiometric UO_2. *J. Am. Ceram. Soc.* **56**, 161–4.

CHAPTER SIXTEEN

A TEM study of dislocation reactions in experimentally deformed chalcopyrite single crystals

Christa Hennig-Michaeli & Jean-Jacques Couderc

16.1 Introduction

Sulphides may contain valuable tectonic information, since they display a variety of deformation features which are particularly sensitive to temperature conditions in the shallow Earth's crust. The understanding of the mechanical behaviour of the common sulphide ore minerals, galena, sphalerite, pyrite, pyrrhotite, and chalcopyrite, has progressed on account of numerous deformation studies in the laboratory over a variety of experimental conditions involving microstructural investigations and texture determinations (for reviews see Clark & Kelly 1976, McClay 1983, Siemes & Hennig-Michaeli 1985). Recently, the flow mechanisms of sulphide minerals in response to various imposed parameters in crustal environments have been comprehensively discussed by Cox (1987) with a view to the development of characteristic microstructures. TEM has been applied to analyse the dislocation substructures in deformed marcasite (Fagot *et al.* 1981), and has provided detailed information on the ductile behaviour of pyrite (Cox *et al.* 1981, Graf *et al.* 1981, Levade *et al.* 1982) and sphalerite (Levade *et al.* 1986). Previous interpretations of the flow mechanisms in experimentally deformed polycrystalline chalcopyrite resulted from texture determinations (Lang 1968) and from metallographic studies of deformation structures (Atkinson 1974, Kelly & Clark 1975, Roscoe 1975). The change from cataclasis to slip and twinning on {112} planes with increasing confining pressure and temperature could be outlined. The rather tentative results need to be revised by means of single-crystal deformation experiments over a similar range of conditions, and by TEM investigation. This chapter is part of such a re-examination.

Chalcopyrite (ccp) is a common accessory mineral in metamorphic and igneous rocks, and it is found in almost all types of ore deposits, apparently representing the widest possible range of conditions for sulphide formation (Stanton 1972). Under tectonic stress chalcopyrite is evidently a relatively weak and ductile mineral. Microscopic observations on grain aggregates in naturally deformed ores reveal a variety of structural features considered to be due to plastic deformation: weakly strained grains with deformation twins; elongated grains being subdivided into more or less organized subgrains; and fine-grained recrystallized aggregates (see, for example, Cox & Etheridge 1984). The dislocation mechanisms, however, are still a matter of speculation, since TEM investigations have only been carried out on crushed powders of few ore samples. Perfect dislocations with unexplored Burgers vectors and partial dislocations terminating extended {112} stacking faults or microtwins (Murr & Lerner 1977, 1978) and a sub-boundary network (Cox & Etheridge 1984) have been reported.

Recent investigations of the glide mechanisms of several ccp single crystals (Couderc & Hennig-Michaeli 1986, 1987; Hennig-Michaeli & Siemes 1987) have established that during experimental deformation at 200°C the behaviour of ccp crystals is more versatile than that of sphalerite-structured crystals. The dislocation mechanisms are in some respect analogous to those in DO_{22}-ordered alloys, with the same type of Bravais lattice (body-centred tetragonal with $c_0/a_0 \simeq 2$). It is the aim of this study to emphasize that not only do the operating glide systems control the deformation, but that also dislocation interactions and dislocation reactions have affected the plastic behaviour of those weakly strained crystals. Already published characteristics of the dislocations will be briefly summarized to provide the basis for this more detailed analysis. The comparison of the dislocation configurations in a crystal compressed at 400°C ($0.58T_m$) with the microstructures of the 200°C crystals ($0.41T_m$) will show that a marked change of the deformation mechanisms occurs in that temperature range.

TEM observations under weak-beam dark-field conditions reveal how partial dislocations are involved in the reactions. Included in the term 'reactions' are dissociations of perfect dislocations and cross-slip processes, since they are associated with transformations of dislocation cores. By means of analysing triple junctions, particular microprocesses, such as the dissociation mode of specific dislocations, dipole interactions, the formation of a microtwin, and the crossing over of dissociated dislocations gliding in conjugate glide planes, become elucidated. Thus the detailed characterization of dislocation reactions leads to a better understanding of the deformation behaviour of the crystals studied.

It would be premature to infer general conclusions on the deformation behaviour of chalcopyrite in crustal environments from the present study.

Electron-transparent foils normal to the axes of compression were prepared by electrochemical polishing of mechanically thinned sections, and were observed in a JEOL 200-CX electron microscope operating at 200 kV (TEM SCAN Service of the Université Paul Sabatier, Toulouse).

16.2 Dislocation reactions in crystals strained at 200°C

16.2.1 Survey of glide modes

In the ccp crystals studied, which were experimentally deformed at 200°C, at 300 MPa confining pressure and with $\dot{\varepsilon} \simeq 4 \times 10^{-6} \text{s}^{-1}$, dislocation glide occurred on three different types of glide planes (Couderc & Hennig-Michaeli 1986, 1987; Hennig-Michaeli & Siemes 1987): on {112} planes, on the (001) plane, and on {100} planes, all being stackings of sulphur layers alternating with CuFe cation layers.

Surface and TEM observations have established that $\{112\}\langle 3\bar{1}\bar{1}\rangle$ slip and $\{112\}\langle\bar{3}11\rangle$ slip are the main glide modes at 200°C. Straight dislocations parallel to low index directions in {112} planes form isolated slip bands with high dislocation densities. Dislocations with $\mathbf{b} = 1/2\langle\bar{1}11\rangle$, which is the shortest lattice vector in {112} planes (cf. Table 16.1), are less numerous than dislocations

Table 16.1 Glide dissociation of perfect dislocations in chalcopyrite crystals deformed at 200°C. Models derived from TEM observations (Couderc & Hennig-Michaeli 1987).

Glide plane	Dissociation reaction		Total dissociation width
	Modulus of **b**	Planar defects	
(A) {112}	$1/2\langle 3\bar{1}\bar{1}\rangle$ 988 pm	$\rightarrow$ $1/3\langle 11\bar{1}\rangle + 1/6\langle 4\bar{2}\bar{1}\rangle + 1/2\langle 1\bar{1}0\rangle$ $\perp$ - - - - - $\perp$ ===== $\perp$ SF APB	20–25 nm
(B)	$1/2\langle\bar{3}11\rangle$ 988 pm	$\rightarrow$ $1/6\langle\bar{1}\bar{1}1\rangle + 1/12\langle\bar{4}21\rangle + 1/2\langle\bar{1}10\rangle + 1/4\langle\bar{2}01\rangle$ $\perp$ - - - - - $\perp$ _____ $\perp$ _____ $\perp$ SF APB$_i$ APB$_i$	20–35 nm
(C)	$1/2\langle\bar{1}\bar{1}1\rangle$ 642 pm	$\rightarrow$ $1/6\langle\bar{1}\bar{1}1\rangle + 1/6\langle\bar{1}\bar{1}1\rangle + 1/6\langle\bar{1}\bar{1}1\rangle$ $\perp$ ===== $\perp$ _ _ _ _ $\perp$ SF ext. SF intr.	~20 nm
(D)	$\langle 1\bar{1}0\rangle$ 748 pm	$\rightarrow$ $1/2\langle 1\bar{1}0\rangle + 1/2\langle 1\bar{1}0\rangle$ $\perp$ ===== $\perp$ APB	~10 nm
(E) (001)	$\langle 110\rangle$ 748 pm	$\rightarrow$ $1/2\langle 110\rangle + 1/2\langle 110\rangle$ $\perp$ _____ $\perp$ APB	~20 nm
(F) {100}	$\langle 010\rangle$ 529 pm	$\rightarrow$ $1/4\langle 021\rangle + 1/4\langle 02\bar{1}\rangle$ $\perp$ ===== $\perp$ APB	~15 nm

SF, geometrical stacking fault; APB, antiphase boundary; APB$_i$, incomplete antiphase boundary. **b**, Burgers vector.

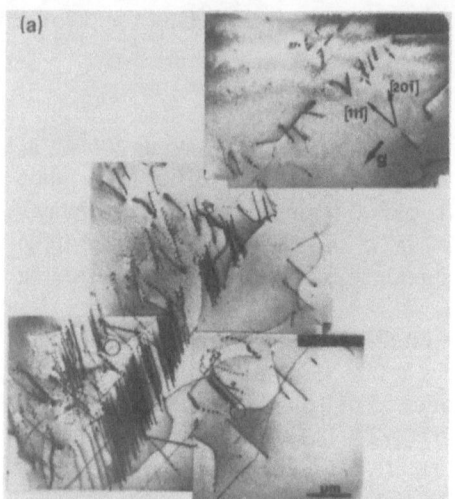
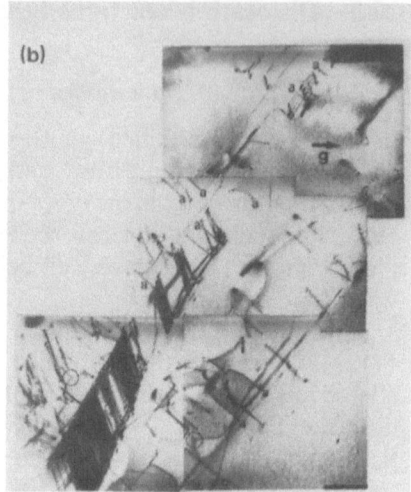

Figure 16.1 (112) slip band in a chalcopyrite crystal, deformed 1.1 per cent at 200°C, showing high-density pile-ups of straight dislocations parallel to [20$\bar{1}$], most of them probably with $\mathbf{b} = 1/2[\bar{3}11]$, and several screw dislocations with $\mathbf{b} = 1/2[\bar{1}\bar{1}1]$ ($\leftarrow$, a). The encircled nodes are presumably junctions of dislocations with $\mathbf{b} = 1/2[\bar{1}\bar{1}1]$, [$\bar{2}$01] and $1/2[\bar{3}11]$. Compression axis (CA) $\simeq$ foil normal (FN) $\simeq$ [110]. Bright field electron micrographs (BF). Electron beam direction (BM) $\simeq$ [221]. (a) $\mathbf{g} = [\bar{2}20]^*$; (b) $\mathbf{g} = [0\bar{2}4]^*$.

with $\mathbf{b} = 1/2\langle\bar{3}11\rangle$. This is illustrated in Figure 16.1 by the slip band in the crystal shortened along [110], in which $1/2[\bar{1}\bar{1}1]$ is the most favoured slip vector in the glide plane (112). Dislocations with $\mathbf{b} = \langle1\bar{1}0\rangle$ can enter {112} planes by cross-slip from the (001) plane, and dislocations with $\mathbf{b}$ parallel to $\langle20\bar{1}\rangle$ are rare in {112} planes. {112}$\langle11\bar{1}\rangle$ deformation twins are characterized by a high density of screw-twin dislocations in their boundaries.

(001)$\langle110\rangle$ slip is a main glide mode in the crystals shortened along [111] and [221]. The slip mode {100}$\langle010\rangle$ has been activated in the vicinity of high-density {112} dislocation bands in the crystal shortened along [110].

16.2.2 Dissociation of perfect dislocations in {112} planes

In previous studies, particular attention has been paid to the dissociation of perfect dislocations, and the kinds of planar defects which are associated with particular types of partial dislocations have been specified (Couderc & Hennig-Michaeli 1986, 1987). Dissociation reactions, which have been inferred from TEM observations in weak-beam conditions are summarized in Table 16.1.

Dislocations with $\mathbf{b} = 1/2\langle3\bar{1}\bar{1}\rangle$, gliding in {112} planes, are observed to be dissociated into up to four non-collinear partials. Due to the overlapping of the elastic strain fields of the partials, analysis by weak-beam methods is difficult. The dissociation width is not constant along the dislocations, and the

394

dissociation mode changes from place to place, indicating that the dislocation core is a changeable configuration. Several dissociation models have been proposed by the authors (Table 16.1), although a definite characterization of the complex splitting has not yet been achieved.

Dislocations with $\mathbf{b} = 1/2\langle\bar{1}\bar{1}1\rangle$ are dissociated into three collinear partials in the {112} planes. A dissociation model, involving dissociation in two adjacent interatomic planes, as proposed by Vanderschaeve & Escaig (1978) for the splitting of $\mathbf{b} = 1/2\{111\}$ in DO_{22}-ordered Ni_3V, is consistent with the observations.

Dislocations with $\mathbf{b} = \langle1\bar{1}0\rangle$ gliding in {112} planes are split into two collinear partials. Four collinear partials are attributes of dislocations with $\mathbf{b}$ parallel to $\langle20\bar{1}\rangle$, which look like double dipoles rather than dissociated perfect dislocations. Dislocations with $\mathbf{b} = \langle20\bar{1}\rangle$ do not seem to be stable. The observation that they can decompose into pairs of perfect dislocations:

$$\langle\bar{2}01\rangle \rightarrow 1/2\langle\bar{1}\bar{1}1\rangle + 1/2\langle\bar{3}11\rangle \tag{16.1}$$

is regarded as an indication that dislocation reactions in {112} planes might be sources for dislocations with the somewhat astonishing Burgers vector $1/2\langle\bar{3}11\rangle$ (Couderc & Hennig-Michaeli 1987).

16.2.3 Dislocation reactions in {112} planes

Several dislocation nodes in Figure 16.1 show that reactions have taken place in the glide plane (112). At the encircled junctions one of the dislocations is out of contrast for $\mathbf{g} = [2\bar{2}0]^*$ (Fig. 16.1a) and has $\mathbf{b} = 1/2[\bar{1}\bar{1}1]$. The nodes are assumed to be joined by dislocations, as indicated in reaction (16.1).

The configuration in Figure 16.2 with six interacting dislocations in the $(\bar{1}12)$ plane manifests the versatile dislocation behaviour in {112} planes. Dislocations 3 and 4 can easily be specified. Dislocation 3, parallel to $\sim [02\bar{1}]$ and out of contrast for $\mathbf{g} = [220]^*$, has $\mathbf{b} = 1/2[\bar{1}1\bar{1}]$ and is dissociated into three collinear partials. Dislocation 7 is of the same type. Dislocation 4, parallel to $\sim [421]$ and out of contrast for $\mathbf{g} = [1\bar{1}2]^*$ and $\mathbf{g} = [008]^*$, has $\mathbf{b} = [\bar{1}\bar{1}0]$ and is split into two collinear partials. Dislocations 1, 2, 5, and 6 remain in contrast whatever the operating reflection, specified by reciprocal lattice vector $\mathbf{g}$.

Dislocation 6 in Figure 16.2, parallel to $\sim [110]$, represents the most frequent dislocation type, with $\mathbf{b} = 1/2\langle3\bar{1}\bar{1}\rangle$ in that sample. It seems to be split into four partials:

$$1/2[\bar{3}\bar{1}\bar{1}] \rightarrow \underbrace{1/6[\bar{1}1\bar{1}] + 1/6[\bar{1}1\bar{1}]}_{1/3[\bar{1}1\bar{1}]} + 1/6[\bar{4}\bar{2}\bar{1}] + 1/2[\bar{1}\bar{1}0] \tag{16.2}$$

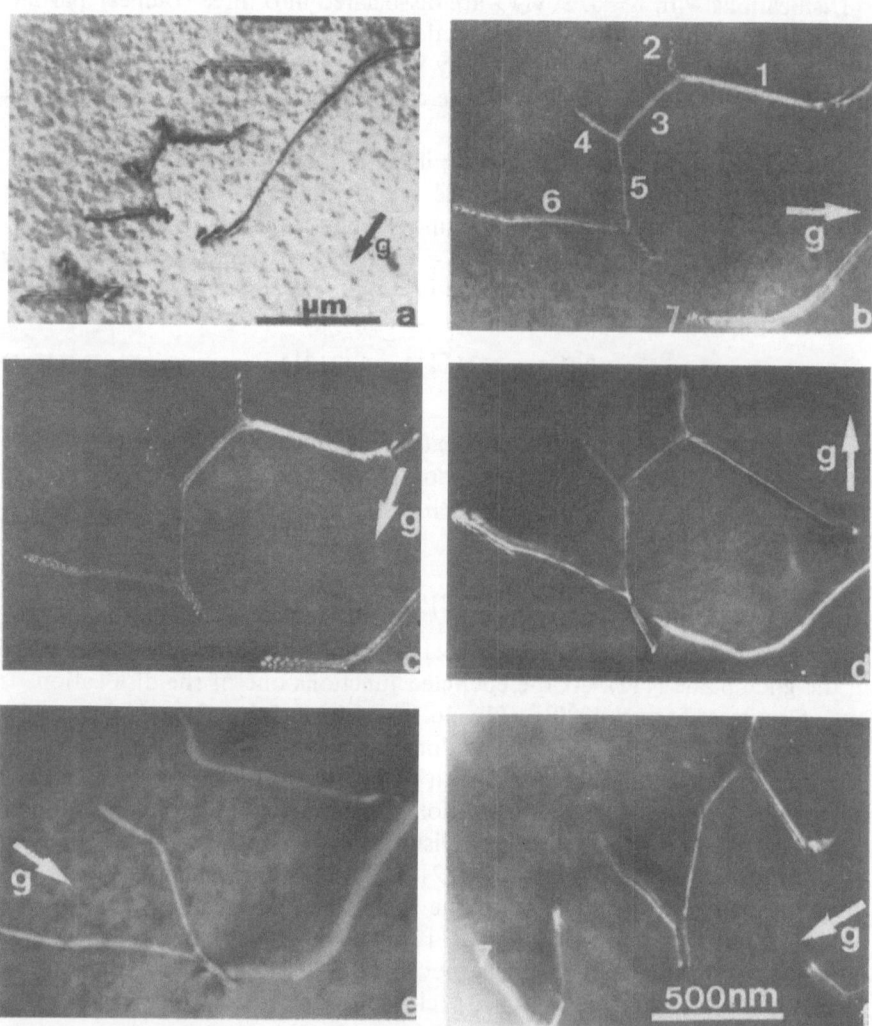

Figure 16.2 Net configuration of dislocations with four different types of Burgers vectors in the $(\bar{1}12)$ plane: 1, 2, $\mathbf{b} = 1/2[\bar{3}\bar{1}\bar{1}]$ or $[\bar{2}0\bar{1}]$; 3, $\mathbf{b} = 1/2[\bar{1}1\bar{1}]$; 4, $\mathbf{b} = [\bar{1}\bar{1}0]$; 5, 6, $\mathbf{b} = 1/2[\bar{3}\bar{1}\bar{1}]$. 1, 2, 3 and 3, 4, 5 form nodes; 5 and 6 join each other in a dipole. 1.0 per cent strain at 200°C. CA ≃ FN ≃ [021]. (a) $\mathbf{g} = [0\bar{2}4]^{*}$, BM ≃ [021], BF. (b) $\mathbf{g} = [11\bar{2}]^{*}$, BM ≃ [021], weak-beam dark-field electron micrograph (DF/WB). (c) Dislocation 4 is out of contrast; $\mathbf{g} = [1\bar{1}2]^{*}$, BM ≃ [021], DF/WB. (d) $\mathbf{g} = [\bar{2}0\bar{4}]^{*}$, BM ≃ [$\bar{2}21$], DF/WB. (e) Dislocation 3 is out of contrast; $\mathbf{g} = [220]^{*}$, BM ≃ [$\bar{2}21$], DF/WB. (f) Dislocation 4 is out of contrast; $\mathbf{g} = [008]^{*}$, BM ≃ [010], DF/WB. (DF, dark field; WB, weak beam.)

Dislocation 5 has to be of the same type, since it results from the reaction:

$$1/2\,[\bar{1}1\bar{1}] + [\bar{1}\bar{1}0] \rightarrow 1/2\,[\bar{3}1\bar{1}] \qquad (16.3)$$

dislocation   (3)   (4)   (5)

Along a short segment the dislocations 5 and 6 form a dipole. That the partial $b = 1/6\,[\bar{4}\bar{2}\bar{1}]$ is the central one in dislocation 5 can be inferred from the configurations of partial dislocations in Figures 16.2c and d:

$$1/2\,[\bar{1}\bar{1}0] + 1/6\,[\bar{1}1\bar{1}] \rightarrow 1/6\,[\bar{4}\bar{2}\bar{1}] \qquad (16.4)$$

partial   (d)   (c)   (f)
of dislocation   (4)   (3)   (5)

The observations at the nodes 3, 4, and 5, as depicted in Figures 16.3b & c, substantiate previously derived dissociation reactions (A, C, and D in Table 16.1). Whereas the partials c, d, e, and f in the glide plane (112) are considered to be situated together in a single interatomic layer, this is not the case for b and c if the model of Vanderschaeve & Escaig (1978) for the splitting of

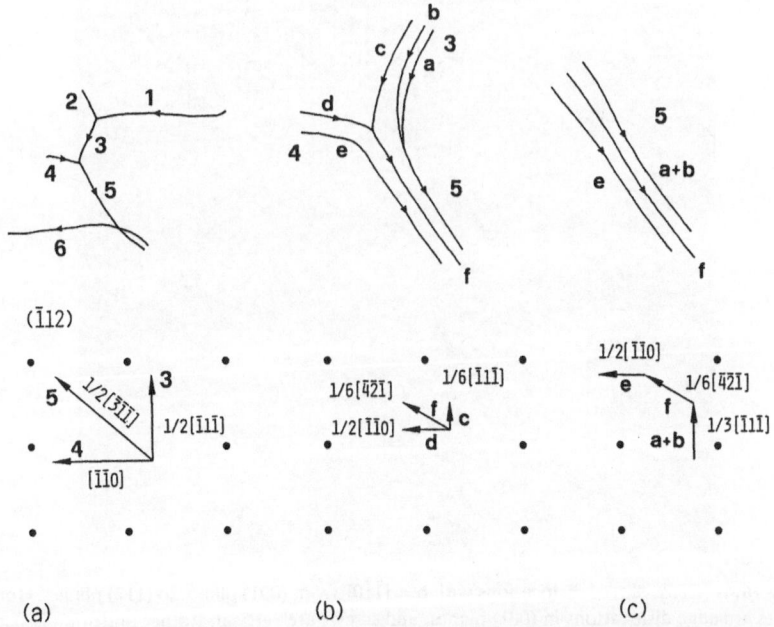

Figure 16.3 Schematic representation of interactions in Figure 16.2. Configuration of dislocations (above), and lattice points and Burgers vectors in the ($\bar{1}$12) plane (below). (a) Reaction of perfect dislocations at the node 3, 4, 5. (b) Reaction of partial dislocations at the node c, d, f. (c) Dissociation of dislocation 5.

$\mathbf{b} = 1/2\langle 111 \rangle$ is maintained. The observations suggest that the core of dislocation 5 has spread over two successive interatomic layers. The splitting of $1/3[\bar{1}1\bar{1}]$ in reaction (16.2) into two collinear partials cannot take place in a single fault plane.

The node between dislocations 1, 2, and 3 might account for the reaction:

$$1/2[\bar{1}1\bar{1}] + 1/2[\bar{3}\bar{1}\bar{1}] \rightarrow [\bar{2}0\bar{1}] \qquad (16.5)$$

dislocation   (3)   (1) or (2)   (1) or (2)

Due to their indistinct contrasts, dislocations 1 and 2 cannot be specified. Nevertheless, the observations show that dislocations with four different types of $\mathbf{b}$ are linked together in one slip plane.

16.2.4   Reactions of dislocations with $\mathbf{b} = \langle 110 \rangle$

Dislocations with $\mathbf{b} = \langle 1\bar{1}0 \rangle$ gliding in their primary slip plane (001) are generally dissociated into two collinear partials (reaction E, Table 16.1) and can

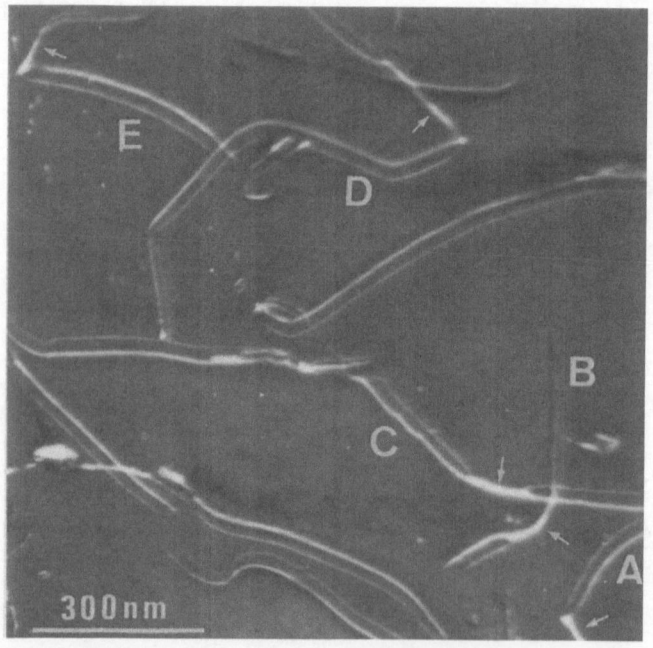

Figure 16.4   Cross-slip of dislocations with $\mathbf{b} = [\bar{1}\bar{1}0]$ from (001) planes to $(\bar{1}12)$ planes. Horizontal lines are edge dislocations in (001) planes, and screws are vertical. Rather uniform dissociation in the (001) planes with only one short less extended segment ($\leftarrow$) at C. The less extended segments at A and D, parallel to $\sim [13\bar{1}]$, and at B, parallel to $\sim [1\bar{1}1]$, lie in $(\bar{1}12)$ planes. The less extended segment at E in the $(\bar{1}12)$ plane, parallel to $\sim [421]$, reorientates to $\sim [1\bar{1}1]$ and redissociates. 1.4 per cent strain at $200°$C. CA $\simeq$ FN $\simeq$ [111], DF/WB, BM $\simeq$ [001], $\mathbf{g} = [220]^{*}$.

easily cross-slip into a {112} plane (Fig. 16.4). Cross-slip can take place without recombination of the partials, since they both are glissile in the two glide planes. A shrinking of the ribbons at the intersection of the primary slip plane with the cross-slip plane has been observed most frequently (Fig. 16.4). The cross-slipped dislocation segments in a {112} plane are always short, even when {112}⟨1$\bar{1}$0⟩ slip is more favoured by the applied stress than (001)⟨1$\bar{1}$0⟩ slip.

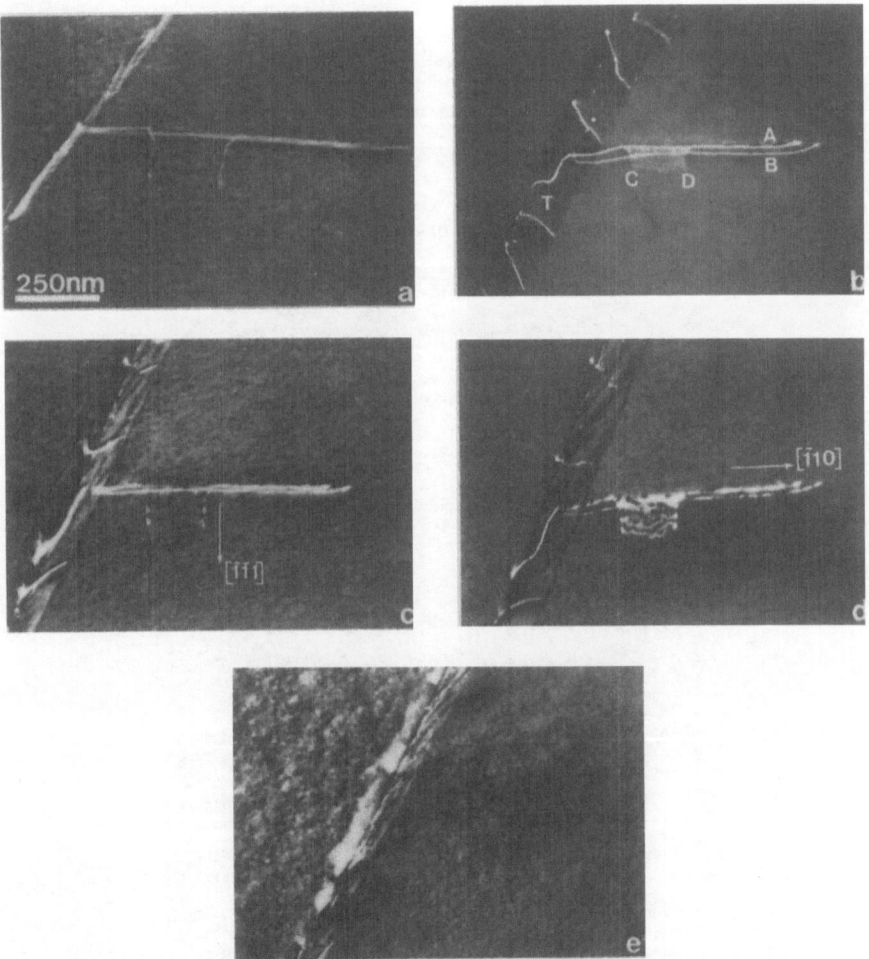

Figure 16.5 Interaction of a dislocation of the glide system (001)[$\bar{1}\bar{1}$0] with a (1$\bar{1}$2) deformation twin (left). The upper partial A with **b** = 1/2[110] in the (001) plane decomposes into a Frank partial 1/6[221] and two Shockley partials ±1/6[11$\bar{1}$] (C, D). The Shockley partials are screws linked by a geometrical SF in the (112) plane showing fringe contrasts in (b) and (d). 1.4 per cut strain at 200°C. CA ≃ FN ≃ [111]. DF/WB. (a) **g** = [$\bar{2}$04]*, BM ≃ [241]; (b) **g** = [220]*, BM ≃ [001]; (c) **g** = [02$\bar{4}$]*, BM ≃ [221]; (d) **g** = [$\bar{1}\bar{1}$2]*, BM ≃ [111]; (e) **g** = [$\bar{2}$20]*, BM ≃ [221].

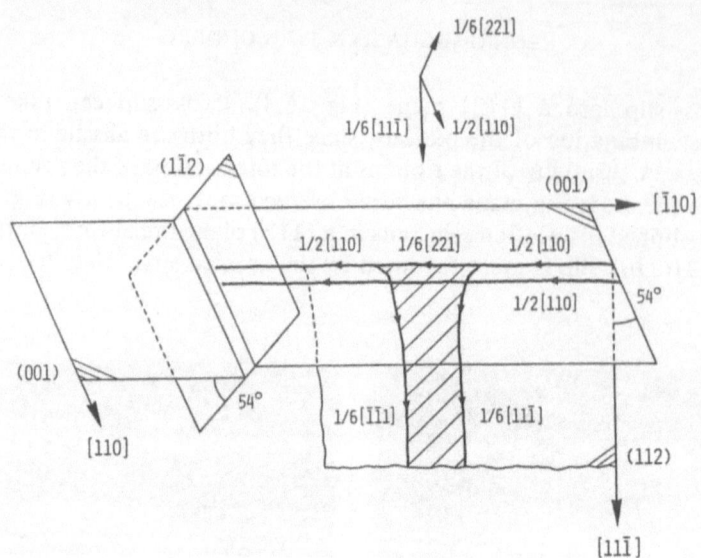

Figure 16.6 Schematic representation of the dislocation reactions in Figure 16.5.

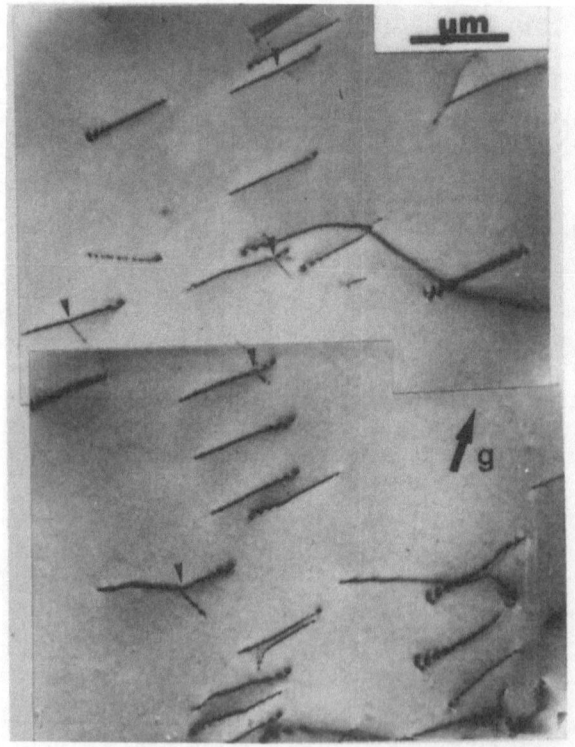

Figure 16.7 Dislocations parallel to [110] in the ($\bar{1}$12) and (001) planes. At triple junctions ($\leftarrow$) several dislocations are joined by [1$\bar{1}$0] directed segments. Most of these junctions result from dipole interactions of dislocations with $\mathbf{b} = [\bar{1}\bar{1}0]$. 1 per cent strain at 200°C. CA $\simeq$ FN $\simeq$ BM $\simeq$ [021], BF, $\mathbf{g} = [02\bar{4}]^*$.

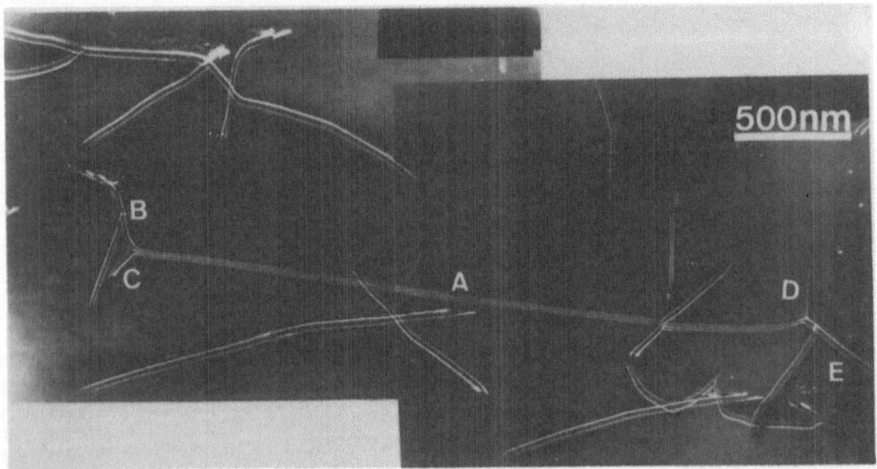

Figure 16.8 Dipole formation of dislocations within **b** = [110] gliding in two adjacent (001) planes. The dislocations B–E and C–D are dissociated into two collinear partials, which are superimposed along an edge dipole A, so that only three or two partials remain visible (cf. Fig. 16.10a). 1.4 per cent strain at 200°C. CA ≃ FN ≃ [111]. DF/WB. **g** = [220]*, BM ≃ [001].

Decomposition reactions of partials with **b** = 1/2⟨110⟩ appear to be sources for microtwin formation. In Figure 16.5 an edge dislocation in the (001) plane is dissociated into two collinear partials as usual. It joins a microtwin in (1Ī2) which acts as an obstacle to dislocation motion in the primary glide plane, and the upper partial spreads into Shockley partials in (112) and into a Frank partial in the way illustrated in Figure 16.6:

$$1/2[110] \rightarrow 1/6[11\bar{1}] + 1/6[221] \tag{16.6}$$

gliding in   (001)   (112)

The Frank partial 1/6[221] is sessile in the (001) plane and in the (112) plane. The Shockley partials ±1/6[11Ī] in the (112) plane bound a pure geometrical stacking fault, a 'one-layer' microtwin. The observed large differences between the twinning shear stresses in ccp crystals of different orientations (Hennig-Michaeli & Siemes 1987) might be explained by processes such as the operation of reaction (16.6), which evidently is induced by local stress concentrations.

Most of the 'triple junctions' in Figure 16.7 are not nodes between dislocations with three different Burgers vectors but result from dipole interactions of dislocations with **b** = ⟨110⟩. It is shown in Figure 16.8 how dislocations dissociated in the (001) plane into two collinear partials with **b** = 1/2[110], and gliding in closely spaced (001) planes, meet and re-orientate in an edge dipole. Along the dipole the four partials in the two glide planes are superimposed so that only three or two remain visible. The dipole immobilizes the edge segments of the dislocations and becomes elongated by the motion of mixed

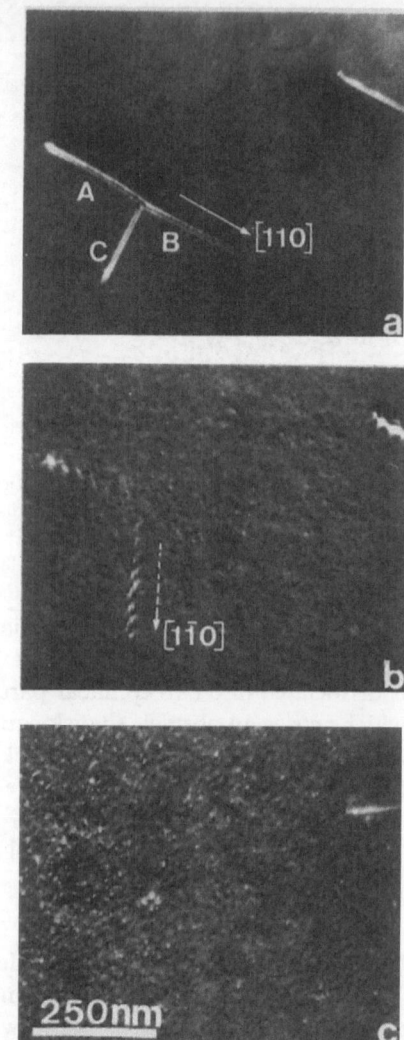

Figure 16.9 Interaction of dissociated dislocations with $\mathbf{b} = [110]$. A and B are screw dislocations and C is an edge dipole of superimposed dislocations in adjacent (001) planes. The upper 'partial dipole' of C is pinched off from the upper partial of A and B. The lower 'partial dipole' of C spreads into mutually opposed screw orientations (cf. Fig. 16.10d). 1.0 per cent strain at 200°C. $CA \simeq FN \simeq [021]$, DF/WB. (a) $\mathbf{g} = [220]^*$, $BM \simeq [\bar{2}21]$; (b) $\mathbf{g} = [1\bar{1}2]^*$, $BM \simeq [021]$; (c) $\mathbf{g} = [2\bar{2}0]^*$, $BM \simeq [221]$.

and screw segments. When screw segments of the two dislocations happen to meet at the end of the dipole (Fig. 16.9) they can join each other by cross-slip, resulting in a jogged-screw dislocation separated from a pinched-off edge dipole (Fig. 16.10, cf. Tetelman 1962). The configuration in Figure 16.9 exemplifies an intermediate state of this process: one partial of dislocation A has cross-slipped so that the 'dislocation' C consists of a pinched-off edge dipole of one partial and a dipole of two partials spreading into mutually opposed screw orientations (cf. Fig. 16.10d). The configuration in Figure 16.10b has also been observed.

402

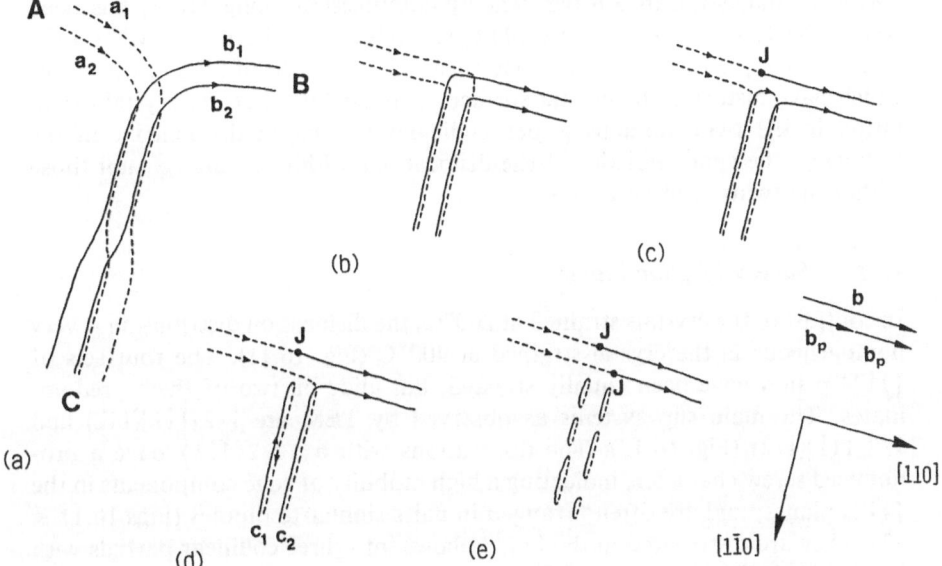

Figure 16.10 Interactions between dissociated dislocations with $\mathbf{b} = [110]$ in two adjacent (001) planes at the end of an edge dipole: (a) superposition of dissociated dislocations (cf. Fig. 16.8); (b) low-energy configuration of pure edge and pure screw dislocations; (c) a_1 and b_1 have joined each other by cross-slip of a partial dislocation containing a jog (J); (d) the dipole c_1 has been pinched off after exchange of partials (cf. Fig. 16.9); (e) c_2 has been pinched off. Note that shortening of c_1 and c_2 can proceed by core diffusion.

16.3 Dislocation reactions in crystals strained at 400°C

16.3.1 Deformation experiments

During experiments at 400°C ($0.58 T_m$) at 300 MPa confining pressure and $\dot{\varepsilon} \approx 5 \times 10^{-6}\,\mathrm{s}^{-1}$, plastic deformation started at low differential stresses which could not be determined accurately due to the external load registration of the high-pressure deformation apparatus. After the experiments the capsules embodying the deformed crystals contain H_2S vapour. The pre-polished surfaces of the samples are corroded, but they show the typical yellow metallic lustre of ccp. Exposed to the atmosphere, the outsides of the crystals and freshly cut surfaces tarnish rapidly. After a couple of months, however, the tarnishing of fresh cuts proceeds less quickly. A crystal that was thermally treated under the experimental conditions but not strained oxidizes only slowly. Obviously, the reactivity of the deformed crystals originates from strain-induced defects and decreases gradually. X-ray diffraction does not reveal any new reflections pointing to phase transitions. No additional spots appear in TEM diffraction patterns.

One crystal, strained 3.8 per cent by compression along [100], has been studied by TEM. Closely spaced {112} slip traces have been observed on the corroded sample surfaces. In a broad shear band two sets of {102} twins occur which have distorted the crystal surfaces only slightly, but the crystal orientation in the twin domains is perpendicular to that in the matrix. In the following, the characteristics of the dislocations within the matrix, not those within the twins, will be outlined.

16.3.2 Survey of glide modes

In contrast to the crystals strained at 200°C, the dislocation distribution is very homogeneous in the crystal strained at 400°C (Fig. 16.11). The four sets of {112} planes have been equally stressed, but glide on two of them predominates. The main slip systems as observed by TEM are $1/2[\bar{1}\bar{1}1](112)$ and $1/2[\bar{1}\bar{1}\bar{1}](11\bar{2})$ (Fig. 16.12). The dislocations with $\mathbf{b} = 1/2\langle\bar{1}\bar{1}1\rangle$ have a pronounced screw character, indicating a high mobility of edge components in the {112} planes, and are often arranged in pairs similar to dipoles (Figs 16.11 & 13). They are dissociated in the {112} planes into three collinear partials with $\mathbf{b} = 1/6\langle\bar{1}\bar{1}1\rangle$ (Figs 16.13 & 16) in the same way as at 200°C and can cross-slip

Figure 16.11 Dislocation arrangement in a chalcopyrite crystal deformed 3.8 per cent at 400°C. Three types of dislocations are in contrast: screws with $\mathbf{b} = 1/2[\bar{1}\bar{1}\bar{1}]$ (~N45°W) in (11$\bar{2}$) planes; screws with $\mathbf{b} = 1/2[\bar{1}\bar{1}1]$ (~N65°E) in (112) planes; and screws with $\mathbf{b} = [1\bar{1}0]$ (~N10°E). CA ≈ FN ≈ [100]. BF, $\mathbf{g} = [024]^*$, BM ≈ [100].

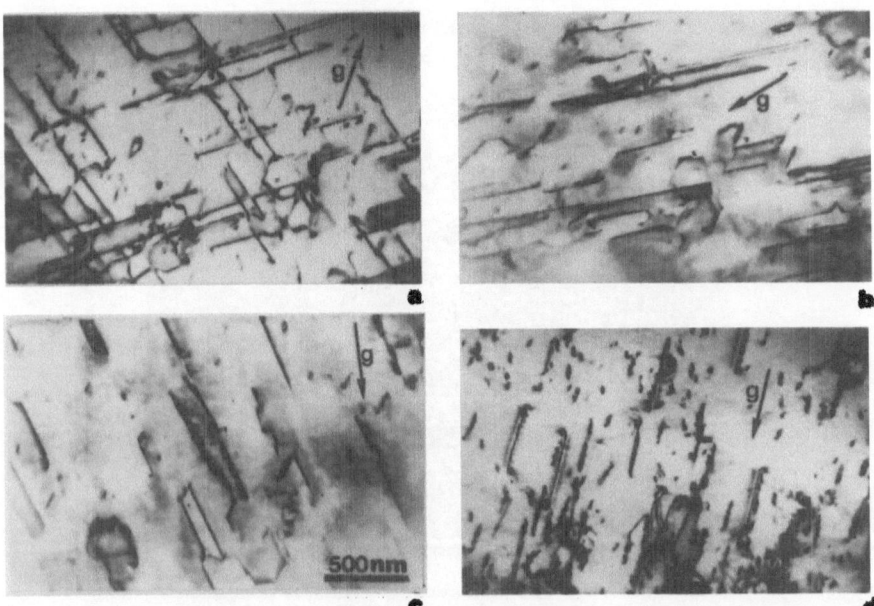

Figure 16.12 Characterization of the dislocations in the 400°C sample. In (a), (b), and (c) the two main glide planes (11$\bar{2}$) (N30°W) and (112) (N80°E) are viewed edge-on. Dislocations with **b** = 1/2[$\bar{1}\bar{1}\bar{1}$] in (11$\bar{2}$) planes are out of contrast in (b), dislocations with **b** = 1/2[$\bar{1}\bar{1}$1] in (112) planes are out of contrast in (c), and both types show numerous cross-slips into (1$\bar{1}$0) planes. Dislocations with **b** = [1$\bar{1}$0] are only visible in (d). 3.8 per cent strain. CA $\simeq$ FN $\simeq$ [100], BF. (a) **g** = [220]*, BM $\simeq$ [1$\bar{1}$0]; (b) **g** = [$\bar{1}\bar{1}$2]*, BM $\simeq$ [1$\bar{1}$0]; (c) **g** = [$\bar{1}\bar{1}\bar{2}$]*, BM $\simeq$ [1$\bar{1}$0]; (d) **g** = [2$\bar{2}$0]*, BM $\simeq$ [110].

into (1$\bar{1}$0). The observation that {112}⟨$\bar{1}\bar{1}$1⟩ slip is the main glide mode in the sample and not the more favoured {112}⟨$\bar{3}$11⟩ slip shows that at 400°C the Burgers vector 1/2⟨$\bar{3}$11⟩ is no longer stable.

Numerous screw dislocations with **b** = [1$\bar{1}$0] occur (Figs 16.11 & 12d), which are split into two collinear partials 1/2[1$\bar{1}$0] (Fig. 16.13). In weak-beam micrographs, the points of emergence of the partials on the foil surface coincide with traces of the (001) plane rather than with traces of {112} planes (cf. Fig. 16.13). The slip mode (001)⟨110⟩ seems to be activated in spite of the zero resolved shear stress acting on it. The slip system (001)[$\bar{1}$10] was activated at undefined low stresses in a crystal strained along [221]. It is probably the easiest glide mode in chalcopyrite at 400°C.

16.3.3   Cross-slip of dislocations with **b** = 1/2⟨$\bar{1}\bar{1}$1⟩

For both main slip systems, (112)[$\bar{1}\bar{1}$1] and (11$\bar{2}$)[$\bar{1}\bar{1}\bar{1}$], the plane of cross-slip is (1$\bar{1}$0), coinciding with the plane of the micrographs a, b, and c in Figure 16.12, so that the lengths of the cross-slipped segments can be seen directly

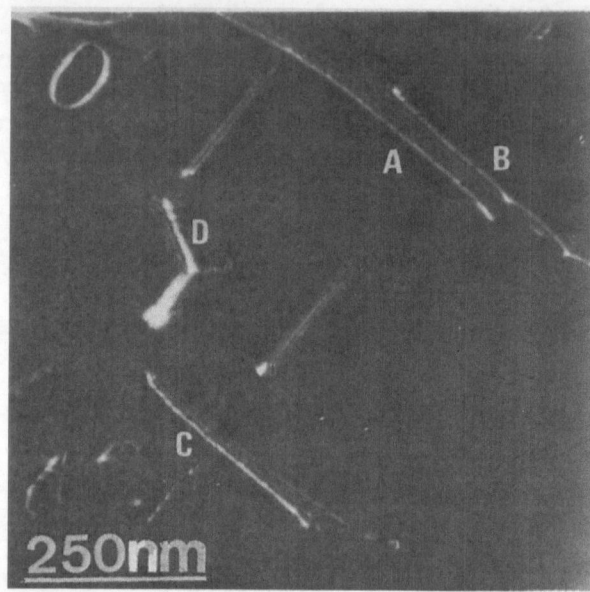

Figure 16.13 Dissociation of dislocations with $b = 1/2[\bar{1}\bar{1}1]$ (A, B, C, D) in (112) planes into three collinear partials. A, B, and C are screws. A and B form a dipole. D joins a node. Dissociation of screw dislocations with $b = [1\bar{1}0]$ (N50°E) into two collinear partials. In the upper left there is a small loop in the (001) plane with b parallel to $[02\bar{1}]$. 3.8 per cent strain at 400°C. $CA \simeq FN \simeq [100]$. DF/WB, $g = [0\bar{2}4]^*$, BM $\simeq [221]$.

(they are several tens of nanometres). The character of the dislocations (55°, 70°) in the (1$\bar{1}$0) plane can be established too, since the Burgers vectors lie also in the planes of the micrographs. The serrated appearance of several dislocations (Figs 16.12b, c) shows that generally only short segments cross-slip, and that the loops usually do not expand in the {112} glide planes which are just reached. The segments in the (1$\bar{1}$0) plane can be described to be long jogs (superjogs) of the screws in {112}, with heights from several b up to 60b.

16.3.4 Interactions between dislocations of different {112}⟨$\bar{1}\bar{1}$1⟩ slip systems

The configuration in Figure 16.14 is due to interactions between dislocations of the two main slip systems, (112)[$\bar{1}\bar{1}$1] (dislocations A, E) and (11$\bar{2}$)[$\bar{1}\bar{1}\bar{1}$] (dislocations B, D). The segment C has $b = [\bar{1}\bar{1}0]$, and is a screw dislocation in the cross-slip plane (1$\bar{1}$0) showing an enhanced contrast in Figure 16.14a where g is parallel to $\pm b$. At the nodes A, B, C and C, D, E the dislocations 1/2⟨$\bar{1}\bar{1}$1⟩ join:

$$1/2[\bar{1}\bar{1}1] + 1/2[\bar{1}\bar{1}\bar{1}] \rightarrow [\bar{1}\bar{1}0] \qquad (16.7)$$

dislocation in $\quad$ (112) $\qquad$ (11$\bar{2}$) $\qquad$ (1$\bar{1}$0)

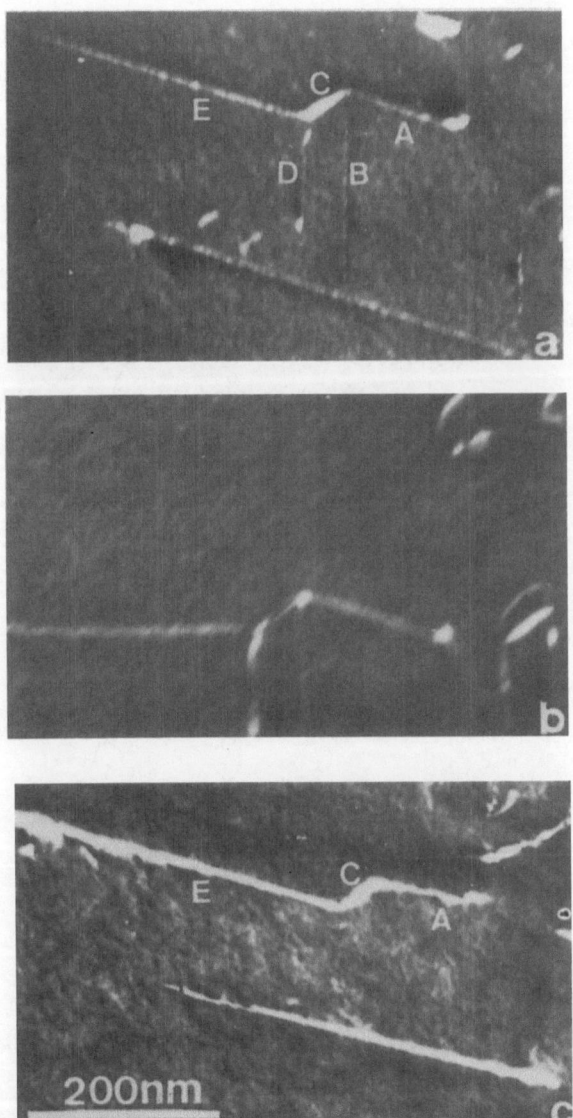

Figure 16.14 Interaction between dislocations of two different $\{112\}\langle\bar{1}\bar{1}1\rangle$ slip systems. E and A are screws with $\mathbf{b} = 1/2[\bar{1}\bar{1}1]$ in (112); D and B are screws with $\mathbf{b} = 1/2[\bar{1}\bar{1}\bar{1}]$ in (11$\bar{2}$); C is a screw with $\mathbf{b} = [\bar{1}\bar{1}0]$. 3.8 per cent strain at 400°C. CA ≃ FN ≃ [100]. DF/WB. (a) $\mathbf{g} = [220]^{*}$, BM ≃ [1$\bar{1}$0]; (b) $\mathbf{g} = [024]^{*}$, BM ≃ [4$\bar{2}$1]; (c) $\mathbf{g} = [\bar{1}\bar{1}2]^{*}$, BM ≃ [1$\bar{1}$0].

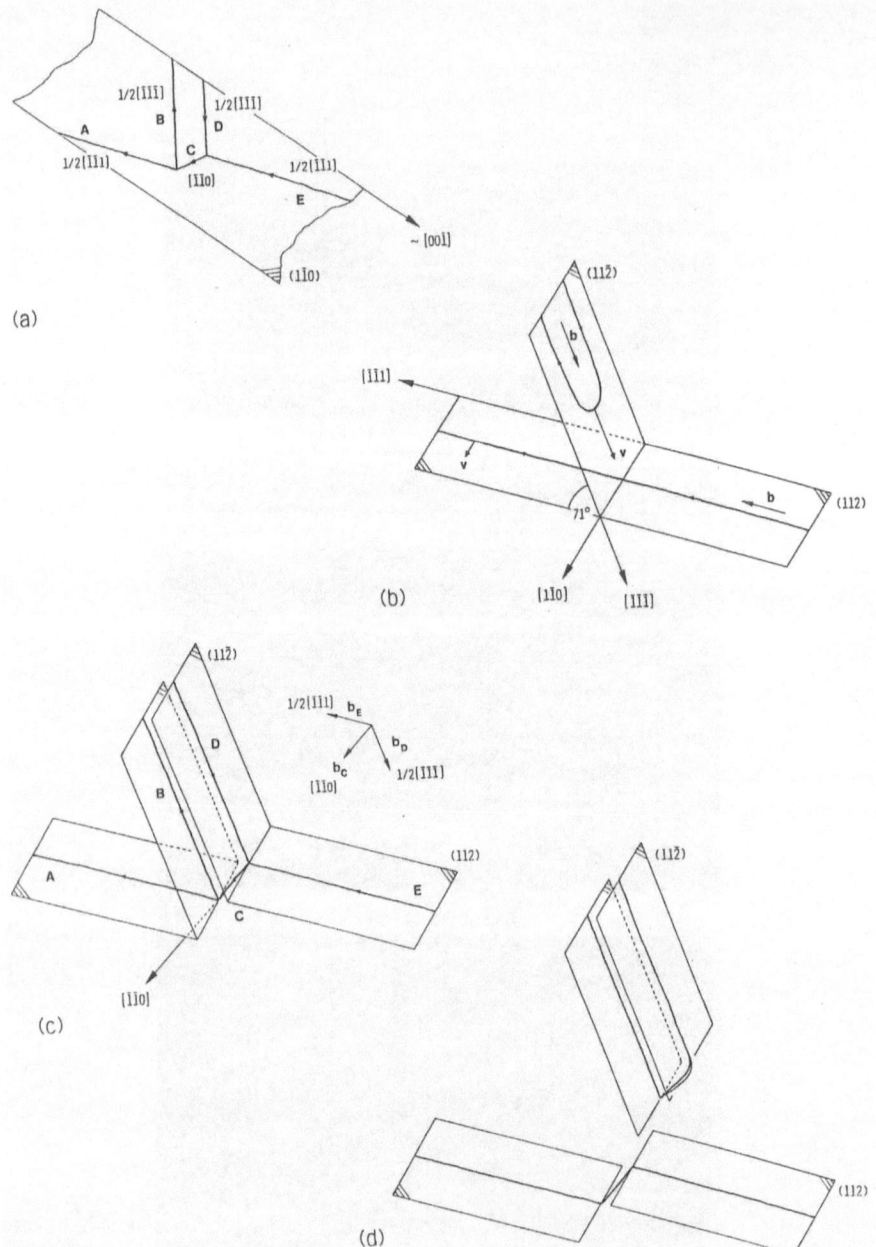

Figure 16.15 Interaction between dislocations of two different {112}⟨$\bar{1}\bar{1}$1⟩ slip systems. (a) Schematic representation of interactions in Figure 16.14. (b) Dislocations of two different 1/2⟨$\bar{1}\bar{1}$1⟩{112} slip systems approach. (c) Joined by formation of a segment in the (1$\bar{1}$0) plane with **b** = [$\bar{1}\bar{1}$0], screw branches in {112} planes are shifted stepwise into successive {112} planes. In (d) they have passed each other and contain segments in (1$\bar{1}$0) planes.

The reaction is energetically favourable since the two $1/2\langle\bar{1}\bar{1}1\rangle$ dislocations ($b_1\hat{\ }b_2 = 109°$) are mutually attractive. In Figure 16.15 it is shown how two dislocations gliding in the conjugate {112} glide planes meet. Because of the shape of the loops, a probable situation will be that edge segments of one loop happen to meet screws of the other one. During the reaction the screws in {112} are shifted stepwise into successive glide planes. The reaction continues until the edge segments of the joining loop are consumed, resulting in the observed coplanar configuration in $(1\bar{1}0)$.

Segment C seems to be split into two collinear partials (Fig. 16.14b):

$$[\bar{1}\bar{1}0] \rightarrow 1/2[\bar{1}\bar{1}0] + 1/2[\bar{1}\bar{1}0] \qquad (16.8)$$

Under an applied stress, segment C can decompose in such a way that the dislocations A–E and D–B can pass each other (Fig. 16.15d). Dislocation A–E will be a screw in two separated (112) planes with a long jog in $(1\bar{1}0)$. Dislocation D–B will look like an elongated 'loop in the $(1\bar{1}0)$ plane'. The microstructure of such a loop with $b = 1/2[111]$ is revealed in Figure 16.16. Whereas the segments A and C are dissociated in $(11\bar{2})$ into three collinear partials, recombination of two partials has taken place along segment B:

$$1/6[111] + 1/6[111] + 1/6[111] \rightarrow 1/6[111] + 1/3[111] \qquad (16.9)$$

$$\text{dissociation in (112)} \qquad\qquad \text{at segment B}$$

The two partials of segment B appear to lie in adjacent $(1\bar{1}0)$ planes. The dissociation mode in $(1\bar{1}0)$ and the wide dissociation as revealed in Figure 16.16c might be assisted by climb processes.

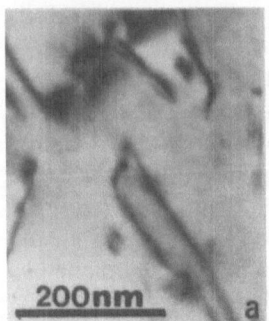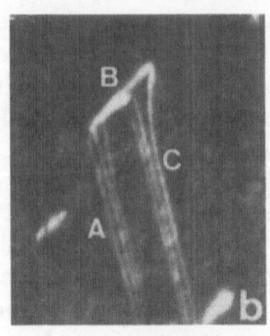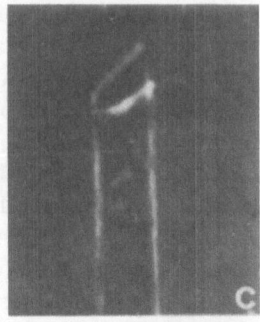

Figure 16.16 Dissociation of a dislocation with $b = 1/2[111]$ probably formed during reactions between dislocations of two different $1/2\langle\bar{1}\bar{1}1\rangle\{112\}$ slip systems (cf. Fig. 16.15). The segments A and C are screw dislocations dissociated into three collinear partials in separated $(11\bar{2})$ glide planes being viewed edge-on in (c). Segment B parallel to $\sim[110]$ in the $(1\bar{1}0)$ plane is dissociated into two collinear partials. 3.8 per cent strain at 400°C. CA $\simeq$ FN $\simeq$ [100]. (a) $g = [024]^*$, BM $\simeq$ [100], BF; (b) $g = [024]$, BM $\simeq$ [100], DF/WB; (c) $g = [0\bar{2}\bar{4}]^*$, BM $\simeq$ [4$\bar{2}$1], DF/WB.

409

The configuration in Figure 16.16 can also be imagined to arise from double cross-slip of a $1/2[111]$ dislocation from a $(11\bar{2})$ plane via the $(1\bar{1}0)$ plane into another $(11\bar{2})$ plane. The joining of serrated $1/2[111]$ screw dislocations gliding in neighbouring $(11\bar{2})$ glide planes might bring about a similar defect structure.

16.4 Discussion

The observation of dislocation reactions in ccp has substantiated the versatile dislocation behaviour at $200°C$ in the main glide plane $\{112\}$. The appearance of dislocations with the Burgers vector $1/2\langle3\bar{1}\bar{1}\rangle$ has been found to result from the recombination of dislocations with $\mathbf{b} = \langle1\bar{1}0\rangle$ and $\mathbf{b} = 1/2\langle11\bar{1}\rangle$. Reactions between dislocations with $\mathbf{b} = \langle\bar{2}01\rangle$ and $1/2\langle\bar{1}\bar{1}1\rangle$ also generate dislocations with $\mathbf{b} = 1/2\langle\bar{3}11\rangle$. Nevertheless, it is very likely that most of the dislocations with $\mathbf{b} = 1/2\langle3\bar{1}\bar{1}\rangle$ are directly activated by the applied stress. The observed node between specific partials (reaction 4, Fig. 16.3b) has confirmed previous dissociation models of the different types of joining perfect dislocations (Couderc & Hennig-Michaeli 1986, 1987): but there is now some indication that the dissociation of $1/2\langle3\bar{1}\bar{1}\rangle$ is spread over two successive interatomic layers, since only one partial $1/6\langle11\bar{1}\rangle$ of the dissociated dislocation $\mathbf{b} = 1/2\langle11\bar{1}\rangle$ participates in reaction (4). The remaining two $1/6\langle11\bar{1}\rangle$ partials which are supposed to glide in successive interatomic layers of the 'glide set' obviously have not undergone reactions at the node and continue along the $1/2\langle3\bar{1}\bar{1}\rangle$ dislocation. It has to be admitted that the splitting of the $1/2\langle3\bar{1}\bar{1}\rangle$ dislocations is not yet understood clearly. In particular, it has to be discovered whether interatomic layers of the 'shuffle set' can be slip planes of specific partials. Most favourable observation conditions are expected in foils parallel to the glide plane $\{112\}$, which are in preparation.

During cross-slip of dislocations with $\mathbf{b} = \langle1\bar{1}0\rangle$ the dislocation core is modified: in the (001) plane the two partials are linked by an APB without any incorrect first neighbours and in $\{112\}$ planes by a complete APB. The energy of the APB in the (001) plane is lower than in $\{112\}$ planes (cf. dissociation widths, Table 16.1) and, therefore, glide in the (001) plane should be favoured. The charge distributions at broken bonds along dislocation cores are not the same in the two glide planes. When an edge dislocation in an (001) plane has two dangling bonds per unit length, there will be three dangling bonds in $\{112\}$ planes if the dislocation lies in an interatomic layer of the 'glide set', and one dangling bond if it lies in an interatomic layer of the 'shuffle set'.

Further significant microprocesses proceed from dislocations of the glide mode $(001)\langle110\rangle$: the decomposition of partial dislocations with $\mathbf{b} = 1/2\langle110\rangle$ supports the formation of microtwins and exchange reactions of these partials at superimposed dipoles generate immobile 'triple junctions' in (001) planes.

The state of the samples deformed at $400°C$ manifests that chemical reactions occurred during the experiments. Small loops in (001) planes with diameters not exceeding 100 nm (Fig. 16.13) with **b** parallel to $\langle 021 \rangle$ indicate diffusive mass transfer. Due to the observed desulphurization it is assumed that anion vacancies have collapsed in the (001) plane which is not the densest-packed plane in ccp.

At 400°C (001)$\langle 110 \rangle$ slip is probably the easiest glide mode. In $\{112\}$ planes dislocations with **b** $= 1/2\langle \bar{3}11 \rangle$ seem not to be stable any longer. They might occur occasionally, as suggested by the node at C in Fig. 16.13, where the following reaction has probably taken place:

$$1/2[\bar{1}\bar{1}1] + [\bar{1}10] \rightarrow 1/2[\bar{3}11] \qquad \text{(cf. reaction (16.3))}$$

$$\text{in (112)} \quad \text{in (001)} \quad \text{in (112)}$$

In the $\{112\}$ slip planes the dislocations with **b** $= 1/2\langle \bar{1}\bar{1}1 \rangle$ can easily move in the glide direction, but the displacement of the screws is determined by the propagation of the long jogs in the cross-slip plane $(1\bar{1}0)$. Dislocations with **b** $= 1/2\langle \bar{1}\bar{1}1 \rangle$ moving in conjugate $\{112\}$ planes can pass each other by combining to **b** $= [\bar{1}\bar{1}0]$ in the $(1\bar{1}0)$ plane and by disjoining under the applied stress. Specific 'loops in the $(1\bar{1}0)$ plane' result from the interactions. Although only short dislocation segments lie in the $(1\bar{1}0)$ plane, the plastic behaviour at 400°C seems to be affected considerably by them. The $\{110\}$ planes in ccp consist of atomic layers with the composition $CuFeS_2$, in contrast with the glide planes at 200°C which are sheets of sulphur anions alternating with sheets of CuFe cations.

16.5 Summary and conclusions

The results of TEM observations of the authors on deformed ccp crystals are summarized in Table 16.2. Although knowledge of the glide behaviour at 200°C is more complete than that at 400°C, the present study has shown that a conspicuous change of the glide mechanisms occurs between the two temperatures. Coarse $1/2\langle \bar{3}11 \rangle \{112\}$ slip bands predominate at 200°C, whereas homogeneously distributed $1/2\langle \bar{1}\bar{1}1 \rangle \{112\}$ dislocations characterize the crystal deformed at 400°C. With rise of temperature (001)$\langle 110 \rangle$ slip evidently becomes more important.

Dislocation reactions and interactions have an effect on the deformation behaviour. At 200°C strong elastic forces in the main glide plane $\{112\}$ due to the long Burgers vector $1/2\langle \bar{3}11 \rangle$ might favour the development of net configurations of dislocations in this glide plane, and might induce dislocation multiplication in neighbouring glide planes, leading to the progressive growth of slip bands. Dipole formation accommodates local stresses which are due to elastic lattice distortions. Dipoles can be immobilized entirely when annihilation of partial dislocations takes place between the superposed dislocations.

411

Table 16.2 Survey of glide modes and dislocation reactions in experimentally deformed chalcopyrite crystals.

	200°C	400°C
Glide modes, main dislocation character		
primary slip modes	{112}⟨3$\bar{1}\bar{1}$⟩ slip, mixed {112}⟨$\bar{3}$11⟩ slip, mixed (001)⟨110⟩ slip, edge	(001)⟨110⟩ slip, screw {112}⟨$\bar{1}\bar{1}$1⟩ slip, screw
secondary slip modes	{112}⟨$\bar{1}\bar{1}$1⟩ slip, screw {112}⟨1$\bar{1}$0⟩ slip, mixed {112}⟨20$\bar{1}$⟩ slip?, 60° {100}⟨010⟩ slip, 45°	{110}⟨$\bar{1}\bar{1}$1⟩ slip, mixed
twinning modes	{112}⟨$\bar{1}\bar{1}$1⟩ twins, screw	{102} twins
Reactions		
dissociation	observed at all dislocation types	observed at all dislocation types
cross-slip processes	(001)⟨1$\bar{1}$0⟩ to {112}⟨1$\bar{1}$0⟩	{112}⟨$\bar{1}\bar{1}$1⟩ to {1$\bar{1}$0}⟨$\bar{1}\bar{1}$1⟩
dipole formation	narrow (001)⟨110⟩ edge dipoles, narrow {112}⟨3$\bar{1}\bar{1}$⟩ mixed dipoles observed	broad {112}⟨$\bar{1}\bar{1}$1⟩ screw dipoles
dipole annihilation	rows of loops	
reactions at dislocation nodes	interactions of dislocations in {112} planes	intersection of {112}⟨$\bar{1}\bar{1}$1⟩ dislocations gliding in conjugate {112} planes
—	formation of a (112) microtwin	formation of elongated 'loops in the (1$\bar{1}$0) plane'
vacancy loops	—	loop plane (001), random distribution

The formation of serrated loops by cross-slip processes at 400°C is assumed to exercise a strong influence on the mobility of the dislocations with $\mathbf{b} = 1/2\langle\bar{1}\bar{1}1\rangle$ in {112} planes. The interaction of dislocations gliding in two conjugate glide planes leaves behind long jogs and elongated loops in the (1$\bar{1}$0) planes, which seem to be rather immobile configurations causing pronounced strain hardening.

The rôle of creep processes during deformation cannot be estimated from the present observations. Annihilation of pure edge dipoles of dislocations gliding in (001) occurs at 200°C. Rows of loops are also common features in the {112} glide planes, where the nature of the loops has not yet been clearly identified. The trails of loops look similar to those observed in experimentally deformed pyrite where they are common attributes of crystals deformed at temperatures higher than 650°C (Graf *et al.* 1981, Cox *et al.* 1981, Levade *et al.* 1982). Narrow dipoles are rare in the ccp crystal deformed at 400°C. Presumably dipole annihilation immediately follows dipole formation. Trails of loops are absent, but numerous prismatic loops in the (001) plane demonstrate condensation of vacancies or interstitials resulting from bulk diffusion. The motion of dislocations with $\mathbf{b} = 1/2\langle \bar{1}\bar{1}1 \rangle$ in the (1$\bar{1}$0) plane is assumed to be controlled by climb.

Perhaps the TEM studies on deformed crystals have provided some information for the interpretation of deformation mechanisms in naturally deformed ccp. The identification of specific types of dislocations and their interactions might be a useful tool with which to distinguish low-temperature mechanisms from high-temperature mechanisms. For this purpose, further deformation experiments on chalcopyrite single crystals over a less restricted range of conditions have to be performed, assisted by detailed TEM work. A future aim will be to learn to understand the defect chemistry of the strain-induced defects and its dependence on stoichiometry and ambient sulphur fugacity.

Due to its extensive distribution in ores and rocks and its temperature-dependent deformation mechanisms, ccp might become an index mineral for deformation temperatures at low stresses. The complex deformation behaviour that sensitively depends on the orientation of the individual crystals is considered, in particular, to be indicative that conclusions can be drawn about palaeostress directions during the final stage of deformation.

Acknowledgements

The authors appreciate helpful discussions with Dr Colette Levade. Thanks are due to Professor David Barber and two anonymous referees for their constructive criticisms of the manuscript. C.H.-M.would like to express her thanks to Professor Heinrich Siemes for his encouragements and thoughtful comments. She is grateful to INSA, Toulouse, for its hospitality, and to the Deutsche Forschungsgemeinschaft and the CNRS for financial support.

References

Atkinson, B. K. 1974. Experimental deformation of polycrystalline galena, chalcopyrite and pyrrhotite. *Trans Inst. Min. Metall.* **83B**, 19–28.

Clark, B. R. & W. C. Kelly 1976. Experimental deformation of common sulphide minerals. In *Physics and chemistry of minerals and rocks*, R. G. J. Strens (ed.), 51–69. New York: Wiley.

Couderc, J.-J. & C. Hennig-Michaeli 1986. Transmission electron microscopy of experimentally deformed chalcopyrite single crystals. *Phys. Chem. Minerals* 13, 393–402.

Couderc, J.-J. & C. Hennig-Michaeli 1987. TEM evidence of various glide modes in experimentally strained CuFeS$_2$ crystals. *Phil Mag. A* 57, 301–25.

Cox, S. F. 1987. Flow mechanisms in sulphide minerals. In *Mechanical and chemical (re)mobilization of metalliferous mineralization*, B. Marshal & L. B. Gilligan (eds). *Ore Geol. Rev.* 2, 133–71.

Cox, S. F. & M. A. Etheridge 1984. Deformation microfabric development in chalcopyrite fault zones, Mt. Lyell, Tasmania. *J. Struct. Geol.* 6, 167–82.

Cox, S. F., M. A. Etheridge & B. E. Hobbs 1981. The experimental ductile deformation of polycrystalline and single crystal pyrite. *Econ. Geol.* 76, 2105–17.

Fagot, M., C. Levade, J.-J. Couderc & J. Bras 1981. Observation of lattice defects in orthorhombic iron disulfide (marcasite). *Phys. Chem. Minerals* 7, 253–9.

Graf, J. L., B. J. Skinner, J. Bras, M. Fagot, C. Levade & J.-J. Couderc 1981. Transmission electron microscopic observation of plastic deformation in experimentally deformed pyrite. *Econ. Geol.* 76, 738–42.

Hennig-Michaeli, C. & H. Siemes 1987. Experimental deformation of chalcopyrite single crystals at 200°C. *Tectonophysics* 135, 217–32.

Kelly, W. C. & B. R. Clark 1975. Sulfide deformation studies: III. Experimental deformation of chalcopyrite to 2000 bars and 500°C. *Econ. Geol.* 70, 431–53.

Levade, C., J.-J. Couderc, J. Bras & M. Fagot 1982. Transmission electron microscopy study of experimentally deformed pyrite. *Phil Mag. A* 46, 307–25.

Levade, C., J.-J. Couderc, I. Dudouit & J. Garigue 1986. The plastic behaviour of natural sphalerite single crystals between 473 and 873 K. *Phil Mag. A* 54, 259–74.

McClay, K. R. 1983. Fabrics of deformed sulphides. *Geol. Rund.* 72, 469–91.

Murr, L. E. & S. L. Lerner 1977. Transmission electron microscopic study of defect structure in natural chalcopyrite (CuFeS$_2$). *J. Mat. Sci.* 12, 1349–54.

Murr, L. E. & S. L. Lerner 1978. Explosive shock deformation of natural chalcopyrite (CuFeS$_2$). *J. Mat. Sci.* 13, 2268–72.

Roscoe, W. E. 1975. Experimental deformation of natural chalcopyrite at temperatures up to 300°C over the strain rate range 10^{-2} to 10^{-6} sec^{-1}. *Econ. Geol.* 70, 454–72.

Siemes, H. & C. Hennig-Michaeli 1985. Ore minerals. In *Preferred orientation in deformed metals and rocks: an introduction to modern texture analysis*, H.-R. Wenk (ed.), 335–60. Orlando, Florida: Academic Press.

Stanton, R. L. 1972. *Ore petrology*. New York: McGraw-Hill.

Tetelman, A. S. 1962. Dislocation dipole formation in deformed crystals. *Acta Metall.* 10, 813–20.

Vanderschaeve, G. & B. Escaig 1978. Dissociation of 1/2⟨112⟩ perfect dislocations and nature of slip dislocations in ordered Ni$_3$V. *J. Phys., Lett.* 39, L74–7.

Index

Because this book has many contributing authors, who also come from diverse fields, the terminology is not entirely uniform over all the chapters. Hence, you may not always find the exact keyword on a particular page, but you should find information relevant to the subject. In approaching this index it may be helpful to seek a general topic (e.g creep) and then look for a specific item under this heading (e.g. recovery-controlled ...). Keywords are indexed in three ways: numbers in ordinary face are page numbers; those in italics represent illustrations; whereas bold face numbers refer either to chapters or sections.